Organization of the Prokaryotic Genome

Organization of the Prokaryotic Genome

Edited by
Robert L. Charlebois

Department of Biology, University of Ottawa,
Ottawa, Ontario, Canada

ASM
PRESS

Washington, D.C.

Library of Congress Cataloging-in-Publication Data

Organization of the prokaryotic genome/edited by Robert L. Charlebois.
    p.   cm.
  Includes bibliographical references and index.
  ISBN 1-55581-151-5 (hardcover)
  1. Microbial genetics.   2. Genomes.   3. Prokaryotes.   I. Charlebois, Robert L.
QH434.074   1999
572.8′629—DC21   99-32407
                                                                                          CIP

CONTENTS

CONTRIBUTORS

Michel Blot
Genomique Bactérienne et Evolution, Université Joseph Fourier-CNRS EP 2029-CEA LRC12, Grenoble 38041 Cedex, France

Mark Borodovsky
Schools of Biology and Mathematics, Georgia Institute of Technology, Atlanta, GA 30332-0230

Ronald Chalmers
Department of Biochemistry, University of Oxford, South Parks Road, Oxford OX1 3QU, United Kingdom

Robert L. Charlebois
Department of Biology, University of Ottawa, Ottawa, Ontario K1N 6N5, Canada, and Canadian Institute for Advanced Research

Stewart T. Cole
Unité de Génétique Moléculaire Bactérienne, Institut Pasteur, 28 rue du Docteur Roux, 75724 Paris Cédex 15, France

Bernard Decaris
Laboratoire de Génétique et Microbiologie UA INRA 952, Université Henri Poincaré, Nancy 1, Faculté des Sciences BP 239, 54506 Vandoeuvre-lès-Nancy, France

Michael Y. Galperin
National Center for Biotechnology Information, National Library of Medicine, National Institutes of Health, Bethesda, MD 20894

William S. Hayes
School of Biology, Georgia Institute of Technology, Atlanta, GA 30332-0230

N. Patrick Higgins
Department of Biochemistry and Molecular Genetics, University of Alabama at Birmingham, Birmingham, AL 35294-2170

Femia G. Hopwood
Australian Proteome Analysis Facility (APAF), Macquarie University, Sydney, NSW 2109, Australia

Diarmaid Hughes
Department of Molecular Biology, Box 590, The Biomedical Center, Uppsala University, S-751 24 Uppsala, Sweden

Eugene V. Koonin
National Center for Biotechnology Information, National Library of Medicine, National Institutes of Health, Bethesda, MD 20894

Bernard Labedan
Institut de Génétique et Microbiologie, CNRS UMR 8621, Bâtiment 409, Université de Paris-Sud, 91405 Orsay Cedex, France

Jeffrey G. Lawrence
Department of Biological Sciences, University of Pittsburgh, Pittsburgh, PA 15260

Pierre Leblond
Laboratoire de Génétique et Microbiologie UA INRA 952, Université Henri Poincaré, Nancy 1, Faculté des Sciences BP 239, 54506 Vandoeuvre-lès-Nancy, France

Shu-Lin Liu
Department of Medical Biochemistry, University of Calgary, Calgary, Alberta, Canada T2N 1N4

Alexander V. Lukashin
School of Biology, Georgia Institute of Technology, Atlanta, GA 30332-0230

Michael McClelland
Sydney Kimmel Cancer Center, Room 300, 3099 Science Park Road, San Diego, CA 92121

Bénédicte Michel
Génétique Microbienne, Institut National de la Recherche Agronomique, Domaine de Vilvert, 78352 Jouy en Josas Cedex France

Roger Milkman
Department of Biological Sciences, The University of Iowa, Iowa City, IA 52242-1324

Amanda S. Nouwens
Australian Proteome Analysis Facility (APAF), Macquarie University, Sydney, NSW 2109, Australia

Theo Odijk
Faculty of Chemical Engineering and Materials Science, Delft University of Technology, P.O. Box 5045, 2600 GA Delft, The Netherlands

Monica Riley
Marine Biological Laboratory, Woods Hole, MA 02543

John R. Roth
Department of Biology, University of Utah, Salt Lake City, UT 84112

Isabelle Saint-Girons
Unité de Bactériologie Moléculaire et Médicale, Institut Pasteur, 28 rue du Docteur Roux, 75724 Paris Cédex 15, France

Kenneth E. Sanderson
Salmonella Genetic Stock Centre, Department of Biological Sciences, University of Calgary, Calgary, Alberta, Canada T2N 1N4

Christoph W. Sensen
National Research Council of Canada, Institute for Marine Biosciences, 1411 Oxford Street, Halifax, Nova Scotia, Canada B3H 3Z1

Andrew St. Jean
Department of Biology, University of Ottawa, Ottawa, Ontario K1N 6N5, Canada

Reginald K. Storms
Centre for Structural and Functional Genomics, Department of Biology, Concordia University, Montreal, Quebec, Canada H3G 1M8

Roman L. Tatusov
National Center for Biotechnology Information, National Library of Medicine, National Institutes of Health, Bethesda, MD 20894

Mathew Traini
Australian Proteome Analysis Facility (APAF), Macquarie University, Sydney, NSW 2109, Australia

Roel Van Driel
E. C. Slater Instituut, BioCentrum Amsterdam, University of Amsterdam, Plantage Muidergracht 12, 1018 TV Amsterdam, The Netherlands

Bradley J. Walsh
Australian Proteome Analysis Facility (APAF), Macquarie University, Sydney, NSW 2109, Australia

Keith L. Williams
Australian Proteome Analysis Facility (APAF), Macquarie University, Sydney, NSW 2109, Australia

Conrad L. Woldringh
Institute for Molecular Cell Biology, BioCentrum Amsterdam, University of Amsterdam, Kruislaan 316, 1098 SM Amsterdam, The Netherlands

PREFACE

Until recently, genetic linkage analysis was restricted to a few model organisms for which genetic tools had been developed. Physical maps soon obviated the need for genetics in determining genome organization, but like their predecessors, these maps were necessarily coarse and incomplete. Genetic maps require genetic markers, and physical maps require discrete probes. Neither could suitably serve to describe a genome in its entirety. Due to this constraint, detailed structural, functional, and evolutionary analyses could only focus on the gene or on relatively tiny regions of the genome. The extensive work on genes as the objects of study has led to certain paradigms in molecular and evolutionary biology which still color our understanding of genomes.

To appreciate the distinction between genes and genomes, it is first necessary to identify which features are in common and which are different between a sequence of a gene and a sequence of genes. Both may suffer point mutations, including insertions, deletions, and replacements. Both contribute a phenotype that is subject to enviromental controls; the gene produces a gene product, and the genome produces a cell. Both evolve with time. The gene, however, is far less modular than is the genome. Although functional domains in genes exist, functional domains in the genome—its genes—are by far its major organizational feature. In addition, structural elements within a gene, often designed to permit the correct placement of functional elements, are fundamentally different in purpose from the structural and functional elements in genomes. The question to be answered then is whether genomic organization contributes significantly to phenotype and to the evolution of phenotype. Linkage facilitates DNA replication as well as partitioning at cell division, and it has important effects in population genetics. But is linkage an important feature in cell and molecular biology? Is the genome an ordered set of genes or is it merely a set of genes?

To better appreciate what resources we have available for answering such fundamental questions, chapters 1 and 2 review the technology of collecting sequence data and finding genes within that data, respectively. Next, a survey of high-level genomic characteristics is presented (chapter 3), followed by a discussion of what distinguishes *Archaea* from *Bacteria* (chapter 4) and eukaryotes from prokaryotes (chapter 5). Chapter 6 addresses the importance of genomic content on phenotype.

The next three chapters describe the major mechanisms by which genomic organization can change. These mechanisms include homologous recombination (chapter 7), illegitimate recombination (chapter 8), and transposition (chapter 9). Then we begin to look at DNA not only as a linear sequence of genes, but as three-dimensional, physical material (chapter 10) whose packaging (chapter 11) ties structure to function (chapter 12).

Next, the mutability of genomes is explored, first by operationally defining stability (chapter 13) and instability (chapter 14) and then by analyzing the causes and effects of horizontal genetic transfer (chapters 15 and 16) and the invention of new functions (chapter 17).

Finally, a more direct link between genomic content and cellular expression is covered, looking at proteomics (chapter 18) and then functional genomics (chapter 19).

Genome sequencing might be regarded as being similar to mountain-climbing expeditions, with surveyors (bioinformaticians) and geologists (biochemists) accompanying the explorers. Once having reached a summit, however, the team leader and resource manager can lose interest. Since the mountain has been climbed and there is nowhere else to go but down, the team sets its sights instead on another (usually higher) peak, dragging the surveyors and geologists away before their respective tasks on the present mountain have been completed.

It might be better to regard genomics metaphorically as akin to civics rather than to mountaineering and cartography. Genome sequencing as a discipline arose from the technology of genome mapping, but genomics should steer away from this physical mindset and return to its biological foundations. Biology is not so much about structure as it is about function; and the cell's function is complex indeed. The analogy to civics may be appropriate in that a city's efficiency is not only a function of the people and services within, but is also a function of their relative arrangement. I hope that the reader of this book learns to appreciate the importance of gene arrangement in the genome and its role in the shaping of life.

ROBERT L. CHARLEBOIS

SEQUENCING MICROBIAL GENOMES

Christoph W. Sensen

1

The introduction of automated fluorescent sequencing methods in 1986 (1, 20) has had a major impact on microbiology, biochemistry, genetics, and related research. Automated sequencing was the key for the transition from genome mapping to genome sequencing and was one of the major factors in the rapid expansion of the new science of genomics. Even though there are basically no new methods or ideas that are applied in genomics research, the combination of methods from various scientific disciplines opens new possibilities for understanding how genomes are organized and how they function. Less than 10 years after the invention of automated sequencing, the first complete microbial genomes were published (9, 10). Three years later, more than a dozen genomes were publicly available and more than 50 were in progress, and by the year 2000, probably more than 100 will be published. For a comprehensive list of microbial-genome-sequencing efforts see http://www.mcs.anl.gov/home/gaasterl/genomes.html.

Even though several different strategies were suggested initially for sequencing microbial genomes completely, in the end, most projects have completed the genomic sequences in very similar ways. It is quite clear now that mathematical models are not very close to the real world of genome sequencing. Biology has the last word in many cases, and statistical equations apply only to a certain degree. A completed genome project will be defined here as one that produces a completely linked sequence, with an error rate of one error per 5 to 10 kbp. This chapter will give an overview of the strategies developed to sequence entire microbial genomes. The advantages and disadvantages of various approaches are discussed.

THREE STRATEGIES FOR GENOME SEQUENCING

Initially, two "pure" strategies were proposed for genome sequencing: total-genome shotgun sequencing (2) (for a viral genome) and primer walking (24) (in the context of the *Saccharomyces cerevisiae* sequencing effort).

Total-Genome Shotgun Sequencing

Total-genome shotgun sequencing was applied initially to the sequencing of viral genomes. It was also proposed initially for microbial genomes, but no genomic sequence has been completed by this method. Still,

Christoph W. Sensen, National Research Council of Canada, Institute for Marine Biosciences, 1411 Oxford Street, Halifax, Nova Scotia, Canada B3H 3Z1.

Organization of the Prokaryotic Genome, Edited by Robert L. Charlebois,
© 1999 American Society for Microbiology, Washington, D.C.

many laboratories use the technique to sequence individual large inserts (cosmids, BACs, etc.).

For total-genome shotgun sequencing, the genomic DNA is fragmented into random pieces and subcloned directly into pUC, M13, or other vectors that accept insert sizes of 1 to 5 kbp. Typically, 6 to 10 genome equivalents are sequenced to cover the DNA molecule completely by using standard primers that prime at the end of the cloning vector. Inexpensive labelled primers, which in most cases are included in the sequencing kits, are sufficient for this method. Sequence assembly is performed at the end of the sequencing phase (see below). Sequence polishing is done only once, after the entire sequence is linked into one contiguous sequence (contig). The major advantages of the method are that no map of the genome is required and large sequencing facilities can be operated with a full machine load at all times. The high redundancy is necessary because of differences in cloning efficiency in different regions of the genome and the gain in accuracy as a result of multiple coverage of each sequenced genomic base pair. Typically, no efforts are made to sequence the entire genome on both strands. Even with a 10-fold coverage of the genome, many regions remain sequenced on only one DNA strand (single stranded) because of cloning biases.

This method is mainly suitable for large sequencing facilities because no thorough analysis of the genomic sequence is possible before the assembly at the end of the sequence production. While the gene set can be identified prior to closure of the genome, most of the overall genome topology cannot be seen initially because of the large number of gaps. A clear picture of the genomic organization develops only during the late phases of the strategy. Major problems with this method occur if large fractions of the genomic DNA cannot be cloned in *Escherichia coli*. The method requires an evenly distributed set of cloned fragments to achieve a linked sequence with 6- to 10-fold coverage. Another major obstacle is

repetitive elements in the genomic sequence. Each repeat that is larger than the largest single sequencing read results in a "branching point" in the genome assembly that needs to be resolved with a different sequencing approach or through mapping experiments. Even though repeats could theoretically be resolved by analyzing their nonredundant ends, this can be very difficult if multiple repeat families exist and the number of repeats is high. Almost all of the genomes for which total-genome shotgun sequencing was chosen as the initial approach have been very small (up to 2 Mbp), with a relatively low content of repeats.

Primer-Walking Strategy and Other Directed-Sequencing Strategies

The primer-walking strategy has been tried primarily in the context of the yeast sequencing project (26). The method requires an ordered library of clones, either an overlapping set of large clones (e.g., a cosmid library) or an ordered set of discrete subclones (e.g., two 6-base cutter restriction digest libraries from a cosmid). Walking primers are calculated at both ends of each clone, and primer walks are repeated until a contig is achieved. Typically, the contigs are covered with very low redundancy (compared to the total-genome shotgun method), and both DNA strands are completely sequenced. Redundancies as low as 2.4-fold have been reported (26). The method requires a source of inexpensive sequencing primers in order to be as cost-effective as the total-genome shotgun method. Because the sequencing primers are almost all custom made, they are typically unlabelled, and thus the sequencing biochemistry applied in a primer-walking project requires either internal labelling techniques (23) or dye terminator biochemistry (17). The advantage of the method is that sequence can be analyzed soon after the project is started. Contigs will rapidly grow to a size of several kilobase pairs, and genes can easily be identified, because the primer-walking method incorporates sequence polishing after each sequencing step.

There are two major problems with the directed-sequencing strategy. The first problem is the lag time between successive sequencing rounds. After a primer is calculated, it takes a day to synthesize and purify the oligonucleotide. If sequencing can be done overnight, the resulting sequence is available the next day for a new primer calculation step. For each day of sequence production, there is another day when the automated sequencers are on hold, unless the project works with interleaved clones (or multiple sequencing projects are being run concurrently). The entire process requires many manual steps that typically result in a lower-than-ideal throughput. The other problem with this method is related to repetitive sequences whose existence is unknown during the primer calculation. A sequencing primer that binds more than once in the contig will result in "double sequence" that is unreadable. Whenever this phenomenon occurs, another, nonredundant primer further away from the end of the subcontig needs to be calculated to continue sequencing. If no satisfactory primer can be found, sequencing of the particular gap can only be continued from one end, which makes the method even slower.

Transposon-mediated sequencing (8) is another directed-sequencing approach. Relatively small inserts (3 to 5 kbp) are mapped, and transposons are inserted into each contig. The known transposon sequence is used as a primer target, and new sequence can be generated with labelled primers, similar to the primary sequencing in random shotgun sequencing strategies. The advantage of the transposon-mediated sequencing strategy is the possibility of using a single custom primer pair, which primes inside the transposon sequence, throughout the entire project. The disadvantage is the relatively slow progress due to the small size of the initial contigs and the complicated procedure that generates target clones for sequencing. No complete genome-sequencing project has been done exclusively with transposon-mediated sequencing.

Like transposon-mediated sequencing, nested deletions have been proposed as a directed approach that is able to use standard primers (24). This method is also quite slow, as it introduces extra steps into the cloning process. The method has never had more than minor applications in total-genome-sequencing projects.

Mixed Strategy

The total-genome shotgun strategy or the primer-walking strategy, if applied alone, would probably work for microbial genomes, but it would be very costly and time-consuming if a completed genomic sequence was required. A third strategy, which combines the two methods, seems to be the approach of choice for almost all groups involved in genome sequencing. Typically, there is an initial random-sequencing phase, which is followed by a directed approach that completes the genomic sequence. Random sequencing generates as many "jump-in sites" as possible for the subsequent primer-walking phase. This allows for an efficient gap closure that avoids most of the negative aspects of the pure primer-walking approach.

There are two different philosophies of how to perform the random-sequencing phase of the mixed approach. It is performed either on the entire genome (mostly by the TIGR [The Institute for Genomic Research] group) (9) or on a set of ordered clones (by almost all other groups) (19).

Sequencing projects that start with a total-genome shotgun approach do not necessarily require an initial genomic map. Only during the gap closure phase are large clones, like lambda, cosmid, phosmid, or BAC clones, necessary to complete the sequence. A complete large-clone map of the genome is not always produced in this kind of genome-sequencing project. Starting with a shotgun library of the genomic DNA has the inherent danger of misassembly of repetitive regions. In the worst case, entire regions of the genome could be left out. If this affects relatively small regions, it may go unrecognized, because there is not necessarily a complete large-clone map against which the genome assembly could be verified.

Most sequencing efforts prefer to have an ordered library with large genomic inserts at the start of the sequencing project. Initially, the map does not need to be complete, but this is preferred to keep the overall redundancy low. Microbial-genome projects dealing with relatively large genomes (more than 2 Mbp in size) (e.g., the *E. coli* [3] and the *Bacillus subtilis* [16] projects) started with complete maps, while smaller genomes with few repetitive elements (e.g., the *Haemophilus influenza* [9] and *Mycoplasma genitalium* [10] projects) were sequenced without initial complete maps. Venter et al. (22) proposed an alternative gap closure strategy that has yet to be proven to work. Working on ordered libraries certainly has the advantage that repetitive regions in the genome can be dealt with on separate clones. Even if these repeats are much longer than a single sequence read, they can be completely characterized on separate clones.

Gap closure in the mixed strategy is performed by primer walking. Typically, the random phase is completed with three- to sixfold random sequence coverage. The primer walks are usually performed on large clones, either lambda or cosmid clones, where gap-spanning clones are identified by end sequencing (TIGR strategy), or on the original clones that were used to map the genome and that represent the minimal tiling path in the sequencing project (19). Recently, some conference contributions have reported direct-sequencing technologies that use entire microbial genomes as the sequencing templates (Ellson Chen's group at Perkin-Elmer, Foster City, Calif., and others). In this method, Applied Biosystems (ABI) big-dye terminator biochemistry is used on DNA templates that can have a size of several megabase pairs. This technique may be applied in the future in the contig-linking process, thus avoiding the mapping phase entirely.

The mixed strategy allows the reduction of the overall redundancy of the genomic sequence to a coverage of four- to sixfold. There is a possibility of sequencing the entire genome on both strands, as primer walks are already applied to link contigs.

THE DIFFERENT PHASES OF GENOME-SEQUENCING PROJECTS

Regardless of the sequencing strategy chosen in a particular project, there are four general phases of the sequencing process. The first phase is the *primary sequencing* phase, where sequencing reads are accumulated in a more or less random fashion. In the primary sequencing phase, the first contigs are assembled from single sequencing reads. These contigs can be considered subcontigs in regard to the final finished sequence, which consists of a single-genome-sized contig. The next project phase is the *linking phase*, where the subcontigs of the primary sequencing phase are sequentially connected into a single contig using directed-sequencing methods. The next phase is the *polishing phase*, where one contig exists but it is not yet completely free of sequence ambiguities and thus needs to be "disambiguated." If desired, regions in the contig that were only sequenced on one DNA strand can be "double stranded" by sequencing the other DNA strand by directed DNA-sequencing techniques. The last phase is the *finished sequence*. This is the sequence that will be analyzed, annotated, and submitted to the databases. The four different phases vary in duration and intensity, based on the chosen sequencing strategy. In the pure total-genome shotgun and primer-walking approaches, some of the four phases would be very short or missing, but in the mixed strategy, all four phases typically occur. Efficient sequence analysis is important as an aid during the various phases to avoid unnecessary work. Screening for potential contaminations (e.g., *E. coli* genomic DNA or cloning vector DNA) can help to highlight clones that need to be excluded from the sequence assembly.

TOOLS FOR GENOME SEQUENCING

Automated Sequencing Machines

Only one genome project, the *E. coli* effort at the University of Wisconsin (5), made substantial progress with radioactive sequencing before changing to automated-sequencing strategies. Initially there were many different approaches to automated sequencing, including multiplexing efforts (7). Almost all projects today are done with slab gel technologies. Two different machine types presently dominate genome sequencing: one-lane, four-dye systems (e.g., ABI 373A and ABI 377) and four-lane, one-dye or four-lane, two-dye systems (Pharmacia ALF, LiCor 4000, LiCor 4200, and others).

Four-lane, one-dye and one-lane, four-dye machine types are both suitable for genome sequencing and have been used to produce completely finished genomes; however, they have different strengths and weaknesses and are ideally used in combination. One-lane, four-dye slab gel systems typically produce read lengths of 650 to 700 bp, with an average error rate of 2% or higher (15) per read. The high error rate is caused by the spectral overlap among the four dyes that are used to label the four different subreactions. The deconvolution process used to separate the different subreactions results in a problem in resolution between the bases C and G. In multiple C and G peaks, the true sequence can only be resolved if both DNA strands are sequenced. The major advantage of the ABI systems is their dye terminator biochemistry, which is available only for this kind of machine (because of patent issues). This biochemistry allows efficient walking with unlabelled primers on templates that can vary widely in size and G+C content. Protocols exist to label large inserts with the dye terminator biochemistry. Primer walks on cosmids and lambda clones are standard technology. The maximal separation distance on ABI machines is restricted to approximately 40 cm, which is about one-third less than the ideal separation distance (23). The read length would probably be 20 to 25% more if the separation distance were increased.

Four-lane, one-dye and four-lane, two-dye systems are very similar in their approaches to radioactive sequencing. Typically, either labelled primers are used or the sequencing reaction is labelled with internal label. (Dye terminators cannot be used unless the patent and licensing situation is changed by ABI or DuPont.) All four subreactions in the sequencing reaction have the same dye as a label; there is no need for a deconvolution process, and the raw data, as read from the gel, can immediately be used for sequence assembly. Sequence with a very low error rate (as low as 0.1%) can be obtained for reads in excess of 1,000 bp. At present, the best machine of this kind appears to be the LiCor 4200, which has two laser detection systems with nonoverlapping wavelengths (around 700 and 800 nm) that can detect two sequencing reactions (ideally, one from the forward reaction and one from the reverse reaction) in the same lane. With this method, up to 2,000 bp can be read from a single sequencing reaction with a very low error rate, saving 50% of the work and sequencing cost because both sequencing reactions are performed in the same reaction tube. This makes the LiCor 4200 an ideal machine for the random-sequencing phase. An average sequence coverage of between three- and fourfold can easily be obtained with LiCor 4200 instruments.

Recently, capillary electrophoresis sequencing machines, which work with essentially the same one-lane, four-dye biochemistry that is used on ABI sequencing machines, have been introduced. Even though these machines are not yet used in production sequencing, it can be expected that they will have a major impact. The formation by Perkin-Elmer and Craig Venter of the Celera company, which will use 200 ABI 3700 capillary electrophoresis sequencing machines to sequence the *Drosophila melanogaster* and human genomes, is an indication of possible future directions in genome sequencing.

Robotic Workstations

Automation is being used in genomics projects to produce very large data sets. This has led to an unprecedented automation of biochemistry laboratories. Over the last few years, the efficiency of laboratory personnel has been raised by a factor of 2 or 3. This trend will likely be steady for some time to come. Sequencing methods are standardized in 96- or 384-well formats, compatible with robotic workstations. It can be expected that sequencing-template preparation, sequencing-reaction setup, and sequencing reactions will be completely automated in many major sequencing laboratories within the next 2 to 3 years. Automated gel-loading techniques are being developed (18), which will allow an increase in the density of lanes on sequencing gels. Capillary electrophoresis machines allow loading directly from the microtiter plate, completely eliminating the need for manual loading procedures.

Primer Synthesis

The cost of synthesizing sequencing primers is still a major factor in genome-sequencing projects. Primer synthesis technology has not made major progress for quite some time. The instrument market has been consolidated by the mergers of several companies that produce these kinds of machines. Commercially available systems still do not synthesize more than 16 oligonucleotides at a time. This is surprising and will certainly change soon. There have been prototypes of oligonucleotide synthesizers that work in a 96-well format, but none of these machines has become commercially available.

Alternative strategies, including Studier's approach (21) of using libraries of short oligonucleotides, have not shown the results that were predicted 10 years ago and have been sidelined along with many other technologies that were proposed to make genome sequencing less expensive.

Software

Next to nonradioactive biochemistry, the biggest breakthrough in automated sequencing was the ability of sequencing machines to produce machine-readable files that can be used almost immediately for sequence assembly. This has had a major impact on the performance of sequencing projects. It is easy today to collaborate in a sequencing project, with automated sequencing machines in various geographical locations, because all of the original data can be shared in a very short time among the laboratories involved in the project. The *Sulfolobus solfataricus* project (19), for example, has automated sequencing machines in two locations in Canada (Halifax and Ottawa) and in Denmark, France, and the Netherlands.

There are several major sequence assembly packages, which were mainly developed by groups associated with large sequencing projects. The Staden package (4) and the TIGR Assembler (9) are examples. All of these packages perform the following tasks in an automated fashion:

- clipping vector from the sequence read
- masking bad sequence at the $5'$ and $3'$ ends of the sequence read
- splitting trace data and sequence character information
- assembling sequence reads
- linking trace data and multiple sequence alignment
- editing
- exporting consensus sequences for sequence analysis.

Directed-sequencing strategies require primer calculation programs to automate the primer selection process. Several packages have been implemented (13, 14). These packages can choose primers for the linking process and also for the final double-stranding sequencing reactions.

Sequence analysis depends on the sequencing strategy. In almost all of the large genome projects, automated analysis tools are used to generate at least an initial overview of the features in the genomic sequence. One example of such a tool is MAGPIE (11), which produces a hierarchical view of genomic features.

An example of the MAGPIE output can be viewed at http://niji.imb.nrc.ca/sulfolobus.

OPTIMAL SEQUENCING STRATEGIES FOR MICROBIAL GENOMES

There is certainly not a single sequencing strategy that could be identified as optimal for all microbial sequencing projects. The multitude of factors that are part of the decision-making process is almost overwhelming. Nonetheless, there are some general factors that need to be considered in all sequencing projects if they are to be successful.

There are two different kinds of sequencing laboratories that produce genomic sequence: sequencing factories (e.g., TIGR, Rockville, Md.; Washington University, St. Louis, Mo.; The Sanger Centre, Cambridge, England; and Genoscope, Paris, France) and smaller laboratories with an output of 2 to 5 Mbp of genomic sequence per year. The sequencing factories operate under conditions very different from those of the smaller laboratories. Sequencing factories typically use fewer biochemistries and protocols, because they can use "brute force" approaches whenever necessary to solve more complicated sequencing problems. Typically, sequencing factories produce sequence with a high coverage (6- to 10-fold per base pair) because redundancy is not an issue for sequence production.

Smaller laboratories face challenges in the sequencing process that are more or less unknown in sequencing factories. They key issue is often personnel. Typically the number of staff is the limiting factor in sequence production. Most of the smaller laboratories are not able to use the available machine capacity completely because they can only operate a single shift per day, 5 days per week. These limits dictate the use of the most efficient sequencing strategy, to be able to complete a genome in as short a time as possible. The learning curve for the techniques in a microbial genome-sequencing laboratory should not be underestimated; almost always, the first year of a new laboratory operation is relatively unproductive due to training requirements.

Small laboratories try to produce sequence with low coverage because the infrastructure for high redundancy is not available.

An ideal sequencing strategy for a smaller laboratory is close to the strategy used in the *S. solfataricus* genome project (6, 19). In this project, mapped clones (cosmids and lambda and BAC clones) are used to obtain an optimal tiling path. To start a project, the map does not need to be complete; the only requirement for the initial set of clones is that they do not overlap much (or at all). The map can be completed during the sequencing project by end sequencing cosmid, BAC, or lambda libraries and mapping the respective end sequences to existing contigs. Each of the initial clones is subcloned after nebulization into the double-stranded vector pUC18. Size selection methods are used to screen for clones with an insert size of 1 to 2 kbp. Primary sequencing is performed on LiCor 4200 automated sequencers. This allows sequencing of both strands of the pUC18 insert in the same sequencing reaction. After a threefold coverage of the initial cosmid, lambda, or BAC clone is achieved, the linking phase is started. First, all contigs that formed in the primary sequencing phase are edited for disambiguation. Next, walking primers are calculated at the ends of the contigs to generate new sequences that link all contigs into a single molecule. All primer-walking reactions are performed with ABI dye terminator technology on the initial templates (e.g., cosmids). After the linking phase is completed, the polishing phase is entered. Walking primers needed to resolve ambiguities or to double-strand single-stranded regions are calculated and applied with the same technology used in the linking phase. The result is a sequence that is completely covered on both strands and has a redundancy of approximately fourfold.

One of the hardest parts of microbial genome sequencing is calculating the costs involved in sequence production. Typically, amounts around 50 cents per finished base pair are cited by the sequencing factories (12), thus achieving the figure that was initially the threshold for the start of the human-genome-sequencing project (25). This figure can certainly be seen as the upper limit for future projects, as more efficient technologies and biochemistries will allow the reduction of sequencing costs. Today, approximately 50% of the sequence production costs are machine usage and consumables; the other half is personnel costs. With increasing levels of automation, these costs will be reduced, and in the future it may be possible to reach 10 cents per finished base pair.

Despite the high initial costs, it is already undeniable that the value of the data that is generated in microbial-genome-sequencing projects far outweighs the production costs. Never before in the history of biology have scientists gained so much insight into the organization of life as they have through the study of complete genomes.

ACKNOWLEDGMENT

This is NRCC publication 42292.

REFERENCES

1. **Ansorge, W., B. Sproat, J. Stegemann, and C. Schwager.** 1986. A non-radioactive automated method for DNA sequence determination. *J. Biochem. Biophys. Methods* **13**:315–323.
2. **Baer, R., A. T. Bankier, M. D. Biggin, P. L. Deininger, P. J. Farrell, T. G. Gibson, G. Hatfull, G. S. Hudson, S. C. Satchwell, C. Sequin, P. S. Tuffnell, and B. G. Barrell.** 1984. DNA sequence and expression of the B95-8 Epstein-Barr virus genome. *Nature* **310**:207–211.
3. **Blattner, F. R., G. Plunkett III, C. A. Bloch, N. T. Perna, V. Burland, M. Riley, J. Collado-Vides, J. D. Glassner, C. K. Rhode, G. F. Mayhew, J. Gregor, N. W. Davis, H. A. Kirkpatrick, M. A. Goeden, D. J. Rose, B. Mau, and Y. Shao.** 1997. The complete genome sequence of *Escherichia coli* K-12. *Science* **277**:1453–1474.
4. **Bonfield, J. K., K. F. Smith, and R. Staden.** 1995. A new DNA sequence assembly program. *Nucleic Acids Res.* **23**:4992–4999.
5. **Burland, V., G. Plunkett III, D. L. Daniels, and F. R. Blattner.** 1993. DNA sequence and analysis of 136 kilobases of the Escherichia coli genome: organizational symmetry around the origin of replication. *Genomics* **16**:551–561.
6. **Charlebois, R. L., T. Gaasterland, M. A. Ragan, W. F. Doolittle, and C. W. Sensen.** 1996. The *Sulfolobus solfataricus* P2 genome project. *FEBS Lett.* **389**:88–91.
7. **Dolan, M., A. Ally, M. S. Purzycki, W. Gilbert, and P. M. Gillevet.** 1995. Large-scale genomic sequencing: optimization of genomic chemical sequencing reactions. *BioTechniques* **19**:264–274.
8. **Elliott, S. J., L. A. Wainwright, T. K. McDaniel, K. G. Jarvis, Y. K. Deng, L.-C. Lai, B. P. McNamara, M. S. Donnenberg, and J. B. Kaper.** 1998. The complete sequence of the locus of enterocyte effacement (LEE) from enteropathogenic *Escherichia coli* E2348/69. *Mol. Microbiol.* **28**:1–4.
9. **Fleischmann, R. D., M. D. Adams, O. White, R. A. Clayton, E. F. Kirkness, A. R. Kerlavage, C. J. Bult, J.-F. Tomb, B. A. Dougherty, J. M. Merrick, K. McKenny, G. Sutton, W. FitzHugh, C. Fields, J. D. Gocayne, J. Scott, R. Shirley, L.-I. Liu, A. Glodek, J. M. Kelley, J. F. Weidman, C. A. Phillips, T. Spriggs, E. Hedblom, M. D. Cotton, T. R. Utterback, M. C. Hanna, D. T. Nguyen, D. M. Saudek, R. C. Brandon, L. D. Fine, J. L. Fritchman, J. L. Fuhrmann, N. S. M. Geoghagen, C. L. Gnehm, L. A. McDonald, K. V. Small, C. M. Fraser, H. O. Smith, and J. C. Venter.** 1995. Whole-genome random sequencing and assembly of *Haemophilus influenzae* Rd. *Science* **269**:496–512.
10. **Fraser, C. M., J. D. Gocayne, O. White, M. D. Adams, R. A. Clayton, R. D. Fleischmann, C. J. Bult, A. R. Kerlavage, G. Sutton, J. M. Kelley, J. L. Fritchman, J. F. Weidman, K. V. Small, M. Sandusky, J. Fuhrmann, D. Nguyen, T. R. Utterback, D. M. Saudek, C. A. Phillips, J. M. Merrick, J.-F. Tomb, B. A. Dougherty, K. F. Bott, P.-C. Hu, T. S. Lucier, S. N. Peterson, H. O. Smith, C. A. Hutchison III, and J. C. Venter.** 1995. The minimal gene complement of *Mycoplasma genitalium*. *Science* **270**:397–403.
11. **Gaasterland, T., and C. W. Sensen.** 1996. Fully automated genome analysis that reflects user needs and preferences. A detailed introduction to the MAGPIE system architecture. *Biochimie* (Paris) **78**:302–310.
12. **Glaser, V.** 1997. TIGR Conference highlights crucial issues of speed and cost of genome sequencing. *Genet. Eng. News* **17**:1.
13. **Gordon, P., et al.** Unpublished data.
14. **Haas, S., M. Vingron, A. Poustka, and S. Wiemann.** 1998. Primer design for large scale sequencing. *Nucleic Acids Res.* **26**:3006–3012.

15. **Koop, B. F., L. Rowan, W.-Q. Chen, P. Deshpande, H. Lee, and L. Hood.** 1993. Sequence length and error analysis of Sequenase and automated Taq cycle sequencing methods. *BioTechniques* **14:**442–447.

16. **Kunst. F., N. Ogasawara, I. Moszer A. M. Albertini, G. Alloni, V. Azevedo, M. G. Bertero, P. Bessières, A. Bolotin, S. Borchert, R. Borriss, L. Boursier, A. Brans, M. Braun, S. C. Brignell, S. Bron, S. Brouillet, C. V. Bruschi, B. Caldwell, V. Capuano, N. M. Carter, S.-K. Choi, J.-J. Codani, I. F. Connerton, N. J. Cummings, R. A. Daniel, F. Denizot, K. M. Devine, A. Düsterhöft, S. D. Ehrlich, P. T. Emmerson, K. D. Entian, J. Errington, C. Fabret, E. Ferrari, D. Foulger, C. Fritz, M. Fujita, Y. Fujita, S. Fuma, A. Galizzi, N. Galleron, S.-Y. Ghim, P. Glaser, A. Goffeau, E. J. Golightly, G. Grandi, G. Giuseppi, B. J. Guy, K. Haga, J. Haiech, C. R. Harwood, A. Hénaut, H. Hilbert, S. Holsappel, S. Hosono, M.-F. Hullo, M. Itaya, L. Jones, B. Joris, D. Karamata, Y. Kashahara, M. Klaerr-Blanchard, C. Klein, Y. Kobayashi, P. Koetter, G. Koningstein, S. Krogh, M. Kumano, K. Kurita, A. Lapidus, S. Lardinois, J. Lauber, V. Lazarevic, S.-M. Lee, A. Levine, H. Liu, S. Masuda, C. Mauël, C. Médigue, N. Medina, R. P. Mellado, M. Mizuno, D. Moestl, S. Nakai, M. Noback, D. Noone, M. O'Reilly, K. Ogawa, A. Ogiwara, B. Oudega, S.-H. Park, V. Parro, T. M. Pohl, D. Portetelle, S. Porwollik, A. M. Prescott, E. Presecan, P. Pujic, B. Purnelle, G. Rapoport, M. Rey, S. Reynolds, M. Rieger, C. Rivolta, E. Rocha, B. Roche, M. Rose, Y. Sadaie, T. Sato, E. Scanlan, S. Schleich, R. Schroeter, F. Scoffone, J. Sekiguchi, A. Sekowska, S. J. Seror, P. Serror, B.-S. Shin, B. Soldo, A. Sorokin, E. Tacconi, T. Takagi, H. Takahashi, K. Takemaru, M. Takeuchi, A. Tamakoshi, T. Tanaka, P. Terpstra, A. Tognoni, V. Tosato, S. Uchiyama, M. Vandenbol, F. Vannier, A. Vassarotti, A. Viari, R. Wambutt, E. Wedler, H. Wedler, T. Weitzenegger, P. Winters, A. Wipat, H. Yamamoto, K. Yamane, K. Yasumoto, K. Yata, K. Yoshida, H.-F. Yoshikawa, E. Zumstein, H. Yoshikawa, and A. Danchin.** 1997. The complete genome sequence of the Gram-positive bacterium *Bacillus subtilis*. *Nature* **390:**249–256.

17. **Lee, L. G., C. R. Cornell, S. L. Woo, R. D. Cheng, B. F. McArdle, C. W. Fuller, N. D. Halloran, and R. K. Wilson.** 1992. DNA sequencing with dye-labelled terminators and T7 polymerase: effect of dyes and dNTPs on the incorporation of dye-terminators and probability analysis of termination fragments. *Nucleic Acids Res.* **20:**2471–2483.

18. **Panussis, D. A., M. W. Cook, L. L. Rifkin, J. E. Snider, J. T. Strong, R. M. McGrane, R. K. Wilson, and E. R. Mardis.** 1998. A pneumatic device for rapid loading of DNA sequencing gels. *Genome Res.* **8:**543–548.

19. **Sensen, C. W., H. P. Klenk, R. K. Singh, G. Allard, C. C.-Y. Chan, Q. Liu, S. Penny, F. Young, M. Schenk, T. Gaasterland, W. F. Doolittle, M. A. Ragan, and R. L. Charlebois.** 1996. Organizational characteristics and information content of an archaeal genome: 156 kbp of contiguous sequence from *Sulfolobus solfataricus* P2. *Mol. Microbiol.* **22:**175–191.

20. **Smith, L. M., J. Z. Sanders, R. J. Kaiser, P. Hughes, C. Dodd, C. R. Connel, C. Heiner, S. B. H. Kent, and L. E. Hood.** 1986. Fluorescence detection in automated DNA sequence analysis. *Nature* **321:**674–679.

21. **Studier, F.** 1989. A strategy for high-volume sequencing of cosmid DNAs: random and directed priming with a library of oligonucleotides. *Proc. Natl. Acad. Sci. USA* **86:**6917–6921.

22. **Venter, J. C., H. O. Smith, and L. Hood.** 1996. A new strategy for genome sequencing. *Nature* **381:**364–366.

23. **Voss, H., S. Wiemann, U. Wirkner, C. Schwager, J. Zimmermann, J. Stegemann, H. Erfle, N. A. Hewitt, T. Rupp, and W. Ansorge.** 1992. Automated DNA sequencing system resolving 1000 bases with fluorescein*-15-dATP as internal label. *Methods Mol. Cell. Biol.* **3:**153–155.

24. **Voss, H., S. Wiemann, D. Grothues, C. Sensen, J. Zimmermann, C. Schwager, J. Stegemann, H. Erfle, T. Rupp, and W. Ansorge.** 1993. Automated low-redundancy large-scale DNA sequencing by primer walking. *BioTechniques* **15:**714–721.

25. **Watson, J. D.** 1990. The human genome project. *Science* **248:**44–49.

26. **Wiemann, S., H. Voss, C. Schwager, T. Rupp, J. Stegemann, J. Zimmermann, D. Grothues, C. Sensen, H. Erfle, A. Banrevi, and W. Ansorge.** 1993. Sequencing and analysis of 51.6 kilobases on the left arm of chromosome XI from *Saccharomyces cerevisiae* reveals 23 open reading frames including the fas1 gene. *Yeast* **9:** 1343–1348.

STATISTICAL PREDICTIONS OF CODING REGIONS IN PROKARYOTIC GENOMES BY USING INHOMOGENEOUS MARKOV MODELS

Mark Borodovsky, William S. Hayes, and Alexander V. Lukashin

2

PROTEIN-CODING GENES AND ORFs IN BACTERIAL GENOMES

Recently the first complete bacterial genome, that of *Haemophilus influenzae*, was sequenced (18). In two and a half years the number of fully sequenced bacterial genomes grew by 10 more, and complete sequences of more than 20 other genomes are expected before the year 2000. The progress in sequencing has dramatically changed the methodology of sequence annotation. Slow and tedious biochemical experiments have been complemented by fast and increasingly reliable methods of computer analysis. Finding genes by computer in bacterial DNA is easy and difficult at the same time. On one hand, the search for protein-coding genes is facilitated by the dense packing of bacterial genes as well as by the absence of introns. On the other hand, it is difficult to locate rather short genes as well as to find the precise starting point of a gene, the position of translation initiation. Dense gene packing leads to gene overlaps; for instance, as many

as 30% of the *Escherichia coli* genes are thought to overlap with their neighbors. The overlaps hamper the detection of exact gene boundaries as well.

A useful concept in bacterial protein-coding gene annotation and in gene-finding algorithms is the open reading frame (ORF). An ORF is a DNA fragment whose 5′ end starts at a start codon. An ORF consists of consecutive triplets (codons) and terminates at the first stop (nonsense) codon encountered. A group of nested ORFs sharing the same 3′ stop codon is usually found in a DNA sequence between two in-frame stop codons. It is convenient to deal with the longest ORF of the group. In this chapter, the term ORF usually refers to the longest ORF. Every bacterial protein-coding gene resides inside an ORF, but far from every ORF observed in bacterial DNA sequence hosts a gene. Nevertheless, ORF detection is a reasonable start for gene hunting. A given DNA sequence carries information about both DNA strands and provides data for six possible reading frames: three frames in the direct DNA strand as well as three frames in the complementary DNA strand read in the opposite direction.

There is significant difference between length distributions of ORFs containing functional protein-coding genes and the "empty"

Mark Borodovsky, Schools of Biology and Mathematics, Georgia Institute of Technology, Atlanta, GA 30332-0230. *William S. Hayes and Alexander V. Lukashin*, School of Biology, Georgia Institute of Technology, Atlanta, GA 30332-0230.

Organization of the Prokaryotic Genome, Edited by Robert L. Charlebois,
© 1999 American Society for Microbiology, Washington, D.C.

ORFs that occur by chance. For the complete *Bacillus subtilis* genome, such distributions were computed with the GenBank annotation and are shown in Fig. 1. The line graphs approximate the histograms. The length frequency distribution of random ORFs has the shape of geometric distribution. The vast majority of random ORFs are shorter than 500 nucleotides (nt), while the mean length of protein-coding ORFs is equal to 893 nt. More precisely, speaking of the *B. subtilis* genome, an ORF longer than 500 nt has a 98% likelihood of hosting a gene. Analyses done for genomes with higher (or lower) GC content produced similar distributions of length frequencies for random ORFs with a bit larger (or smaller) average length value. The average length of protein-coding ORFs stayed close to 900 nt. Such an observation suggests that to identify a "long" gene-hosting ORF should be a rather simple matter. However, the exact 5′ boundary of the gene has yet to be determined. Therefore, there are at least two important things to learn: (i) how to accurately identify short bacterial genes (shorter than 500 nt) and (ii) how to determine the translation start position.

DETECTING PROTEIN-CODING POTENTIAL

There are two approaches to discovering whether an ORF contains a gene. The intrinsic, or statistical, approach examines ORF length and/or oligonucleotide frequencies within the ORF sequence to determine whether these values are typical for a true pro-

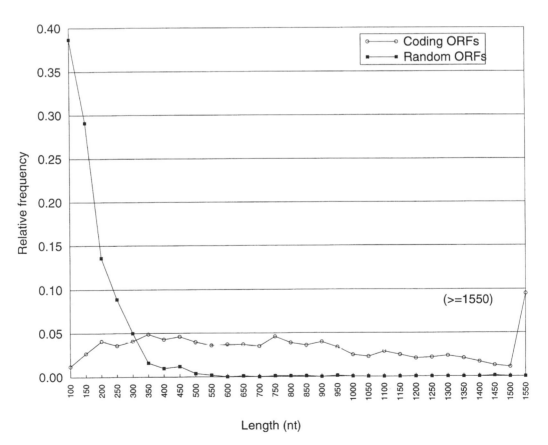

FIGURE 1 Distributions of ORF lengths in the *B. subtilis* genome for protein-coding ORFs and for random ORFs found in intergenic regions.

tein-coding sequence or for a true noncoding sequence. Figure 2 shows coding and noncoding patterns of nucleotide frequencies revealed by multiple alignment. A large number of nucleotide sequences from the B. subtilis genome containing the translation start were aligned against the start codons that occupy positions 1 to 3. The pattern observed in noncoding regions, upstream of the start codon position, is clearly more random than the three-periodic pattern observed in the coding region, downstream from the start codon position. The ribosome binding site (RBS) position, −19 to −4, is marked by a strong abundance of purines. The difference in position-specific nucleotide frequency patterns of coding and noncoding regions (Fig. 2), as well as more pronounced differences that exist in position-specific oligonucleotide frequencies (not shown), has been efficiently used in

statistical gene-finding algorithms. The extrinsic, or similarity search, approach seeks to determine whether the protein translation of a given ORF reveals a significant similarity to a known protein sequence (1). Normally, such a finding indicates the presence of a functional gene and provides an insight into the possible function of the protein product. The capacity of similarity searches is high but limited by the fact that up to 30% of newly discovered bacterial proteins, the "pioneers," do not show any similarity to proteins already stored in the databases (11, 22, 31–33). The two approaches to gene identification are complementary, and a researcher using the statistical (intrinsic) approach should keep in mind the additional extrinsic opportunities and vice versa (8, 20).

The first characterization of nucleotide compositional bias related to the DNA pro-

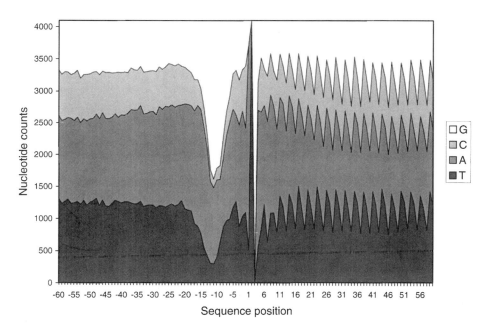

FIGURE 2 Positional nucleotide frequency patterns observed in 4,076 B. subtilis sequences, with those that surrounded annotated gene starts aligned against the start codon positions. Frequency values for T, A, C, and G nucleotides in each position are shown superimposed. The three-periodic pattern is clearly seen within the protein-coding region, downstream from the start codon position at positions 1 to 3. A rather random pattern is seen upstream of the start codon position, with the RBS region, positions −20 to −5, shown as a sharp decrease of pyrimidins T and C and abundance of purines G and A.

tein-coding function was apparently done by Erickson and Altman in 1979 (16). In the 1980s a number of DNA coding-potential assessment algorithms based on various statistically detectable features of protein-coding sequences were suggested (see Fickett and Tung [17] for a review and comparison of 21 different measures). These measures include codon position-specific frequencies of oligonucleotides, starting from mononucleotides and, notably, including triplets, particularly codons (codon usage pattern).

Experimenting with oligonucleotide statistics has led to a natural attempt at using probabilistic Markov chain models whose parameters are defined by oligonucleotide statistics. A Markov model theory provided a natural basis for mathematical treatment of DNA sequence (15). Ordinary Markov models have been used since the earliest studies of DNA sequences (21). Later, with a larger amount of sequence data available, three-periodic inhomogeneous Markov models were proven to be more informative and more useful for protein-coding-sequence modeling and recognition (9, 10, 46). Interestingly, the most discriminative protein-coding-potential measures as assessed by Fickett and Tung, the hexamer frequencies specific to three codon positions, could be integrated into the fifth-order three-periodic Markov model statistics. Subsequently, Markov models of different types, ordinary (homogeneous) and inhomogeneous, the ones necessary to describe functionally distinct regions of DNA (see below), could be integrated within a hidden Markov model (HMM) architecture. Recent applications of the HMM approach to bacterial gene identification (40) are described in later sections.

THE GeneMark CORE ALGORITHM

The GeneMark core algorithm identifies an anonymous DNA sequence fragment [$S = (b_1, b_2, \ldots, b_n)$] as a sequence of one of three possible pure types: (i) protein-coding sequence (gene), (ii) DNA sequence complementary to a protein-coding gene (gene shadow), and (iii) noncoding sequence whose complement is a noncoding sequence as well (6). In both cases i and ii, the algorithm also indicates the reading frame for genetic code reading.

Here we need to introduce Markov models in more detail. An order k Markov model for DNA sequence defines the probability of observing a nucleotide, b, following a given k-tuple (a word, W, of length k). The Markov model is characterized by the set of transition probabilities, $P(b|W)$, for $W \in$ (words of length k) and $b \in (T, C, A, G)$ and initial probabilities, $P_0(W)$, for $W \in$ (words of length k). In ordinary Markov models, transition probabilities, $P(b|W)$, do not depend on the position of W in the sequence and are estimated by using the counts of words observed in a training sequence (4):
$P(b|W) = n(W\&b)/\{n(W\&T) + n(W\&C) + n(W\&A) + n(W\&G)\}$, $P_0(W) = n(W)/N$.
Here $n(W\&b)$ denoted the count of events when the word W is observed and is followed by nucleotide b and N denotes the total count of k-tuples. Ordinary Markov models are used to describe noncoding DNA sequence. A Markov model of a protein-coding region should reproduce the three-periodic pattern in nucleotide frequencies (Fig. 2). Therefore, probabilities $P(b|W)$ and $P_0(W)$ should depend on the location (positional phase) of the word W in the sequence. This positional phase is characterized, for instance, by the frame (codon position) where the word W begins. The three-periodic Markov model is described by three sets of probabilities, $P_i(b|W)$, for $W \in \{$words of length $k\}$ and $b \in \{T, C, A, G\}$ and three sets of initial probabilities, $P_{0i}(W)$, for $W \in \{$words of length $k\}$, where $i = 1, 2, 3$ designates the frame of the initial nucleotide of W. The values of the probabilities are estimated from counts of $(k + 1)$-tuples in a sample of sequences of a known type (either a sample of protein-coding sequences or a sample of sequences complementary to protein-coding ones) as follows:
$P_i(b|W) = n_i(W\&b)/\{n_i(W\&T) + n_i(W\&C) + n_i(W\&A) + n_i(W\&G)\}$, where $i = 1, 2, 3$, and $P_{0i}(W) = n_i(W)/N_i$, where $i = 1, 2, 3$.

Here $n_i(W\&b)$ denotes the count of occasions when the word W is observed to begin in frame position i and is followed by the nucleotide b and N_i, is the total count of k-tuples observed to begin in frame position i in the training sequence. Now the Markov models are ready to be used in the algorithm. For an anonymous fragment, S, we calculate the probabilities $P(S|\text{coding type } g) = P^g_{f(j)}(W_1)$ $\prod P^g_{f(j)}(b_{j+k}|W_j)$, where $j = 1, 2, \ldots, n - k$ and $g = d$ or r and $P(S|\text{noncoding}) = P(W_1)$ $\prod P(b_{j+k}|W_j)$, where $j = 1, 2, \ldots, n - k$. Here, g indicates the type of protein-coding model (direct coding or reversed coding), $f(j)$ is the frame position of the jth nucleotide, and W_j is the word of length k starting at the jth nucleotide. The start of a sequence S could fall into any frame; therefore, the event that the sequence S is of a particular coding type (direct or reversed) splits into three subevents defined by the frame of the initial nucleotide, $f(1)$. Therefore, seven categories, six coding and one noncoding, can be considered for a sequence S. Probabilities $P(S|\text{category } i)$, where $i = 1, 2, \ldots 7$ of the events where S appears within a sequence of a pure category i are computed with the Markov models defined above. Assuming that a priori probability $P(\text{category } i)$ for S to belong to a category i is known, the a posteriori probabilities are defined as follows:

$$P(\text{category } i|S)$$
$$= \frac{P(S|\text{category } i)\,P(\text{category } i)}{\Sigma\,P(S|\text{category } i)\,P(\text{category } i)} \quad (1)$$

where $i - 1, 2, \ldots 7$

The a priori probability to pick up a noncoding sequence S, $P(\text{category } 7)$, is assumed to be equal to 1/2, while the a priori probability to pick up a sequence S that belongs to each one of six coding categories, $P(\text{category } i)$, where $i = 1, 2, \ldots 6$, is assumed to be equal to $[1 - P(\text{category } 7)]/6 = 1/12$ (this set of a priori values is a default in the GeneMark program).

The value $P(\text{category } i|S) = p_i$, where $i = 1, \ldots, 7$, $(\Sigma p_i = 1)$, defines the a posteriori probability for the hypothesis that the sequence S carries genetic code in a particular reading frame, or that the sequence S carries no genetic code, given the observed sequence S. If one p_i, where $i = 1, \ldots, 6$, is greater than T, the threshold, then the fragment S is identified as a protein-coding region in the respective reading frame. If p_7 is greater than T or if no p_i, where $i = 1, \ldots, 6$, is greater than T, S is identified as noncoding. Note that formula 1 treats the protein-coding events in six possible frames as mutually exclusive; therefore, strictly speaking, it is not applicable for overlapping genes. The main algorithm parameters are the values of the threshold and the a priori probability values, as well as the Markov model orders and the initial and transition probabilities. The length, n, of the input sequence S is an important parameter also.

THE CORE ALGORITHM SENSITIVITY TO PARAMETER VARIATIONS

A change of algorithm parameters causes increase or decrease in the prediction accuracy, which could be characterized by false-negative or false-positive error rates or by values of sensitivity or specificity. These terms are defined as follows. Let us consider a set of rather short sequences, S_i, where $i = 1, 2, \ldots N$, consisting of n_1 pure coding sequences and n_2 pure noncoding sequences. Let us assume that the core algorithm applied to these sequences has identified m_1 coding sequences as noncoding and m_2 noncoding sequences as coding. The false-negative error rate is defined, then, as $100(m_1/n_1)\%$ and the false-positive error rate is defined as $100(m_2/n_2)\%$. The sensitivity with regard to protein-coding prediction is defined as $100[(n_1 - m_1)/n_1]\%$, the percentage of correctly predicted coding sequences among all tested coding sequences. The specificity with regard to protein-coding prediction is defined as $100[(n_1 - m_1)/(n_1 - m_1 + m_2)]\%$, the percentage of correct coding predictions among all coding predictions made. It was observed that the greater the length of fragment S, the higher the sensitivity of the algorithm's coding region prediction became. However, the longer fragment S is more likely to contain a boundary between coding and noncoding regions. Therefore, the increase of the fragment S length is undesirable under the algorithm assumption that it deals with a pure-type sequence. In the discussion below, if not specified otherwise, we will assume that a fragment S is a rather short sequence fragment, about 100 nt long. The threshold, T, affects the balance between sensitivity and specificity of prediction. Larger T values make the sensitivity lower and the specificity higher. Smaller T values may raise sensitivity to an extreme of 100%, but the specificity will drop dramatically. A similar effect on sensitivity and specificity was induced by making larger or smaller another parameter, the value of the a priori probability of the noncoding sequence category in formula 1.

Parameters of the Markov models of coding and noncoding regions, such as model orders and initial and transition probabilities, are implicitly present in formula 1. Note that the orders of coding and noncoding sequence models should not necessarily be the same. Using a high order of the Markov model of coding sequence along with a low order of the noncoding model, even zero order, did not produce any noticeable loss of accuracy (23a). This observation corresponds to the fact that a noncoding sequence does not reveal any strong pattern of nucleotide ordering (e.g., Fig. 2). In the following discussion, we focus on variations of the parameters of the protein-coding-sequence model.

First of all, it may be true that a complete set of bacterial genes can be accurately modeled by a single Markov model, the model of a certain order trained on a randomly chosen gene subset. Most likely such a model does not exist. For instance, according to Médigue et al. (42), there are three classes of genes in *E. coli*. Class I genes possess intermediate codon usage bias. These genes maintain a low or intermediate level of expression, although some genes may occasionally be expressed at a very high level in environmentally triggered (rare) conditions. Class II genes have a high codon usage bias. These genes are highly expressed under exponential growth conditions. Genes from class III, with low codon usage bias, mainly belong to plasmids and insertion sequences. This class also includes genes coding for fimbriae, major pili, many membrane proteins, restriction endonucleases, and lambdoid phage Iysogeny control proteins. Many class III genes can be expressed at a fairly high level, but their weakly biased codon usage pattern does not reflect the proportions in the distribution of *E. coli* tRNAs under exponential growth conditions. To separate class I, class II, and class III genes, Médigue et al. used clusterization of the *E. coli* gene sequences characterized as vectors of 61 codon frequencies (42). Lawrence and Ochman (38) performed further analysis of the *E. coli* gene classes and presented a model of amelioration of horizontally transferred gene sequences within the *E. coli* genome. The nucleotides in the third positions of codons in newly transferred genes were assumed to be the most convenient targets for a mutation process that adjusted the GC content of the transferred gene to the average GC content of the host genome. However, the rate of this change was not high enough to completely adopt the codon usage pattern of the host. The remaining difference may present a difficulty for detecting coding potential in horizontally transferred gene sequence by the Markov model trained on the bulk gene set.

PREDICTION ACCURACY OF THE CORE ALGORITHM (PURE-TYPE DNA FRAGMENTS)

To evaluate the prediction accuracy of the GeneMark core algorithm, we used updated sets of genes of the *E. coli* classes I, II, and III containing 1,234, 231, and 156 genes, respectively (reference 14a; see reference 7 for earlier results). By the rules of cross-validation, the testing for each class was performed seven times in such a way that six-sevenths of the set was used to train the gene model and the remaining one-seventh was used for the testing (29). A control set of coding sequences was divided into nonoverlapping fragments of identical lengths, 96 nt, and predictions for the sample of pure-type fragments were made by the core algorithm. The prediction accuracy characteristics observed for seven nonoverlapping test sets were averaged. The cross-class prediction error rates were defined as well. The model derived from all sequences of one gene class was tested on the set of 96-nt coding fragments derived from sequences of another class. The error rates produced by the class-specific models employed by the core algorithm are shown in Fig. 3. The observed error rate typically decreased sharply as the model order increased from 0 to 2. Then, after an almost flat segment, the error rate started to increase. The same general trend was observed for coding-potential detection error rates in DNA sequences of other species (23a). Notably, the second-order model is derived from triplet statistics, including codon frequencies.

A model trained on a certain class of genes could be expected to be the most accurate predictor for the sequences selected from the same class. For instance, the class II model performed well for class II sequences, recognized sequences from class I genes poorly, and failed to identify sequences from class III (Fig. 3b). However, the model trained on class I genes identified class II gene sequences with lower error rates than those observed for class I sequences themselves (Fig. 3a). Therefore, the class I model can be used to identify both class I and class II genes, comprising almost all native E. coli genes. Similarly, the class III model of order 2 recognized sequences from both class I and class II genes with better accuracy than sequences from class III (Fig. 3c). This observation, surprisingly, may indicate that the class III model is the only one needed for finding genes in E. coli. This idea has been further explored for the case when whole gene sequences, not the short fixed-length fragments, are considered (see below). The phenomenon

of more accurately detecting coding potential in a gene sequence of another class, as compared to the prediction of gene sequence from the training class, could be explained in terms of Kullback-Liebler (K-L) distance. The K-L distance, a concept frequently used in information theory, is a measure of distance between two statistical models (14). It can be shown that the protein-coding-sequence Markov model that "misfits" the sequence data may actually have a larger "efficient" K-L distance toward the noncoding model than the K-L distance defined in the normal case where the gene model fits the sequence data (10a). The greater distance facilitates better discrimination. The results presented above led us to the conclusion that the set of models for gene finding may be reduced to the gene models of class I and class III genes, called below typical and atypical gene models, respectively.

Interestingly, the dependence of prediction accuracy on the model order was greatly relaxed when, instead of DNA fragments of fixed short lengths, we analyzed genomic sequences containing multiple genes. In terms of accuracy of gene prediction versus annotation (see below), the first- and the zero-order Markov models are much closer to the prediction accuracy of the second- and higher-order models than was seen above (Fig. 3). In this context, the E. coli class III model alone produced nearly as accurate predictions as class I and class III models used in concert (40a).

WHICH ORDER MARKOV MODEL SHOULD BE USED?

Somewhat surprisingly (Fig. 3), the highest-order Markov model of protein-coding DNA did not necessarily produce the highest prediction accuracy. The reason for this observation is as follows. The amount of statistics (counts of an oligonucleotide of a particular length) available in a given training sequence for estimating a parameter of the Markov model of order k decreases geometrically as k increases. The larger the model order, the higher the chance that a given training set

a) Testing Class I model

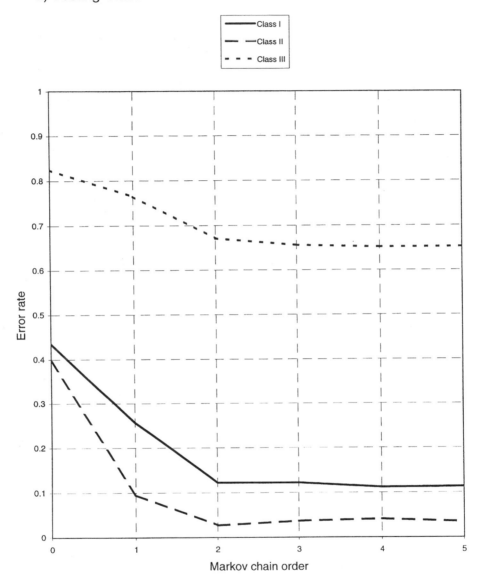

FIGURE 3 False-positive and false-negative rates produced by the GeneMark core algorithm for three sets of *E. coli* gene fragments derived from genes of class I (a), class II (b), and class III (c).

does not have enough information for reliable estimation of Markov transition probabilities. What minimal-size training set is suitable to determine a sufficiently accurate model of order k? The accuracy of the estimation of the model parameter $P(b|W)$ can be defined in terms of $100(1 - \alpha)\%$ confidence intervals for $P(b|W)$. The larger the observed number of counts of the word W, the narrower the width of the confidence interval becomes. If the maximum error of the $P(b|W)$ point estimate $p = n(W\&b)/n(W)$ is required to be less than

b) Testing Class II model

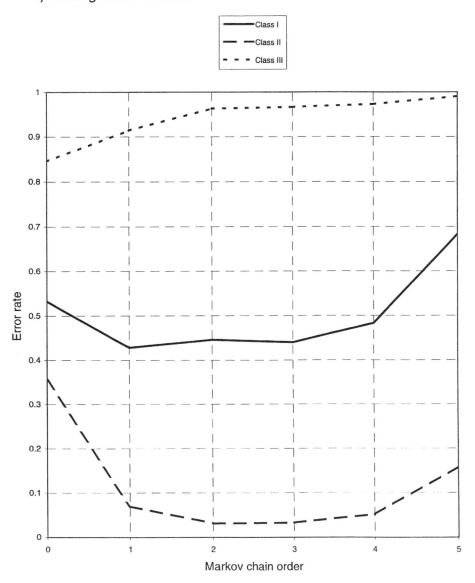

FIGURE 3 (*Continued*)

ε, then the W count, n_s, should be at least $z_{\alpha/2}^2/4\varepsilon^2$, where $z_{\alpha/2}$ is the z score for $\alpha/2$ (26). For ε equal to 0.02 and α equal to 0.05, the number n_s is equal to 400. There are 4^k- possible k-tuples, and let us assume that the sequence composition uniformity condition is fulfilled in the sense that in a DNA sequence of length L the observed counts of k words

are close to $L/4^k$. Obviously, if equally high estimation accuracy is required for each model parameter, then a bias in sequence composition will increase the necessary size of the training set. Under an assumption of uniform composition, the minimum size of the training set required for estimating parameters of an ordinary Markov model of order k is $L =$

c) Testing Class III model

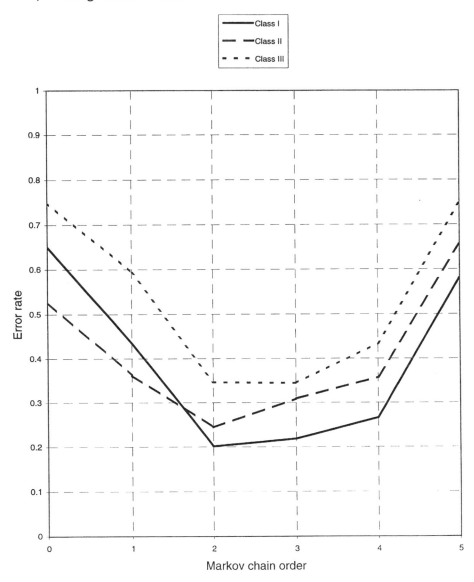

FIGURE 3 (*Continued*)

$4^{(k+1)} \cdot 400$, where $k = 0,1, \ldots$ (Table 1, middle row). For estimating parameters of the three-periodic Markov model, a training set three times as large is needed (Table 1, bottom row). An alternative approach argues that the most accurate model to be obtained from a given training sequence does not necessarily belong to the class of Markov models of fixed order. If a Markov model of fixed order k is used in formula 1, the transition probability $P(b_i|W_k)$ for a nucleotide b in a given position i is defined by the ratio $n(W_k\&b)/n(W_k)$, where $n(W_k)$ and $n(W_k\&b)$ are the reference set word counts for the word W_k of length k preceding position i and for the word $W_k\&b$ of length $k + 1$. If the value $n(W_k)$ is larger

TABLE 1 Sizes of sequence sets recommended for training a Markov model of a given order

Model type	Size (kb) for model of order:								
	0	1	2	3	4	5	6	7	8
Ordinary	0.4	1.6	6.4	25.6	102.4	409.6	1,638	6,553	26,214
Three periodic	1.2	4.8	19.2	76.8	307.2	1,229.0	4,915	19,661	78,643

than n_s, the transition probability estimate is considered statistically well defined. Let us now consider a set of words W_1, W_2, . . . , W_k, . . . preceding b_i such that W_1 is the nucleotide preceding b_i, W_2 is the dinucleotide preceding b_i, and so on. By looking up the counts of these words in the reference set, we should define the integer k^* such that $n(W_{k^*})$ is greater than n_s and $n(W_{K^*+1})$ is less than n_s. Then $P(b_i|W_{k^*})$ is used in formula 1 for nucleotide b_i. Selecting $k^* = k^*(b_i)$ and using $P(b_i|W_{k^*(bi)})$ for each b_i where $i = 2, . . . , n$ means that the order of the model varies in formula 1 when an a posteriori probability value $P(\text{category } j|S)$ is determined. An approach of this sort or an even further refined approach, called an interpolated Markov model (27), when transition probability $P(b_i|\text{set of preceding words } W_k)$ is defined as a linear combination of $P(b_i|W_k)$, where $k = 0, 1, 2, . . . $, has a potential to improve the prediction accuracy (44). In fact, the variable-order model involves a larger number of parameters and thus possesses a higher potential for higher resolution if a large enough training set is available. In practice, the actual difference in results produced by a fixed-order model versus a variable-order model depends on the degree of asymmetry of word distributions in a particular training sequence. For instance, the asymmetry can be characterized by the percentage of the words of length k whose counts in the training set exceed a critical number, such as an n_s value of 400. To give an example, counts of words of different length were done in the set of coding sequences from the *Helicobacter pylori* genome (Table 2). The word counts were obtained for each reading frame as required for deriving parameters of a three-periodic Markov model.

For small word lengths, k, the counts of all possible words W_k, $n(W_k)$, are larger than 400 (100% entries in Table 2), while for large k, almost all counts $n(W_k)$ are smaller than 400 (0% entries). Transitions from 100 to 0% occur sharply at word lengths of 4 and 5. Similar sharp transitions have been observed for other bacterial genomes (23a). This property of word distributions, a rather quick transition of word counts into a critical region n_s greater than 400 as k decreases, indicates that the prediction accuracy of a variable-order model may not differ much from the accuracy of the fixed-order model. Particularly in the *H. pylori* case, the prediction accuracy of a variable-order model, operating with word lengths l of ≥ 5, should be compared with the accuracy of a fixed model of order 4. The actual figures, close indeed, are given below.

GENE FINDING IN GENOMIC SEQUENCES

The assumption used in the core algorithm that S is a sequence of a pure type is unlikely to be true for a sufficiently long sequence. There are three possibilities for dealing with long DNA sequences. The first option is to divide a long sequence into short, fixed-length, overlapping sequences (the sliding-window technique). The majority of these sequences will be of pure types, and the local analyses will produce reasonable results. An advantage of this type of approach is that it lends itself readily to visualization of a local coding potential. The graphical representation per se is indicative of gene boundaries, frameshifts, and other irregularities of the coding-potential distribution (41). The statistical score F_{stat} for an ORF is obtained by averaging a posteriori probability values computed for a

TABLE 2 Percentage of words of given length k (for *H. pylori* genome) in which counts exceed the critical number, n_s, of 400[a]

Frame	% for k of:							
	1	2	3	4	5	6	7	8
1	100.0	100.0	95.3	83.6	36.5	6.8	0.4	0.0
2	100.0	100.0	100.0	84.4	36.4	6.4	0.4	0.0
3	100.0	100.0	100.0	86.3	38.4	5.9	0.4	0.0

[a] Sufficient to accurately determine the transition probability from the k word to the $k + 1$ word, the parameter of the Markov model of order k.

set of sequence fragments covering the ORF. These fragments are selected by a sliding window of length W moving with a step D, a multiple of 3. The fragments are chosen to belong to the same frame as the ORF. The scoring function F_{stat} was shown to be insensitive to variations of parameter D (6). The GeneMark program defaults for W and D are 96 and 12 nt, respectively. An ORF is predicted to contain a gene if F_{stat} is greater than T_g, the threshold; otherwise, if F_{stat} is less than T_g, an ORF is predicted to be noncoding. By computational experiments with known genes and random ORFs, one may define a value of T_g providing desirable prediction sensitivity and specificity values. Raising T_g leads to increasing specificity and decreasing sensitivity. Lowering T_g may raise sensitivity to an extreme of 100%, but the specificity will drop significantly unless some filtering is used to exclude unrealistic predictions of very short ORFs.

The fixed window length and fixed step of window sliding seem to impose some artificial restrictions on the sequence analysis procedure. Therefore, the second option is to select and analyze by the core GeneMark algorithm some potential pure-type sequences, such as ORFs of various lengths. For instance, the core algorithm of the GeneMark program (6a) has an option to evaluate a whole ORF as a single sequence S. A suitable T_g value has to be defined for this case. Also, the analysis of every ORF taken in isolation requires a special postprocessing to reconcile the results obtained for the ORFs that overlap in actual ge-

nomic sequence. Both options described so far, especially the second, do not identify the 5′ gene boundary with enough precision. The global score of the whole ORF is not sensitive enough to variations of the position of the translation start within a range that is small in comparison with the ORF size.

The third option is to modify formula 1 to be applicable to a sequence S that may contain subsequences of different genetic categories. Such an approach was implemented in the GeneMark.hmm algorithm modeling switches between coding and noncoding regions as transitions between states of the HMM (40).

PREDICTION OF GENE-CONTAINING ORFs IN WHOLE GENOMES

We used data on 10 complete bacterial genomes available from the GenBank database: *H. influenzae* (18) (GenBank accession no. L42023), *Mycoplasma genitalium* (19) (L43967), *Methanococcus jannaschii* (11) (L77117), *Mycoplasma pneumoniae* (25) (U00089), *Synechocystis* sp. strain PCC6803 (28), *E. coli* (5) (U00096), *H. pylori* (47) (AE000511), *Methanobacterium thermoautotrophicum* (45) (AE000666), *B. subtilis* (37) (AL009126), and *Archaeoglobus fulgidus* (30) (AE000782). The GeneMark gene predictions obtained for each genome were compared with the GenBank annotation. The accuracy of predictions was characterized by the values of sensitivity (Sn) and specificity (Sp). The sensitivity value was defined as the fraction of annotated genes whose location was predicted correctly in terms of the prediction and the annotation belonging to the same group of nested ORFs. Specificity was defined as the fraction of all predictions matching annotated genes. Obviously, the parameters of a gene-finding program can be adjusted so that the sensitivity would be brought close to 100% by making an ORF protein coding for the slightest reason. However, the specificity of the method would drop significantly due to a large number of false positives. To adequately characterize a gene-finding system, both sensitivity and specificity must be presented. In the core GeneMark algorithm, the balance between sensitivity and specificity is efficiently controlled by the threshold parameter, T_g (see above). The default value of the threshold (0.5) provides the sensitivity-specificity balance for GeneMark prediction accuracy (Table 3). By lowering T_g to 0.1, a higher sensitivity can be achieved while the specificity is kept high by filtering out unrealistic short overlapping ORFs (Table 3). The recently developed algorithm GLIMMER uses interpolated Markov models to compute the a posteriori probability of coding function (44). The GLIMMER program, kindly provided by Steven Salzberg, was used to predict genes in eight complete genomes (the nonstandard genetic code of two *Myco-*

plasma species precluded running GLIMMER for these genomes at that time [February 1998]). This test produced the total average characteristics of Sn, 0.98, and Sp, 0.84. Note that the averages of Sn and Sp values for Gene-Mark ($T_g = 0.1$) for the same eight genomes are exactly the same (as obtained from Table 3).

MORE ACCURATE LOCATION OF GENE BOUNDARIES

As we have seen, an ORF that carries a protein-coding gene can be detected with high accuracy by the statistical methods described above. However, the exact position of the translation initiation codon might remain uncertain. The range of uncertainty for the gene start, if GeneMark is used, is the size of the algorithm sliding window, i.e., about 100 nt. As a palliative, the GeneMark program indicates several possible start codons and scores them (6a). Obviously, the accuracy of prediction of the beginning of the gene needs to be improved. The algorithm predicting the exact position of the translation start should have variables related to gene boundaries. This requirement might be satisfied as follows. (Note that DNA sequence annotation identifies the

TABLE 3 GeneMark performance for 10 complete genomes[a]

Genome	Value			
	$T_g = 0.5$		$T_g = 0.1$	
	Sn	Sp	Sn	Sp
A. fulgidus	0.93	0.95	0.98	0.88
B. subtilis	0.93	0,97	0.98	0.74
E. coli	0.88	0.99	0.97	0.83
H. influenzae	0.96	0.96	0.99	0.85
H. pylori	0.93	0.96	0.97	0.84
M jannaschii	0.99	0.96	0.99	0.84
M. thermoautotrophicum	0.93	0.99	0.96	0.91
Synechocystis	0.91	0.93	0.97	0.79
M. genitalium	0.96	0.89	0.99	0.66
M. pneumoniae	0.92	0.92	0.97	0.70
Average	0.93	0.96	0.98	0.81

[a] Sn, sensitivity (the fraction of annotated genes for which the ORF has been correctly predicted); Sp, specificity (the fraction of predicted genes for which the ORF is identical to an ORF of an annotated gene).

functional role of each nucleotide in the sequence.) For a DNA sequence of length L
designated as $S = \{b_1, b_2, \ldots, b_L\}$, the functional role of each nucleotide could be indi-
cated by a "functional" sequence, $A = \{a_1, a_2, \ldots, a_L\}$, where a_i represents a function of b_i
and its complement. For instance, a_i may take
the integer value 0 if nucleotide b_i as well as
its complement are residing in noncoding
regions, it may take the value 1 if b_i is a part
of a gene residing in the direct strand, and it
may take a value of 2 if the b_i complement is
involved in encoding a protein in the complementary strand. Further possibilities for b_i,
such as residing within a region of gene overlap, are not considered at the moment. Identifying protein-coding regions in sequence S
is equivalent to finding a true functional sequence A related to S. Let us assume that for
given sequences A and S the algorithm is
available to compute the probability, $P(A|S) =
P(a_1, a_2, \ldots, a_L|b_1, b_2, \ldots, b_L)$, that A describes
the biological function of S. Such an algo-
rithm could be used to find A^*, providing the
largest value $P(A^*|S)$ among all possible A.
Such an A^*, the most likely one among all
possible functional sequences, is supposed to
define the true annotation of protein-coding
genes in the sequence S.

The problem of the $P(A|S)$ computation
and maximization could be formulated in
terms of HMMs. The HMM technique di-
vides the object of interest into interrelated
levels of observable states and hidden states,
each one with its own dynamics (15). This
technique is quite general and could be used
in a variety of applications and where the input information has to be processed to reveal
some underlying meanings. The HMM theory
was successfully applied in speech recognition
(see reference 43 for a review). Applications
of HMM to DNA and protein sequence analysis were described by several groups (2, 12,
13, 24, 34–36, 48). The first HMM-based
gene-finding algorithm, ECOPARSE, was
developed by Krogh et al. (34) for the *E. coli*
genome analysis.

Interestingly, the HMM approach to gene
finding allows a variety of HMM architectures

to be used. In the recently developed
GeneMark.hmm algorithm, the scheme of
transitions between hidden Markov states
shown in Fig. 4 is used. The Markov models
of coding (note the two types of protein-
coding models: typical and atypical) and noncoding regions used in the GeneMark algorithm were essential parts of the new
algorithm. These models were incorporated
into the HMM framework to evaluate
stretches of DNA sequence emitted by a particular hidden state. This type of HMM architecture is known as HMM with duration
(43). A sequence of hidden states with their
duration defines nucleotide positions where
one function (noncoding) is switching into
another (protein coding) and vice versa. The
sequence of hidden states in this model is
equivalent to the previously introduced functional sequence A with the addition of splitting coding states into typical and atypical.
Thus, the problem is reduced to finding the
A^* providing the largest value of $P(A|S)$, the
conditional probability of A given S. The
computation of $P(A|S)$ makes use of statistical
models of length distributions of coding and
noncoding regions and initial and transition
probabilities of hidden states, as well as Markov models of coding and noncoding regions.
The core GeneMark.hmm procedure is the
Viterbi-type algorithm (43) that finds the sequence A^*. Notably, this core procedure does
not take into account the possibility of gene
overlaps. The observed overlaps, though frequent, were not extensive enough to provide
sufficient data for deriving statistical models of
overlapping genes in several possible orientations.

To further improve prediction of the translation start position, we used a postprocessing
step utilizing the RBS model, similar to one
described previously (23). The accuracy of the
GeneMark.hmm program was assessed by using several test sets, including the 10 complete
bacterial genomes mentioned above. The
GeneMark.hmm predictions were compared
with GeneBank annotations (Table 4). It was
shown that the frequency of exact gene predictions is clearly higher than that of

GeneMark.hmm

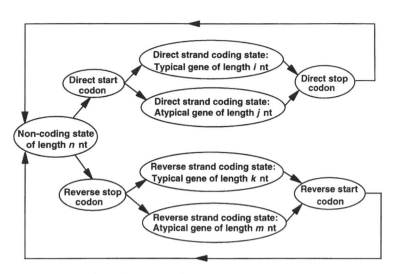

FIGURE 4   The architecture of the Hidden Markov model used in the GeneMark.hmm algorithm. The diagram shows the hidden states and possible transitions between them. Note that the architecture does not allow for a gene overlap. (Reprinted from *Nucleic Acids Research* [40] with permission of the publisher.)

GeneMark, the version which also used the RBS model. Nevertheless, the GeneMark program correctly identified the ORFs where 98.8% of all genes predicted by Gene-Mark.hmm resided. It is worth mentioning, however, that the evaluation of the algorithm performance by comparison with the database annotation may not be sufficiently conclusive, since in only a few cases is the precise position of the translation initiation codon known from an experiment. Nevertheless, the database annotation of the initiation codon posi-

TABLE 4   GeneMark.hmm performance[a]

Set no.	No. of genes	Prediction method	No. (%)		
			Exact prediction	Only 3'-end prediction	Missing genes
1	4,288	VA	2,483 (58)	1,592 (37)	213 (5)
1	4,288	PP	3,233 (75)	842 (20)	213 (5)
2	2,821	VA	2,017 (71)	750 (27)	54 (2)
2	2,821	PP	2,268 (80)	499 (18)	54 (2)
3	325	VA	255 (78)	64 (20)	6 (2)
3	325	PP	289 (89)	30 (9)	6 (2)
4	204	VA	156 (76.5)	47 (23)	1 (0.5)
4	204	PP	177 (87.5)	26 (12)	1 (0.5)

[a] The four control sets of annotated genes selected for the comparison are described in the text. The numbers in the rows designated VA correspond to predictions made by the GeneMark.hmm program with the Viterbi algorithm only. The rows designated PP show the results of predictions with postprocessing (the RBS identification procedure). The "Exact prediction" column contains the numbers of genes with both the 3' and 5' ends predicted exactly. The numbers of genes predicted with the 5' end misplaced are shown in the column "Only 3'-end prediction." The genes annotated but not correctly predicted at the 5' or 3' end fall into the category of missing genes. The percentage (shown in parentheses) is the fraction relative to the total number of annotated genes. (Reprinted from *Nucleic Acids Research* [40] with permission of the publisher.)

tion represents the expert decision summarizing all available direct and indirect evidence and is assumed to be close to the real one.

Gene overlaps make is difficult to precisely locate the translation initiation start. In spite of frequent opinion that gene overlaps are likely to happen only in phage and virus genomes where requirements for tight gene packing are vitally important, the complete bacterial genomes demonstrate quite a few gene overlaps. The overlap regions are of special interest because of their double genetic code load. The distributions of overlap length are different for overlapping genes residing in the same strand and for genes residing in opposite strands. The overlaps in the same strand are more common, with the trivial overlaps of length 1 (TGA/ATG) or 4 (ATGA) constituting the majority of same-strand overlaps. Overlaps with lengths greater than 48 nt, half the default length of the GeneMark window, were observed for 4% of the *E. coli* genes. Overlaps between genes residing in opposite strands are less common. The majority of them fell into the category of trivial overlaps of length 4 (TTAA, TTAG, CTAA, and CTAG). However, 2% of the *E. coli* genes are involved in opposite-strand overlaps with lengths greater than 48 nt. A similar distribution of frequencies of overlap with different lengths has been observed in *B. subtilis* (Fig. 5).

There has been a tendency for bacterial-genome annotators to prefer a longer ORF to a shorter one provided there is no convincing evidence in favor of the shorter one. Statistically, this tendency is justified by the expectation that in about 75% of cases the actual gene occupies the longest ORF. This figure can be explained as follows. Let us consider the set of four triplets ATG, TAA, TAG, and TGA, assuming for simplicity that these triplets are equally frequent in a noncoding DNA sequence. Consider the intergenic region located upstream of the true initiation codon of a gene, X. Let us start from the initiation codon and read the triplets in the 5′ direction in the same reading frame until the first triplet

from the set specified above is encountered. If this triplet is ATG, the event expected to happen in 25% of cases, then the gene X does not occupy the longest ORF. Otherwise, in 75% of cases, one of the stop codons is met; thus, the gene X does occupy the longest ORF. The assumption that the four triplets occur with equal frequencies and that ATG is the only possible initiation codon can be easily modified for the case of a real genome with a given oligonucleotide composition. Interestingly, the assessment of the accuracy of methods of translation start prediction frequently leads to figures in the 70 to 80% range. Such a figure may be interpreted as showing that the method is indeed highly accurate (>95% correct predictions), but the annotation using the "longest-ORF rule" is actually correct in about 70 to 80% of cases.

ACCURACY OF EXACT GENE FINDING

The performance of the GeneMark.hmm program was tested with several control sets, including 10 complete bacterial genomes. Our focus was on the *E. coli* genome. The complete genomic sequence of *E. coli* consists of 4,639,221 nt, with 4,288 genes annotated (5). The GeneMark.hmm program identified 4,440 genes in the *E. coli* genomic sequence. Each predicted gene was also characterized as typical or atypical, depending on the type of underlying coding (hidden) state. Twenty percent of the predicted genes were identified as atypical. The gene-finding accuracy was evaluated with four control sets of genes annotated in the *E. coli* genome (Table 4). Control set 1 contained all annotated *E. coli* genes. Set 2 was compiled from nonoverlapping *E. coli* genes. The *E. coli* genes whose RBSs were annotated in GenBank constituted set 3. The genes coding for proteins with experimentally verified amino termini (39) were included in set 4.

The evaluation results (Table 4) show that the Viterbi-type algorithm alone was able to exactly predict 58% of the *E. coli* genes in set 1. The gene overlap seems to be an important factor indeed, since the percentage of exact

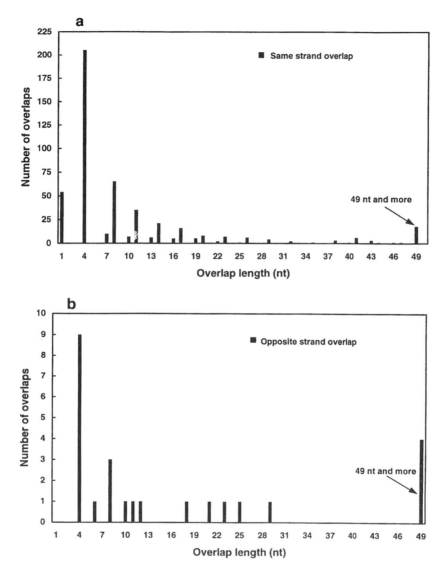

FIGURE 5 Distribution of *B. subtilis* gene overlaps over the overlap length as annotated in GenBank. (a) Same-strand overlap; (b) Opposite-strand overlap.

gene predictions jumped to 71% when the overlapping genes were eliminated (set 2). It is worth mentioning that the 58 and 71% figures may not be consistent estimates of the algorithm's real performance, since the majority of annotated translation initiation codons in control sets 1 and 2 were not verified in experiments. In control sets 3 and 4, the core algorithm exactly predicted 78 and 76.5% of the genes, respectively. These two close fig-ures give a more realistic estimation of the core algorithm's predictive power for genes with no overlaps. The percentages of the *E. coli* genes predicted either exactly or with misplaced translation starts around 95, 98, 98, and 99.5% for sets 1, 2, 3, and 4, respectively. These figures did not change when the RBS prediction was combined with the core algorithm prediction at the postprocessing step (Table 4). However, for many genes partially

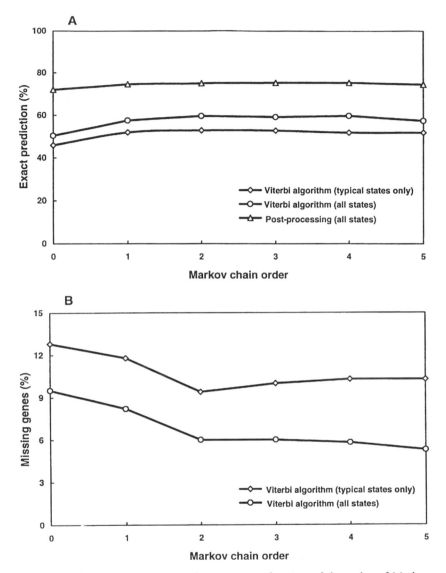

FIGURE 6 GeneMark.hmm performance as a function of the order of Markov model used in the algorithm. The results of comparison between the predicted and annotated parses are shown for the sequence of the first 500,000 nt taken from the entire *E. coli* genomic sequence. This contig contains 468 annotated genes. (A) Exact prediction: the fraction of annotated genes with both the 5′ and 3′ ends predicted exactly. ◇, the predicted parse was generated by the Viterbi algorithm with the Markov models for typical genes only; ○, the Markov model for atypical genes was included in the GeneMark.hmm algorithm; △, the parse was corrected by postprocessing with the RBS model. (B) Missing genes: the fraction of annotated genes with neither the 5′ nor the 3′ end predicted exactly (the postprocessing procedure did not change the number of missing genes [Table 2]). The graph notations are the same as in panel A. (Reprinted from *Nucleic Acids Research* [40] with permission of the publisher.)

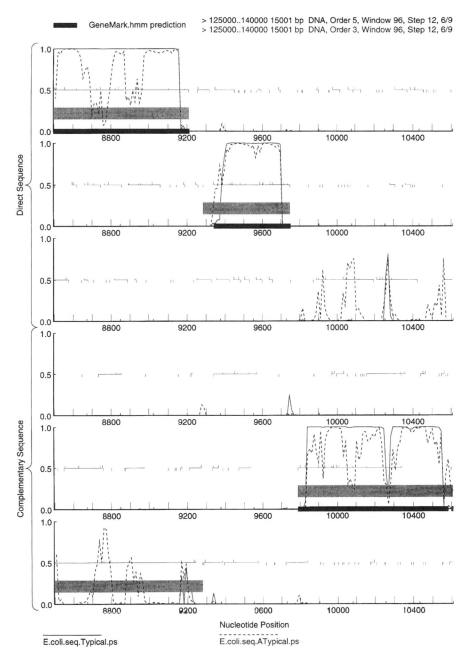

FIGURE 7 Graphical output of the combined GeneMark and GeneMark.hmm programs (6a). The ORFs are indicated by the thin lines in the middle of each panel. The coding-potential graphs, obtained by plotting local a posteriori probabilities of protein-coding as defined by GeneMark, are shown in each panel. The probability values generated by using the typical-gene model are shown by solid lines; those obtained by using atypical-gene models are shown by dashed lines. The genes predicted by GeneMark.hmm are shown by bold face solid lines at the bottoms of the panels. Hatched rectangles mark the regions bound by two in-frame stop codons and possessing high enough coding potential.

predicted by the core algorithm, the correct position of the translation start was found. The fraction of exact predictions increased from 58 to 75% for set 1, from 71 to 80% for set 2, from 78 to 89% for set 3, and from 76.5 to 87.5% for set 4. One may conclude that RBS correction produces a 10% increase in the percentage of exactly predicted genes under nonoverlap conditions. Also, it appears from the results of program testing on set 1 that gene overlaps were responsible for about 10% of nonexact predictions. A new gene-finding algorithm with HMM architecture addressing the possibility of gene overlaps has recently been developed (44a).

HIGHER-ORDER MODELS AND MODELS OF TYPICAL AND ATYPICAL GENES

The second-order Markov models of coding and noncoding regions were used by GeneMark.hmm to produce the results presented in Table 4. The percentage of exact predictions as a function of the model order is shown in Fig. 6A. Surprisingly, even the zero-order models yield high enough accuracy. This observation could be explained by the fact that the GeneMark.hmm algorithm accumulates detectable signal within the rather long bacterial gene even if the relatively weak zero-order model is used. This does not happen in the case where the GeneMark core algorithm is applied to a short DNA sequence restricted by the sliding window. For short sequences, the higher-order models are known to be more accurate in coding-potential detection (6, 41). In fact, as the analysis of GeneMark.hmm accuracy shows, the number of missing genes, presumably short genes, decreased as the model's order increased (Fig. 6B). Using high-order models is also well justified for analysis of eukaryotic DNA with ubiquitous short coding exons. The role of the atypical-gene model is illustrated in Fig. 6. Switching off the atypical model produced a decrease in the number of exact predictions (Fig. 6A) and an increase in the number of missing genes (Fig. 6B).

A sample graphical output of the combined GeneMark and GeneMark.hmm program (6a) is shown in Fig. 7 for a fragment of the *E. coli* genome. Interestingly, the gene predicted in the opposite strand (Fig. 7, second panel from the bottom) has a nonstandard start codon, TTG; thus, the ORF indicated for this gene falls short. However, the GeneMark and GeneMark.hmm programs both agree on predicting the longer gene.

As was mentioned above, GeneMark.hmm may use the atypical gene model only as a quite viable option for finding bacterial genes (40a). However, using both typical and atypical gene models provides specific information on a gene type. Genes predicted as atypical are likely to be horizontally transferred genes, a category of special interest for evolutionary studies, as well as for studies of pathogenic bacteria whose pathogenicity islands or antibiotic-resistance genes could be relatively recent additions to the whole genome.

ACKNOWLEDGMENTS

The valuable programming assistance of John Besemer is greatly appreciated.

The work was supported in part by the National Institutes of Health and the Isaac Newton Institute for Mathematical Sciences (Cambridge, United Kingdom).

REFERENCES

1. **Altshul, S. F., T. L. Madden, A. A. Schaffer, J. Zhang, Z. Zhang, W. Miller, and D. J. Lipman.** 1997. Gapped BLAST and PSI-BLAST: a new generation of protein database search programs. *Nucleic Acids Res.* **25:**3389–3402.
2. **Baldi, P., Y. Chauvin, T. Hunkapiller, and M. A. McClure.** 1994. Hidden Markov models of biological primary sequence information. *Proc. Natl. Acad. Sci. USA* **9:**1059–1063.
3. **Berg, O. G., and P. H. von Hippel.** 1988. Selection of DNA binding sites by regulatory proteins. *Trends Biochem. Sci.* **13:**207–211.
4. **Billingsley, P.** 1961. Statistical methods in Markov chains. *Ann. Math. Stat.* **82:**12–40.
5. **Blattner, F. R., G. Plunkett III, C. A. Bloch, N. T. Perna, V. Burland, M. Riley, J. Collado-Vides, J. D. Glasner, C. K. Rode, G. F. Mayhew, J. Gregor, N. W. Davis, H. A. Kirkpatrick, M. A. Goeden, D. J. Rose, B.

Mau, and Y. Shao. 1997. The complete genome sequence of *Escherichia coli* K-12. *Science* **277**:1453–1462.

6. **Borodovsky, M., and J. McIninch.** 1993. GeneMark: parallel gene recognition for both DNA strands. *Comput. Chem.* **18**:259–268.

6a. **Borodovsky, M., and J. McIninch.** 1996. [Online.] http://genemark.biology.gatech.edu/GeneMark.

7. **Borodovsky, M., J. McIninch, E. Koonin, K. Rudd, C. Médigue, and A. Danchin.** 1995. Detection of new genes in a bacterial genome using Markov models for three gene classes. *Nucleic Acids Res.* **23**:3554–3562.

8. **Borodovsky, M., K. Rudd, and E. Koonin.** 1994. Intrinsic and extrinsic approaches for detecting genes in a bacterial genome. *Nucleic Acids Res.* **22**:4756–4767.

9. **Borodovsky, M., Y. A. Sprizhitsky, E. I. Golovanov, and A. A. Alexandrov.** 1986. Statistical features in the *Escherichia coli* genome functional primary structure. II. Non-homogeneous Markov chains. *Mol. Biol.* **20**:833–840.

10. **Borodovsky, M., Y. A. Sprizhitsky, E. I. Golovanov, and A. A. Alexandrov.** 1986. Statistical features in the *Escherichia coli* genome functional primary structure. III. Computer recognition of protein coding regions. *Mol. Biol.* **20:** 1144–1150.

10a. **Borodovsky, M.** Unpublished data.

11. **Bult, C. J., O. White, G. J. Olsen, L. Zhou, R. D. Fleischmann, G. G. Sutton, J. A. Blake, L. M. FitzGerald, R. A. Clayton, J. D. Gocayne, A. R. Kerlavage, B. A. Dougherty, J. Tomb, M. D. Adams, C. I. Reich, R. Overbeek, E. F. Kirkness, K. G. Weinstock, J. M. Merrick, A. Glodek, J. D. Scott, N. S. Geoghagen, J. F. Weidman, J. L. Fuhrmann, D. T. Nguyen, T. Utterback, J. M. Kelley, J. D. Peterson, P. W. Sadow, M. C. Hanna, M. D. Cotton, M. A. Hurst, K. M. Roberts, B. B. Kaine, M. Borodovsky, H. P. Klenk, C. M. Fraser, H. O. Smith, C. R. Woese, and J. C. Venter.** 1996. Complete genome sequence of the methanogenic archeon *Methanococcus jannaschii*. *Science* **273:** 1058–1073.

12. **Burge, C., and S. Karlin.** 1997. Prediction of complete gene structures in human genomic DNA. *J. Mol. Biol.* **268**:78–94.

13. **Churchill, G. A.** 1989. Stochastic models for heterogeneous DNA sequences. *Bull. Math. Biol.* **51**:79–94.

14. **Clover, T. M., and J. A. Thomas.** 1991. *Elements of Information Theory*. John Wiley & Sons, New York, N.Y.

14a. **Danchin, A.** Personal communication.

15. **Durbin, R., S. Eddy, A. Krogh, and G. Mitchison.** 1998. *Biological Sequence Analysis. Probabilistic Models of Proteins and Nucleic Acids*. Cambridge University Press, Cambridge, United Kingdom.

16. **Erickson, J. W., and G. G. Altman.** 1979. A search for patterns in the nucleotide sequence of the MS2 genome. *J. Math. Biol.* **7**:219–230.

17. **Fickett, J. W., and C. S. Tung.** 1992. Assessment of protein coding measures. *Nucleic Acids Res.* **20**:6441–6450.

18. **Fleischmann, R. D., M. D. Adams, O. White, R. A. Clayton, E. F. Kirkness, A. R. Kerlavage, C. J. Bult, J.-F. Tomb, B. A. Dougherty, J. M. Merrick, K. McKenney, G. Sutton, W. FitzHugh, C. A. Fields, J. D. Gocayne, J. D. Scott, R. Shirley, L.-I. Liu, A. Glodek, J. M. Kelley, J. F. Weidman, C. A. Phillips, T. Spriggs, E. Hedblom, M. D. Cotton, T. R. Utterback, M. C. Hanna, D. T. Hguyen, D. M. Saudek, R. C. Brandon, L. D. Fine, J. L. Fritchman, J. L. Fuhrmann, N. S. M. Geoghagen, C. L. Gnehm, L. A. McDonald, K. V. Small, C. M. Fraser, H. O. Smith, and J. C. Venter.** 1995. Whole-genome random sequencing and assembly of *Haemophilus influenzae*. *Science* **269:** 496–512.

19. **Fraser, C. M., J. D. Gocayne, O. White, M. D. Adams, R. A. Clayton, R. D. Fleischmann, C. J. Bult, A. R. Kerlavage, G. Sutton, J. M. Kelley, J. L. Fritchman, J. F. Weidman, K. V. Small, M. Sandusky, J. L. Fuhrmann, D. T. Nguyen, T. R. Utterback, D. M. Saudek, C. A. Phillips, J. M. Merrick, J.-F. Tomb, B. A. Dougherty, K. F. Bott, P.-C. Hu, T. S. Lucier, S. N. Peterson, H. O. Smith, C. A. Hutchison III, and J. C. Venter.** 1995. The minimal gene complement of *Mycoplasma genitalium*. *Science* **270**:397–403.

20. **Frishman, D., A. Mironov, H. W. Mewes, and M. Gelfand.** 1998. Combining diverse evidence for gene recognition in completely sequenced bacterial genomes. *Nucleic Acids Res.* **26**:2941–2947. (Erratum, **26**:3870.)

21. **Gatlin, L.** 1972. *Information Theory and Living Systems*. Columbia University Press, New York, N.Y.

22. **Green, P., D. Lipman, L. Hillier, R. Waterston, D. States, and J. M. Claverie.** 1993. Ancient conserved regions in new gene sequences and the protein databases. *Science* **259**:1711–1716.

23. **Hayes, W. S., and M. Borodovsky.** 1998. Deriving ribosome binding site (RBS) statistical models from unannotated DNA sequences and the use of the RBS model for N-terminal pre-

diction, p. 279–290. *In Proceedings of the Pacific Symposium on Biocomputing 1998.* World Scientific, Maui, Hawaii.

23a. Hayes, W. S., and M. Borodovsky. Unpublished data.

24. Henderson, J., S. Salzberg, and K. H. Fasman. 1997. Finding genes in DNA with a hidden Markov model. *J. Comp. Biol.* 4:127–141.

25. Himmelreich, R., H. Hilbert, H. Plagens, E. Pirkl, B.-C. Li, and R. Herrmann. 1996. Complete sequence of the genome of the bacterium *Mycoplasma pneumoniae. Nucleic Acids Res.* 24:4420–4449.

26. Hogg, R. V., and E. A. Tanis. 1997. *Probability and Statistical Inference.* Prentice Hall, Englewood Cliffs, N.J.

27. Jelinek, F., and R. I. Mercer. 1980. Interpolated estimation of Markov source parameters from sparse data p. 252–260. *In* E. S. Gelsema and L. N. Kanak (ed.), *Pattern Recognition in Practice.* Elsevier/North Holland, New York, N.Y.

28. Kaneko, T., A. Tanaka, S. Sato, H. Kotani, T. Sazuka, N. Miyajima, M. Sugiura, and S. Tabata. 1995. Sequence analysis of the genome of the unicellular cyanobacterium *Synechocystis* sp. strain PCC6803. I. Sequence features in the 1 Mb region from map positions 64% to 92% of the genome. *DNA Res.* 2:153–166.

29. Kleffe, J., K. Hermann, and M. Borodovsky. 1995. Statistical analysis of GeneMark performance by cross-validation. *Comput. Chem.* 20:123–134.

30. Klenk, H. P., R. A. Clayton, J. Tomb, O. White, K. E. Nelson, K. A. Ketchum, R. J. Dodson, M. Gwinn, E. K. Hickey, J. D. Peterson, D. L. Richardson, A. R. Kerlavage, D. E. Graham, N. C. Kyrpides, R. D. Fleischmann, J. Quackenbush, N. H. Lee, G. G. Sutton, S. Gill, E. F. Kirkness, B. A. Dougherty, K. McKenney, M. D. Adams, B. Loftus, S. Peterson, C. I. Reich, L. K. McNeil, J. H. Badger, A. Glodek, L. Zhou, R. Overbeek, J. D. Gocayne, J. F. Weidman, L. McDonald, T. Utterback, M. D. Cotton, T. Spriggs, P. Artiach, B. P. Kaine, S. M. Sykes, P. W. Sadow, K. P. D'Andrea, C. Bowman, C. Fujii, S. A. Garland, T. M. Mason, G. J. Olsen, C. M. Fraser, H. O. Smith, C. R. Woese, and J. C. Venter. 1997. The complete genome sequence of the hyperthermophilic, sulphate-reducing archaeon *Archaeoglobus fulgidus. Nature* 390:364–370.

31. Koonin, E. V., P. Bork, and C. Sander. 1994. Yeast chromosome III: new gene functions. *EMBO J.* 13:493–503.

32. Koonin E. V., and M. Y. Galperin. 1997. Prokaryotic genomes: the emerging paradigm of genome-based microbiology. *Curr. Opin. Genet. Dev.* 7:757–763.

33. Koonin, E. V., A. R. Mushegian, M. Y. Galperin, and D. R. Walker. 1997. Comparison of archaeal and bacterial genomes: computer analysis of protein sequences predicts novel functions and suggests a chimeric origin for the archaea. *Mol. Microbiol.* 25:619–637.

34. Krogh, A., M. Brown, I. S. Mian, K. Sjolander, and D. Haussler. 1994. Hidden Markov models in computational biology. Applications to protein modeling. *J. Mol. Biol.* 235:1501–1531.

35. Krogh, S., I. S. Mian, and D. Haussler. 1994. A hidden Markov model that finds genes in *E. coli* DNA. *Nucleic Acids Res.* 22:4768–4778.

36. Kulp, D., D. Haussler, M. G. Reese, and F. H. Eeckman. 1996. *In* D. J. States, P. Agarwal, T. Gaasterland, L. Hunter, and R. F. Smith (ed.), *Proceedings of the Fourth International Conference on Intelligent Systems for Molecular Biology (ISMB-96)*, p. 134–142. AAAI Press, Menlo Park, Calif.

37. Kunst, F., N. Ogasawara, I. Moszer, A. M. Albertini, G. Alloni, V. Azevedo, M. G. Bertero, P. Bessieres, A. Bolotin, S. Borchert, R. Borriss, L. Boursier, A. Brans, M. Braun, S. C. Brignell, S. Bron, S. Brouillet, C. V. Bruschi, B. Caldwell, V. Capuano, N. M. Carter, S. K. Choi, J. J. Codani, I. F. Connerton, N. J. Cummings, R. A. Daniel, F. Denizot, K. M. Devine, A. Dusterhoft, S. D. Ehrlich, P. T. Emmerson, K. D. Entian, J. Errington, C. Fabret, E. Ferrari, D. Foulger, C. Fritz, M. Fujita, Y. Fujita, S. Fuma, A. Galizzi, N. Galleron, S. Y. Ghim, P. Glaser, A. Goffeau, E. J. Golightly, G. Grandi, G. Guiseppi, B. J. Guy, K. Haga, J. Haiech, C. R. Harwood, A. Henaut, H. Hilbert, S. Holsappel, S. Hosono, M. F. Hullo, M. Itaya, L. Jones, B. Joris, D. Karamata, Y. Kasahara, M. Klaerr-Blanchard, C. Klein, Y. Kobayashi, P. Koetter, G. Koningstein, S. Krogh, M. Kumano, K. Kurita, A. Lapidus, S. Lardinois, J. Lauber, V. Lazarevic, S. M. Lee, A. Levine, H. Liu, S. Masuda, C. Mauel, C. Medigue, N. Medina, R. P. Mellado, M. Mizuno, D. Moestl, S. Nakai, M. Noback, D. Noone, M. O'Reilly, K. Ogawa, A. Ogiwara, B. Oudega, S. H. Park, V. Parro, T. M. Pohl, D. Portetelle, S. Porwollik, A. M. Prescott, E. Presecan, P. Pujic, B. Purnelle, G. Rapoport, M. Rey, S. Reynolds, M. Rieger, C. Rivolta, E. Rocha, B. Roche, M. Rose, Y. Sadaie, T. Sato, E. Scanlan, S. Schleich, R. Schroeter, F. Scoffone, J. Sekiguchi, A. Sekowska, S. J. Seror,

P. Serror, B. S. Shin, B. Soldo, A. Sorokin, E. Tacconi, T. Takagi, H. Takahashi, K. Takemaru, M. Takeuchi, A. Tamakoshi, T. Tanaka, P. Terpstra, A. Tognoni, V. Tosato, S. Uchiyama, M. Vandenbol, F. Vannier, A. Vassarotti, A. Viari, R. Wambutt, E. Wedler, H. Wedler, T. Weitzenegger, P. Winters, A. Wipat, H. Yamamoto, K. Yamane, K. Yasumoto, K. Yata, K. Yoshida, H. F. Yoshikawa, E. Zumstein, H. Yoshikawa, and A. Danchin. 1997. The complete genome sequence of the gram-positive bacterium *Bacillus subtilis*. *Nature* **390**:249–256.

38. **Lawrence, J. G., and H. Ochman.** 1997. Amelioration of bacterial genomes: rates of change and exchange. *J. Mol. Evol.* **44**:383–397.

39. **Link, A. J., K. Robison, and G. M. Church.** 1997. Comparing the predicted and observed properties of proteins encoded in the genome of *Escherichia coli* K-12. *Electrophoresis* **18**:1259–1313.

40. **Lukashin, A. V., and M. Borodovsky.** 1998. GeneMark.hmm: new solutions for gene finding. *Nucleic Acids Res.* **26**:1107–1115.

40a.**Lukashin, A. V., and M. Borodovsky.** Unpublished data.

41. **McIninch, J., W. Hayes, and M. Borodovsky.** 1996. Application of GeneMark in multispecies environment, p. 176–188. *In* D. J. States, P. Agarwal, T. Gaasterland, L. Hunter, and R. F. Smith (ed.), *Proceedings of the Fourth International Conference on Intelligent Systems for Molecular Biology (ISMB-96)*. AAAI Press, Menlo Park, Calif.

42. **Médigue, C., T. Rouxel, P. Vigier, A. Henaut, and A. Danchin.** 1991. Evidence for horizontal gene transfer in *Escherichia coli* speciation. *J. Mol. Biol.* **222**:851–856.

43. **Rabiner, L. R.** 1989. Tutorial on hidden Markov models and selected applications in speech recognition. *Proc. IEEE* **77**:257–286.

44. **Salzberg, S. L., A. L. Delcher, S. Kasif, and O. White.** 1998. Microbial gene identification using interpolated Markov models. *Nucleic Acids Res.* **26**:544–548.

44a.**Shmatkov, A. M., A. A. Melikyan, F. L. Chernousko, and M. Borodovsky.** Unpublished data.

45. **Smith, D. R., L. A. Doucette-Stamm, C. Deloughery, H.-M. Lee, J. Dubois, T. Aldredge, R. Bashirzadeh, D. Blakely, R. Cook, K. Gilbert, D. Harrison, L. Hoang, P. Keagle, W. Lumm, B. Pothier, D. Qiu, R. Spadafora, R. Vicare, Y. Wang, J. Wierzbowski, R. Gibson, N. Jiwani, A. Caruso, D. Bush, H. Safer, D. Patwell, S. Prabhakar, S. McDougall, G. Shimer, A. Goyal, S. Pietrovski, G. M. Church, C. J. Daniels, J.-I. Mao, P. Rice, J. Nolling, and J. N. Reeve.** 1997. Complete genome sequence of *Methanobacterium thermoautotrophicum* ΔH: functional analysis and comparative genomics. *J. Bacteriol.* **179**:7135–7155.

46. **Tavare, S., and B. Song.** 1989. Codon preference and primary sequence structure in protein-coding regions. *Bull. Math. Biol.* **51**:95–115.

47. **Tomb, J.-F., O. White, A. R. Kerlavage, R. A. Clayton, G. G. Sutton, R. D. Fleischmann, K. A. Ketchum, H. P. Klenk, S. Gill, B. A. Dougherty, K. Nelson, J. Quackenbush, L. Zhou, E. F. Kirkness, S. Peterson, B. Loftus, D. Richardson, R. Dodson, H. G. Khalak, A. Glodek, K. McKenney, L. M. Fitzegerald, N. Lee, M. D. Adams, E. K. Hickey, D. E. Berg, J. D. Gocayne, T. R. Utterback, J. D. Peterson, J. M. Kelley, M. D. Cotton, J. M. Weidman, C. Fujii, C. Bowman, L. Watthey, E. Wallin, W. S. Hayes, M. Borodovsky, P. D. Karp, H. O. Smith, C. M. Fraser, and J. C. Venter.** 1997. The complete genomic sequence of the gastric pathogen *Helicobacter pylori*. *Nature* **388**:539–548.

48. **Yada, T., and M. Hirosawa.** 1996. *In* D. J. States, P. Agarwal, T. Gaasterland, L. Hunter, and R. F. Smith (ed.), *Proceedings of the Fourth International Conference on Intelligent Systems for Molecular Biology (ISMB-96)*, p. 252–260. AAAI Press, Menlo Park, Calif.

BACTERIAL GENOMES—ALL SHAPES AND SIZES

Stewart T. Cole and Isabelle Saint-Girons

3

One of the most fascinating aspects of bacteria is their remarkable diversity, which is often manifested in pronounced phenotypic differences, such as colony morphology, pigmentation, cell shape and size, metabolic capacity, and longevity. Of course, all of these properties are mere reflections of genotypic differences and programmed gene expression. After a decade of research it is now clear that there is much more variation at the level of the microbial genome than was initially thought and that bacteria display great variety in the size, number, organization, and topology of their chromosomes. The aim of this chapter is to summarize the recent findings of bacterial genomics and to comment on the themes and trends which are emerging.

APPROACHES USED FOR PHYSICAL MAPPING OF BACTERIAL GENOMES

The strategy used to physically map a bacterial genome depends largely on the degree of resolution required, the future use for the map,

and the resources available. Clearly, now that whole-genome sequencing is becoming more widespread, the most detailed and informative maps will be obtained by this approach. However, as not all projects require this depth of information, physical mapping techniques will still have an important role to play in establishing genome shape and size, in comparative studies with different strains of a given species or genus, and in epidemiology and taxonomy.

A variety of techniques and methods are available to construct physical maps, and those most commonly employed involve pulsed-field gel electrophoresis (PFGE) of macrorestriction fragments generated by digesting intact genomic DNA, immobilized in agarose plugs, with rare-cutting enzymes (185, 186). Samples may be subjected to complete or partial digestion with one or more enzymes and then separated by electrophoresis in one or two dimensions (144, 175, 176, 178). As these approaches have been described extensively in earlier reviews (48, 67), they will be discussed only briefly here, although it should be noted that the most informative and reliable maps have generally been constructed by a combination of methods (Tables 1 to 4).

Hybridization techniques are often used to construct a map and to deduce the positions of genetic markers. Among the more com-

Stewart T. Cole, Unité de Génétique Moléculaire Bactérienne, Institut Pasteur, 28 rue du Docteur Roux, 75724 Paris Cédex 15, France. *Isabelle Saint-Girons*, Unité de Bactériologie Moléculaire et Médicale, Institut Pasteur, 28 rue du Docteur Roux, 75724 Paris Cédex 15, France.

Organization of the Prokaryotic Genome, Edited by Robert L. Charlebois,
© 1999 American Society for Microbiology, Washington, D.C.

TABLE 1 Physical and genetic maps from purple bacteria

Organism	Chromosome(s) (kbp)	Plasmid(s)[a] (kbp)	Genome size (kbp)	rRNA[b]	Marker[c]	Method[d]	Strategy[e]	Reference[f]
Acinetobacter sp.	3,780	–	3,780	7	26	PFGE	NS, Tf	83
Agrobacterium tumefaciens	3,000 + 2,100 (**L**)[g]	200, 450	5,750	>1 + 1	2 + 2	PFGE	CH	2
Bartonella bacilliformis	1,600	–	1,600	2	2	PFGE	2D, DD	114
Bordetella pertussis	3,750	–	3,750	ND[g]	21	PFGE	DD, NS	197
Bradyrhizobium japonicum	8,700	–	8,700	1	63	PFGE	CH, NS	115, 116
Brucella melitensis	2,100 + 1,150	–	3,250	2 + 1	6	PFGE	CH	146
Brucella melitensis, B. abortus, B. suis, B. canis, B. ovis, B. neotomae	2,100 + 1,150	–	3,250	2 + 1	14	PFGE	DD, PD, NS	147
Burkholderia cepacia ATCC 25416	3,650 + 3,170 + 1,070	200	8,090	4 + 1 + 1	1	PFGE	2D, DD	173
Burkholderia cepacia ATCC 17616	3,400 + 2,500 + 900	170	6,970	3 + 1 + 1	4	PFGE	CH, DD, NS	41
Campylobacter coli	1,700	–	1,700	ND	2	PFGE	CH, PD, Tf	220
Campylobacter fetus TK (+)	1,690	–	1,690	3*	14	PFGE	CH, PD	207
Campylobacter fetus subsp. fetus	2,015	–	2,015	ND	1	PFGE	CH, DD	76
Campylobacter jejuni TGH9011	1,160	–	1,160	3	8	PFGE	CH	181
Campylobacter jejuni	1,810	–	1,810	3	6	PFGE	CH	106
Campylobacter jejuni 8116	1,810	–	1,810	3	18	PFGE	GM	107
	1,700	–	1,700	3	1	PFGE	CH	155
Campylobacter jejuni UA580	1,720	–	1,720	ND	ND	PFGE	PD	37
	1,720	–	1,720	3*	11	PFGE	CH, PD	207
	1,725	–	1,725	3*	22	PFGE	CH, DD	153
Campylobacter upsaliensis	2,005	–	2,005	ND	13	PFGE	CH	21
Caulobacter crescentus	3,800	–	3,800	1	ND	PFGE	CH, NS, Tn	60
	4,000	–	4,000	1	35	PFGE	CH, NS, Tn	59
Chromatium vinosum	3,675	–	3,675	3	ND	PFGE	CH	79

Continued on following page

Organism								
Dichelobacter nodosus	1,540	—	1,540	3	18	PFGE	CH	121
Escherichia coli K12 W3110	4,700	100	4,800	ND	ND	OL	EMBL4	109
Escherichia coli K12 MG1655	4,600	—	4,600	ND	53	PFGE	LC, PD, Tn	161
	4,595	—	4,595	ND	53	PFGE	LC, NS	89
	4,640	—	4,640	7	4,288	GS	SG	19
Escherichia coli K12 EMG2	4,700	100	4,800	7	48	PFGE	LC, PD	192
Escherichia coli K12 (eight strains)	4,545	100	4,800	ND	ND	PFGE	LC, NS	89
	4,800	100	4,800	ND	ND	PFGE	CH, LC	52
Escherichia coli K12 (six strains)	4,600	—	4,600	ND	ND	PFGE	CH, DD, Tn	162
Haemophilus ducreyi	1,760	—	1,760	6	16	PFGE	CH, DD, PD	92
Haemophilus influenzae b	2,100	—	2,100	6	15	PFGE	CH, DD	26
Haemophilus influenzae Rd	1,900	—	1,900	6	13	PFGE	CH, DD, Tf	128
	1,980	—	1,980	ND	ND	PFGE	CH, DD, NS	104
	1,830	—	1,830	6	1,743	GS	SG	64
Haemophilus parainfluenzae	2,340	—	2,340	ND	3	PFGE	CH, DD, Tf	103
Helicobacter pylori NCTC11638	1,800	—	1,800	2*	30	OL/PFGE	Lorist 6/CH, LC	24
Helicobacter pylori UA802	1,710	—	1,710	3*	3	CH, PD	PFGE	206
Helicobacter pylori (five strains)	1,670–1,740	—	>1,670	2*	15	PFGE	CH, DD, PD	98
Helicobacter pylori 26695	1,667	—	1,667	2*	1,590	GS	SG	211
Micrococcus sp. strain Y-1	4,060	—	4,060	ND	ND	PFGE	DD, PD	160
Moraxella catarrhalis	1,940	—	1,940	ND	12	PFGE	LC, PD	77
Myxococcus xanthus	9,455	—	9,455	ND	ND	PFGE	DD, LC, NS, PD	39
	9,455	—	9,455	ND	ND	OL	YAC	119
	9,200	—	9,200	4	>100	OL/PFGE	YAC/CH, NS	88

TABLE 1 (*Continued*)

Organism	Chromosome(s) (kbp)	Plasmid(s)[a] (kbp)	Genome size (kbp)	rRNA[b]	Marker[c]	Method[d]	Strategy[e]	Reference[f]
Neisseria gonorrhoeae FA1090	2,220	—	2,220	4	68	PFGE	DD, GM, LC	55, 56
Neisseria gonorrhoeae MS11	2,330	+	>2,330	4	60	PFGE	2D, DD, PD	17
Neisseria meningitidis B1940	2,300	—	2,300	4	35	PFGE	2D	78
	2,300	—	2,300	ND	ND	PFGE	2D	15
Neisseria meningitidis Z2491	2,225	—	2,225	4	75	PFGE	DD, GM	57
Pseudomonas aeruginosa PAO	5,900	—	5,900	4	11	PFGE	2D, PD	174
	5,900	—	5,900	4	27	PFGE	2D, PD	178
	5,860	—	5,860	ND	35	PFGE	LC, PD, Tn	171
	5,900	—	5,900	ND	40	PFGE	GM	130
Pseudomonas aeruginosa C	6,500	—	6,500	4	17	PFGE	CH, 2D, PD	184
Pseudomonas aeruginosa C (21 isolates)	6,345–6,605	—	>6,345	4	26	PFGE	2D	177
Pseudomonas fluorescens	6,630	—	6,630	5	26	PFGE	2D, DD, PD	170
Rhizobium meliloti	3,400	1,340, 1,700	6,440	3	17	PFGE	DD, NS, Tn	94
Rhodobacter capsulatus	3,800	134	3,934	4	31	OL/PFGE	Lorist 6/CH, IE	66
Rhodobacter sphaeroides	3,600	134	3,634	4	54	OL/PFGE	Lorist 6/CH, DD	69, 70
	3,045 + 915	31, 42, 63, 97, 105, 110	4,408	1 + 2	25 + 6	PFGE	DD, LC	201, 202
				ND + 2	ND + 14	OL/PFGE	pLA1297/Tn	42
				ND + 2	ND + 144	GS (13%)		43

Organism								
Salmonella enteritidis	4,500	60	4,660	60	7	PFGE	DD, NS, Tn, Td	133
Salmonella paratyphi B	4,555	—	4,655	61	7	PFGE	DD, NS, Tn, Td	132
Salmonella paratyphi A	4,595	—	4,595	81	7	PFGE	DD, NS, Tn, Td	136
Salmonella typhi	4,780	—	4,780	37	7	PFGE	DD, NS, Tn, Td	137
Salmonella typhimurium	4,300	90	4,890	102	7	PFGE	NS, Tn	135
	4,780	90	4,870	38	7	PFGE	NS, Tn	218
	4,300	90	4,890	109	7	PFGE	CH, DD	134
Shigella flexneri	4,590	230	4,820	52	ND	PFGE	LC, NS, PD	157
Stigmatella aurantiaca	9,350	—	9,350	37	5	PFGE	CH, DD, IE, LC	152
Thermus thermophilus HB8	1,740	+	>1,740	21	2*	PFGE	CH, PD	20
Thermus thermophilus HB27	1,320	250	2,070	12	2*	PFGE	CH, LC	205
	1,320	250	2,070	59	2*	PFGE	CH, LC	204
Thiobacillus cuprinus	3,800	50 (**L**)	3,850	2	1	PFGE	CH	143
Thiobacillus ferrooxidans	2,900	8.6	2,908	1	2	PFGE	CH, DD, PD	95
Vibrio cholerae	3,200	—	3,200	20	7	PFGE	CH, DD, LC, PD	142

[a] — , no extrachromosomal element has been reported; +, plasmid of unknown size present.

[b] *, organization of the ribosomal genes is different from the *rrs* (16S)-*rrl* (23S)-*rrs* (5S) operon. When there are several chromosomes, the number of *rrn* genes is shown as 4 + 1 + 1, meaning four on chromosome 1, one on chromosome 2, etc.

[c] The number excludes random probes and does not include the ribosomal genes, indicated in the previous column.

[d] The methods are as follows: GS, whole-genome sequencing; OL, ordered library with vector, e.g., EMBL4.

[e] The strategies are as follows: CH, cross-hybridization; 2D, two-dimensional gel electrophoresis; DD, double digestion; GM, gene mapping; IE, indirect end labeling; LC, linking clones; PD, partial digests; NS, insertion of new restriction sites; SG, shotgun; Td, transduction; Tf, transformation; Tn, transposition.

[f] Physical maps for 38 species were redrawn in reference 54, with minor modifications relative to the original publications (essentially, with additional markers).

[g] **L**, linear; ND, not done.

TABLE 2 Physical and genetic maps from gram-positive bacteria

Organism	Chromosome(s) (kbp)	Plasmid(s)[a] (kbp)	Genome size (kbp)	rRNA[b]	Marker[c]	Method[d]	Strategy[e]	Reference[f]
Adoleplasma oculi	1,630	–	1,630	2 *rrs*	ND	PFGE	DD, NS	208
Bacillus cereus ATCC 10987	5,700	–	5,700	ND[g]	3	PFGE	PD, DD	110
Bacillus cereus (three strains)	5,485–6,270	–	>5,485	ND	6	PFGE	PD, DD	29
Bacillus cereus F0837/76	2,400	40, 230, 260, 360, 760, 960	5,010	ND	30	PFGE	CH, DD, PD	33
Bacillus sp. strain C-125	3,700	–	3,700	4	9	PFGE	CH, DD, LC, PD	199
Bacillus subtilis	4,190	–	4,190		40	PFGE	PD, LC	3
	4,200	–	4,200	10	ND	PFGE	DD, LC	210
	4,165	–	4,165	ND	29	PFGE	DD, LC, PD	97
	4,165	–	4,165	10	ND	PFGE	GM	96
	4,165	–	4,165	ND	65	OL	YAC, GW	6
	4,215	–	4,215	10	4,100	GS	PCR	117
Bacillus thuringiensis	5,400	+	>5,400	ND	6	PFGE	PD	32
Bacillus thuringiensis subsp. *berliner*	5,700	60, 60, 100, 130, 270, 600	6,920	3	14	PFGE	CH, DD, PD	31
Bacillus thuringiensis subsp. *canadiensis*	4,300	–	4,300	ND	15	PFGE	CH	30
Clostridium acetobutylicum	4,145	210	4,355	11	40	PFGE	2D, IE	50
Clostridium beijerinckii	6,685	–	6,685	13	32	PFGE	CH, NS, Tn	217
Clostridium perfringens CPN50	3,600	+	>3,600	10	28	PFGE	IE, DD, LC	27
Clostridium perfringens	3,600	+	>3600	10	97	PFGE	DD, GM	102
Clostridium perfringens (eight strains)	3,070–3,650	+	>3070	10	11	PFGE	IE, DD, LC	28
Corynebacterium glutamicum	3,080	–	3,080	>4	39	PFGE	CH, LC, PD	13
Enterococcus faecalis	2,825	–	2,825	>4	12	PFGE	CH, PD, Tn	150
Lactococcus lactis DL11	2,580	+	>2,580	6	ND	PFGE	2D, DD, PD	212
Lactococcus lactis IL1403	2,420	–	2,420	6	77	PFGE	IE, NS, Tn	125, 126
Lactococcus lactis MG1363	2,560	–	2,560	6	77	PFGE	IE, NS, Tn	126
Listeria monocytogenes	3,150	–	3,150	6	9	PFGE	DD, LC, PD	148
Mycobacterium leprae	2,800	–	2,800	1	71	OL	Lorist 6/GW	58
	2,800	–	2,800	1	1,046	GS (66%)	SG	194
Mycobacterium tuberculosis H37Rv	4,400	–	4,400	1	165	OL/PFGE/GS	pYUB18/2D, LC	47, 165
Mycobacterium bovis BCG Pasteur	4,350	–	4,350	1	18	PFGE	2D, LC	164
Mycoplasma capricolum	1,155	–	1,155	ND	ND	PFGE	2D, DD, PD	149
	1,070	–	1,070	2	17	PFGE	DD	216

Organism	Genome size	Plasmid[a]	Size	rrs[b]	No.[c]	Method[d]	Strategy[e]	Reference[f]
Mycoplasma gallisepticum S6	1,055	—	1,055	2*	4	PFGE	DD	82
Mycoplasma gallisepticum (two strains)	1,000–1,040	—	>1000	2 *rrs*	ND	PFGE	NS	209
Mycoplasma genitalium	600	—	600	1	3	PFGE	2D, DD	49
	580	—	580	ND	ND	OL	pcos RW2, GW	140
	580	—	580	1	70	OL	GM	163
	580	—	580	1	470	GS	SG	72
Mycoplasma hominis (five strains)	695–760	—	>695	2	10	PFGE	CH, DD, LC, PD	120
Mycoplasma mobile	780	—	780	ND	ND	PFGE	2D	14
Mycoplasma mycoides subsp. *myc* Y	1,200	—	1,200	2	2	PFGE	DD	168
Mycoplasma mycoides subsp. *myc* (five strains)	1,040–1,280	—	>1,040	2	9	PFGE	DD	169
Mycoplasma pneumoniae	775	—	775	1	3	PFGE	DD, LC, PD	113
	816	—	816	1	677	GS	CW	90
Rickettsiella melolonthae	1,720	—	1,720	ND	ND	PFGE	PD	75
Spiroplasma citri	1,780	—	1,780	1	33	PFGE	CH, DD, IE	221
Streptococcus mutans MT8148	2,165	—	2,165	ND	5	PFGE	LC, PD	158
Streptococcus mutans GS-5	2,125	—	2,125	ND	28	PFGE	CH	86
Streptococcus mutans GS-5	2,125	—	2,125	5	53	PFGE	GM	223
Streptococcus pneumoniae	2,270	—	2,270	6	24	PFGE	CH, Tf	80
Streptococcus pyogenes	1,920	—	1,920	6	32	PFGE	DD, LC	200
Streptococcus thermophilus A054	1,825	—	1,825	6	15	PFGE	DD, PD	180
Streptococcus thermophilus (three strains)	1,825–1,865	—	>1,825	5–6	10	PFGE	DD, PD	179
Streptomyces ambofaciens (three strains)	8,000 (L)[g]	+ (L or C)[g]	>8,000	6	13	PFGE	CH, DD, LC	122
Streptomyces coelicolor	8,000 (L)	350 (L)	8,300	6	170	OL	Supercos-1	172
	8,000 (L)	350 (L)	8,350	ND	36	PFGE	CH, DD, LC	105
	8,000 (L)	350 (L)	8,350	6	16	PFGE	LC	123
Streptomyces griseus	7,800 (L)	350 (L)	8,150	ND	ND	PFGE	CH, DD, LC	129
Streptomyces lividans	8,000 (L)		8,050	6	15	PFGE	DD, LC	123
Ureaplasma urealyticum	900	50	900	2	ND	PFGE	DD, PD	45

[a] —, no extrachromosomal element has been reported; +, plasmid of unknown size present.

[b] *, organization of the ribosomal genes is different from the *rrs* (16S)-*rrl* (23S)-*rrs* (5S) operon.

[c] The number excludes random probes and does not include the ribosomal genes, indicated in the previous column.

[d] The methods are as follows: GS, whole-genome sequencing; OL, ordered library with vector, e.g., pcos RW2.

[e] The strategies are as follows: CH, cross-hybridization; GW, genome walking; 2D, two-dimensional gel electrophoresis; DD, double digestion; GM, gene mapping; IE, indirect end labeling; LC, linking clones; PD, partial digests; NS, insertion of new restriction sites; SG, shotgun; Tf, transformation; Tn, transposition.

[f] Physical maps for 38 species were recrawn in reference 54, with minor modifications relative to the original publications (essentially, with additional markers).

[g] C, circular; L, linear; ND, not done.

TABLE 3 Physical and genetic maps from other bacteria: chlamydia, cyanobacteria, green sulfur bacteria, planctomycetes, and spirochetes

Organism	Chromosome(s) (kbp)	Plasmid(s)[a] (kbp)	Genome size (kbp)	rRNA[b]	Marker[c]	Method[d]	Strategy[e]	Reference[f]
Chlamydia trachomatis	1,045	–	1,045	2	12	PFGE	CH, DD, LC, PD	18
Anabaena sp. strain PCC 7120	6,370	110, 190, 410	7,080	2	28	PFGE	CH, DD	7
	6,420		7,130	2	39	PFGE	DD, NS, PD, Tn	118
Synechococcus sp. strain PCC 7002	2,700	Several	>2,700	2	19	PFGE	CH, DD, LC	40
Synechocystis sp. strain PCC 6803	3,820	2.2, 5.2, 50, 100	3,977	2	30	PFGE	CH, LC	44, 118
	3,573			2	3,168	GS	SG	100
Chlorobium tepidum	2,100	–	2,100	ND[a]	15	PFGE	DD, PD	151
Planctomyces limnophilus	5,205	+	>5,205	2*	5	PFGE	CH	215
Borrelia afzelii VS461	950 (**L**)[g]	9–58 (**L** and **C**)[g]	>1,250	2*	30	PFGE	CH, 2D, DD	156
Borrelia afzelii (three strains)	935–955 (**L**)	9–58 (**L** and **C**)	>1,250	2*	15	PFGE	CH	34
Borrelia andersonii (two strains)	935–955 (**L**)	9–175 (**L** and **C**)	>1,250	2*	15	PFGE	CH	34
Borrelia burgdorferi 212	945 (**L**)	9–58 (**L** and **C**)	>1,250	2*	ND	PFGE	2D	53
	945 (**L**)	9–58 (**L** and **C**)	>1,250	2*	38	PFGE	2D	156
	945 (**L**)	9–58 (**L** and **C**)	>1,250	2*	9	PFGE	CH	159

Borrelia burgdorferi B31	910 (**L**)	9–53 (**L** and **C**)	>1,440	2*	853 + 430	GS	SG	71
Borrelia burgdorferi SH2-82	950 (**L**)	9–53 (**L** and **C**)	>1,250	2*	19	PFGE	CH	35
Borrelia burgdorferi (six strains)	935–955 (**L**)	9–53 (**L** and **C**)	>1,250	2*	15	PFGE	CH	34
Borrelia burgdorferi sensu lato (five strains)	935–955 (**L**)	9–53 (**L** and **C**)	>1,250	2*	15	PFGE	CH	156
Borrelia garinii 20047	950 (**L**)	9–58 (**L** and **C**)	>1,250	2*	36	PFGE	CH, 2D, DD	156
Borrelia garinii (three strains)	935–955 (**L**)	9–58 (**L** and **C**)	>1,250	2*	15	PFGE	CH	34
Borrelia japonica (two strains)	935–955 (**L**)	9–140 (**L** and **C**)	>1,250	2*	15	PFGE	CH	34
Leptospira interrogans RZ11	4,400 + 350	—	4,750	2*	14	PFGE	2D, LC, PD	224
Leptospira interrogans Verdun	4,600 + 350	—	4,950	2*	ND	PFGE	2D, DD, PD, LC	10
Leptospira interrogans (RZ11 and Verdun)	See above	—	>4,750	2*	13 + 1	PFGE	CH	225
Leptospira interrogans (RZ11 and Verdun)	See above	—	>4,750	2*	27 + 2	PFGE	GM	22
Serpulina hyodysenteriae	3,200	—	3,200	1*	10	PFGE	CH, DD	226
Treponema denticola	3,000	2.6	3,002	2	5	PFGE	CH, 2D, DD, LC	141
Treponema pallidum	1,080	—	1,080	1	12	PFGE	CH, DD	213
	1,137	—	1,137	1		GS	SG	73

[a] —, no extrachromosomal element has been reported; +, plasmid of unknown size present.

[b] *, organization of the ribosomal genes is different from the *rrs* (16S)-*rrl* (23S)-*rrs* (5S) operon.

[c] The number excludes random probes and does not include the ribosomal genes, indicated in the previous column.

[d] GS, whole-genome sequencing.

[e] The strategies are as follows: CH, cross-hybridization; 2D, two-dimensional gel electrophoresis; DD, double digestion; GM, gene mapping; LC, linking clones; PD, partial digests; NS, insertion of new restriction sites; SG, shotgun; Tn, transposition.

[f] Physical maps for 38 species were redrawn in reference 54, with minor modifications relative to the original publications (essentially, with additional markers).

[g] C, circular; **L**, linear; ND, not done

TABLE 4 Physical and genetic maps from archaea

Organism	Chromosome(s) (kbp)	Plasmid(s)[a] (kbp)	Genome size (kbp)	rRNA[b]	Marker[c]	Method[d]	Strategy[e]	Reference[f]
Aquifex pyrophilus	1,620	–	1,620	6	1	PFGE	CH, LC	188
Halobacterium salinarium GRB	2,040	1.8, 37, 90, 305	2,473	1	20	OL/PFGE	Lorist M/CH	198
Halobacterium salinarium (three strains)	2,000	Several	>2,000	ND[g]	ND	PFGE	2D	85
Haloferax mediterranei	2,900	130, 320, 490	3,840	2	5	PFGE	CH, LC, PD	138
	2,900	130, 320, 490	3,840	2	15	PFGE	CH, 2D	4
	2,900	130, 320, 490	3,840	2	24	PFGE	GM	139
Haloferax volcanii	2,920	6.4, 86, 442, 690	4,144	2	24	OL	Lorist M, GW	38
	2,920	6.4, 86, 442, 690	4,144	2	139	PFGE	CH, Tf	46
Methanobacterium thermoautotrophicum Marburg	1,620	4.5	1,625	2	46	PFGE	CH, DD, PD	195, 196
Methanobacterium thermoautotrophicum ΔH	1,751	–	1,751	2	1,855	GS	SG	193
Methanococcus jannaschii	1,660	16, 58	1,734	2*	1,738	GS	SG	25
Methanococcus voltae	1,880	–	1,880	1*	37	PFGE	CH, PD	191
Methanococcus wolfei	1,730	–	1,730	2	26	PFGE	CH, DD	195
Sulfolobus tokodai (previously acidocaldarius)	3,000	–	3,000	ND	ND	PFGE	LC	219
Thermococcus celer	2,760	–	2,760	1	ND	PFGE	CH, DD, LC	112
	1,890	–	1,890	1*	ND	PFGE	DD, PD	154

[a] –, extrachromosomal element has been reported; +, plasmid of unknown size present.
[b] *, organization of the ribosomal genes is different from the *rrs* (16S)-*rrl* (23S)-*rrs* (5S) operon.
[c] The number excludes random probes and does not include the ribosomal genes, indicated in the previous column.
[d] The methods are as follows: GS, whole-genome sequencing; OL, ordered library with vector, e.g., Lorist M.
[e] The strategies are as follows: CH, cross-hybridization; GW, genome walking; 2D, two-dimensional gel electrophoresis; DD, double digestion; GM, gene mapping; LC, linking clones; PD, partial digests; SG, shotgun; Tf, transformation.
[f] Physical maps for 38 species were redrawn in reference 54, with minor modifications relative to the original publications (essentially, with additional markers).
[g] ND, not done.

monly used methods are cross-hybridization and linking-clone analysis. Cross-hybridization, involving the isolation and labelling of a genomic fragment, detects neighboring fragments but does not always determine their order. This can be readily achieved by linking-clone analysis, in which probes spanning sites for the mapping enzyme are used to demonstrate the contiguity of particular fragments. Analysis of products obtained by partial digestion is also useful for determining the order and relatedness of restriction fragments, and it is particularly powerful when combined with two-dimensional gel electrophoresis, although this is a technically demanding technique (175). Details of the different approaches used to derive the genomic organization of numerous bacteria and archaea are summarized in Tables 1 to 4, where it can be seen that cross-hybridization in conjunction with PFGE analysis is the most popular approach. Another technique that is finding wider application is the transposon-mediated introduction of novel or unique sites for rare-cutting enzymes into bacterial genomes, and this has facilitated mapping studies of, among others, *Brucella* and *Lactococcus* spp. (99, 124, 147).

With the growing awareness of the multiplicity of chromosomes in microrganisms, it has become increasingly important to distinguish between the chromosomal and plasmid components of the genome. One means of achieving this objective is to cleave genomic DNA with the intron-encoded nuclease I-*Ceu*I, as this cuts uniquely in the *rrl* genes, encoding the 23S rRNA, of essentially all bacteria. Since *rrn* genes are confined to chromosomes by definition, plasmid DNA is never cleaved by this enzyme. This strategy enabled Katayama et al. (101) to demonstrate that many of the toxin genes in *Clostridium perfringens* were plasmid borne by comparing the hybridization profiles of samples treated with I-*Ceu*I with those of untreated DNA. Digestion with I-*Ceu*I also enables the orientations and the numbers of *rrn* operons to be established, as each operon gives rise to one fragment. In this way, *Clostridium acetobutylicum* and *Clos-*

tridium beijerinckii were shown to have 11 and 14 copies of the *rrn* operon, respectively (50, 217). Digestion with I-*Ceu*I is obviously of greater usefulness in studies of bacteria with large numbers of operons, like members of the family *Enterobacteriaceae* or *Bacillaceae*, and it works successfully for the majority of bacterial genera, with the notable exception of the mycobacteria, where the recognition site of I-*Ceu*I has incurred a single base deletion. I-*Ceu*I has proved to be a powerful tool for genome analysis of *Salmonella* spp. (132, 133, 136) and was especially useful for characterizing chromosomal rearrangements in *Salmonella typhi* and *Salmonella paratyphi*, where inversions resulted from recombination between *rrn*H and *rrn*G (137).

In recent years significant effort has been devoted to developing direct-mapping techniques for large DNA molecules that do not require gel electrophoresis. Among the more promising of these are two new methods known as DNA combing and optical mapping, both of which make use of fluorescence microscopy and image analysis to visualize single DNA molecules. In the DNA-combing technique (16), DNA molecules attached to a glass support by one end are straightened or "combed" by a receding meniscus and then used as a substrate for fluorescent in situ hybridization. While combing has not yet been used to map a bacterial genome, its potential has been validated in pilot studies in which prophage genomes were localized and measured. Optical mapping (187, 214) is similar to the combing procedure and could be used to construct maps by visualization of individual DNA molecules that have been cleaved with restriction enzymes, since the cleavage points are clearly apparent on microscopic examination. In both cases, inspection of a statistically significant sample is required. Given the increasing availability of fluorescence microscopy facilities in microbiology laboratories, it is conceivable that these approaches will find wider application in the years to come.

Physical maps of bacterial genomes have also been derived by "bottom-up" approaches, in which large numbers of clones based on lambda, cosmids, or yeast artificial chromosomes (YAC) are assembled into contigs in piecemeal fashion. The ordered-library strategy was pioneered by Kohara et al. for *Escherichia coli* (109) and has also been applied to the analysis of about a dozen other prokaryotic genomes (Tables 1 to 4), often in conjunction with genome-sequencing projects. A major advantage of this approach is the generation of an immortalized source of DNA in the form of a minimally redundant set of clones. Ordered libraries are also referred to as encyclopedias, as they have the potential to impart a large body of information (see reference 68 for a recent review). However, constructing maps in this way is labor-intensive and relatively expensive. Furthermore, it is generally difficult to achieve complete coverage of the genome, and other methods are required to obtain gap closure. The development of the bacterial artificial chromosome (BAC) cloning system (190) represented a significant breakthrough in the construction of comprehensive ordered libraries and has been successfully applied to *Mycobacterium tuberculosis* and *Mycobacterium bovis* BCG (23), from which it had previously proved difficult to obtain fully representative cosmid libraries. The BAC cloning system is based on the *E. coli* F factor, whose replication is strictly controlled, thus ensuring stable maintenance of large constructs (190). BAC libraries are likely to find wider application in bacterial-genome mapping and sequencing in the coming years.

DO WE STILL NEED TO MAP?

The ease with which microbial genome sequences can be obtained by conventional methods (64) or, in the near future, through the use of "DNA chips" (65), raises the question of whether genome mapping has a future, given the relative paucity of information obtained in comparison to that derived from a genome sequence. We believe that there are several reasons to continue with genome mapping. Firstly, not only does it provide valuable confirmatory data about size and topology, it also reflects more accurately the situation of a bacterial genome in vivo, as no cloning or in silico assembly steps are necessarily involved. In this respect, it is particularly encouraging to note the excellent agreement in size for genomes that have been both mapped and completely sequenced, such as those from *Haemophilus influenzae*, *Helicobacter pylori*, *E. coli*, and *Mycobacterium tuberculosis* (19, 47, 64, 211). Secondly, for reasons of cost it is unlikely that either genome sequencing or hybridization chip-based sequencing will be available or affordable to many microbiologists in the next few years. Thirdly, some forms of chromosomal modifications are difficult or even impossible to detect by hybridization methods involving DNA chips or microarrays. In consequence, it is most probable that PFGE profiling and mapping will continue to be employed for comparative studies of bacterial genomes in the immediate future and for characterizing chromosomal rearrangements, such as deletions, insertions, duplications, and inversions. In contrast, it seems increasingly unlikely that any new linkage maps of bacterial chromosomes will be produced by classical genetic mapping procedures.

GENOME STATISTICS FOR >100 BACTERIA AND ARCHAEA

Genome-mapping studies have been published for well over 100 species of bacterial descent and at least a dozen archaea. Table 1 presents the genome statistics derived from the maps of a large variety of purple bacteria. The chromosomes range in size from 1,100 to 9,400 kb, and the number of *rrn* operons present varies from one to seven. Details of mapping studies performed with many grampositive bacteria are summarized in Table 2, where it can be seen that genome size ranges from 580 kb in the case of *Mycoplasma genitalium* to around 8 Mb for the *Streptomyces* species. The number of *rrn* operons in this group varies from one in the slow-growing mycobacteria to 14 in *C. beijerinckii*. Physical and genetic maps from the remaining bacteria, chlamydiae, green sulfur bacteria, planctomycetes, rickettsiae, and spirochetes are described in Table 3, where it is apparent that genome sizes range from 935 to 6,420 kb. Members of this group typically have low *rrn* copy numbers (1 or 2).

Details of genome characterization studies conducted with archaea are outlined in Table 4. The genome sizes range from 1.62 to 4.15 Mb. It is of interest that plasmids make a significant contribution to the total genome size of the halophilic species (18% for *Halobacterium* sp. strain GRB to 30% for *Haloferax volcanii*), in sharp contrast to the situation for the methanogens and extreme thermophiles.

Overall, bacterial genomes range in size from about 0.6 to 9.4 Mb. In a recent review (189), it was suggested that there may be a relationship between the genome sizes and the lifestyles of bacteria. Specialist prokaryotes, which occupy restricted ecological niches, tend to have smaller genomes, <1.5 Mb in size, and many of them are extremely fastidious (e.g., *Borrelia burgdorferi*) or behave as obligate intracellular parasites (e.g., *Chlamydia trachomatis*). Generalist bacteria that have broad metabolic potential and few organic growth requirements in vitro possess larger genomes (~3 to 5 Mb). The largest genomes are found in bacteria that have complex developmental cycles, such as the myxobacteria and *Streptomyces*.

LINEAR CHROMOSOMES IN PROKARYOTES: THE OTHER SIDE OF THE COIN?

Until the end of the 1980s, bacteria were thought to have a single, circular chromosome. This paradigm, based on studies with *E. coli* and *Bacillus subtilis*, was difficult to shake. However, in 1989, it was demonstrated convincingly that the *B. burgdorferi* chromosome was linear (11, 61). Since then, the linearity of the bacterial chromosome has been demonstrated in many members of the genus *Streptomyces* (131) and in *Rhodococcus fascians* (51).

Linear and circular replicons (plasmid or chromosome) can be found in all possible combinations within a bacterium. Linear and circular replicons coexisting in the same bacterium have been reported—*Agrobacterium tumefaciens* has both a linear and a circular chromosome (2). Furthermore, a linear plasmid, N15 (203), was shown to lysogenize *E. coli*. Linear plasmids (20 to 320 kbp in size) can be found in some species of *Mycobacterium* (*Mycobacterium xenopi*, *Mycobacterium celatum*, and *Mycobacterium branderi*), which all contain a circular chromosome (166). It should be emphasized that the first report of a linear plasmid in prokaryotes was nearly 20 years ago (reference 87; for a review, see reference 145) and concerned *Streptomyces rochei*, which, as mentioned above, has a linear chromosome. Both linear and circular plasmids have been found in all species of the genus *Borrelia* (8, 9, 167).

One of the major issues which had to be resolved was the structure of the telomeres (or termini of the chromosome). One can distinguish two major classes of telomeres in prokaryotes, blunt-ended extremities with a protein covalently attached and covalently closed hairpin loops. The latter, in which a DNA strand loops around to become the complementary strand, was exhibited by *Borrelia* plasmids (9, 91) and was recently demonstrated for the *Borrelia* chromosome (36). The former, in which the telomeres carry covalently bound proteins, is exhibited by the plasmids (108) and chromosomes of seven species of *Streptomyces*, among which *Streptomyces coelicolor*, *Streptomyces ambofaciens*, *Streptomyces lividans*, and *Streptomyces griseus* are the best studied (122, 123, 129, 131, 172).

Several models for the replication of linear molecules have been proposed for eukaryotic genomes. They can be adapted to prokaryotes by analogy. Models of replication of telomeric structures involving covalently closed ends were given for the genomes of vaccinia virus and other poxviruses (12) and African swine fever virus (81). The model implies a concatemeric replicative intermediate, which was shown to occur in vaccinia virus. Models involving a protein covalently attached to blunt ends have been well characterized for adenovirus and phage Φ29 (182). The telomere is the origin of replication. Other models, for example, involving a circular intermediate (see below), could be proposed.

The distinction between circular and linear replicons is perhaps not so simple, since conversion does occur between them. It has been shown that the linear chromosome of S. lividans can be circularized by joining the two extremities artificially or through spontaneous deletions (131). Furthermore, an origin of replication, a full equivalent of oriC typical of a circular chromosome, was shown to be functional when cloned on a circular plasmid. This oriC is located in the center of the Streptomyces chromosome (222). The replication of the Streptomyces linear chromosomes could proceed either from oriC or from the telomeres. The ability to use an origin of replication typical of a circular replicon is not restricted to the Streptomyces chromosome. Replication of the pSLA2 linear plasmid from S. rochei occurs bidirectionally, extending toward the extremities of the plasmid from a centrally located origin (37). The linear chromosome of B. burgdorferi also contains an oriC-like region near its middle (159), which carries the characteristic genes, gyrA, gyrB, dnaA, and dnaN. The conversion of a linear plasmid to a circular plasmid has been shown to occur in the relapsing fever agent, Borrelia hermsii (62). This suggests that circular intermediates may be involved in the replication of linear replicons in species of Borrelia.

Large inverted repeats are found at the ends of the Streptomyces chromosome (24 to 210 kbp) and plasmids (0.04 to 80 kbp), while in contrast, small imperfect inverted repeats were reported for the ends of the Borrelia chromosome (26 bp) and plasmids (19 bp). In both cases there is homology between the telomeres of the bacterial chromosome and plasmids. There is homology between one end of the SLP2 linear plasmid and the telomeres of S. lividans, while the telomeres of the B. burgdorferi chromosome are homologous to those of its linear plasmids. Casjens et al. (36) hypothesized a possible exchange between telomeres from the Borrelia chromosome and those from linear plasmids. Such a replacement of the linear chromosomal telomeres by telomeres from a linear plasmid (pPGZ101) has been shown in Streptomyces rimosus (84). Exchanges of extremities might be a common property among linear replicons, the mechanism of which has to be further studied.

ONE CHROMOSOME OR MORE?

As mentioned above, until the advent and widespread application of genome-mapping studies, it was commonly believed that all bacteria possessed a single circular chromosome and that, in some cases, this was accompanied by extrachromosomal genetic elements such as plasmids, independent replicons that are often capable of promoting their own transfer (127). However, as the technology for examining both chromosomes and plasmids improved, it became apparent that some bacteria harbored very large plasmids that are sometimes referred to as "megaplasmids," and early examples were provided by Rhizobium or Agrobacterium spp. As more bacterial species were investigated, it seemed possible that some of these large plasmids might represent second chromosomes, as they carried genes that were known, or believed, to be essential. This led to much reflection on the true nature of the bacterial chromosome and the possible reasons for the evolution of additional chromosomes. DNA sequence analysis, combined with attempts to obtain derivatives lacking some of the additional chromosomes, will undoubtedly resolve the issue of whether a plasmid corresponds to a second chromosome or is simply another replicon. It is currently difficult to determine whether the smaller replicons present in bacterial cells correspond to second chromosomes or to plasmids. For the sake of clarity and simplicity, the definition of a chromosome used here is a replicon which carries an essential gene, such as an rRNA or housekeeping gene, although it should be remembered that conditions might exist where these genes are inessential. Conversely, plasmids should not be required for cell viability under optimal growth conditions in the laboratory, although they almost certainly carry genes that confer a competitive advantage on the host under certain circumstances.

It is now clear from genome-mapping and genetic studies that *A. tumefaciens*, *Leptospira interrogans*, *Brucella melitensis*, and *Rhodobacter sphaeroides* have two distinct chromosomes (1, 2, 42, 225), whereas *Burkholderia cepacia* has three (41, 173). It is probable that as more species are examined new variations on this theme will be found, although in the overwhelming majority of cases a single chromosome can be expected (Tables 1 to 4).

L. interrogans shows the simplest genomic organization of the "multichromosomal" bacteria, as it contains two circular chromosomes, CI and CII, the smaller of which was initially thought to be a plasmid (10, 224). Genome maps are available for two distinct serovars, *L. interrogans* serovar icterohaemorrhagiae and *L. interrogans* serovar pomona, and in both cases the size of CII has been estimated at 0.35 Mb. In contrast, CI of *L. interrogans* serovar icterohaemorrhagiae, at 4.61 Mb, is 200 kb larger than its counterpart in *L. interrogans* serovar pomona (225). Map comparison indicated that CI may have diverged significantly during evolution, as it shows a "mosaic" type organization that may reflect insertion sequence (IS)-mediated rearrangements. In both instances, the genes *dnaA*, *gidA*, *gyrA*, and *gyrB* are localized in a narrow interval of CI, and this probably defines *oriC*. Although the smaller size of CII was suggestive of its being a large plasmid, mapping studies showed the 0.35-Mb circular species of both serovars to carry the single-copy gene *asd*, encoding aspartate semialdehyde dehydrogenase, whose function is indispensable for bacterial viability (22). These findings, together with the observation that strains of *L. interrogans* harboring only the 4.4- or 4.6-Mb chromosome have never been observed, led to the 0.35-Mb species being designated a bacterial chromosome.

The genome of *B. melitensis*, like that of *L. interrogans*, comprises two circular chromosomes of 2.05 and 1.13 Mb. Sixteen genetic loci have been mapped to the larger chromosome, which bears two *rrn* operons. The smaller chromosome also harbors an *rrn* operon, as well as the single-copy genes for the GroE chaperones and bacterioferritin, thus ruling out the possibility that it corresponds to an exceptionally large plasmid.

Perhaps the best characterized of the multichromosomal bacteria is the photosynthetic microorganism *R. sphaeroides*, which has two circular chromosomes, CI (~3 Mb) and CII (~0.9 Mb) (42), as extensive DNA sequence information has been published for CII (43). The genomic structure of *R. sphaeroides* is well conserved; multiple independent isolates have been examined, and all have been shown to contain two chromosomes, although some variation in size and *rrn* operon copy number was observed (described in reference 43). Like those of *B. melitensis*, both chromosomes of *R. sphaeroides* carry *rrn* operons, as well as duplicate copies of genes encoding a number of isofunctional enzymes that are structurally similar but subject to different regulatory controls. A further 146 genes and open reading frames (ORFs) have been localized through "sequence-skimming" studies to CII, and most of them do not appear to have counterparts on CI. As many of them (e.g., *bioA*, *bioD*, *proA*, and *proB*) mediate basic metabolic functions, it is again clear that attribution of chromosomal status to CII is legitimate. Furthermore, the indistinguishable codon usages, G+C contents, and di- and trinucleotide frequencies of CI and CII are all indicative of CII having coexisted with CI for a very long time, and they rule out the possibility of recent acquisition. Furthermore, the fact that genes involved in the same biochemical pathways also localize to one or the other chromosome eliminates the possibility of chromosomal specialization.

The *A. tumefaciens* genome is most unusual, as it comprises a 3-Mb circular chromosome, a 2.1-Mb linear chromosome, and two plasmids of 450 and 200 kb, respectively (Table 1). The 200-kb plasmid is the well-characterized, tumor-inducing pTi, while the

450-kb species is the cryptic plasmid pAtC58. The large linear and circular replicons both carry *rrn* genes, as well as other metabolically essential genes, thus confirming their identities as chromosomes (2)

The genome of *B. cepacia*, which is one of the largest described among bacteria (Table 1), shows the most complex organization of the multichromosomal bacteria, comprising three circular chromosomes (41, 173). Maps are available for the genomes of two strains of *B. cepacia*, 17616 and ATCC 25416, and although both genomes consist of three chromosomes, they display significant differences in their sizes, organizations, and *rrn* copy numbers (Table 1). At 8.15 Mb, the genome of strain ATCC 25416 is some 1.35 Mb bigger than that of strain 17616, and it includes a plasmid of 200 kb that has not been described in the latter strain. All three chromosomes of both strains of *B. cepacia* carry *rrn* operons, and *recA* genes have been located on CI and CII in strain ATCC 25416 (41, 173).

An interesting example of functional genomic plasticity is provided by the nitrogen-fixing symbiotic bacteria belonging to the related genera *Azorhizobium*, *Bradyrhizobium*, and *Rhizobium*. Detailed maps (Table 1) are available for *Bradyrhizobium japonicum*, whose genome comprises a single circular chromosome of 8.7 Mb (116), and *Rhizobium meliloti* 1021, whose 6.6-Mb genome consists of three circular replicons, a 3.54-Mb chromosome, and two megaplasmids of 1.34 and 1.7 Mb (94). Generally speaking, genes important for symbiosis are localized on plasmids in *Rhizobium* spp. but on the chromosomes of *Azorhizobium* and *Bradyrhizobium*. The recent systematic DNA sequence analysis of pNGR234*a*, the symbiotic plasmid from *Rhizobium* sp. strain NGR234, has been immensely instructive (74). This 536,165-bp replicon shows a significant degree of functional organization and carries clusters of genes involved in amino acid and peptide catabolism, rhamnose synthesis, nitrogen fixation, and plasmid replication and transfer. The vast majority of the proteins necessary for bacteroid

development and production of the nitrogen-fixing complex are apparently encoded by pNGR234*a*, although some essential *fix* genes are chromosomally localized (74). No genes required for housekeeping or primary metabolism are located on pNGR234*a*, and this is consistent with the fact that strain NGR234 can be cured of its plasmid (74). Of particular importance is the finding that the locus containing the replication and conjugal transfer genes of pNGR234*a* is highly similar, in both sequence and organization, to those of several other plasmids from *Agrobacterium* and *Rhizobium* spp., and this is strongly suggestive of recent lateral transfer of genetic information. Taken together, these combined observations suggest that pNGR234*a* corresponds to a true plasmid rather than a second chromosome, such as CII of *R. sphaeroides*. Or does it?

FROM ONE CHROMOSOME TO TWO?

The genome size of *Bacillus cereus* strains varies from 5.5 to 6.3 Mb, and great diversity is seen in the number and organization of the chromosomes (Table 2). In some isolates, there is a single circular chromosome of >5.5 Mb (110), whereas in others, such as strain F0837/76, the chromosome is only 2.4 Mb in size (33) but is accompanied by six "plasmids" ranging in size from 40 to 960 kb that represent a further 2.2 Mb of genetic material (Table 2). Comparison of the genome maps of six strains of *B. cereus* led to the identification of a segment of the chromosome that was frequently found in the form of a separate replicon (33). In three of the strains, there is a 2-Mb replicon that carries the essential pyruvate dehydrogenase gene, in addition to the 2.4-Mb chromosome. Kolstö has proposed the terms primary and secondary chromosomes to describe these species (111).

Since the vast majority of bacteria are content with a single chromosome, why have some organisms opted to divide their genomes into two or more chromosomes? If one accepts the reasonable principle that initially all bacteria were unichromosomal, then two possibilities spring to mind. Firstly, it is conceivable that a single large chromosome split into several smaller replicons or, alternatively, that the second chromosome was obtained from another organism by horizontal transfer, as might have been the case with pNGR234a in *Rhizobium* sp. strain NGR234. Support for the first hypothesis is provided by the comparative mapping studies performed with *B. cereus*, which suggest not only that chromosomal splitting occurs but that it may be a relatively recent event in evolutionary terms, as contemporary strains with one or more chromosomes coexist. In turn, this raises the question of whether the process is reversible; can a single large chromosome be re-formed from two smaller ones by means of homologous recombination or a similar mechanism? Clearly *B. cereus* represents an ideal system in which to study this issue. It also seems likely that the two chromosomes present in *L. interrogans* may be the descendants of a single ancestral chromosome, possibly as a result of the extensive rearrangements that led to the current mosaic organization. However, as the *L. interrogans* genome appears to be much less dynamic than that of *B. cereus*, and all strains have two chromosomes, this must have been an ancient event.

One immediate consequence of this putative chromosome fission event is the necessity for origins of replication for each of the chromosomal progeny. Apart from *L. interrogans*, in which it is clear that the large chromosome carries a typical *oriC* (225), little is known about the nature and distribution of the origins of replication in other multichromosomal bacteria. While essential functions, such as the DnaA-mediated ability to initiate replication, can be furnished in *trans*, the second chromosome has to have its own origin of replication. It is conceivable that multiple chromosomes arose in some bacteria as an attempt at achieving faster overall rates of replication, although this would also require the existence of an efficient system for partitioning all chromosomes.

In *E. coli* it is known that alternative replication systems exist, but these are generally only active under very specific conditions (5). Alternatively, a replication origin provided by an integrated genetic element, such as a phage or plasmid, could be activated following the imprecise excision of the element, leading to a segment of the bacterial chromosome changing replicon. If this excised segment contained genes that were essential for bacterial growth or survival, this would ensure a strong selective pressure for maintenance of the second chromosome, since its loss would result in cell death. In the same vein, second chromosomes may have arisen as the result of an increased requirement for a specific gene product. Higher levels of gene expression might result from altered topological constraints acting on the DNA or from a more favorable location with respect to the origin of replication, leading to an increase in gene dosage.

As is so often the case in microbiology, this reasoning leads one back to *E. coli*, the molecular genetic paradigm, and raises the question of whether the F-prime factor (183) was the prototype for the second bacterial chromosome. F-prime factors arise as the result of integration of the F factor (63), which carries three mobile elements, IS2, IS3, and Tn1000 ($\gamma\delta$), into the bacterial chromosome via homologous recombination between chromosomal and F factor-borne IS elements or by transpositional recombination. Reversal of these events, or recombination between duplicated copies of IS elements, leads to excision of the F factor together with a flanking segment of DNA, which can represent up to 30% (~1.5 Mb) of the *E. coli* chromosome (93). The formation of primes has been described for several plasmids, including the broad-host-range plasmids ColV and R68, and in a variety of bacterial hosts, such as *E. coli*, *Salmonella* spp., *Klebsiella pneumoniae*, *Proteus mirabilis*, and *Pseudomonas aeruginosa* (93). In brief, two components are required to produce a second replicon from the bacterial chromosome, a stable plasmid replicon and an insertion sequence. It is most noteworthy that pNGR234a of *Rhizobium* sp. strain NGR234 is exceptionally rich in IS, and this, together with the fact that many of the genes carried by pNGR234a are chromosomal in *Azorhizobium* and *Bradyrhizobium*, suggests that pNGR234a may well be of chromosomal descent. Undoubtedly, as more sequence data become available from multichromosomal bacteria, further insight into genome dynamics will be obtained.

ACKNOWLEDGMENTS

Work in our laboratories is supported by the Institut Pasteur and the Association Française Raoul Follereau.

REFERENCES

1. **Allardet-Servent, A., M.-J. Carles-Nurit, G. Bourg, S. Michaux, and M. Ramuz.** 1991. Physical map of the *Brucella melitensis* 16 M chromosome. *J. Bacteriol.* **173:**2219–2224.

2. **Allardet-Servent, A., S. Michaux-Charachon, E. Jumas-Bilak, L. Karayan, and M. Ramuz.** 1993. Presence of one linear and one circular chromosome in the *Agrobacterium tumefaciens* C58 genome. *J. Bacteriol.* **175:**7869–7874.

3. **Amjad, M., J. M. Castro, H. Sandoval, J.-J. Wu, M. Yang, D. J. Henner, and P. J. Piggot.** 1990. An *Sfi*I restriction map of the *Bacillus subtilis* 168 genome. *Gene* **101:**15–21.

4. **Anton, J., P. Lopez-Garcia, J. P. Abad, C. L. Smith, and R. Amila.** 1994. Alignment of genes and *Swa*I restriction sites to the *Bam*HI genomic map of *Haloferax mediterranei*. *FEMS Microbiol. Lett.* **117:**53–60.

5. **Asai, T., and T. Kogoma.** 1994. D-loops and R-loops: alternative mechansims for the initiation of chromosome replication in *Escherichia coli*. *J. Bacteriol.* **176:**1807–1812.

6. **Azevedo, V., E. Alvarez, E. Zumstein, G. Damiani, V. Sgaramella, S. D. Ehrlich, and P. Serror.** 1993. An ordered collection of *Bacillus subtilis* DNA segments cloned in yeast artificial chromosomes. *Proc. Natl. Acad. Sci. USA* **90:**6047–6051.

7. **Bancroft, I., P. Wolk, and E. V. Oren.** 1989. Physical and genetic maps of the genome of the heterocyst-forming cyanobacterium *Anabaena* sp. strain PCC7120. *J. Bacteriol.* **171:**5940–5948.

8. **Barbour, A. G.** 1988. Plasmid analysis of *Borrelia burgdorferi*, the Lyme disease agent. *J. Clin. Microbiol.* **26:**475–478.

9. **Barbour, A. G., and C. F. Garon.** 1987. Linear plasmids of the bacterium *Borrelia burgdorferi* have covalently closed ends. *Science* **237:**409–411.

10. **Baril, C., J. L. Herrmann, C. Richaud, D. Margarita, and I. Saint Girons.** 1992. Scattering of the rRNA genes on the physical map of the circular chromosome of *Leptospira interrogans* serovar icterohaemorrhagiae. *J. Bacteriol.* **174:**7566–7571.

11. **Baril, C., C. Richaud, G. Baranton, and I. Saint Girons.** 1989. Linear chromosome of *Borrelia burgdorferi*. *Res. Microbiol.* **140:**507–516.

12. **Baroudy, B. M., S. Venkatesan, and B. Moss.** 1982. Incompletely base-paired flip-flop terminal loops link the two DNA strands of the vaccinia virus genome into one uninterrupted polynucleotide chain. *Cell* **28:**315–324.

13. **Bathe, B., J. Kalinowski, and A. Pühler.** 1996. A physical and genetic map of the *Corynebacterium glutamicum* ATCC 13032 chromosome. *Mol. Gen. Genet.* **252:**255–265.

14. **Bautsch, W.** 1988. Rapid mapping of the *Mycoplasma mobile* genome by two dimensional field inversion gel electrophoresis techniques. *Nucleic Acids Res.* **16:**11461–11467.

15. **Bautsch, W.** 1993. A *Nhe*I macrorestriction map of *Neisseria meningitidis* B1940 genome. *FEMS Microbiol. Lett.* **107:**191–198.

16. **Bensimon, A., A. Simon, A. Chiffaudel, V. Croquette, F. Heslot, and D. Bensimon.** 1994. Alignment and sensitive detection of DNA by a moving interface. *Science* **265:**2096–2098.

17. **Bihlmaier, A., U. Römling, T. F. Meyer, B. Tümmler, and C. P. Gibbs.** 1991. Physical and genetic map of the *Neisseria gonorrhoeae* strain MS11–N198 chromosome. *Mol. Microbiol.* **5:**2529–2539.

18. **Birkelund, S., and R. S. Stephens.** 1992. Construction of physical and genetic maps of *Chlamydia trachomatis* serovar L2 by pulsed-field gel electrophoresis. *J. Bacteriol.* **174:**2742–2747.

19. **Blattner, F. R., G. R. Plunkett, C. A. Bloch, N. T. Perna, V. Burland, M. Riley, J. Collado-Vides, J. D. Glasner, C. K. Rode, G. F. Mayhew, J. Gregor, N. W. Davis, H. A. Kirkpatrick, M. A. Goeden, D. J. Rose, B. Mau, and Y. Shao.** 1997. The complete genome sequence of *Escherichia coli* K-12. *Science* **277:**1453–1474.

20. **Borges, K. M., and P. L. Bergquist.** 1993. Genomic restriction map of the extremely thermophilic bacterium *Thermus thermophilus* HB8. *J. Bacteriol.* **175:**103–110.

21. **Bourke, B., P. Sherman, H. Louie, E. Hani, P. Islur, and V. L. Chan.** 1995. Physical and genetic map of the genome of *Campylobacter upsaliensis*. *Microbiology* **141:**2417–2424.

22. **Boursaux-Eude, C., I. Saint Girons, and R. Zuerner.** 1998. *Leptospira* genomics. *Electrophoresis* **19:**589–592.

23. **Brosch, R., S. V. Gordon, A. Billault, T. Garnier, K. Eiglmeier, C. Soravito, B. G. Barrell, and S. T. Cole.** 1998. Use of a *Mycobacterium tuberculosis* H37Rv bacterial artificial chromosome library for genome mapping, sequencing, and comparative genomics. *Infect. Immun.* **66:**2221–2229.

24. **Bukanov, N. O., and D. E. Berg.** 1994. Ordered cosmid library and high-resolution physical-genetic map of *Helicobacter pylori* strain NCTC11638. *Mol. Microbiol.* **11:**509–523.

25. **Bult, C. J., O. White, G. J. Olsen, L. Zhou, R. D. Fleischmann, G. G. Sutton, J. A. Blake, L. M. FitzGerald, R. A. Clayton, J. D. Gocayne, A. R. Kerlavage, B. A. Dougherty, J. F. Tomb, M. D. Adams, C. I. Reich, R. Overbeek, E. F. Kirkness, K. G. Weinstock, J. M. Merrick, A. Glodek, J. L. Scott, N. S. M. Geoghagen, and J. C. Venter.** 1996. Complete genome sequence of the methanogenic archaeon, *Methanococcus jannaschii*. *Science* **273:**1058–1073.

26. **Butler, P. D., and E. R. Moxon.** 1990. A physical map of the genome of *Haemophilus influenzae* type b. *J. Gen. Microbiol.* **136:**2333–2342.

27. **Canard, B., and S. T. Cole.** 1989. Genome organization of the anaerobic pathogen *Clostridium perfringens*. *Proc. Natl. Acad. Sci. USA* **86:**6676–6680.

28. **Canard, B., B. Saint-Joanis, and S. T. Cole.** 1992. Genomic diversity and organization of virulence genes in the pathogenic anaerobe *Clostridium perfringens*. *Mol. Microbiol.* **6:**1421–1429.

29. **Carlson, C. R., A. Grönstad, and A.-B. Kolstö.** 1992. Physical maps of the genomes of three *Bacillus cereus* strains. *J. Bacteriol.* **174:**3750–3756.

30. **Carlson, C. R., T. Johansen, and A.-B. Kolstö.** 1996. The chromosome map of *Bacillus thuringiensis* subsp. canadensis HD224 is highly similar to that of the *Bacillus cereus* type strain ATCC 14579. *FEMS Microbiol. Lett.* **141:**163–167.

31. **Carlson, C. R., T. Johansen, M.-M. Lecadet, and A.-B. Kolstö.** 1996. Genomic organization of the entomopathogenic bacterium *Bacillus thuringiensis* subsp. berliner 1715. *Microbiology* **142:**1625–1634.

32. **Carlson, C. R., and A.-B. Kolstö.** 1993. A complete physical map of a *Bacillus thuringiensis* chromosome. *J. Bacteriol.* **175:**1053–1060.

33. **Carlson, C. R., and A.-B. Kolstö.** 1994. A small (2.4 Mb) *Bacillus cereus* chromosome corresponds to a conserved region of a larger (5.3 Mb) *Bacillus cereus* chromosome. *Mol. Microbiol.* **13:**161–169.

34. **Casjens, S., M. Delange, H. L. Ley III, P. Rosa, and W. M. Huang.** 1995. Linear chromosomes of Lyme disease agent spirochetes: genetic diversity and conservation of gene order. *J. Bacteriol.* **177:**2769–2780.

35. **Casjens, S., and W. M. Huang.** 1993. Linear chromosomal physical and genetic map of *Borrelia burgdorferi*, the Lyme disease agent. *Mol. Microbiol.* **8:**967–980.

36. **Casjens, S., M. Murphy, M. DeLange, L. Sampson, R. vanVugt, and W. M. Huang.** 1997. Telomeres of the linear chromosomes of Lyme disease spirochaetes: nucleotide sequence and possible exchange with linear plasmid telomeres. *Mol. Microbiol.* **26:**581–596.

37. **Chang, P. C., and S. N. Cohen.** 1994. Bidirectional replication from an internal origin in a linear *Streptomyces* plasmid. *Science* **265:**952–954.

38. **Charlebois, R. L., L. C. Schalkwyk, J. D. Hofman, and W. F. Doolittle.** 1991. Detailed physical map and set of overlapping clones covering the genome of the archaebacterium *Haloferax volcanii* DS2. *J. Mol. Biol.* **222:**509–524.

39. **Chen, H., A. Kuspa, I. M. Keseler, and L. J. Shimkets.** 1991. Physical map of the *Myxococcus xanthus* chromosome. *J. Bacteriol.* **173:**2109–2115.

40. **Chen, X., and W. R. Widger.** 1993. Physical genome map of the unicellular cyanobacterium *Synechococcus* sp. strain PCC 7002. *J. Bacteriol.* **175:**5106–5116.

41. **Cheng, H.-P., and T. G. Lessie.** 1994. Multiple replicons constituting the genome of *Pseudomonas cepacia* 17616. *J. Bacteriol.* **176:**4034–4042.

42. **Choudhary, M., C. Mackenzie, K. Nereng, E. Sodergren, G. M. Weinstock, and S. Kaplan.** 1994. Multiple chromosomes in bacteria: structure and function of chromosome II of *Rhodobacter sphaeroides* 2.4.1T. *J. Bacteriol.* **176:**7694–7702.

43. **Choudhary, M., C. Mackenzie, K. Nereng, E. Sodergren, G. M. Weinstock, and S. Kaplan.** 1997. Low-resolution sequencing of *Rhodobacter sphaeroides* 2.4.1^T chromosome II is a true chromosome. *Microbiology* **143:**3085–3099.

44. **Churin, Y. N., I. N. Shalak, T. Börner, and S. V. Shestakov.** 1995. Physical and genetic map of the chromosome of the unicellular cyanobacterium *Synechocystis* sp. strain PCC 6803. *J. Bacteriol.* **177:**3337–3343.

45. **Cocks, B. G., L. E. Pyle, and L. R. Finch.** 1989. A physical map of the genome of *Ureaplasma urealyticum* 960^T with ribosomal RNA loci. *Nucleic Acids Res.* **17:**6713–6719.

46. **Cohen, A., W. L. Lam, R. L. Charlebois, W. F. Doolittle, and L. C. Schalkwyk.** 1992. Lo-

calizing genes on the map of the genome of *Haloferax volcanii*, one of the Archaea. *Proc. Natl. Acad. Sci. USA* **89:**1602–1606.

47. **Cole, S. T., R. Brosch, J. Parkhill, T. Garnier, C. Churcher, D. Harris, S. V. Gordon, K. Eiglmeier, S. Gas, C. E. Barry III, F. Tekaia, K. Badcock, D. Basham, D. Brown, T. Chillingworth, R. Connor, R. Davies, K. Devlin, T. Feltwell, S. Gentles, N. Hamlin, S. Holroyd, T. Hornsby, K. Jagels, A. Krogh, A. McLean, S. Moule, L. Murphy, K. Oliver, J. Osborne, M. A. Quail, M.-A. Rajandream, J. Rogers, S. Rutter, K. Seeger, J. Skelton, R. Squares, S. Squares, J. E. Sulston, K. Taylor, S. Whitehead, and B. G. Barrell.** 1998. Deciphering the biology of *Mycobacterium tuberculosis* from the complete genome sequence. *Nature* **393:**537–544.

48. **Cole, S. T., and I. Saint Girons.** 1994. Bacterial genomics. *FEMS Microbiol. Rev.* **14:**139–160.

49. **Colman, S. D., P.-C. Hu, W. Litaker, and K. F. Bott.** 1990. A physical map of the *Mycoplasma genitalium* genome. *Mol. Microbiol.* **4:**683–687.

50. **Cornillot, E., C. Croux, and P. Soucaille.** 1997. Physical and genetic map of the *Clostridium acetobutylicum* ATCC 824 chromosome. *J. Bacteriol.* **179:**7426–7434.

51. **Crespi, M., E. Messens, A. B. Caplan, M. Van Montagu, and J. Desomer.** 1992. Fasciation induction by the phytopathogen *Rhodococcus fascians* depends upon a linear plasmid encoding a cytokinin synthase gene. *EMBO J.* **11:**795–804.

52. **Daniels, D. L.** 1990. The complete *Avr*II restriction map of the *Escherichia coli* genome and comparisons of several laboratory strains. *Nucleic Acids Res.* **18:**2649–2651.

53. **Davidson, B. E., J. MacDougall, and I. Saint Girons.** 1992. Physical map of the linear chromosome of the bacterium *Borrelia burgdorferi* 212, a causative agent of Lyme disease, and localization of rRNA genes. *J. Bacteriol.* **174:**3766–3774.

54. **de Bruijn, F. J., J. R. Lupski, and G. M. Weinstock.** 1997. *Bacterial Genomes: Physical Structure and Analysis,* p. 585–775. Chapman and Hall, New York, N.Y.

55. **Dempsey, J. A., and J. G. Cannon.** 1994. Locations of genetic markers on the physical map of the chromosome of *Neisseria gonorrhoeae* FA 1090. *J. Bacteriol.* **176:**2055–2060.

56. **Dempsey, J. A. F., W. Litaker, A. Madhure, T. L. Snodgrass, and J. G. Cannon.** 1991. Physical map of the chromosome of *Neisseria gonorrhoeae* FA1090 with locations of genetic markers, including *opa* and *pil* genes. *J. Bacteriol.* **173:**5476–5486.

57. **Dempsey, J. A. F., A. B. Wallace, and J. G. Cannon.** 1995. The physical map of the chromosome of a serogroup A strain of *Neisseria meningitidis* shows complex rearrangements relative to the chromosomes of the two mapped strains of the closely related species *N. gonorrhoeae*. *J. Bacteriol.* **177:**6390–6400.

58. **Eiglmeier, K., N. Honoré, S. A. Woods, B. Caudron, and S. T. Cole.** 1993. Use of an ordered cosmid library to deduce the genomic organisation of *Mycobacterium leprae*. *Mol. Microbiol.* **7:**197–206.

59. **Ely, B., T. W. Ely, C. J. Gerardot, and A. Dingwall.** 1990. Circularity of the *Caulobacter crescentus* chromosome determined by pulsed-field gel electrophoresis. *J. Bacteriol.* **172:**1262–1266.

60. **Ely, B., and C. J. Gerardot.** 1988. Use of pulsed-field-gradient gel electrophoresis to construct a physical map of the *Caulobacter crescentus* genome. *Gene* **68:**323–333.

61. **Ferdows, M. S., and A. G. Barbour.** 1989. Megabase-sized linear DNA in the bacterium *Borrelia burgdorferi*, the Lyme disease agent. *Proc. Natl. Acad. Sci. USA* **86:**5969–5973.

62. **Ferdows, M. S., P. Serwer, G. A. Griess, S. J. Norris, and A. G. Barbour.** 1996. Conversion of a linear to a circular plasmid in the relapsing fever agent *Borrelia hermsii*. *J. Bacteriol.* **178:**793–800.

63. **Firth, N., K. Ippen-Ihler, and R. A. Skurray.** 1996. Structure and function of the F factor and mechanism of conjugation, p. 2377–2395. *In* F. C. Neidhardt, R. Curtiss III, J. L. Ingraham, E. C. C. Lin, K. B. Low, B. Magasanik, W. S. Reznikoff, M. Riley, M. Schaechter, and H. E. Umbarger (ed.), Escherichia coli *and* Salmonella: *Cellular and Molecular Biology*, 2nd ed. American Society for Microbiology, Washington, D.C.

64. **Fleischmann, R. D., M. D. Adams, O. White, R. A. Clayton, E. F. Kirkness, A. R. Kerlavage, C. J. Bult, J.-F. Tomb, B. A. Dougherty, J. M. Merrick, K. McKenney, G. Sutton, W. FitzHugh, C. Fields, J. D. Gocayne, J. Scott, R. Shirley, L.-I. Liu, A. Glodek, J. M. Kelley, J. F. Weidman, C. A. Phillips, T. Spriggs, E. Hedblom, M. D. Cotton, T. R. Utterback, M. C. Hanna, D. T. Nguyen, D. M. Saudek, R. C. Brandon, L. D. Fine, J. L. Fritchman, J. L. Fuhrmann, N. S. M. Geoghagen, C. L. Gnehm, L. A. McDonald, K. V. Small, C. M. Fraser, H. O. Smith, and J. C. Venter.** 1995. Whole-genome random sequencing and assembly of *Haemophilus influenzae* Rd. *Science* **269:**496–512.

65. **Fodor, S. P. A., J. L. Read, M. C. Pirrung, L. Stryer, A. T. Lu, and D. Solas.** 1996.

Light-directed, spatially addressable parallel chemical synthesis. *Science* **251:**767–773.

66. **Fonstein, M., and R. Haselkorn.** 1993. Chromosomal structure of *Rhodobacter capsulatus* strain SB1003: cosmid encyclopedia and high-resolution physical and genetic map. *Proc. Natl. Acad. Sci. USA* **90:**2522–2526.

67. **Fonstein, M., and R. Haselkorn.** 1995. Physical mapping of bacterial genomes. *J. Bacteriol.* **177:**3361–3369.

68. **Fonstein, M., and R. Haselkorn.** 1997. Encyclopedias of bacterial genomes, p. 348–361. *In* F. J. de Bruijn, J. R. Lupski, and G. M. Weinstock (ed.), *Bacterial Genomes: Physical Structure and Analysis.* Chapman and Hall, New York, N.Y.

69. **Fonstein, M., E. G. Koshy, T. Nikolskaya, P. Mourachov, and R. Haselkorn.** 1995. Refinement of the high-resolution physical and genetic map of *Rhodobacter capsulatus* and genome surveys using blots of the cosmid encyclopedia. *EMBO J.* **14:**1827–1841.

70. **Fonstein, M., S. Zheng, and R. Haselkorn.** 1992. Physical map of the genome of *Rhodobacter capsulatus* SB 1003. *J. Bacteriol.* **174:**4070–4077.

71. **Fraser, C. M., S. Casjens, W. M. Huang, G. G. Sutton, R. Clayton, R. Lathigra, O. White, K. A. Ketchum, R. Dodson, E. K. Hickey, M. Gwinn, B. Dougherty, J.-F. Tomb, R. D. Fleischmann, D. Richardson, J. Peterson, A. R. Kerlavage, J. Quackenbush, S. Salzberg, M. Hanson, R. van Vugt, N. Palmer, M. D. Adams, J. Gocayne, J. Weidman, T. Utterback, L. Watthey, L. McDonald, P. Artiach, C. Bowman, S. Garland, C. Fujii, M. D. Cotton, K. Horst, K. Roberts, B. Hatch, H. O. Smith, and J. C. Venter.** 1997. Genomic sequence of a Lyme disease spirochaete, *Borrelia burgdorferi. Nature* **390:** 580–586.

72. **Fraser, C. M., J. D. Gocayne, O. White, M. D. Adams, R. A. Clayton, R. D. Fleischmann, C. J. Bult, A. R. Kerlavage, G. Sutton, J. M. Kelley, J. L. Fritchman, J. F. Weidman, K. V. Small, M. Sandusky, J. Fuhrmann, D. Nguyen, T. R. Utterback, D. M. Saudek, C. A. Phillips, J. M. Merrick, J.-F. Tomb, B. A. Dougherty, K. F. Bott, P.-C. Hu, T. S. Lucier, S. N. Peterson, H. O. Smith, C. A. Hutchison III, and J. C. Venter.** 1995. The minimal gene complement of *Mycoplasma genitalium. Science* **270:**397–403.

73. **Fraser, C. M., S. J. Norris, G. M. Weinstock, O. White, G. G. Sutton, R. Dodson, M. Gwinn, E. K. Hickey, R. Clayton, K. A. Ketchum, E. Sodergren, J. M. Hardham, M. P. McLeod, S. Salzberg, J. Peterson, H. Khalak, D. Richardson, J. K. Howell, M. Chidambaram, T. Utterback, L. McDonald, P. Artiach, C. Bowman, M. D. Cotton, C. Fujii, S. Garland, B. Hatch, K. Horst, K. Roberts, M. Sandusky, J. Weidman, H. O. Smith, and J. C. Venter.** 1998. Complete genome sequence of *Treponema pallidum,* the syphilis spirochete. *Science* **281:**375–388.

74. **Freiberg, C., R. Fellay, A. Bairoch, W. J. Broughton, A. Rosenthal, and X. Perret.** 1997. Molecular basis of symbiosis between *Rhizobium* and legumes. *Nature* **387:**394–401.

75. **Frutos, R., M. Pages, M. Bellis, G. Roizes, and M. Bergoin.** 1989. Pulsed-field gel electrophoresis determination of the genome size of obligate intracellular bacteria belonging to the genera *Chlamydia, Rickettsiella,* and *Porochlamydia. J. Bacteriol.* **171:**4511–4513.

76. **Fujita, M., and K. Amako.** 1994. Localization of the *sapA* gene on a physical map of *Campylobacter fetus* chromosomal DNA. *Arch. Microbiol.* **162:**375–380.

77. **Furihata, K., K. Sato, and H. Matsumoto.** 1995. Construction of a combined *Not*I/*Sma*I physical and genetic map of *Moraxella* (*Branhamella*) *catarrhalis* strain ATCC25238. *Microbiol. Immunol.* **39:**745–751.

78. **Gäher, M., K. Einsiedler, T. Crass, and W. Bautsch.** 1996. A physical and genetic map of *Neisseria meningitidis* B1940. *Mol. Microbiol.* **19:** 249–259.

79. **Gaju, N., V. Pavon, I. Marin, I. Esteve, R. Guerrero, and R. Amils.** 1995. Chromosome map of the phototrophic anoxygenic bacterium *Chromatium vinosum. FEMS Microbiol. Lett.* **126:** 241–248.

80. **Gasc, A.-M., L. Kauc, P. Barraillé, M. Sicard, and S. Goodgal.** 1991. Gene localization, size, and physical map of the chromosome of *Streptococcus pneumoniae. J. Bacteriol.* **173:**7361–7367.

81. **Gonzales, A., A. Talavera, J. M. Almendral, and E. Vinuela.** 1986. Hairpin loop structure of African swine fever virus DNA. *Nucleic Acids Res.* **14:**6835–6844.

82. **Gorton, T. S., M. S. Goh, and S. J. Geary.** 1995. Physical mapping of the *Mycoplasma gallisepticum* S6 genome with localization of selected genes. *J. Bacteriol.* **177:**259–263.

83. **Gralton, E. M., A. L. Campbell, and E. L. Neidle.** 1997. Directed introduction of DNA cleavage sites to produce a high-resolution genetic and physical map of the *Acinetobacter sp.* strain ADP1 (BD413UE) chromosome. *Microbiology* **143:**1345–1357.

84. **Gravius, B., D. Glocker, J. Pigac, K. Pandza, D. Hranueli, and J. Cullum.** 1994. The 387 kb linear plasmid pPZ101 of *Streptomyces*

rimosus and its interaction with the chromosome. *Microbiology* **140:**2271–2277.

85. **Hackett, N. R., Y. Bobovnikova, and N. Heyrovska.** 1994. Conservation of chromosomal arrangement among three strains of the genetically unstable archeon *Halobacterium salinarium. J. Bacteriol.* **176:**7711–7718.

86. **Hantman, M. J., S. Sun, P. J. Piggot, and L. Daneo-Moore.** 1993. Chromosome organization of *Streptococcus mutans* GS-5. *J. Gen. Microbiol.* **139:**67–77.

87. **Hayakawa, T., T. Tanaka, K. Sakaguchi, N. Otake, and H. Yonehara.** 1979. A linear plasmid-like DNA in *Streptomyces* sp. producing lankacidin group antibiotics. *J. Gen. Appl. Microbiol.* **25:**255–260.

88. **He, Q., H. Chen, A. Kuspa, Y. Cheng, D. Kaiser, and L. J. Shimkets.** 1994. A physical map of the *Myxococcus xanthus* chromosome. *Proc. Natl. Acad. Sci. USA* **91:**9584–9587.

89. **Heath, J. D., J. D. Perkins, B. Sharma, and G. M. Weinstock.** 1992. *Not*I genomic cleavage map of *Escherichia coli* K-12 strain MG1655. *J. Bacteriol.* **174:**558–567.

90. **Himmelreich, R., H. Hilbert, H. Plagens, E. Pirkl, B. C. Li, and R. Herrmann.** 1996. Complete sequence analysis of the genome of the bacterium *Mycoplasma pneumoniae. Nucleic Acids Res.* **24:**4420–4449.

91. **Hinnebusch, J., S. Bergström, and A. G. Barbour.** 1990. Cloning and sequence analysis of linear plasmids of the bacterium *Borrelia burgdorferi. Mol. Microbiol.* **4:**811–820.

92. **Hobbs, M. M., M. J. Leonardi, F. R. Zaretzky, T. H. Wang, and T.-H. Kawula.** 1996. Organization of the *Haemophilus ducreyi* 35000 chromosome. *Microbiology* **142:**2587–2594.

93. **Holloway, B., and K. B. Low.** 1996. F-prime and R-prime factors, p. 2413–2420. *In* F. C. Neidhardt, R. Curtiss III, J. L. Ingraham, E. C. C. Lin, K. B. Low, B. Magasanik, W. S. Reznikoff, M. Riley, M. Schaechter, and H. E. Umbarger (ed.), Escherichia coli *and* Salmonella*: Cellular and Molecular Biology,* 2nd ed. American Society for Microbiology, Washington, D.C.

94. **Honeycutt, R. J., M. McClelland, and B. W. Sobral.** 1993. Physical map of the genome of *Rhizobium meliloti. J. Bacteriol.* **175:**6945–6952.

95. **Irazabal, N., I. Marin, and R. Amils.** 1997. Genomic organization of the acidophilic chemolithoautotrophic bacterium *Thiobacillus ferrooxidans* ATCC 21834. *J. Bacteriol.* **179:**1946–1950.

96. **Itaya, M.** 1993. Construction of the *Bacillus subtilis* chromosome physical map and the strategy for mapping newly isolated genes in one

membrane filter for hybridization. *Nucleic Acids Symp. Series* **29:**145–146.

97. **Itaya, M., and T. Tanaka.** 1991. Complete physical map of the *Bacillus subtilis* 168 chromosome constructed by a gene-directed mutagenesis method. *J. Mol. Biol.* **220:**631–648.

98. **Jiang, Q., K. Hiratsuka, and D. E. Taylor.** 1996. Variability of gene order in different *Helicobacter pylori* strains contributes to genome diversity. *Mol. Microbiol.* **20:**833–842.

99. **Jumas-Bilak, E., C. Maugard, S. Michaux-Characlion, A. Allardet-Servent, A. Perrin, D. O'Callaghan, and M. Ramuz.** 1995. Study of the organization of the genomes of *Escherichia coli, Brucella melitensis* and *Agrobacterium tumefaciens* by insertion of a unique restriction site. *Microbiology* **141:**2425–2432.

100. **Kaneko, T., T. Matsubayashi, M. Sugita, and M. Sugiura.** 1996. Physical and gene maps of the unicellular cyanobacterium *Synechococcus* sp. strain PCC6301 genome. *Plant Mol. Biol.* **31:**193–201.

101. **Katayama, S., B. Dupuy, G. Daube, B. China, and S. T. Cole.** 1996. Genome mapping of *Clostridium perfringens* strains with I-*Ceu*I shows many virulence genes to be plasmidborne. *Mol. Gen. Genet.* **251:**720–726.

102. **Katayama, S., B. Dupuy, T. Garnier, and S. T. Cole.** 1995. Rapid expansion of the physical and genetic map of the chromosome of *Clostridium perfringens* CPN50. *J. Bacteriol.* **177:**5680–5685.

103. **Kauc, L., and S. H. Goodgal.** 1989. The size and physical map of *Haemophilus parainfluenzae. Gene* **83:**377–380.

104. **Kauc, L., M. Mitchell, and S. H. Goodgal.** 1989. Size and physical map of the chromosome of *Haemophilus influenzae. J. Bacteriol.* **171:**2474–2479.

105. **Kieser, H. M., T. Kieser, and D. A. Hopwood.** 1992. A combined genetic and physical map of the *Streptomyces coelicolor* A3(2) chromosome. *J. Bacteriol.* **174:**5496–5507.

106. **Kim, N. W., H. Bingham, R. Khawaja, H. Louie, E. Hani, K. Neote, and V. L. Chan.** 1992. Physical map of *Campylobacter jejuni* TGH9011 and localization of 10 genetic markers by use of pulsed field gel electrophoresis. *J. Bacteriol.* **174:**3494–3498.

107. **Kim, N. W., R. Lombardi, H. Bingham, E. Hani, H. Louie, D. Ng, and V. L. Chan.** 1993. Fine mapping of the three rRNA operons on the updated genomic map of *Campylobacter jejuni* TGH9011 (ATCC 43431). *J. Bacteriol.* **175:**7468–7470.

108. **Kinashi, H., and M. Shimaji-Murayama.** 1991. Physical characterization of SCP1, a giant

linear plasmid from *Streptomyces coelicolor*. *J. Bacteriol.* **173**:1523–1529.

109. **Kohara, Y., K. Akiyama, and K. Isono.** 1987. The physical map of the whole *E. coli* chromosome: application of a new strategy for rapid analysis and sorting of a large genomic library. *Cell* **50**:495–508.

110. **Kolstö, A.-B., A. Grönstad, and H. Oppegaard.** 1990. Physical map of the *Bacillus cereus* chromosome. *J. Bacteriol.* **172**:3821–3825.

111. **Kolstö, A. B.** 1997. Dynamic bacterial genome organization. *Mol. Microbiol.* **24**:241–248.

112. **Kondo, S., A. Yamagishi, and T. Oshima.** 1993. A physical map of the sulfur-dependent archaebacterium *Sulfolobus acidocaldarius* 7 chromosome. *J. Bacteriol.* **175**:1532–1536.

113. **Krause, D. C., and C. B. Mawn.** 1990. Physical analysis and mapping of the *Mycoplasma pneumoniae* chromosome. *J. Bacteriol.* **172**:4790–4797.

114. **Krueger, C. M., K. L. Marks, and G. M. Ihler.** 1995. Physical map of the *Bartonella bacilliformis* genome. *J. Bacteriol.* **177**:7271–7274.

115. **Kündig, C., C. Beck, H. Hennecke, and M. Göttfert.** 1995. A single rRNA gene region in *Bradyrhizobium japonicum*. *J. Bacteriol.* **177**:5151–5154.

116. **Kündig, C., H. Hennecke, and M. Göttfert.** 1993. Correlated physical and genetic map of the *Bradyrhizobium japonicum* 110 genome. *J. Bacteriol.* **175**:613–622.

117. **Kunst, F., N. Ogasawara, I. Moszer, A. M. Albertini, G. Alloni, V. Azevedo, M. G. Bertero, P. Bessières, A. Bolotin, S. Borchert, R. Borris, L. Boursier, A. Brans, M. Braun, S. C. Brignell, S. Bron, S. Brouillet, C. V. Bruschi, B. Caldwell, V. Capuano, N. M. Carter, S.-K. Choi, J.-J. Codani, I. F. Connerton, N. J. Cummings, R. A. Daniel, F. Denizot, K. M. Devine, A. Düsterhöft, S. D. Ehrlich, P. T. Emmerson, K. D. Entian, J. Errington, C. Fabret, E. Ferrari, D. Foulger, C. Fritz, M. Fujita, Y. Fujita, S. Fuma, A. Galizzi, N. Galleron, S.-Y. Ghim, P. Glaser, A. Goffeau, E. J. Golightly, G. Grandi, G. Guiseppi, B. J. Guy, K. Haga, J. Haiech, C. R. Harwood, A. Hénaut, H. Hilbert, S. Holsappel, S. Hosono, M.-F. Hullo, M. Itaya, L. Jones, B. Joris, D. Karamata, Y. Kasahara, M. Klaerr-Blanchard, C. Klein, Y. Kobayashi, P. Koetter, G. Koningstein, S. Krogh, M. Kumano, K. Kurita, A. Lapidus, S. Lardinois, J. Lauber, V. Lazarevic, S.-M. Lee, A. Levine, H. Liu, S. Masuda, C. Mauël, C. Médigue, N. Medina, R. P. Mellado, M. Mizuno, D. Moestl, S. Nakai, M. Noback, D. Noone, M. O'Reilly, K. Ogawa, A. Ogiwara, B. Oudega, S.-H. Park, V. Parro, T. M. Pohl, D. Portetelle, S. Porwollik, A. M. Prescott, E. Presecan, P. Pujic, B. Purnelle, G. Rapoport, M. Rey, S. Reynolds, M. Rieger, C. Rivolta, E. Rocha, B. Roche, M. Rose, Y. Sadaie, T. Sato, E. Scanlan, S. Schleich, R. Schroeter, F. Scoffone, J. Sekiguchi, A. Sekowska, S. J. Seror, P. Serror, B.-S. Shin, B. Soldo, A. Sorokin, E. Tacconi, T. Takagi, H. Takahashi, K. Takemaru, M. Takeuchi, A. Tamakoshi, T. Tanaka, P. Terpstra, A. Tognoni, V. Tosato, S. Uchiyama, M. Vandenbol, F. Vannier, A. Vassarotti, A. Viari, R. Wambutt, E. Wedler, H. Wedler, T. Weitzenegger, P. Winters, A. Wipat, H. Yamamoto, K. Yamane, K. Yasumoto, K. Yata, K. Yoshida, H.-F. Yoshikawa, E. Zumstein, H. Yoshidawa, and A. Danchin.** 1997. The complete genome sequence of the Gram-positive bacterium *Bacillus subtilis*. *Nature* **390**:249–256.

118. **Kuritz, T., A. Ernst, T. A. Black, and P. Wolk.** 1993. High-resolution mapping of genetic loci of *Anabaena* PCC 7120 required for photosynthesis and nitrogen fixation. *Mol. Microbiol.* **8**:101–110.

119. **Kuspa, A., D. Vollrath, Y. Cheng, and D. Kaiser.** 1989. Physical mapping of the *Myxococcus xanthus* genome by random cloning in yeast artificial chromosomes. *Proc. Natl. Acad. Sci. USA* **86**:8917–8921.

120. **Ladefoged, S. A., and G. Christiansen.** 1992. Physical and genetic mapping of the genomes of five *Mycoplasma hominis* strains by pulsed-field gel electrophoresis. *J. Bacteriol.* **174**:2199–2207.

121. **La Fontaine, S., and J. I. Rood.** 1997. Physical and genetic map of the chromosome of *Dichelobacter nodosus* strain A198. *Gene* **184**:291–298.

122. **Leblond, P., G. Fischer, F.-X. Francou, F. Berger, M. Guérineau, and B. Decaris.** 1996. The unstable region of *Streptomyces ambofaciens* includes 210 kb terminal inverted repeats flanking the extremities of the linear chromosomal DNA. *Mol. Microbiol.* **19**:261–271.

123. **Leblond, P., M. Redenbach, and J. Cullum.** 1993. Physical map of the *Streptomyces lividans* 66 genome and comparison with that of the related strain *Streptomyces coelicolor* A 3(2). *J. Bacteriol.* **175**:3422–3429.

124. **Le Bourgeois, P., M. Lautier, M. Mata, and P. Ritzenthaler.** 1992. New tools for physical and genetic mapping of *Lactococcus* species. *Gene* **111**:109–114.

125. **Le Bourgeois, P., M. Lautier, M. Mata, and P. Ritzenthaler.** 1992. Physical and genetic map of the chromosome of *Lactococcus lactis* subsp. *lactis* IL1403. *J. Bacteriol.* **174:**6752–6762.

126. **Le Bourgeois, P., M. Lautier, L. van der Berghe, M. J. Gasson, and P. Ritzenthaler.** 1995. Physical and genetic map of the *Lactococcus lactis* subsp. *cremoris* MG1363 chromosome: comparison with that of *Lactococcus lactis* subsp. *lactis* IL 1403 reveals a large genome inversion. *J. Bacteriol.* **177:**2840–2850.

127. **Lederberg, J.** 1952. Cell genetics and hereditary symbiosis. *Physiol. Rev.* **32:**403.

128. **Lee, J. J., H. O. Smith, and R. J. Redfield.** 1989. Organization of the *Haemophilus influenzae* Rd genome. *J. Bacteriol.* **171:**3016–3024.

129. **Lezhava, A., T. Mizukami, T. Kajitani, D. Kameoka, M. Redenbach, H. Shinkawa, O. Nimi, and H. Kinashi.** 1995. Physical map of the linear chromosome of *Streptomyces griseus*. *J. Bacteriol.* **177:**6492–6498.

130. **Liao, X., I. Charlebois, C. Ouellet, M.-J. Morency, K. Dewar, J. Lightfoot, J. Foster, R. Siehnel, H. Schweizer, J. S. Lam, R. E. Hancock, and R. C. Levesque.** 1996. Physical mapping of 32 genetic markers on the *Pseudomonas aeruginosa* PAO1 chromosome. *Microbiology* **142:**79–86.

131. **Lin, Y.-S., H. M. Kieser, D. A. Hopwood, and C. W. Chen.** 1993. The chromosomal DNA of *Streptomyces lividans* 66 is linear. *Mol. Microbiol.* **10:**923–933.

132. **Liu, S.-L., A. Hessel, H.-Y. Cheng, and K. E. Sanderson.** 1994. The *Xba*I-*Bln*I-*Ceu*I genomic cleavage map of *Salmonella paratyphi* B. *J. Bacteriol.* **176:**1014–1024.

133. **Liu, S.-L., A. Hessel, and K. E. Sanderson.** 1993. The *Xba*I-*Bln*I-*Ceu*I genomic cleavage map of *Salmonella enteritidis* shows an inversion relative to *Salmonella typhimurium* LT2. *Mol. Microbiol.* **10:**655–664.

134. **Liu, S.-L., A. Hessel, and K. E. Sanderson.** 1993. The *Xba*I-*Bln*I-*Ceu*I genomic cleavage map of *Salmonella typhimurium* LT2 determined by double digestion, end labeling, and pulsed-field gel electrophoresis. *J. Bacteriol.* **175:**4104–4120.

135. **Liu, S.-L., and K. E. Sanderson.** 1992. A physical map of the *Salmonella typhimurium* LT2 genome made by using *Xba*I analysis. *J. Bacteriol.* **174:**1662–1672.

136. **Liu, S.-L., and K. E. Sanderson.** 1995. The chromosome of *Salmonella paratyphi* A is inverted by recombination between *rrn*H and *rrn*G. *J. Bacteriol.* **177:**6585–6592.

137. **Liu, S.-L., and K. E. Sanderson.** 1995. Rearrangements in the genome of the bacterium *Salmonella typhi*. *Proc. Natl. Acad. Sci. USA* **92:**1018–1022.

138. **Lopez-Garcia, P., J. P. Abad, C. Smith, and R. Amils.** 1992. Genomic organization of the halophilic archaeon *Haloferax mediterranei*: physical map of the chromosome. *Nucleic Acids Res.* **20:**2459–2464.

139. **Lopez-Garcia, P., A. St. Jean, R. Amils, and R. L. Charlebois.** 1995. Genomic stability in the archaeae *Haloferax volcanii* and *Haloferax mediterranei*. *J. Bacteriol.* **177:**1405–1408.

140. **Lucier, T. S., P.-Q. Hu, S. N. Peterson, X.-Y. Song, L. Miller, K. Heitzman, K. F. Bott, C. A. R. Hutchison, and P.-C. Hu.** 1994. Construction of an ordered genomic library of *Mycoplasma genitalium*. *Gene* **150:**27–34.

141. **MacDougall, J., and I. Saint Girons.** 1995. Physical map of the *Treponema denticola* circular chromosome. *J. Bacteriol.* **177:**1805–1811.

142. **Majumber, R., S. Sengupta, G. Khetawat, R. K. Bhadra, S. Roychoudhury, and J. Das.** 1996. Physical map of the genome of *Vibrio cholerae* 569B and localization of genetic markers. *J. Bacteriol.* **178:**1105–1112.

143. **Marin, I., R. Amils, and J. P. Abad.** 1997. Genomic organization of the metal-mobilizing bacterium *Thiobacillus cuprinus*. *Gene* **187:**99–105.

144. **McClelland, M., K. K. Wong, and K. Sanderson.** 1997. Physical mapping and fingerprinting of bacterial genomes using rare cutting restriction enzymes, p. 253–311. *In* F. J. de Bruijn, J. R. Lupski, and G. M. Weinstock (ed.), *Bacterial Genomes: Physical Structure and Analysis*. Chapman and Hall, New York, N.Y.

145. **Meinhardt, F., R. Schaffrath, and M. Larsen.** 1997. Microbial linear plasmids. *Appl. Microbiol. Biotechnol.* **47:**329–336.

146. **Michaux, S., J. Paillisson, M.-J. Carles-Nurit, G. Bourg, A. Allardet-Servent, and M. Ramuz.** 1993. Presence of two independent chromosomes in the *Brucella melitensis* 16M genome. *J. Bacteriol.* **175:**701–705.

147. **Michaux-Charachon, S., G. Bourg, E. Jumas-Bilak, P. Guigue-Talet, A. Allardet-Servent, D. O'Callaghan, and M. Ramuz.** 1997. Genome structure and phylogeny in the genus *Brucella*. *J. Bacteriol.* **179:**3244–3249.

148. **Michel, E., and P. Cossart.** 1992. Physical map of the *Listeria monocytogenes* chromosome. *J. Bacteriol.* **174:**7098–7103.

149. **Miyata, M., L. Wang, and T. Fukumura.** 1991. Physical mapping of the *Mycoplasma capricolum* genome. *FEMS Microbiol. Lett.* **63:**329–334.

150. **Murray, B. F., K. V. Singh, R. P. Ross, J. D. Heath, G. M. Dunny, and G. M. Wein-**

stock. 1993. Generation of restriction map of *Enterococcus faecalis* OG1 and investigation of growth requirements and regions encoding biosynthetic function. *J. Bacteriol.* **175:**5216–5223.

151. **Naterstad, K., A. B. Kolstö, and R. Sirevag.** 1995. Physical map of the genome of the green phototrophic bacterium *Chlorobium tepidum. J. Bacteriol.* **177:**5480–5484.

152. **Neumann, B., A. Pospiech, and H. U. Schairer.** 1993. A physical and genetic map of the *Stigmatella aurantiaca* DW4/3.1 chromosome. *Mol. Microbiol.* **10:**1087–1099.

153. **Newnham, E., N. Chang, and D. E. Taylor.** 1996. Expanded genomic map of *Campylobacter jejuni* UA580 and localization of 23S ribosomal rRNA genes by I-*Ceu*I restriction endonuclease digestion. *FEMS Microbiol. Lett.* **142:**223–229.

154. **Noll, K. M.** 1989. Chromosome map of the thermophilic archaebacterium *Thermococcus celer. J. Bacteriol.* **171:**6720–6725.

155. **Nuijten, P. J. M., C. Bartels, N. M. C. Bleumink-Pluym, W. Gaastra, and B. A. M. Van der Zeijst.** 1990. Size and physical map of the *Campylobacter jejuni* chromosome. *Nucleic Acids Res.* **18:**6211–6214.

156. **Ojaimi, C., B. E. Davidson, I. Saint Girons, and I. G. Old.** 1994. Conservation of gene arrangement and an unusual organization of rRNA genes in the linear chromosomes of the Lyme disease spirochaetes *Borrelia burgdorferi, B. garinii* and *B. afzelii. Microbiology* **140:**2931–2940.

157. **Okada, N., C. Sasakawa, T. Tobe, K. A. Talukder, K. Komatsu, and M. Yoshikawa.** 1991. Construction of a physical map of the chromosome of *Shigella flexneri* 2a and the direct assignment of nine virulence-associated loci identified by Tn5 insertions. *Mol. Microbiol.* **5:**2171–2180.

158. **Okahashi, N., C. Sasakawa, N. Okada, M. Yamada, M. Yoshikawa, M. Tokuda, I. Takahashi, and T. Koga.** 1990. Construction of a *Not*I restriction map of the *Streptococcus mutans* genome. *J. Gen. Microbiol.* **136:**2217–2223.

159. **Old, I. G., J. MacDougall, I. Saint Girons, and B. E. Davidson.** 1992. Mapping of genes on the linear chromosome of the bacterium *Borrelia burgdorferi*—possible locations for its origin of replication. *FEMS Microbiol. Lett.* **99:**245–250.

160. **Park, J. H., J.-C. Song, M. H. Kim, D.-S. Lee, and C. H. Kim.** 1994. Determination of genome size and a preliminary physical map of an extreme alkaliphile, *Micrococcus sp.* Y-1, by pulsed-field gel electrophoresis. *Microbiology* **140:**2247–2250.

161. **Perkins, J. D., J. D. Heath, B. R. Sharma, and G. M. Weinstock.** 1992. *Sfi*I genomic cleavage map of *Escherichia coli* K-12 strain MG 1655. *Nucleic Acids Res.* **20:**1129–1137.

162. **Perkins, J. D., J. D. Heath, B. R. Sharma, and G. M. Weinstock.** 1993. *Xba*I and *Bln*I genomic cleavage maps of *Escherichia coli* K-12 strain MG1655 and comparative analysis of other strains. *J. Mol. Biol.* **232:**419–445.

163. **Peterson, S. N., T. Lucier, K. Heitzman, E. A. Smith, K. F. Bott, P. C. Hu, and C. A. R. Hutchison.** 1995. Genetic map of the *Mycoplasma genitalium* chromosome. *J. Bacteriol.* **177:**3199–3204.

164. **Philipp, W. J., S. Nair, G. Guglielmi, M. Lagranderie, B. Gicquel, and S. T. Cole.** 1996. Physical mapping of *Mycobacterium bovis* BCG Pasteur reveals differences from the genome map of *Mycobacterium tuberculosis* H37Rv and from *M. bovis. Microbiology* **142:**3135–3145.

165. **Philipp, W. J., S. Poulet, K. Eiglmeier, L. Pascopella, V. Balasubramanian, B. Heym, S. Bergh, B. R. Bloom, W. R. Jacobs, Jr., and S. T. Cole.** 1996. An integrated map of the genome of the tubercle bacillus, *Mycobacterium tuberculosis* H37Rv, and comparison with *Mycobacterium leprae. Proc. Natl. Acad. Sci. USA* **93:**3132–3137.

166. **Picardeau, M., and V. Vincent.** 1997. Characterization of large linear plasmids in mycobacteria. *J. Bacteriol.* **179:**2753–2756.

167. **Plasterk, R. H. A., M. I. Simon, and A. G. Barbour.** 1985. Transposition of structural genes to an expression sequence on a linear plasmid causes antigenic variation in the bacterium *Borrelia hermsii. Nature* **318:**257–263.

168. **Pyle, L. E., and L. R. Finch.** 1988. A physical map of the genome of *Mycoplasma mycoides* subspecies *mycoides* Y with some functional loci. *Nucleic Acids Res.* **16:**6027–6039.

169. **Pyle, L. E., T. Taylor, and L. R. Finch.** 1990. Genomic maps of some strains within the *Mycoplasma mycoides* cluster. *J. Bacteriol.* **172:**7265–7268.

170. **Rainey, P. B., and M. J. Bailey.** 1996. Physical and genetic map of the *Pseudomonas fluorescens* SBW25 chromosome. *Mol. Microbiol.* **19:**521–533.

171. **Ratnaningsih, E., S. Dharmsthiti, V. Krishnapillai, A. Morgan, M. Sinclair, and B. W. Holloway.** 1990. A combined physical and genetic map of *Pseudomonas aeruginosa* PAO. *J. Gen. Microbiol.* **136:**2351–2357.

172. **Redenbach, M., H. M. Kieser, D. Denapaite, A. Eichner, J. Cullum, H. Kinashi, and D. A. Hopwood.** 1996. A set of ordered cosmids and a detailed genetic and physical map

for the 8 Mb *Streptomyces coelicolor* A 3(2) chromosome. *Mol. Microbiol.* **21**:77–96.

173. **Rodley, P. D., U. Römling, and B. Tümmler.** 1995. A physical genome map of the *Burkholderia cepacia* type strain. *Mol. Microbiol.* **17**:57–67.

174. **Römling, U., D. Grothues, W. Bautsch, and B. Tümmler.** 1989. A physical genome map of *Pseudomonas aeruginosa* PAO. *EMBO J.* **8**:4081–4089.

175. **Römling, U., K. Schmidt, and B. Tümmler.** 1997. One-dimensional pulsed-field gel electrophoresis, p. 312–325. *In* F. J. de Bruijn, J. R. Lupski, and G. M. Weinstock (ed.), *Bacterial Genomes: Physical Structure and Analysis.* Chapman and Hall, New York, N.Y.

176. **Römling, U., K. Schmidt, and B. Tümmler.** 1997. Two-dimensional pulsed-field gel electrophoresis, p. 326–336. *In* F. J. de Bruijn, J. R. Lupski, and G. M. Weinstock (ed.), *Bacterial Genomes: Physical Structure and Analysis.* Chapman and Hall, New York, N.Y.

177. **Römling, U., K. D. Schmidt, and B. Tümmler.** 1997. Large genome rearrangements discovered by the detailed analysis of 21 *Pseudomonas aeruginosa* clone C isolates found in environment and disease habitats. *J. Mol. Biol.* **271**:386–404.

178. **Römling, U., and B. Tümmler.** 1991. The impact of two-dimensional pulsed-field gel electrophoresis techniques for the consistent and complete mapping of bacterial genomes: refined physical map of *Pseudomonas aeruginosa* PAO. *Nucleic Acids Res.* **19**:3199–3206.

179. **Roussel, Y., F. Bourgoin, G. Guédon, M. Pébay, and B. Decaris.** 1997. Analysis of the genetic polymorphism between three *Streptococcus thermophilus* strains by comparing their physical and genetic organization. *Microbiology* **143**:1335–1343.

180. **Roussel, Y., M. Pebay, G. Guedon, J. M. Simonet, and B. Decaris.** 1994. Physical and genetic map of *Streptococcus thermophilus* A054. *J. Bacteriol.* **176**:7413–7422.

181. **Salama, S. M., E. Newnham, N. Chang, and D. E. Taylor.** 1995. Genome map of *Campylobacter fetus* subsp. fetus ATCC 27374. *FEMS Microbiol. Lett.* **132**:239–245.

182. **Salas, M.** 1991. Protein-priming of DNA replication. *Annu. Rev. Biochem.* **60**:39–71.

183. **Scaife, J.** 1966. F-prime factor formation in *E. coli* K12. *Genet. Res.* **8**:189.

184. **Schmidt, K. D., B. Tümmler, and U. Römling.** 1996. Comparative genome mapping of *Pseudomonas aeruginosa* PAO with *P. aeruginosa* C, which belongs to a major clone in cystic fibrosis patients and aquatic habitats. *J. Bacteriol.* **178**:85–93.

185. **Schwartz, D. C., and C. R. Cantor.** 1984. Separation of yeast chromosome-sized DNAs by pulsed-field gel electrophoresis. *Cell* **37**:67–75.

186. **Schwartz, D. C., W. Saffran, J. Welsh, R. Haas, M. Goldenberg, and C. R. Cantor.** 1983. New techniques for purifying large DNAs and studying their properties and packing. *Cold Spring Harbor Symp. Quant. Biol.* **47**:189–195.

187. **Schwartz, D. C., and A. Samad.** 1997. Optical mapping approaches to molecular genomics. *Curr. Opin. Biotechnol.* **8**:70–74.

188. **Shao, Z., W. Mages, and R. Schmitt.** 1994. A physical map of the hyperthermophilic bacterium *Aquifex pyrophilus* chromosome. *J. Bacteriol.* **176**:6776–6780.

189. **Shimkets, L. J.** 1997. Structure and sizes of genomes of the archaea and bacteria. p. 5–11. *In* F. J. de Bruijn, J. R. Lupski, and G. M. Weinstock (ed.), *Bacterial Genomes: Physical Structure and Analysis.* Chapman and Hall, New York, N.Y.

190. **Shizuya, H., B. Birren, U. J. Kim, V. Mancino, T. Slepak, Y. Tachiiri, and M. Simon.** 1992. Cloning and stable maintenance of 300-kilobase-pair fragments of human DNA in *Escherichia coli* using an F-factor-based vector. *Proc. Natl. Acad. Sci. USA* **89**:8794–8797.

191. **Sitzman, J., and A. Klein.** 1991. Physical and genetic map of the *Methanococcus voltae* chromosome. *Mol. Microbiol.* **5**:505–513.

192. **Smith, C. L., J. G. Econome, A. Schutt, S. Klco, and C. R. Cantor.** 1987. A physical map of the *Escherichia coli* K12 genome. *Science* **236**:1448–1453.

193. **Smith, D. R., L. A. Doucette-Stamm, C. Deloughery, H. Lee, J. Dubois, T. Aldredge, R. Bashirzadeh, D. Blakely, R. Cook, K. Gilbert, D. Harrison, L. Hoang, P. Keagle, W. Lumm, P. Pothier, D. Qiu, R. Spadafora, R. Vicaire, Y. Wang, J. Wierzbowski, R. Gibson, N. Jiwani, A. Caruso, D. Bush, H. Safer, D. Patwell, S. Prabhakar, S. McDougall, G. Shimer, A. Goyal, S. Pietrokovski, G. M. Church, C. J. Daniels, J.-I. Mao, P. Rice, J. Nölling, and J. N. Reeve.** 1997. Complete genome sequence of *Methanobacterium thermoautotrophicum* ΔH: functional analysis and comparative genomics. *J. Bacteriol.* **179**:7135–7155.

194. **Smith, D. R., P. Richterich, M. Rubenfield, P. W. Rice, C. Butler, H.-M. Lee, S. Kirst, K. Gundersen, K. Abendschan, Q. Xu, M. Chung, C. Deloughery, T. Aldredge, J. Maher, R. Lundstrom, C. Tulig, K. Falls, J. Imrich, D. Torrey, M. Engel-**

stein, G. Breton, D. Madan, R. Nietupski, B. Seitz, S. Connelly, S. McDougall, H. Safer, R. Gibson, L. Doucette-Stamm, K. Eiglmeier, S. Bergh, S. T. Cole, K. Robison, L. Richterich, J. Johnson, G. M. Church, and J.-I. Mao. 1997. Multiplex sequencing of 1.5 Mb of the *Mycobacterium leprae* genome. *Genome Res.* **7**:802–819.

195. Stettler, R., G. Erauso, and T. Leisinger. 1995. Physical and genetic map of *Methanobacterium wolfei* genome and its comparison with the updated genomic map of *Methanobacterium thermoautotrophicum* Marburg. *Arch. Microbiol.* **163**:205–210.

196. Stettler, R., and T. Leisinger. 1992. Physical map of the *Methanobacterium thermoautotrophicum* Marburg chromosome. *J. Bacteriol.* **174**:7227–7234.

197. Stibitz, S., and T. L. Garletts. 1992. Derivation of a physical map of the chromosome of *Bordetella pertussis* Tohama I. *J. Bacteriol.* **174**:7770–7777.

198. St. Jean, A., B. A. Trieselmann, and R. L. Charlebois. 1994. Physical map and set of overlapping cosmid clones representing the genome of the archaeon *Halobacterium sp.* GRB. *Nucleic Acids Res.* **22**:1476–1483.

199. Sutherland, K. J., M. Hashimoto, T. Kudo, and K. Horikoshi. 1993. A partial physical map for the chromosome of alkalophilic *Bacillus sp.* strain C-125. *J. Gen. Microbiol.* **139**:661–667.

200. Suvorov, A. M., and J. J. Ferretti. 1996. Physical and genetic chromosomal map on an M type 1 strain of *Streptococcus pyogenes*. *J. Bacteriol.* **178**:5546–5549.

201. Suwanto, A., and S. Kaplan. 1989. Physical and genetic mapping of the *Rhodobacter sphaeroides* 2.4.1 genome: presence of two unique circular chromosomes. *J. Bacteriol.* **171**:5850–5859.

202. Suwanto, A., and S. Kaplan. 1989. Physical and genetic mapping of the *Rhodobacter sphaeroides* 2.4.1 genome: genome size, fragment identification and gene localization. *J. Bacteriol.* **171**:5840–5849.

203. Svarchevsky, A. N., and V. N. Rybchin. 1984. Characterization of plasmid properties of bacteriophage N15. *Mol. Genet. Mikrobiol. Virusol.* **5**:34–39.

204. Tabata, K., and T. Hoshino. 1996. Mapping of 61 genes on the refined physical map of the chromosome of *Thermus thermophilus* HB27 and comparison of genome organization with that of *T. thermophilus* HB8. *Microbiology* **142**:401–410.

205. Tabata, K., T. Kosuge, T. Nakahara, and T. Hoshino. 1993. Physical map of the extremely thermophilic bacterium *Thermus thermophilus* HB27 chromosome. *FEBS Lett.* **331**:81–85.

206. Taylor, D. E., M. Eaton, N. Chang, and S. M. Salama. 1992. Construction of a *Helicobacter pylori* genome map and demonstration of diversity at the genome level. *J. Bacteriol.* **174**:6800–6806.

207. Taylor, D. E., M. Eaton, W. Yan, and N. Chang. 1992. Genome maps of *Campylobacter jejuni* and *Campylobacter coli*. *J. Bacteriol.* **174**:2332–2337.

208. Tigges, E., and F. C. Minion. 1994. Physical map of the genome of *Acholeplasma oculi* ISM1499 and construction of a Tn*4001* derivative for macrorestriction chromosomal mapping. *J. Bacteriol.* **176**:1180–1183.

209. Tigges, E., and F. C. Minion. 1994. Physical map of *Mycoplasma gallisepticum*. *J. Bacteriol.* **176**:4157–4159.

210. Toda, T., and M. Itaya. 1995. I-*Ceu*I recognition sites in the *rrn* operons of the *Bacillus subtilis* 168 chromosome: inherent landmarks for genome analysis. *Microbiology* **141**:1937–1945.

211. Tomb, J.-F., O. White, A. R. Kerlavage, R. A. Clayton, G. G. Sutton, R. D. Fleischmann, K. A. Ketchum, H. P. Klenk, S. Gill, B. A. Dougherty, K. Nelson, J. Quackenbush, L. Zhou, E. F. Kirkness, S. Peterson, B. Loftus, D. Richardson, R. Dodson, H. G. Khalak, A. Glodek, K. McKenney, L. M. Fitzegerald, N. Lee, M. D. Adams, E. K. Hickey, D. E. Berg, J. D. Gocayne, T. R. Utterback, J. D. Peterson, J. M. Kelley, M. D. Cotton, J. M. Weidman, C. Fujii, C. Bowman, L. Watthey, E. Wallin, W. S. Hayes, M. Borodovsky, P. D. Karp, H. O. Smith, C. M. Fraser, and J. C. Venter. 1997. The complete genome sequence of the gastric pathogen *Helicobacter pylori*. *Nature* **388**:539–547.

212. Tulloch, D. L., L. R. Finch, A. J. Hillier, and B. E. Davidson. 1991. Physical map of the chromosome of *Lactococcus lactis* subsp. *lactis* DL11 and localization of six putative rRNA operons. *J. Bacteriol.* **173**:2768–2775.

213. Walker, E. M., J. K. Howell, Y. You, A. R. Hoffmaster, J. D. Heath, G. M. Weinstock, and S. J. Norris. 1995. Physical map of the genome of *Treponema pallidum* subsp. pallidum (Nichols). *J. Bacteriol.* **177**:1797–1804.

214. Wang, Y.-K., E. J. Huff, and D. C. Schwartz. 1995. Optical mapping of site-directed cleavages on single DNA molecules by the RecA-assisted restriction endonuclease technique. *Proc. Natl. Acad. Sci. USA* **92**:165–169.

215. Ward-Rainey, N., F. A. Rainey, E. M. H. Wellington, and E. Stackebrandt. 1996.

Physical map of the genome of *Planctomyces lim-nophilus*, a representative of the phylogenetically distinct planctomycete lineage. *J. Bacteriol.* **178:** 1908–1913.

216. **Whitley, J. C., A. Muto, and L. R. Finch.** 1990. A physical map for *Mycoplasma capricolum* Cal. kid with loci for all known tRNA species. *Nucleic Acids Res.* **19:**399–400.

217. **Wilkinson, S. R., and M. Young.** 1995. Physical map of the *Clostridium beijerinckii* (formerly *Clostridium acetobutylicum*) NCIMB 8052 chromosome. *J. Bacteriol.* **177:**439–448.

218. **Wong, K. K., and M. McClelland** 1992. A *Bln*I restriction map of the *Salmonella typhimurium* LT2 genome. *J. Bacteriol.* **174:**1656–1661.

219. **Yamagishi, A., and T. Oshima.** 1990. Circular chromosomal DNA in the sulfur-dependent archaebacterium *Sulfolobus acidocaldarius*. *Nucleic Acids Res.* **18:**1133–1136.

220. **Yan, W., and D. E. Taylor.** 1991. Sizing and mapping of the genome of *Campylobacter coli* strain UA417R using pulsed-field gel electrophoresis. *Gene* **101:**117–120.

221. **Ye, F., F. Laigret, J. C. Whitley, C. Citti, L. R. Finch, P. Carle, J. Renaudin, and J. M. Bove.** 1992. A physical and genetic map of the *Spiroplasma citri* genome. *Nucleic Acids Res.* **20:**1559–1565.

222. **Zakrzewska-Czerwinska, J., and H. Schrempf.** 1992. Characterization of an autonomously replicating region from the *Streptomyces lividans* chromosome. *J. Bacteriol.* **147:**2688–2693.

223. **Zuccon, F. M., M. J. Hantmann, L. A. Sechi, S. Sun, P. J. Piggot, and L. Daneo-Moore.** 1995. Physical map of *Streptococcus mutans* GS-5 chromosome. *Dev. Biol. Stand.* **85:** 403–407.

224. **Zuerner, R. L.** 1991. Physical map of chromosomal and plasmid DNA comprising the genome of *Leptospira interrogans*. *Nucleic Acids Res.* **19:**4857–4860.

225. **Zuerner, R. L., J. L. Herrmann, and I. Saint Girons.** 1993. Comparison of genetic maps for two *Leptospira interrogans* serovars provides evidence for two chromosomes and intraspecies heterogeneity. *J. Bacteriol.* **175:**5445–5451.

226. **Zuerner, R. L., and T. B. Stanton.** 1994. Physical and genetic map of the *Serpulina hyodysenteriae* B78[T] chromosome. *J. Bacteriol.* **176:**1087–1092.

ARCHAEA: WHOSE SISTER LINEAGE?

Robert L. Charlebois

4

Archaeal genomes resemble those from bacteria in structure and in style (63): (i) protein-coding genes are virtually uninterrupted by introns, and inteins are uncommon; (ii) genes whose products coordinate a common function often occur in a tight, operonic linkage; (iii) the chromosome itself is circular, typically between 1 and 3 Mbp, with plasmids contributing up to another 1 Mbp; (iv) the genome is not enveloped by a nuclear membrane; and (v) expression of the genome leads to a cell whose biological strategy is much like that of the bacteria.

Given the similarity in cellular and genetic organization, it has been argued (12, 79, 81) that Woese's three-domain view of the biological world (119) exaggerates the distinction between *Archaea* and *Bacteria*. In that argument, an obvious line can be drawn between the eukaryotic cell and genome on the one hand and the prokaryotic cell and genome on the other. Although prokaryotes were originally defined by the characters which they did not possess (115), phylogenetic analysis now strongly suggests a common ancestor for all

prokaryotes (3, 8, 72). Eucarya do not fit comfortably within this tree, being derived from at least two lineages of prokaryotes in some sort of chimeric ménage (45, 46, 52–54, 79, 80, 106, 120). The deepest node in the universal tree separates *Bacteria* from *Archaea*, but soon after this bifurcation, other divergences—not quite as old, but almost—are observed. Considering the sequential bifurcations and radiations in the evolution of lineages, the three-domain view may be arbitrary, especially given the chimeric and uncertain origins of one of the domains.

If *Archaea* are nothing more than deep-branching *Procarya*, what then is unique about archaeal genomes relative to bacterial genomes—enough so to warrant a separate chapter in this book? There are, after all, other deeply branching prokaryotic lineages. Is there something unique about *Archaea* beyond what one would expect from deep divergence? Or is the privilege earned by the apparent phylogenetic kinship between *Archaea* and *Eucarya*, elevating this group of prokaryotes to a status of eukaryotic ancestor?

In this chapter, following a review of archaeal genomics, I wish to scrutinize the convenient though perhaps misleading construct that is organismic phylogeny. In so doing, I will address theories of the origin of eukary-

Robert L. Charlebois, Department of Biology, University of Ottawa, Ottawa, Ontario K1N 6N5, Canada, and Canadian Institute for Advanced Research.

Organization of the Prokaryotic Genome, Edited by Robert L. Charlebois,
© 1999 American Society for Microbiology, Washington, D.C.

otes, theories which look to *Archaea* for answers, thanks to the (largely) accepted rooting (3, 8, 72) of the tree of life in the bacterial branch. Since their recognition as a unified clade, *Archaea* have been favorite subjects of molecular evolutionists interested in early life (62). They provide a third perspective from which biology can be studied and understood, a perspective from which one might extrapolate to the last common ancestor (the "cenancestor" [28, 33]) or even back to the progenote (116, 117).

What I hope this chapter contributes to the understanding of genomic organization is the notion of bottom-up construction. The archaea especially have contributed a significant database from which to generate ideas of chimerism, to the point that it seems unreasonable to continue believing in strictly linear models of genomic inheritance. Perhaps the gene is truly the unit of selection as Dawkins (22) proposed, with only the gene's phylogeny being valid. Thinking that *Archaea*—or any other lineage—represent a unified group of organisms sharing a common ancestor, whose evolution can accurately be drawn as a tree, might in fact be largely incorrect. The components which now assemble into a given cell may well have followed quite independent evolutionary paths. This would be especially true of modular components, whose gene products may be unconstrained by a precise dependence on other specific gene products.

In order to address such conjectural topics scientifically, sequence analysis will be important in the years to come and will have much to reveal about genomic construction and evolution. My hypothesis is that genomic rearrangement will be found not to be confined to the space bounded by a cytoplasmic membrane but will draw instead (with some constraints) from a global pool of dynamic genetic systems. The analysis of archaeal genomes will be useful in addressing such hypotheses, as will the analysis of other ancient lineages whose sequences are rapidly being deciphered.

MAPPING AND SEQUENCING OF ARCHAEAL GENOMES

Interest in archaeal genomes, as judged by volume in the literature, has focused mainly on the extremely halophilic archaea (reviewed in references 14, 15, 21, and 88). Haloarchaea possess among the largest of known archaeal genomes, typically including numerous plasmids of various sizes (30, 55, 76, 90). The DNA is often inhomogeneous in composition, separable into two fractions differing in moles percent G + C (83, 92) or oligonucleotide bias (17). Much of the minor, lower-GC fraction of haloarchaeal DNA takes an extrachromosomal form (30, 92), but islands of low-GC DNA can be found interrupting chromosomal regions as well (17, 29, 89). Insertion sequences are common, abundant, and active in some strains (reviewed in references 15, 21, and 88), and they concentrate in this minor fraction (19, 85, 89, 92). Despite their disruptive effect on genes, however, the insertion sequences seem not to be potent instigators of chromosomal rearrangement (56, 75), though they seem to be important in plasmid construction and evolution (85). The principal mode of persistent chromosomal rearrangement in this lineage, or in any other for that matter, remains unclear. Some of this evolution may be driven by intercellular exchange: evidence is accumulating that the haloarchaeal genome is chimeric, with hints of recruitment of cyanobacterial and gram-positive bacterial genes (26, 48, 95, 113). Finally, haloarchaeal DNA is interesting in yet another respect: the extent of gene duplication in the genome (2, 109). As argued elsewhere (14), it will be of obvious benefit to examine the sequence of several haloarchaeal genomes in order to understand their structures, functions, and evolution. Mapping efforts, generating as much detail as the method allowed (17, 56, 110), have been insufficient to resolve many critical issues (109).

Of particular interest to students of genomic organization, genetic plasticity in archaeal genomes has been well documented. This plasticity takes the form of frequent insertional inactivation of genes, with rates as high as 10^{-4} to 10^{-2} per cell per generation (91, 101, 114), and significant insertion sequence-mediated rearrangement (93, 94). The genetic instability appears to be prevalent not only among members of the family *Halobacteriaceae* (15, 21, 88) but also among members of the order *Sulfolobales* (39a, 101, 102), though the insertion sequences responsible are apparently unrelated (102). As discussed above, however, the frequent genomic rearrangements in the haloarchaea may be lateral to their phylogenetic continuity, with genome structure apparently conserved at least in intermediate evolutionary time (75). Still, these rearrangements, along with sequence "tinkering" (24), may provide the necessary variation to allow evolution, especially within inconsistent environments (16, 56). Whether the insertion sequences have in hindsight proven useful or whether they have only burdened their hosts as parasites is, however, a difficult question to answer. Unfortunately, evolutionary literature is rich in "probable" answers to experimentally difficult questions such as this one, the answers often being individually plausible but collectively incompatible.

Although they have been the principal focus, haloarchaeal genomes are not the only archaeal genomes which have been characterized. Representative maps from both the euryarchaeotal (*Halobacterium salinarum* GRB [110], *Halobacterium salinarum* NRC-1 and S9 [56], *Haloferax mediterranei* [2, 74], *Haloferax volcanii* DS2 [17, 19], *Methanobacterium thermoautotrophicum* Marburg [107, 108], *Methanobacterium wolfei* [108], *Methanococcus voltae* [103], *Pyrococcus* sp. strain KOD1 [42], and *Thermococcus celer* [86]) and crenarchaeotal (*Pyrobaculum aerophilum* [34] and *Sulfolobus tokodaii* 7 [67]) branches are available and have been found to conform to the general prokaryotic paradigm. Four archaeal genomes have been completely sequenced, all euryarchaeotes: *Methanococcus jannaschii* (10), *M. thermoautotrophicum* ∆H (97, 104), *Archaeoglobus fulgidus* (65), and *Pyrococcus horikoshii* (61, 61a). Several other genomes are on the way to completion,

including the euryarchaeotes *Pyrococcus abyssi* (Genoscope, Évry, France) and *Pyrococcus furiosus* (http://www.genome.utah.edu) and the crenarchaeotes *Aeropyrum pernix* (Biotechnology Center, National Institute of Technology and Evaluation, Tokyo, Japan), *Pyrobaculum aerophilum* (34), and *Sulfolobus solfataricus* P2 (18). Projects have been initiated for *H. salinarum* (Max-Planck-Institute for Biochemistry, Martinsried, Germany, and University of Washington), *H. volcanii* (University of Scranton), *Methanosarcina mazei* Gö1 (University of Goettingen, Goettingen, Germany), and *Thermoplasma acidophilum* (Max-Planck-Institute for Biochemistry) as well. (For the current status of archaeal genome projects, follow the links found at http://www.tigr.org/tdb/mdb/mdb.html.) The available sequences largely confirm the distinctness of *Archaea*, although as discussed later in this chapter, not necessarily the distinctness of individual archaeal genes.

PHYSICAL ORGANIZATION OF THE ARCHAEAL NUCLEOID

Physical analysis of archaeal nucleoids (77, 96, 111) lags behind the efforts of genetic characterization. One of the more interesting findings has been that of histone homologs in the euryarchaeotes, in addition to other types of nucleoid DNA-binding proteins in both euryarchaeotes and crenarchaeotes (reviewed in references 47 and 99). The fact that the histones have been found in only one branch of *Archaea* and not (yet?) in the other (references 34, 47, and 99 and unpublished data) may reflect missing data, but more likely it reflects a different style of genomic compaction. Even within the euryarchaeotes, several distinct families of nucleoid proteins are used (47): the MC type in *Methanosarcina* and *Methanothrix*, the bacterial-HU type in *Thermoplasma*, and the histone type in *Archaeoglobus* (65), *Methanobacterium*, *Methanococcus*, *Methanothermus*, and *Pyrococcus/Thermococcus* (100). The diversity of known nucleoid proteins will undoubtedly increase as they are characterized from other species.

The study of topoisomerases has also attracted attention to the archaea (39), especially to the extremely thermophilic archaea which possess reverse gyrase—a type I topoisomerase and ATP-dependent helicase (20) that overwinds DNA. Reverse gyrase is, however, not present in all archaea, nor is it limited to archaea (5, 6, 23). The positive supercoiling introduced by reverse gyrase has important effects on DNA topology at high temperatures (reviewed in reference 39), alone or in combination with other nucleoid proteins (78). Reverse gyrase is found only in hyperthermophiles.

THE ARCHAEAL PROMISE

The first indication that *Archaea* were different from other prokaryotes came with Woese and colleagues' analysis of rRNA-based oligonucleotide sequence catalogs (40, 41, 118), soon confirmed by full ribosomal DNA sequencing (50, 73). Although unrooted, the small-subunit rRNA tree grouped a variety of disparate prokaryotes together and apart from other prokaryotes. It was soon realized that members of this new lineage—then called the archaebacteria—had other characteristics in common besides their rRNA sequences, such as membrane lipids based on glycerol ethers of isoprenoids (43), the use of unformylated tRNA^{Met} for initiation of translation and distinctive tRNA modification patterns (49, 82), insensitivity to many common antibiotics otherwise effective against bacteria or eucarya (1, 57), an RNA polymerase subunit structure different from the classic bacterial $\alpha_2\beta\beta'\sigma$ and more similar to the eukaryotic RNA polymerase (70), and complete lack of *N*-acetylmuramic acid and hence of (true) peptidoglycan (60).

Rooting of the tree, using anciently duplicated paralogous sequences (44, 59), further divided *Archaea* from *Bacteria* by positioning the root in the bacterial branch. The resulting archaeal-eucaryal clade gave molecular evolutionists further incentive to look to *Archaea* for eukaryotic origins. Subsequently, the similarity between archaeal and eucaryal information processing systems came to light (reviewed in references 9, 62, and 70), although archaeal metabolism still seemed to resemble that in bacteria more than that in eucarya (10, 87). Given the penchant of molecular biologists for transcription and translation—the central dogma—the conflict between styles of gene expression and of biochemistry may have biased the former in terms of greater evolutionary weight. Interestingly, the root itself is most supported by genes involved in processing information (8, 44, 59), though not exclusively (72). Metabolic genes, for reasons not fully understood, tend to produce confusing phylogenies (9).

The extent of this discrepancy concerning the alliance of shared archaeal genes either with *Bacteria* or with *Eucarya* has been best illustrated by Gupta and Golding (45, 51, 52). They found, using the accumulating public sequence data, that archaeal genes tree with either eucaryal nuclear genes or gram-positive ("Monoderm" [51]) bacterial genes, eucaryal nuclear genes match gram-negative ("Diderm" [51]) bacterial genes or archaeal genes, and gram-negative bacterial genes group with those from gram-positive bacteria. Brown and Doolittle (9) further resolved the gram-negative grouping, finding support for secondary relationships between archaea and cyanobacteria and between gram-positive bacteria and cyanobacteria. One interpretation of these findings is that the premitochondrial eukaryotic cell was already a chimera combining a primitive bacterium with an early archaeon, as first proposed by Zillig (120) and illustrated in Fig. 1. However, for archaea to bear more similarity to gram-positive bacteria and cyanobacteria implies a common ancestor more recent than the split between these and other *Bacteria*, a conclusion which conflicts with the unity of *Bacteria* (9). The conflict might be resolved if we decide that the early bacterial branches in the 16S rRNA tree are artifactually early (7, 23, 64) and if we posit that the early archaeal lineage went through a period of rapid sequence evolution, distancing *Archaea* from most gram-negative bacteria more

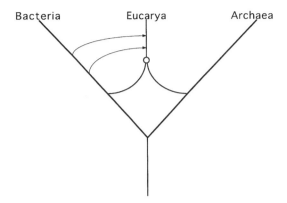

FIGURE 1 Eucaryal origin from the fusion of a bacterium with an archaeon (120). The arrows indicate the two most important endosymbiotic events to occur since the origin of *Eucarya*: mitochondria and chloroplasts.

rapidly than the latter were diverging from gram-positive bacteria and cyanobacteria (Fig. 2). As an alternative to Gupta's (51) conclusions, the apparent kinship between gram-positive bacteria and archaea may instead be explained by assuming horizontal genetic transfer, a strong case having recently been made by Gribaldo et al. (48) for HSP70.

By analyzing entire genomic sequences, Koonin et al. (68) arrived at yet different conclusions, thanks to different assumptions.

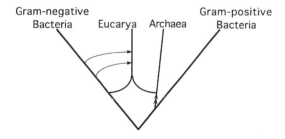

FIGURE 2 Eucaryal origin from the fusion of a gram-negative bacterium with an archaeon (45, 52). Although this modification of Fig. 1 helps to explain why *Archaea* are more closely related to gram-positive bacteria than to gram-negative bacteria, one must assume that a burst of evolutionary sequence change occurred shortly after the inception of *Archaea* and that early-branching *Bacteria* are artifactually early branching (see text).

Their argument is that for some archaeal genes to resemble bacterial sequences, and for others to resemble eucaryal nuclear sequences, two periods of rapid evolution followed by a return to normal rates would be needed: one distancing bacterial gene expression mechanisms and the other distancing eucaryal metabolism. Since such a scenario is implausible, the paper's conclusion is that the *Archaea* themselves are chimeric, representing a mixture of bacterial genes and what the 16S rRNA tree calls early archaeal genes. Meanwhile, the nonchimeric cousin of the early archaeon would meet with its α-proteobacterial symbiont, initiating the chimeric eukaryotic cell (Fig. 3). The deductions of Koonin et al. do not require the root of the universal tree to be grossly misplaced or early-branching *Bacteria* to be erroneously early branching, but they do require extensive interkingdom horizontal DNA transfer.

From a metabolic rather than a genetic point of view, Martin and Müller (80) recently proposed a syntrophic origin of the eukaryotic cell, driven by an autotroph's dependence on H_2 provided in the waste of another cell's fermentative activities. The partnership is further specified as that between a methanogen and a proteobacterium, with good argument, although a methanogenic candidate for the archaeal partner contradicts the controversial (9) eocyte hypothesis (3, 98). Methanogen or not, the model of syntrophic origins is testable and

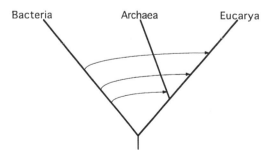

FIGURE 3 Model of archaeal chimerism (68) in which *Archaea* owe their metabolic similarity with *Bacteria* to a large-scale lateral transfer of bacterial genes.

worth testing with genomic sequences. The model's only solution to the origin of cytoskeletal components is to suppose that after two genomes merge, genetic redundancy enables rapid evolution.

What all of these models do is assume that today's chimeric cells are chimerae of DNA extracted from still-extant lineages. Other primary lineages contemporary with early *Bacteria* and *Archaea*, lineages which have since gone extinct, are excluded from most such models, though they have been proposed as the ancestors of viruses (35). Assuming that such microbial lineages existed, which amounts to assuming that microbial biodiversity is roughly constant, resulting in a failure for every success, I formulated my own model of chimeric origins of eukaryotes (Fig. 4). This model is based on Sogin's (106), except that I assume genomic, not progenomic, parentage. Extrapolation of cellular physiology, molecular genetics, and biochemistry to the tree's root has suggested that the last common ancestor of all extant life was already a sophisticated cell (11, 37) and not a progenote (but see Woese [116] for counterarguments). This common ancestor, dubbed the cenancestor (28, 33), is not the same creature in Fig. 4 as it is in currently published rooted trees, but this is a consequence of my assumption that one should not

be constrained by today's highly pruned tree in searching for ancient genetic partnerships. The model in Fig. 4 provides time for the urkaryotic (cytotrophic) lineage to develop what are recognized as eukaryotic-specific traits, such as actin and tubulin (4, 25, 26); it allows time for eukaryotic metabolism to diverge from prokaryotic metabolism; it permits the cytotroph to capture an archaeal version of prokaryotic gene expression and combine this innovation with its own gene expression mechanisms; and it generates a mix of different trees relating *Archaea* either to *Bacteria* or to *Eucarya*, depending on what the archaeal-cytotrophic chimera discarded or kept. It helps to suppose that the rate of protein evolution was greater (25) in the "primitive" cytotrophic lineage, prior to its merging with *Archaea* and its adoption of the archaeal family B DNA polymerases (*Escherichia coli*'s DNA polymerase II is a family B DNA polymerase involved in repair, but it is the only known example of a family B DNA polymerase in bacteria [32]; genome sequences from a variety of bacterial phyla published since the work of Edgell et al. [32] have failed to turn up any other examples.) The model in Fig. 4 does not explain why archaeal genes might be more closely related to gram-positive bacterial genes, although as mentioned above, a gram-

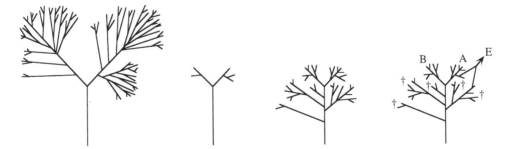

FIGURE 4 Fourth Domain model of eucaryal origins (right), based on the hypothesis that *Eucarya* (E) are chimerae of *Archaea* (A) and an extinct (daggers) lineage of cytotrophs. If one trims the current phylogenetic tree (left) to the time in question (second from left), one suggests that the world supported an implausibly small microbial biodiversity, unless one accepts that other lineages existed (second from right) which have since gone extinct. Such a lineage may have engulfed an early archaeon, adopting its information processing mechanisms but retaining many of its own genes for fermentative metabolism.

positive ancestor spawning gram-negative bacteria and then later *Archaea*, followed by a burst of archaeal sequence evolution, could turn the tree inside out while keeping it right side up (Fig. 5). Is a period of such rapid evolution plausible? If it followed the invention of the archaeal style of gene expression machinery, or an adaptation to hyperthermal environments, it might be. Interestingly, Forterre and colleagues (35, 36, 38) argue for a thermoreductive origin of prokaryotes from a more complex mesophilic cell, based on topoisomerase evolution and the thermoinstability of mRNA. In that scenario, too, *Eucarya* are outgroups to *Procarya*. Should the ancestor

of all life not have been a prokaryote, we will have to rewrite our textbooks on early evolution.

Unless one is a strict and obstinate neutralist, it is difficult not to accept that genes may evolve at different rates in different lineages, at least sometimes (27). The observation by Brown and Doolittle (9) that the rate of a protein's evolution is the same across domains (*Archaea, Bacteria,* and *Eucarya*) suffers both from averaging across sublineages and from only including genes which are universal— and which therefore may have been "perfected" very early in evolution so that they now evolve neutrally. Although variable ev-

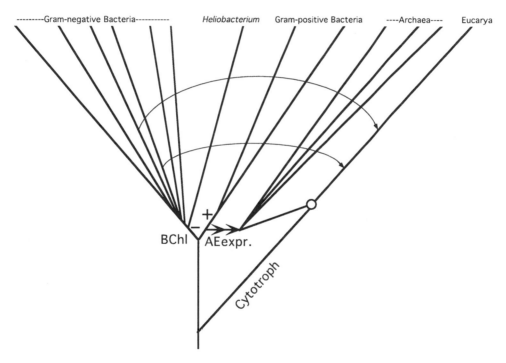

FIGURE 5 Fourth Domain model, adjusted to take into account the sequence similarities among archaea, gram-positive bacteria, and cyanobacteria. Here, one must suppose that after the divergence of *Archaea* from gram-positive bacteria, *Archaea* redesigned their gene expression mechanisms, resulting in rapid sequence evolution. In this scenario, eucaryal genes are more closely related to archaeal genes or to gram-negative bacterial genes, since any affinity with gram-positive bacterial genes would be outweighed by the closer similarity to archaeal genes. *Heliobacterium* is a gram-negative member of the gram-positive lineage, with the simplest of bacteriochlorophyll (BChl)-based photosynthetic apparatuses. +, addition of depth to the otherwise sheet-like murein sacculus; −, loss of the bacterial outer membrane, characteristic of gram-negative bacteria and believed to be ancestral (11); AEexpr., invention of the archaeal-eucaryal style of gene expression and loss of peptidoglycan.

olutionary rates gum up the phylogenetic machinery as much as do lateral genetic transfers, they cannot be ignored or necessarily minimized. Invoking such explanations often comes as a last resort in explaining why one's tree does not look as expected. It might be more of a challenge to explain why a tree does look as expected: why might there never have been lateral genetic transfer or directed sequence evolution? Is the gene in question co-dependent on others so that they would have to have moved at the same time? Was the gene product perfected very early on so that it cannot improve its characteristics? These are interesting questions!

There are several models, based on different assumptions. Whose assumptions are more reasonable? One hopes that the resources expended today in sequencing genomes will help to answer these important evolutionary questions. The application of an organism's genomic sequence to the understanding of its own biology is not in question, but many sequences are currently being generated mainly to address phylogenetic or other evolutionary issues (31). Although we are conscious of the breadth and importance of lateral genetic transfer, the concepts of lineage and species are still prevalent in many evolutionary discussions. What genomic sequencing should teach us is not how to make lists of exceptions to the true phylogeny but rather to discard the notion of there being any globally coherent, tree-like organismic phylogeny. More data and increasingly robust statistics will do little to support a unified tree.

Chimerism in eukaryotic evolution is generally accepted, in the form of the endosymbiotic theory (46), and is proposed for the creation of the premitochondrial eukaryote (52, 54, 79, 106, 120). The occurrence of lateral genetic transfer among prokaryotes, too, is not in question (69, 71), but this type of chimerism may be more gradually developed than is envisioned in the cellular fusion leading to eukaryotes. Models such as those from Tchelet and Mevarech (112) and Martin and Müller (80), however, make even prokaryotic fusions

plausible without a need for phagocytosis. And although tempo and mode may differ among all of these genetic transactions, the end result of chimerism is the same: a multiply grafted tree.

COMPARATIVE GENOMIC ORGANIZATION

As argued in the previous section, the underlying, unwritten assumption inherent in current models of global evolution has been that genes are inherited by descent. Although this may be true in principle, the cumulative effect over geological time of rare lateral genetic transfers must be significant. This is analogous to sequence divergence, where within a recent lineage, (orthologous) sequences are mostly the same, but with increasing time of divergence, these sequences become largely randomized relative to one another. If import of a gene into a genome is viewed as equivalent to the incorporation of a mutation within a gene, both being rare, a genome is just as likely to lose all semblance of clonality as a gene is to lose all semblance of identity.

The biological species will tend to restrict lateral transfers to specific groups. However, movement of genes from more distantly related organisms is not precluded, even between *Bacteria* and *Eucarya* (105). The end result of species-specific gene pools is that genetic exchange will not be random across the biosphere, but neither can it be fully restricted. The effect should also change with time: as lineages go extinct, extant lineages should tend to become more isolated from one another, creating more genetically isolated clades (Fig. 6). When lineages were still new in the early days of life on Earth, however, one might expect this genetic isolation to be proportionately less, since genetic distances were necessarily shorter and barriers to lateral transfer may not yet have been fully erected (80, 116).

What can comparative genomics tell us about the tempo and mode of genomic evolution? Clearly, gene order is rather plastic over large phylogenetic distances (66), al-

FIGURE 6 General model of the history of life, in which genetic traits are largely inherited by descent except for common intraphyletic lateral genetic transfer—hence the thick branches—and occasional interphyletic lateral genetic transfer. The shaded bands represent contemporary gene pools, and the degree of shading indicates both the ease with which cross-kingdom exchanges might have occurred and the impact on all of extant life of such exchanges. The daggers indicate extinct lineages.

though forces exist that counter the pressure to rearrange in the short term (13, 16). If gene order is meaningless across the tree of life, one can turn instead to genomic content as a measure of genomic structure. Comparative inventories have been generated for currently sequenced genomes, and they have revealed a very sparse set of universally shared genes (26, 42a). Sequences may have diverged to the point that they are no longer confidently recognized across lineages (109a), alternative genes and pathways might be utilized in order to accomplish the same biochemistry (84), or many genes may be unnecessary for life (58, 84). It is a surprising conclusion that "256 genes are close to the minimal gene set that is necessary and sufficient to sustain the existence of a modern-type cell" (84).

Archaea have their own collections of genes, some unique to the archaeal lineage and many shared with either *Bacteria* or *Eucarya* or present in both (reviewed in reference 9). Specific genomes have subsets of these collections and typically possess a number of open reading frames not found anywhere else. Obviously,

the cenancestor could not have held a repository of hundreds of thousands of genes, selectively lost in this lineage and that. Sequence evolution is responsible for the unmatched open reading frames, having erased the evidence of their homologies. Confounding things further are lateral genetic transfer and gene replacement, especially from extinct lineages. Only with a much greater sampling of genomic sequences might we ever be able to trace the histories of genomes. Then we would be in a position to better appreciate the extent of drift, selective innovation, and genetic trading that has gone on.

It might be unwise to try to force genomes into some kind of consensus phylogeny when such a tree might be artifactual. Genes will retain their phylogenies (though not necessarily), but I predict that genomes will be found to represent collections of genes, recently vertical but ultimately horizontal. Comparative genomic analyses have begun, and they will undoubtedly transform the method of molecular evolutionary study. What we should not do is attempt to encapsulate genomic inventories within cytoplasmic membranes; instead, we should identify genes or gene sets which are interdependent and trace their fates through the history of the world. These functional units, be they lone genes or small partnerships, should represent the way that the global genome is truly organized.

ACKNOWLEDGMENTS

For support, I thank the Canadian Institute for Advanced Research, the Natural Sciences and Engineering Research Council of Canada, and the Canadian Genome Analysis and Technology program.

REFERENCES

1. **Amils, R., L. Ramírez, J. L. Sanz, I. Marín, A. G. Pisabarro, and D. Ureña.** 1989. The use of functional analysis of the ribosome as a tool to determine archaebacterial phylogeny. *Can. J. Microbiol.* **35:**141–147.
2. **Antón, J., P. López-García, J. P. Abad, C. L. Smith, and R. Amils.** 1994. Alignment of genes and *Swa*I restriction sites to the *Bam*HI genomic map of *Haloferax mediterranei. FEMS Microbiol. Lett.* **117:**53–60.

3. **Baldauf, S. L., J. D. Palmer, and W. F. Doolittle.** 1996. The root of the universal tree and the origin of eukaryotes based on elongation factor phylogeny. *Proc. Natl. Acad. Sci. USA* **93:** 7749–7754.

4. **Baumann, P., and S. P. Jackson.** 1996. An archaebacterial homologue of the essential eubacterial cell division protein FtsZ. *Proc. Natl. Acad. Sci. USA* **93:**6726–6730.

5. **Bouthier de la Tour, C., C. Portemer, M. Nadal, K. O. Stetter, P. Forterre, and M. Duguet.** 1990. Reverse gyrase, a hallmark of the hyperthermophilic archaebacteria. *J. Bacteriol.* **172:**6803–6808.

6. **Bouthier de la Tour, C., C. Portemer, R. Huber, P. Forterre, and M. Duguet.** 1991. Reverse gyrase in thermophilic eubacteria. *J. Bacteriol.* **173:**3921–3923.

7. **Brendel, V., L. Brocchieri, S. J. Sandler, A. J. Clark, and S. Karlin.** 1997. Evolutionary comparisons of RecA-like proteins across all major kingdoms of living organisms. *J. Mol. Evol.* **44:**528–541.

8. **Brown, J. R., and W. F. Doolittle.** 1995. Root of the universal tree of life based on ancient aminoacyl-tRNA synthetase gene duplications. *Proc. Natl. Acad. Sci. USA* **92:**2441–2445.

9. **Brown, J. R., and W. F. Doolittle.** 1997. Archaea and the prokaryote-to-eukaryote transition. *Microbiol. Mol. Biol. Rev.* **61:**456–502.

10. **Bult, C. J., O. White, G. J. Olsen, L. Zhou, R. D. Fleischmann, G. G. Sutton, J. A. Blake, L. M. FitzGerald, R. A. Clayton, J. D. Gocayne, A. R. Kerlavage, B. A. Dougherty, J.-F. Tomb, M. D. Adams, C. I. Reich, R. Overbeek, E. F. Kirkness, K. G. Weinstock, J. M. Merrick, A. Glodek, J. L. Scott, N. S. M. Geoghagen, J. F. Weidman, J. L. Fuhrmann, D. Nguyen, T. R. Utterback, J. M. Kelley, J. D. Peterson, P. W. Sadow, M. C. Hanna, M. D. Cotton, K. M. Roberts, M. A. Hurst, B. P. Kaine, M. Borodovsky, H.-P. Klenk, C. M. Fraser, H. O. Smith, C. R. Woese, and J. C. Venter.** 1996. Complete genome sequence of the methanogenic archaeon, *Methanococcus jannaschii. Science* **273:** 1058–1073.

11. **Cavalier-Smith, T.** 1990. Microorganism megaevolution: integrating the fossil and living evidence. *Rev. Micropaléontologie* **33:**145–154.

12. **Cavalier-Smith, T.** 1992. Bacteria and eukaryotes. *Nature* **356:**570.

13. **Charlebois, R. L.** 1996. The modern science of bacterial genomics. *ASM News* **62:**255–259.

14. **Charlebois, R. L.** 1999. Evolutionary origins of the haloarchaeal genome, p. 309–317. *In* A. Oren (ed.), *Microbiology and Biogeochemistry of Hypersaline Environments.* CRC Press, Boca Raton, Fla.

15. **Charlebois, R. L., and W. F. Doolittle.** 1989. Transposable elements and genome structure in halobacteria, p. 297–307. *In* M. Howe and D. Berg (ed.), *Mobile DNA.* American Society for Microbiology, Washington, D.C.

16. **Charlebois, R. L., and A. St. Jean.** 1995. Supercoiling and map stability in the bacterial chromosome. *J. Mol. Evol.* **41:**15–23.

17. **Charlebois, R. L., L. C. Schalkwyk, J. D. Hofman, and W. F. Doolittle.** 1991. A detailed physical map and set of overlapping clones covering the genome of the archaebacterium *Haloferax volcanii* DS2. *J. Mol. Biol.* **222:**509–524.

18. **Charlebois, R. L., T. Gaasterland, M. A. Ragan, W. F. Doolittle, and C. W. Sensen.** 1996. The *Sulfolobus solfataricus* P2 genome project. *FEBS Lett.* **389:**88–91. (Erratum, **398:**343.)

19. **Cohen, A., W. L. Lam, R. L. Charlebois, W. F. Doolittle, and L. C. Schalkwyk.** 1992. Localizing genes on the map of the genome of *Haloferax volcanii,* one of the Archaea. *Proc. Natl. Acad. Sci. USA* **89:**1602–1606.

20. **Confalonieri, F., C. Elie, M. Nadal, C. Bouthier de la Tour, P. Forterre, and M. Duguet.** 1993. Reverse gyrase: a helicase-like domain and a type I topoisomerase in the same polypeptide. *Proc. Natl. Acad. Sci. USA* **90:** 4753–4757.

21. **DasSarma, S.** 1989. Mechanisms of genetic variability in *Halobacterium halobium*: the purple membrane and gas vesicle mutations. *Can. J. Microbiol.* **35:**65–72.

22. **Dawkins, R.** 1976. *The Selfish Gene.* Oxford University Press, Oxford, United Kingdom.

23. **Deckert, G., P. V. Warren, T. Gaasterland, W. G. Young, A. L. Lenox, D. E. Graham, R. Overbeek, M. A. Snead, M. Keller, M. Aujay, R. Huber, R. A. Feldman, J. M. Short, G. J. Olsen, and R. V. Swanson.** 1998. The complete genome of the hyperthermophilic bacterium *Aquifex aeolicus. Nature* **392:**353–358.

24. **Dennis, P. P., and L. C. Shimmin.** 1997. Evolutionary divergence and salinity-mediated selection in halophilic archaea. *Microbiol. Mol. Biol. Rev.* **61:**90–104.

25. **Doolittle, R. F.** 1995. The origins and evolution of eukaryotic proteins. *Phil. Trans. R. Soc. Lond. B* **349:**235–240.

26. **Doolittle, R. F.** 1998. Microbial genomes opened up. *Nature* **392:**339–342.

27. **Doolittle, R. F., D.-F. Feng, S. Tsang, G. Cho, and E. Little.** 1996. Determining divergence times of the major kingdoms of living organisms with a protein clock. *Science* **271:**470–477.

28. **Doolittle, W. F., and J. R. Brown.** 1994. Tempo, mode, the progenote, and the universal root. *Proc. Natl. Acad. Sci. USA* **91:**6721–6728.

29. **Ebert, K., and W. Goebel.** 1985. Conserved and variable regions in the chromosomal and extrachromosomal DNA of halobacteria. *Mol. Gen. Genet.* **200:**96–102.

30. **Ebert, K., W. Goebel, and F. Pfeifer.** 1984. Homologies between heterogeneous extrachromosomal DNA populations of *Halobacterium halobium* and four new halobacterial isolates. *Mol. Gen. Genet.* **194:**91–97.

31. **Edgell, D. R., and W. F. Doolittle.** 1997. Archaebacterial genomics: the complete genomic sequence of *Methanococcus jannaschii*. *Bioessays* **19:**1–4.

32. **Edgell, D. R., H.-P. Klenk, and W. F. Doolittle.** 1997. Gene duplications in evolution of archaeal family B DNA polymerases. *J. Bacteriol.* **179:**2632–2640.

33. **Fitch, W. M., and K. Upper.** 1987. The phylogeny of tRNA sequences provides evidence for ambiguity reduction in the origin of the genetic code. *Cold Spring Harbor Symp. Quant. Biol.* **52:**759–767.

34. **Fitz-Gibbon, S., A. J. Choi, J. H. Miller, K. O. Stetter, M. I. Simon, R. Swanson, and U.-J. Kim.** 1997. A fosmid-based genomic map and identification of 474 genes of the hyperthermophilic archaeon *Pyrobaculum aerophilum*. *Extremophiles* **1:**36–51.

35. **Forterre, P.** 1992. New hypotheses about the origins of viruses, prokaryotes and eukaryotes, p. 221–234. *In* I. K. Trân Thanh Vân, J. C. Mounolou, J. Schneider, and C. McKay (ed.), *Frontiers of Life*. Éditions Frontières, Gif-sur-Yvette, France.

36. **Forterre, P.** 1995. Thermoreduction, a hypothesis for the origin of prokaryotes. *C. R. Acad. Sci. III* **318:**415–422.

37. **Forterre, P., N. Benachenhou-Lahfa, F. Confalonieri, M. Duguet, C. Elie, and B. Labedan.** 1993. The nature of the last universal ancestor and the root of the tree of life, still open questions. *BioSystems* **28:**15–32.

38. **Forterre, P., F. Confalonieri, F. Charbonnier, and M. Duguet.** 1995. Speculations on the origin of life and thermophily: review of available information on reverse gyrase suggests that hyperthermophilic procaryotes are not so primitive. *Orig. Life Evol. Biosph.* **25:**235–249.

39. **Forterre, P., A. Bergerat, and P. López-García.** 1996. The unique DNA topology and DNA topoisomerases of hyperthermophilic Archaea. *FEMS Microbiol. Rev.* **18:**237–248.

39a.**Fortier, A., and R. L. Charlebois.** Unpublished data.

40. **Fox, G. E., L. J. Magrum, W. E. Balch, R. S. Wolfe, and C. R. Woese.** 1977. Classification of methanogenic bacteria by 16S ribosomal RNA characterization. *Proc. Natl. Acad. Sci. USA* **74:**4537–4541.

41. **Fox, G. E., E. Stackebrandt, R. B. Hespell, J. Gibson, J. Maniloff, T. A. Dyer, R. S. Wolfe, W. E. Balch, R. S. Tanner, L. J. Magrum, L. B. Zablen, R. Blakemore, R. Gupta, L. Bonen, B. J. Lewis, D. A. Stahl, K. R. Luehrsen, K. N. Chen, and C. R. Woese.** 1980. The phylogeny of prokaryotes. *Science* **209:**457–463.

42. **Fujiwara, S., S. Okuyama, and T. Imanaka.** 1996. The world of archaea: genome analysis, evolution and thermostable enzymes. *Gene* **179:**165–170.

42a.**Gaasterland, T., and M. A. Ragan.** 1998. Microbial genescapes: phyletic and functional patterns of ORF distribution among prokaryotes. *Microb. Comp. Genomics* **3:**199–217.

43. **Gambacorta, A., A. Trincone, B. Nicolaus, L. Lama, and M. De Rosa.** 1994. Unique features of lipids of Archaea. *Syst. Appl. Microbiol.* **16:**518–527.

44. **Gogarten, J. P., H. Kibak, P. Dittrich, L. Taiz, E. J. Bowman, B. J. Bowman, M. F. Manolson, R. J. Poole, T. Date, T. Oshima, J. Konishi, K. Denda, and M. Yoshida.** 1989. Evolution of the vacuolar H$^+$-ATPase: implications for the origin of eukaryotes. *Proc. Natl. Acad. Sci. USA* **86:**6661–6665.

45. **Golding, G. B., and R. S. Gupta.** 1995. Protein-based phylogenies support a chimeric origin for the eukaryotic genome. *Mol. Biol. Evol.* **12:**1–6.

46. **Gray, M. W., and W. F. Doolittle.** 1982. Has the endosymbiont hypothesis been proven? *Microbiol. Rev.* **46:**1–42.

47. **Grayling, R. A., K. Sandman, and J. N. Reeve.** 1994. Archaeal DNA binding proteins and chromosome structure. *Syst. Appl. Microbiol.* **16:**582–590.

48. **Gribaldo, S., V. Lumia, R. Creti, E. Conway de Macario, A. Sanangelantoni, and P. Cammarano.** 1999. Discontinuous occurrence of the *hsp70* (*dnaK*) gene among *Archaea* and sequence features of HSP70 suggest a novel outlook on phylogenies inferred from this protein. *J. Bacteriol.* **181:**434–443.

49. **Gupta, R.** 1984. *Halobacterium volcanii* tRNAs. Identification of 41 tRNAs covering all amino acids, and the sequences of 33 class I tRNAs. *J. Biol. Chem.* **259:**9461–9471.

50. **Gupta, R., J. M. Lanter, and C. R. Woese.** 1983. Sequence of the 16S ribosomal RNA from

Halobacterium volcanii, an archaebacterium. *Science* **221**:656–659.

51. **Gupta, R. S.** 1998. Protein phylogenies and signature sequences: a reappraisal of evolutionary relationships among archaebacteria, eubacteria, and eukaryotes. *Microbiol. Mol. Biol. Rev.* **62**:1435–1491.

52. **Gupta, R. S., and G. B. Golding.** 1993. Evolution of HSP70 gene and its implications regarding relationships between archaebacteria, eubacteria, and eukaryotes. *J. Mol. Evol.* **37**:573–582.

53. **Gupta, R. S., and B. Singh.** 1994. Phylogenetic analysis of 70 kD heat shock protein sequences suggests a chimeric origin for the eukaryotic cell nucleus. *Curr. Biol.* **4**:1104–1114.

54. **Gupta, R. S., K. Aitken, M. Falah, and B. Singh.** 1994. Cloning of *Giardia lamblia* heat shock protein HSP70 homologs: implications regarding origin of eukaryotic cells and of endoplasmic reticulum. *Proc. Natl. Acad. Sci. USA* **91**:2895–2899.

55. **Gutiérrez, M. C., M. T. García, A. Ventosa, J. J. Nieto, and F. Ruiz-Berraquero.** 1986. Occurrence of megaplasmids in halobacteria. *J. Appl. Bacteriol.* **61**:67–71.

56. **Hackett, N. R., Y. Bobovnikova, and N. Heyrovska.** 1994. Conservation of chromosomal arrangement among three strains of the genetically unstable archaeon *Halobacterium salinarium. J. Bacteriol.* **176**:7711–7718.

57. **Hilpert, R., J. Winter, W. Hammes, and O. Kandler.** 1981. The sensitivity of archaebacteria to antibiotics. *Zentbl. Bakteriol. Hyg. Abt. 1 Orig. C* **2**:11–20.

58. **Itaya, M.** 1995. An estimation of minimal genome size required for life. *FEBS Lett.* **262**:257–260.

59. **Iwabe, N., K. Kuma, M. Hasegawa, S. Osawa, and T. Miyata.** 1989. Evolutionary relationship of archaebacteria, eubacteria, and eukaryotes inferred from phylogenetic trees of duplicated genes. *Proc. Natl. Acad. Sci. USA* **86**:9355–9359.

60. **Kandler, O.** 1994. Cell wall biochemistry and the three-domain concept of life. *Syst. Appl. Microbiol.* **16**:501–509.

61. **Karawabayasi, Y., M. Sawada, H. Horikawa, Y. Haikawa, Y. Hino, S. Yamamoto, M. Sekine, S. Baba, H. Kosugi, A. Hosoyama, Y. Nagai, M. Sakai, K. Ogura, R. Otsuka, H. Nakazawa, M. Takamiya, Y. Ohfuku, T. Funahashi, T. Tanaka, Y. Kudoh, J. Yamazaki, N. Kushida, A. Oguchi, K. Aoki, and H. Kikuchi.** 1998. Complete sequence and gene organization of the genome of a hyper-thermophilic archaebacterium, *Pyrococcus horikoshii* OT3. *DNA Res.* **5**:55–76.

61a.**Karawabayasi, Y., M. Sawada, H. Horikawa, Y. Haikawa, Y. Hino, S. Yamamoto, M. Sekine, S. Baba, H. Kosugi, A. Hosoyama, Y. Nagai, M. Sakai, K. Ogura, R. Otsuka, H. Nakazawa, M. Takamiya, Y. Ohfuku, T. Funahashi, T. Tanaka, Y. Kudoh, J. Yamazaki, N. Kushida, A. Oguchi, K. Aoki, and H. Kikuchi.** 1998. complete sequence and gene organization of the genome of a hyper-thermophilic archaebacterium, *Pyrococcus horikoshii* OT3 (supplement). *DNA Res.* **5**:147–155.

62. **Keeling, P. J., and W. F. Doolittle.** 1995. Archaea: narrowing the gap between prokaryotes and eukaryotes. *Proc. Natl. Acad. Sci. USA* **92**:5761–5764.

63. **Keeling, P. J., R. L. Charlebois, and W. F. Doolittle.** 1994. Archaebacterial genomes: eubacterial form and eukaryotic content. *Curr. Opin. Genet. Dev.* **4**:816–822.

64. **Klenk, H.-P., P. Palm, and W. Zillig.** 1994. DNA-dependent RNA polymerases as phylogenetic marker molecules. *Syst. Appl. Microbiol.* **16**:638–647.

65. **Klenk, H.-P., R. A. Clayton, J.-F. Tomb, O. White, K. E. Nelson, K. A. Ketchum, R. J. Dodson, M. Gwinn, E. K. Hickey, J. D. Peterson, D. L. Richardson, A. R. Kerlavage, D. E. Graham, N. C. Kyrpides, R. D. Fleischmann, J. Quackenbush, N. H. Lee, G. G. Sutton, S. Gill, E. F. Kirkness, B. A. Dougherty, K. McKenney, M. D. Adams, B. Loftus, S. Peterson, C. I. Reich, L. K. McNeil, J. H. Badger, A. Glodek, L. Zhou, R. Overbeek, J. D. Gocayne, J. F. Weidman, L. McDonald, T. Utterback, M. D. Cotton, T. Spriggs, P. Artiach. B. P. Kaine, S. M. Sykes, P. W. Sadow, K. P. D'Andrea, C. Bowman, C. Fujii, S. A. Garland, T. M. Mason, G. J. Olsen, C. M. Fraser, H. O. Smith, C. R. Woese, and J. C. Venter.** 1997. The complete genome sequence of the hyper-thermophilic sulphate-reducing archaeon *Archaeoglobus fulgidus. Nature* (London) **390**:364–370.

66. **Kolstø, A.-B.** 1997. Dynamic bacterial genome organization. *Mol. Microbiol.* **24**:241–248.

67. **Kondo, S., A. Yamagishi, and T. Oshima.** 1993. A physical map of the sulfur-dependent archaebacterium *Sulfolobus acidocaldarius* 7 chromosome. *J. Bacteriol.* **175**:1532–1536.

68. **Koonin, E. V., A. R. Mushegian, M. Y. Galperin, and D. R. Walker.** 1997. Comparison of archaeal and bacterial genomes: computer analysis of protein sequences predicts novel functions

and suggests a chimeric origin for the archaea. *Mol. Microbiol.* **25**:619–637.

69. **Lan, R., and P. R. Reeves.** 1996. Gene transfer is a major factor in bacterial evolution. *Mol. Biol. Evol.* **13**:47–55.

70. **Langer, D., J. Hain, P. Thuriaux, and W. Zillig.** 1995. Transcription in Archaea: similarity to that in Eucarya. *Proc. Natl. Acad. Sci. USA* **92**: 5768–5772.

71. **Lawrence, J. G., and H. Ochman.** 1997. Amelioration of bacterial genomes: rates of change and exchange. *J. Mol. Evol.* **44**:383–397.

72. **Lawson, F. S., R. L. Charlebois, and J. R. Dillon.** 1996. Phylogenetic analysis of carbamoylphosphate synthetase genes: complex evolutionary history includes an internal duplication within a gene which can root the tree of life. *Mol. Biol. Evol.* **13**:970–977.

73. **Leffers, H., J. Kjems, L. Østergaard, N. Larsen, and R. A. Garrett.** 1987. Evolutionary relationships amongst archaebacteria. A comparative study of 23S ribosomal RNAs of a sulphur-dependent extreme thermophile, an extreme halophile and a thermophilic methanogen. *J. Mol. Biol.* **195**:43–61.

74. **López-García, P., J. P. Abad, C. Smith, and R. Amils.** 1992. Genomic organization of the halophilic archaeon *Haloferax mediterranei*: physical map of the chromosome. *Nucleic Acids Res.* **20**: 2459–2464.

75. **López-García, P., A. St. Jean, R. Amils, and R. L. Charlebois.** 1995. Genomic stability in the archaea *Haloferax volcanii* and *Haloferax mediterranei*. *J. Bacteriol.* **177**:1405–1408.

76. **López-García, P., R. Amils, and J. Antón.** 1996. Sizing chromosomes and megaplasmids in haloarchaea. *Microbiology* **142**:1423–1428.

77. **Lurz, R., M. Grote, J. Dijk, R. Reinhardt, and B. Dobrinski.** 1986. Electron microscopic study of DNA complexes with proteins from the archaebacterium *Sulfolobus acidocaldarius*. *EMBO J.* **5**:3715–3721.

78. **Mai, V. Q., X. Chen, R. Hong, and L. Huang.** 1998. Small abundant DNA binding proteins from the thermoacidophilic archaeon *Sulfolobus shibatae* constrain negative DNA supercoils. *J. Bacteriol.* **180**:2560–2563.

79. **Margulis, L.** 1996. Archaeal-eubacterial mergers in the origin of Eukarya: phylogenetic classification of life. *Proc. Natl. Acad. Sci. USA* **93**:1071–1076.

80. **Martin, W., and M. Müller.** 1998. The hydrogen hypothesis for the first eukaryote. *Nature* **392**: 37–41.

81. **Mayr, E.** 1990. A natural system of organisms. *Nature* **348**:491.

82. **McCloskey, J. A.** 1986. Nucleoside modification in archaebacterial transfer RNA. *Syst. Appl. Microbiol.* **7**:246–252.

83. **Moore, R. L., and B. J. McCarthy.** 1969. Characterization of the deoxyribonucleic acid of various strains of halophilic bacteria. *J. Bacteriol.* **99**:248–254.

84. **Mushegian, A. R., and E. V. Koonin.** 1996. A minimal gene set for cellular life derived by comparison of complete bacterial genomes. *Proc. Natl. Acad. Sci. USA* **93**:10268–10273

85. **Ng, W.-L. V., S. A. Ciufo, T. M. Smith, R. E. Bumgarner, D. Baskin, J. Faust, B. Hall, C. Loretz, J. Seto, J. Slagel, L. Hood, and S. DasSarma.** 1998. Snapshot of a large dynamic replicon in a halophilic archaeon: megaplasmid or minichromosome? *Genome Res.* **8**: 1131–1141.

86. **Noll, K. M.** 1989. Chromosome map of the thermophilic archaebacterium *Thermococcus celer*. *J. Bacteriol.* **171**:6720–6725.

87. **Olsen, G. J., and C. R. Woese.** 1997. Archaeal genomes: an overview. *Cell* **89**:991–994.

88. **Pfeifer, F.** 1986. Insertion elements and genome organization of *Halobacterium halobium*. *Syst. Appl. Microbiol.* **7**:36–40.

89. **Pfeifer, F., and M. Betlach.** 1985. Genome organization in *Halobacterium halobium*: a 70 kb island of more (AT) rich DNA in the chromosome. *Mol. Gen. Genet.* **198**:449–455.

90. **Pfeifer, F., G. Weidinger, and W. Goebel.** 1981. Characterization of plasmids in halobacteria. *J. Bacteriol.* **145**:369–374.

91. **Pfeifer, F., G. Weidinger, and W. Goebel.** 1981. Genetic variability in *Halobacterium halobium*. *J. Bacteriol.* **145**:375–381.

92. **Pfeifer, F., K. Ebert, G. Weidinger, and W. Goebel.** 1982. Structure and functions of chromosomal and extrachromosomal DNA in halobacteria. *Zentbl. Bakteriol. Hyg. Abt. 1 Orig. C* **3**:110–119.

93. **Pfeifer, F., U. Blaseio, and P. Ghahraman.** 1988. Dynamic plasmid populations in *Halobacterium halobium*. *J. Bacteriol.* **170**:3718–3724.

94. **Pfeifer, F., U. Blaseio, and M. Horne.** 1989. Genome structure of *Halobacterium halobium*: plasmid dynamics in gas vacuole deficient mutants. *Can. J. Microbiol.* **35**:96–100.

95. **Pfeifer, F., J. Griffig, and D. Oesterhelt.** 1993. The *fdx* gene encoding the [2Fe-2S] ferredoxin of *Halobacterium salinarium* (*H. halobium*). *Mol. Gen. Genet.* **239**:66–71.

96. **Poplawski, A., and R. Bernander.** 1997. Nucleoid structure and distribution in thermophilic Archaea. *J. Bacteriol.* **179**:7625–7630.

97. **Reeve, J. N., J. Nölling, R. M. Morgan, and D. R. Smith.** 1997. Methanogenesis: genes, ge-

nomes, and who's on first? *J. Bacteriol.* **179:** 5975–5986.

98. **Rivera, M. C., and J. A. Lake.** 1992. Evidence that eukaryotes and eocyte prokaryotes are immediate relatives. *Science* 257:74–76.

99. **Ronimus, R. S., and D. R. Musgrave.** 1995. A comparison of the DNA binding properties of histone-like proteins derived from representatives of the two kingdoms of the Archaea. *FEMS Microbiol. Lett.* **134:**79–84.

100. **Ronimus, R. S., and D. R. Musgrave.** 1996. A gene, *han1A*, encoding an archaeal histone-like protein from the *Thermococcus* species AN1: homology with eukaryal histone consensus sequences and the implications for delineation of the histone fold. *Biochim. Biophys. Acta* **1307:**1–7.

101. **Schleper, C., R. Röder, T. Singer, and W. Zillig.** 1994. An insertion element of the extremely thermophilic archaeon *Sulfolobus solfataricus* transposes into the endogenous β-galactosidase gene. *Mol. Gen. Genet.* **243:**91–96.

102. **Sensen, C. W., H.-P. Klenk, R. K. Singh, G. Allard, C. C.-Y. Chan, Q. Y. Liu, S. L. Penny, F. Young, M. E. Schenk, T. Gaasterland, W. F. Doolittle, M. A. Ragan, and R. L. Charlebois.** 1996. Organizational characteristics and information content of an archaeal genome: 156 kb of sequence from *Sulfolobus solfataricus* P2. *Mol. Microbiol.* **22:**175–191.

103. **Sitzmann, J., and A. Klein.** 1991. Physical and genetic map of the *Methanococcus voltae* chromosome. *Mol. Microbiol.* **5:**505–513.

104. **Smith, D. R., L. A. Doucette-Stamm, C. Deloughery, H. Lee, J. Dubois, T. Aldredge, R. Bashirzadeh, D. Blakely, R. Cook, K. Gilbert, D. Harrison, L. Hoang, P. Keagle, W. Lumm, P. Pothier, D. Qiu, R. Spadafora, R. Vicaire, Y. Wang, J. Wierzbowski, R. Gibson, N. Jiwani, A. Caruso, D. Bush, H. Safer, D. Patwell, S. Prabhakar, S. McDougall, G. Shimer, A. Goyal, S. Pietrokovski, G. M. Church, C. J. Daniels, J.-I. Mao, P. Rice, J. Nölling, and J. N. Reeve.** 1997. Complete genome sequence of *Methanobacterium thermoautotrophicum* ΔH: functional analysis and comparative genomics. *J. Bacteriol.* **179:**7135–7155.

105. **Smith, M. W., D.-F. Feng, and R. F. Doolittle.** 1992. Evolution by acquisition: the case for horizontal gene transfers. *Trends Biochem. Sci.* **17:**489–493.

106. **Sogin, M. L.** 1991. Early evolution and the origin of eukaryotes. *Curr. Opin. Genet. Dev.* **1:**457–463.

107. **Stettler, R., and T. Leisinger.** 1992. Physical map of the *Methanobacterium thermoautotrophicum* Marburg chromosome. *J. Bacteriol.* **174:**7227–7234.

108. **Stettler, R., G. Erauso, and T. Leisinger.** 1995. Physical and genetic map of the *Methanobacterium wolfei* genome and its comparison with the updated genomic map of *Methanobacterium thermoautotrophicum* Marburg. *Arch. Microbiol.* **163:**205–210.

109. **St. Jean, A., and R. L. Charlebois.** 1996. Comparative genomic analysis of the *Haloferax volcanii* DS2 and *Halobacterium salinarium* GRB contig maps reveals extensive rearrangement. *J. Bacteriol.* **178:**3860–3868.

109a. **St. Jean, A., and R. L. Charlebois.** Unpublished data.

110. **St. Jean, A., B. A. Trieselmann, and R. L. Charlebois.** 1994. Physical map and set of overlapping cosmid clones representing the genome of the archaeon *Halobacterium* sp. GRB. *Nucleic Acids Res.* **22:**1476–1483.

111. **Takayanagi, S., S. Morimura, H. Kusaoke, Y. Yokoyama, K. Kano, and M. Shioda.** 1992. Chromosomal structure of the halophilic archaebacterium *Halobacterium salinarium*. *J. Bacteriol.* **174:**7207–7216.

112. **Tchelet, R., and M. Mevarech.** 1994. Interspecies genetic transfer in halophilic archaebacteria. *Syst. Appl. Microbiol.* **16:**578–581.

113. **Walsby, A. E.** 1994. Gas vesicles. *Microbiol. Rev.* **58:**94–144.

114. **Weidinger, G., G. Klotz, and W. Goebel.** 1979. A large plasmid from *Halobacterium halobium* carrying genetic information for gas vacuole formation. *Plasmid* **2:**377–386.

115. **Woese, C. R.** 1987. Bacterial evolution. *Microbiol. Rev.* **51:**221–271.

116. **Woese, C. R.** 1998. The universal ancestor. *Proc. Natl. Acad. Sci. USA* **95:**6854–6859.

117. **Woese, C. R., and G. E. Fox.** 1977. The concept of cellular evolution. *J. Mol. Evol.* **10:** 1–6.

118. **Woese, C. R., and G. E. Fox.** 1977. Phylogenetic structure of the prokaryotic domain: the primary kingdoms. *Proc. Natl. Acad. Sci. USA* **74:**5088–5090.

119. **Woese, C. R., O. Kandler, and M. L. Wheelis.** 1990. Towards a natural system of organisms: proposal for the domains Archaea, Bacteria, and Eucarya. *Proc. Natl. Acad. Sci. USA* **87:**4576–4579.

120. **Zillig, W.** 1991. Comparative biochemistry of Archaea and Bacteria. *Curr. Opin. Genet. Dev.* **1:**544–551.

THE EUKARYOTIC PERSPECTIVE: SIMILARITIES AND DISTINCTIONS BETWEEN PRO- AND EUKARYOTES

Conrad L. Woldringh and Roel Van Driel

5

The presence of a membrane-bound nucleus and the nucleosomal organization of the DNA in eukaryotes represent important differences from prokaryotic cells. Another fundamental difference lies in the mechanism of chromosome segregation. Because of the size of prokaryotic cells, their replicated chromosomes have to be displaced over only relatively small distances (<2 μm). This segregation occurs along with cell growth. In contrast, eukaryotic cells keep their replicated sister chromatids together until segregation during mitotic anaphase. In this case, chromosomes are displaced with the help of the microtubular spindle apparatus over relatively large distances (>10 μm).

This mitotic mechanism of segregation has such a strong appeal that it is often considered to represent a uniform mechanism, also applicable to prokaryotes. This has been suggested by several authors in their studies of the bacterial cell cycle (15, 63, 65, 76). However, eubacteria, as well as archaebacteria, seem to lack a cytoskeletal segregation apparatus, although they do have cytoskeleton-like proteins (like FtsZ [40]; for a review, see reference 53), "motor" proteins (like MukB [22]), and condensing proteins (like Smc [17]). Moreover, the size and complexity of the eukaryotic mitotic apparatus are such that a similar mechanism is difficult to accommodate in a bacterial cell, as illustrated in Fig. 1A.

A prerequisite for mitotic segregation is the cohesion of sister chromatids, which is needed for the bipolar attachment of kinetochores to microtubules coming from opposite poles. The cohesion lasts until anaphase, when it is abolished by proteolysis and the chromatids are pulled apart by the action of the spindle microtubuli. The mechanism of maintenance of cohesion is not known. It is either the result of DNA strand entanglement or due to specific protein bridges (19, 32). However, in metaphase chromosomes, the sister chromatids are largely disentangled and most DNA regions have separated. When we consider the separation of sister chromatids at the scale of a typical bacterial cell, the average distance between pairs of identical genes in sister chromatids during metaphase (1 to 2 μm) is similar to that of DNA segregation in prokaryotes

Conrad L. Woldringh, Institute for Molecular Cell Biology, BioCentrum Amsterdam, University of Amsterdam, Kruislaan 316, 1098 SM Amsterdam, The Netherlands. *Roel Van Driel*, E. C. Slater Instituut, BioCentrum Amsterdam, University of Amsterdam, Plantage Muidergracht 12, 1018 TV Amsterdam, The Netherlands.

Organization of the Prokaryotic Genome, Edited by Robert L. Charlebois,
© 1999 American Society for Microbiology, Washington, D.C.

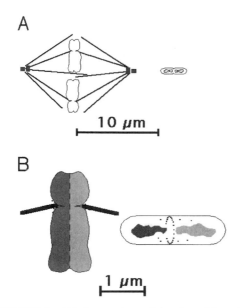

A

10 μm

B

1 μm

FIGURE 1   (A) Comparison of the size of the spindle apparatus in a HeLa cell and the size of an *E. coli* cell grown in glucose minimal medium. (B) The displacement of DNA between the sister chromatids in a metaphase chromosome is comparable to that of the segregated nucleoids in *E. coli*. The spots indicate tubulin or, in *E. coli*, the tubulin-like FtsZ proteins, which form a ring structure prior to division (40, 53).

(Fig. 1B; cf. the pairs of fluorescent spots shown in Fig. 3A and B of reference 35).

Segregation is defined here as the process that moves the replicated and separated chromosomes apart toward opposite poles and into the prospective daughter cells (49). Before or during this movement, the DNA daughter strands have to be disentangled in a process called separation.

This chapter begins by comparing the organization of DNA in bacteria and eukaryotes. Next, the possibility that similar mechanisms are utilized by both eukaryotes and prokaryotes to separate or disentangle sister chromatids or daughter strands is outlined. It is suggested that transcriptional activity and resulting RNA–protein (hnRNP) particles may play a role in DNA strand separation in eukaryotic cells, just as coupled transcription-translation does in prokaryotes. It is implied that genome organization is important not only for gene expression but also for the pro-

cess of chromosome separation. First, we will discuss how the primary separation of daughter strands in bacteria leads to a hierarchical mode of segregation (12), which differs fundamentally from the mitotic mode of segregation of chromosomes in the eukaryotic cell.

HIERARCHICAL DNA SEGREGATION IN BACTERIA

Two characteristics of a bacterial cell, such as *Escherichia coli*, may contribute to the difficulty in understanding its mechanism of genome segregation: first, the occurrence of DNA synthesis throughout the whole cell cycle during rapid growth (8), and second, the lack of a unique centromere sequence, which is characteristic of the eukaryotic chromosome. In contrast to the multiple replication origins in each eukaryotic chromosome, bacterial chromosomes sustain a unique origin, from which replication initiates and proceeds bidirectionally. Although the replication forks move bidirectionally at a maximal rate (about 50 kb per min [31]), the duration of one round of replication (40 min [8]) is long compared to the shortest generation time possible, i.e., 20 min. Despite a low rate of fork movement (0.5 to 5 kb per min [31]), eukaryotic cells, such as *Saccharomyces cerevisiae*, can have an S phase of 30 min, because replication starts at about 400 origins (55), whereas their shortest cycle time is still about 60 min.

The bacterial cell is able to make its cycle time shorter than the duration of the DNA replication period by making use of multifork replication and overlapping cycles (8). It should be noted that in bacteria, multifork replication refers to replication forks that have initiated in previous cell cycles and that therefore belong to different genealogical orders. In eukaryotic chromosomes, multiple replication forks have all initiated during the same cycle and are thus of the same genealogical order. Under conditions of rapid growth, when the cycle time is shorter than the replication period, bacterial DNA can be synthesized continuously throughout the cycle, while once per cycle a new round of replication is initiated and one round is completed (for a review, see reference 21). Bacterial multifork replication is a form of rereplication and implies the presence of several genealogical orders of origins and of replicated daughter strand regions. How are these multiple orders of daughter strands disentangled and segregated?

In the case of eukaryotic chromosomes, with many origins of replication (but no reinitiations during a single cell cycle), segregation is related to the buildup of two back-to-back kinetochores on the unique centromere. These juxtaposed structures are then attached to microtubuli of the mitotic spindle coming from opposite poles. Subsequently, the chromatids are transported to the poles. Inherent in this mechanism is the moving apart of the two recently replicated sister chromatids.

In haploid cells, as depicted schematically in Fig. 2A, segregation seems to take place in the same way in prokaryotes and eukaryotes. If a haploid cell would skip a division, it would contain two chromosomes, as shown in Fig. 2B. The mitotic segregation mechanism will, after one round of replication, segregate each of the sister chromatids to opposite poles and to the respective daughter cells (Fig. 2B). In contrast, in bacteria the replicated daughter strands of one chromosome (sister chromatids) will remain in the same compartment and will be located in the same daughter cell after division; they will only become distributed during the next division cycle.

In principle, a bacterial cell could segregate in a mitotic fashion. If the bacterial chromosome contained a centromere at its terminus of replication, it could always segregate the fully replicated chromosomes, even if reinitiation and multifork replication had produced different genealogical orders of origins and daughter strand regions. Such a mitotic segregation mechanism in *E. coli* was proposed by Hiraga (22) based on the characterization of a kinesin-like motor protein (MukB) and by Begg and Donachie (2) based on the observation that nucleoids move rapidly apart after termination of replication. However, cytological observations (77) have indicated that DNA segregation is a continuous process and

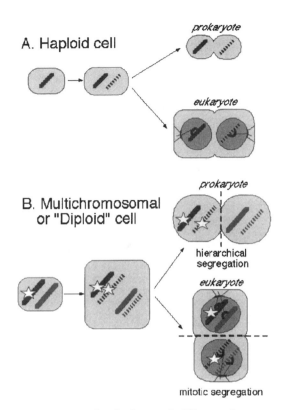

FIGURE 2 The fundamental difference between the modes of segregation in prokaryotes and eukaryotes becomes evident in the case of multichromosomal cells. (A) In a schematic haploid cell, the chromosome, depicted as a solid rod, is semiconservatively replicated, generating two daughter chromosomes, depicted as a solid rod and a dashed rod. In the prokaryotic cell, these chromosomes are segregated into the daughter cells as indicated. Likewise, in the eukaryotic cell, they are segregated as sister chromatids pulled by spindle microtubuli to opposite poles. (B) In a multichromosomal, or diploid, cell, replication of the two chromosomes results in two pairs of chromosomes (chromatids), depicted schematically as solid and shaded rods and dashed rods. In the prokaryotic cell, each pair of daughter chromosomes is located in a prospective daughter cell (as in eukaryotic meiosis I); these daughter chromosomes become segregated only in the next cycle. This has been called hierarchical segregation by Donachie et al. (12). In the eukaryotic cell, each pair of daughter chromosomes or sister chromatids becomes segregated into different daughter cells because of the cohesion between chromatids and because of the attachment of microtubuli from opposite poles to the kinetochores. If a mutation occurs in one of the chromosomes, as indicated by the white stars, the cell becomes heterozygous. As proposed by Donachie et al. (12), hierarchical segregation does not maintain this heterozygosity whereas mitotic segregation does.

that the replicating nucleoid increases gradually in size along the major cell axis in such a way that, immediately upon termination, the two daughter nucleoids are already separated (67).

It thus seems that the origins and subsequently replicated DNA regions are continuously being separated and segregated as each pair of replication forks proceeds from the origin to the terminus of replication. As emphasized by Donachie et al. (12), this situation differs fundamentally from that in eukaryotes. If an *E. coli* cell postpones a division, the cell will contain two chromosomes and it will become multichromosomal (or diploid). After replication of the chromosomes, the next division will not take place between the recently replicated daughter strands (sister chromatids) but between the two pairs of chromosomes (Fig. 2B). In eukaryotic cells this situation is obtained during meiosis I, when the replicated homologous chromosomes pair and their sister kinetochores become associated. Because the connected kinetochores attach to microtubuli from the same pole (49), the chromosomes separate from one another as pairs. In prokaryotes, however, such specific pairing, as during meiosis I, or cohesion, as during mitosis, does not occur. Nevertheless, division occurs normally between the pairs of chromosomes. This phenomenon, called hierarchical segregation by Donachie et al. (12), explains why multichromosomal bacteria cannot maintain heterozygosity. Thus, they will always develop to homozygosity. This is because, when a mutated gene is replicated (Fig. 2B), the gene pair is not segregated over the daughter cells but remains in one cell because of the mechanism of segregation. This mechanism represents a gradual and continuous movement of the nucleoids (see below). A lateral movement of the nucleoids, as in filaments (68) and as depicted in Fig. 2B, suggests that cell division will separate the pairs of chromosomes. However, this lateral movement is not the only reason for the separation of chromosome pairs, because the same hierarchical mode of segregation occurs when the cells assume an ovoid shape

(12). The reason probably lies in a relationship between separated nucleoids and the induction of a division plane, as suggested by Donachie et al. (reference 12; see also reference 80), but that is outside the scope of this review.

COMPARISON OF THE EUKARYOTIC NUCLEUS WITH THE BACTERIAL CELL

Before discussing mechanisms for the separation of replicated daughter strands in prokaryotes and eukaryotes, we will first compare the organization of DNA in both kinds of organisms. In vitro studies have shown that phenomena like phase separation (71), monomolecular collapse, and intermolecular aggregation of the DNA into a condensed state (5, 36, 51) can be induced by high concentrations of proteins (macromolecular crowded solution) and ions. In bacterial cells, high plasmid concentrations have been observed to result in liquid crystalline structures (62). In growing bacteria, a phase separation between nucleoid and cytoplasm is observed by phase contrast microscopy (46) and was also predicted on the basis of physicochemical calculations (59).

Merely on a basis of volume, the bacterial cell can be compared with, for instance, the nucleus of a yeast cell. Table 1 gives the DNA concentrations in both compartments. If the DNA of *E. coli* were dispersed throughout the whole volume of the cell, the DNA concentration would be about 9 mg/ml (Table 1). This is of the same order of magnitude as the DNA concentration in the nucleus of a diploid yeast cell, 13 mg/ml (Table 1). The DNA concentration in the nucleus of a mammalian cell depends on the estimate of the volume of the nucleus and seems to be in the range of 1 to 100 mg/ml (Table 1). The highest concentration for a human cell nucleus, with a radius of 2.5 μm, would be 100 mg/ml, which is similar to the DNA concentration within the volume of the bacterial nucleoid. As discussed in chapter 10, estimates of the volume of the nucleoid visualized with a confocal scanning

TABLE 1 Comparison of the bacterial cell with the eukaryotic nucleus: concentration of macromolecules

Parameter	Value		
	E. coli	*S. cerevisiae*	Human cell
Cell volume (μm^3)	1.06[a]	55[b]	1,200 ($r = 6.5$ μm)[c]
Nucleiod or nuclear volume (μm^3)	0.14[d]	2.5 ($r = 0.75$ μm)[b]	65.5 ($r = 2.5$ μm)[e]
			524 ($r = 5$ μm)[c]
Genome size (kbp)	4×10^3	13.5×10^3 (diploid)	6×10^6 (diploid)
Amount of DNA per cell (pg)	0.01	0.032	6.7
DNA concn (mg/ml)			
In nucleoid	65[d]		
In cell	9		
In nucleus		13	1–100[f]
In metaphase chromosome			120[e]

[a] Calculated from average cell length (2.5 μm) and diameter (0.5 μm), assuming the shape of the cell to be a right cylinder with hemispherical polar caps.
[b] See reference 79.
[c] See references 1 and 43.
[d] See chapter 10.
[e] Calculated for the haploid human chromosome 1 in metaphase (300 Mb) with dimensions given by Manuelidis and Chen (43).
[f] See reference 6.

light microscope operating at a lateral resolution of 130 μm resulted in a DNA concentration of 65 mg/ml. Bohrmann et al. (6) estimated the DNA concentration in the *E. coli* nucleoid to be 80 to 100 mg/ml by applying low-dose ratio contrast imaging with the scanning transmission electron microscope on resin-embedded sections. This probably represents an upper limit, considering their use of dehydrated and embedded cells that may have suffered shrinkage. Thus, on a chemical basis, considering the amounts of DNA in the eukaryotic nucleus and the bacterial cell (Fig. 3A), the DNA concentrations are similar (Table 1).

Within the bacterial cell, the concentrations of protein and RNA are about 20 and 6 mg/ml, respectively (54, 78). In the eukaryotic nucleus, the synthesis of mRNA is a process involving co- and posttranscriptional processing of pre-mRNA, including 5' capping, 3' cleavage, polyadenylation, and splicing. Subsequently, the processed RNA is transported as hnRNPs (heterogeneous ribonucleoprotein particles) by an as-yet-obscure mechanism (see reference 61 for a review). It seems well established that RNA synthesis in the nucleus largely exceeds the need for cytoplasmic mRNA and rRNA, as 80 to 90% of all the pre-mRNA is degraded without export to the cytoplasm (1). The synthetic machineries that transcribe and replicate DNA and process and transport RNA are so extensive that it may be expected that the concentrations of RNA and protein in the eukaryotic nucleus and in the bacterial cell are similar.

Another striking similarity between the prokaryotic cell and the eukaryotic nucleus is the occurrence of cotranscriptional processes (Fig. 3B). In prokaryotes, ribosomes bind immediately to mRNA for the synthesis of proteins in the so-called coupled transcription-translation process. In eukaryotes, the pre-mRNA becomes associated during its synthesis with proteins involved in splicing and transport of the transcripts (10, 30, 47), forming large spliceosome structures of a size similar to that of the bacterial ribosome (30).

However, as illustrated in Fig. 3B, there is a fundamental difference in the conformation of the DNA between pro- and eukaryotes. In the bacterial nucleoid the DNA occurs in the

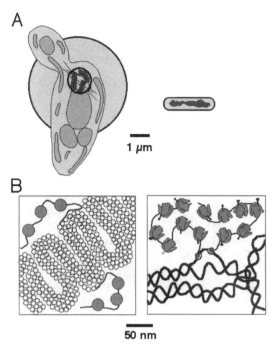

FIGURE 3 (A) Comparison of a diploid yeast cell, its nucleus, and the diameter of a HeLa cell nucleus with an *E. coli* cell. See Table 1 for a calculation of the DNA concentrations in these compartments. (B) Regions (200 nm wide) of a yeast nucleus and an *E. coli* cell. The chromatin in the yeast nucleus (left panel) is depicted as a folded 130-nm-wide thread (4) formed by folding of the 30-nm-diameter fiber (see the text). Nucleosomes are depicted as 10-nm-diameter circles. The large circles represent hnRNP particles with diameters of 20 nm (30) involved in cotranscriptional processing of pre-mRNA. The DNA in the *E. coli* nucleoid is drawn as plectonemic supercoils with diameters of 20 nm (see chapter 10). Ribosomes, with diameters of about 30 nm, are involved in cotranscriptional translation.

form of plectonemic supercoils (see chapter 10) with relatively few bound proteins. In the eukaryotic nucleus the nucleosomal organization involves a 10-fold-increased mass of proteins bound to DNA (6). Viewed on the small scale of the double helix, the DNA (145 bp) is wrapped around a histone octamer in two almost-complete left-handed superhelical turns, forming a thread of nucleosomes with a diameter of 10 nm. Subsequently, this 10-nm-diameter thread is folded as a solenoid into a 30-nm-diameter fiber (42). Minsky et al. (48) proposed that this assembly into nucleosomes is necessary to prevent condensation of the large amounts of DNA in the eukaryotic nucleus, allowing accessibility to RNA polymerases throughout the chromatin. Extremely condensed DNA molecules occur, for instance, in sperm cells, where the DNA is compacted by basic proteins rather than structured in nucleosomes. After fertilization, these proteins are exchanged for acidic nucleoplasmin and histones stored in the egg, resulting in a decondensation of the DNA and in the assembly of nucleosomes (60).

In order to fit into the nucleus, the second-order level of helical folding, the 30-nm-diameter fiber, must be packed into fibers with a larger diameter. Because the higher-order folding is very sensitive to variations in the ionic conditions of nuclear isolation buffers (3), different methods of preparation have resulted in different concepts, e.g., that of the formation of radial loops (26, 45) and that of further superhelical arrangements (42). By visualizing chromosome structures both in living cells (using optical sectioning microscopy and image deconvolution) and in dehydrated and embedded cells (using electron microscopy), Belmont et al. (3, 4) came to the conclusion that the 30-nm-diameter fiber must occur in threadlike domains with diameters of about 130 nm. Such a thread, or folded chromonema, has been schematically depicted in the yeast nucleus of Fig. 3B.

Application of the techniques of whole-chromosome "painting" and fluorescence in situ hybridization support the concept that the eukaryotic nucleus is partitioned into irregularly shaped chromosome territories (9, 14, 32). Visualization of RNA transcription and DNA replication has shown that these two processes occur in hundreds of different domains scattered throughout the S-phase nucleus (11, 25, 72, 73, 75). The chromosome territory model predicts that in interphase interchromosomal channels (34, 52, 73, 82) occur between the individual chromosomes, containing RNA and protein complexes involved in replication, transcription, and RNA processing and transport steps.

Before considering how the chromatids are separated after replication, we will first discuss possible mechanisms used in bacteria for movement of DNA and for nucleoid segregation.

SEPARATION OF DAUGHTER STRANDS IN BACTERIA

As previously proposed (51) and as discussed in chapter 10, the combined effect of supercoiling and macromolecular depletion forces lead to DNA compaction and phase separation between nucleoid and cytoplasm (71). How does the segregation mechanism displace the DNA against this force of compaction?

During segregation, the nucleoids have been observed to move gradually along with the elongating cell, both in normal *E. coli* cells and in bacterial filaments (67, 69). This movement has been shown to be independent of DNA replication because, upon inhibition of DNA synthesis, the nucleoids were pulled out into small lobules dispersed throughout the filament. This dispersion was not due to a physical fragmentation, as it was reversible; inhibition of protein synthesis caused the small DNA regions to coalesce again into a single region (reference 80; see also Fig. 1B and C in chapter 10). Likewise, if protein synthesis in normal cells is inhibited with either rifampin or chloramphenicol, the often lobular structure of the nucleoid is abolished and the DNA becomes confined in smooth, sometimes spherical structures (28, 81). In the case of filaments, already segregated nucleoids are even rapidly pushed together, forming multinucleoid bodies (68). Upon removal of the chloramphenicol and after recovery of protein synthesis, the nucleoids resegregate (69). These observations have led to the concept that separation of daughter strands and segregation of nucleoids can only occur during active transcription and translation, and thus during active growth of the cell.

As a working hypothesis, it has been proposed (69) that the driving force for nucleoid movement results from the random expression of numerous genes coding for membrane proteins. Like that in eukaryotic cells, translocation of inner membrane proteins in *E. coli* has been shown to require a signal recognition particle and its receptor (39, 66). By the formation of an mRNA-ribosome-signal recognition particle-translocon complex, these genes form DNA loops indirectly attached to the plasma membrane, as depicted in Fig. 4A (41, 57) (for a more detailed representation of this complex, see Fig. 2 in reference 81). Although each individual loop is only formed transiently, collectively, the many loops are able to expand and move the nucleoid in the growing cell. How are loops belonging to one daughter strand distinguished from those emanating from the other strand? To achieve such a directionality, it is necessary that an initial displacement occur between the replicated origins. Once the origin regions have separated over some distance, loops belonging to the same strand will, on average, attach to that membrane region of the cell to which the origin has moved. Subsequently, replicated DNA regions will be pulled apart as new loops attach to the membrane at either side of the cell (Fig. 4A).

In sporulating *B. subtilis* cells, it could be demonstrated that one-third of the chromosome centered around the origin is specifically compacted in the forespore (70), suggesting an orientation of the origin toward the cell pole. This orientation appeared dependent on the Spo0J protein (64), which was found to be associated with the origin and also positioned toward the cell pole (37, 38). Similar results were obtained with a homologous protein in *Caulobacter crescentus* (50). A specific positioning was also confirmed by following the movement of the origin region in living *B. subtilis* cells (74) and in *E. coli* (16). For visualization, green fluorescent protein-LacI fusion proteins that bind to an array of *lacO* sequences integrated near the origin were used. All of these observations suggest the existence of a dedicated mechanism for the displacement of replicated DNA sequences.

A

B

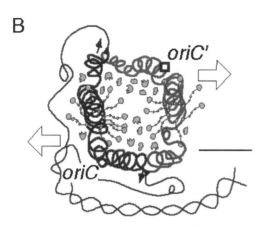

FIGURE 4 (A) Working model of transcription-mediated segregation assuming (i) a dedicated mechanism for the initial displacement of the two replicated origins (*oriC* and *oriC'*, indicated by the round and square symbols, respectively) and (ii) the expansion of the nucleoid by the transient attachment of DNA loops through cotranscriptional and cotranslational translocation of membrane proteins, forming two membrane growth zones that cause cell elongation. Solid and shaded loops attached to the membrane represent DNA from the two replicated daughter strands. See Fig. 2 in reference 81 for a more detailed representation of this attachment complex. (B) Initial displacement of the origins could occur by the formation of a hypothetical ribosome assembly compartment formed at rRNA genes near the unique origin of replication. Through this cotranscriptional assembly the origins are pushed apart (open arrows). The movement of this initial displacement is subsequently taken over and enhanced by DNA loops pulled to the membrane during the cotranscriptional and cotranslational translocation of membrane proteins (transertion [57]). This segregation mechanism is also suitable for multifork replication. When, after reinitiation, the origins of each pair are moved apart along the cell's long axis, the subsequently replicated daughter strands will again generate loops that attach to the membrane, now forming two pairs of growth zones. (Bar, 150 nm.)

Alternative to a displacement by the above-mentioned specific proteins, which are homologous to the products of the *parA* and *parB* families of plasmid-encoded genes (for a review, see reference 76), the initial displacement of the replicated origins could again be obtained through the process of transcription. For example, in the case of the high transcriptional activity of rRNA genes, combined with the assembly of the ribosomal subunits, it can be envisaged that the newly replicated daughter strands are pushed apart by the formation of a ribosome assembly compartment, as schematized in Fig. 4B (78, 80). It should be noted that rRNA genes occur close to the origin of replication in *E. coli* and *B. subtilis* but not in *C. crescentus*. Future studies will have to elucidate the mechanism(s) by which bacteria displace their replicated origins and give directionality to the movement of their daughter nucleoids.

SEPARATION OF SISTER CHROMATIDS IN EUKARYOTES

In contrast to prokaryotes, eukaryotic cells have to perform two different processes after DNA replication. The replicated sister chromatids must not only be separated and disentangled, but they also have to be compacted, while remaining paired for bipolar attachment to the mitotic spindle during metaphase. Even in the yeast nucleus some mitotic condensation is necessary to prevent the lagging ends of large chromosomes from being damaged by the process of cytokinesis (18).

Because of the termination of many converging replication forks, and because of the various orders of helical folding, DNA replication results in a complex plectonemic intertwining of the daughter strands. Duplantier et al. (13) propose that the disentanglement of the sister chromatids occurs in two steps. The first is a separation that generates two catenated daughter chromosomes. Remarkably, this partial separation of the chromatids occurs also when the mitotic spindle has been disrupted through treatment with colchicine (49). The second step is a pulling at the start of anaphase by the spindle, which is coupled with and gives direction to the strand-passing reaction of topoisomerase II. In this second step, the chromatids are pulled and moved over relatively large distances.

It has been suggested that the driving force for an initial disentanglement of the DNA strands could be given by the process of DNA condensation as such (44) and that as compaction proceeds the dissolution of the sister chromatids becomes more efficient (24). The condensation could be obtained with the help of force-generating systems, possibly involving SMC (for structural maintenance of chromosome) proteins that act as chromatin motors (32) and topoisomerase II (23). In addition, the generation of superhelical tension by wrapping the DNA around multisubunit protein complexes (condensin [29]) has been suggested as a mechanism to compact the chromatin fiber. It is interesting that SMC proteins have a nucleotide binding motif and extended coiled-coil domains that bear structural similarity to proteins occurring in the cytoplasm of *Salmonella enterica* (TlpA [33]), *E. coli* (MukB [56]), and *Mycoplasma hyorhinis* (P115 [58]).

For the disentanglement of chromatin fibers, the strand-passing activity of topoisomerase II is essential. However, this enzyme can only be functional if it is given a direction for strand passing (for a theoretical analysis, see reference 27). What mechanism can give force and direction to the activity of these enzymes if not the traction force of the spindle? How can mere condensation of entangled fibers lead to a dissolution of the sister chromatids? Here, an analogy with the transcription-translation-mediated separation of daughter strands in bacteria could suggest a possible mechanism.

The continuing process of transcription during G_2 or even during S phase and a similar phenomenon of phase separation between nucleosomal DNA and hnRNP particles as observed in the bacterial cell (Fig. 3B) and as suggested by Cremer et al. (9) could provide such a driving force. The formation of a domain of transcriptional activity on one repli-

cated daughter strand and the fusion of this domain with other transcriptional domains at the border of a chromosome territory could result in a pulling that gives directionality to the strand-passing activity of topoisomerases and separate local entanglements between the daughter strands (Fig. 5A). At a later stage, the condensation of the daughter strand could again exert a pulling force, with the hnRNP particles locally separating the chromatids and allowing them to slide past each other (Fig. 5B). An indication of a structural role for continuous transcriptional activity in nuclear organization was obtained from experiments with transcriptional inhibitors which induced a dispersion of chromosome territories (20). The occurrence in mammalian nuclei of sep-

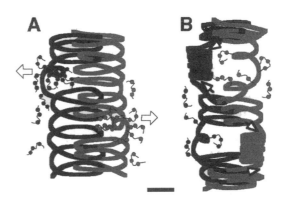

FIGURE 5 (A) Recently replicated sister chromatids, drawn as solid and shaded 30-nm-diameter fibers, are locally disentangled by topoisomerases and a pulling and pushing force exerted by transcriptional microcompartments. These are formed around one daughter strand by the activities of cotranscriptional splicing and transport of pre-mRNA. Compare this with the initial separation of replicated *oriC* in a bacterium (Fig. 4B). The force is directed perpendicularly to the long axis of the chromatids (open arrows). (B) Condensation of the DNA in two separated (euchromatin) regions exerts a pulling force on the chromatids (small arrows) and gives directionality for further disentanglement by topoisomerases. Transcription may separate the strands locally, allowing them to slide past each other to the regions of condensation. The cohesion between sister chromatids is shown here as the result of residual entanglements, but it could also be achieved by specific binding proteins (32) (Bar, 150 nm.)

arate replication and transcription sites grouped into distinct clusters has been described by Wansink et al. (73) and Wei et al. (75). The dynamic interdigitation of these transcription and replication domains may reflect the transition from one functional state to the other and serve to separate and disentangle the replicated DNA strands.

CONCLUSION

In a bacterial cell, chromosomal DNA occurs in a compact nucleoid region separated from the cytoplasmic phase. During segregation, the geometrical centers of the duplicated nucleoids have to move apart in the elongating cell only over a distance equal to the length increase of the cell during its growth cycle. Thus, the doubling in cell length already suffices to move the compact nucleoids apart, assuming that they move the maximum distance. In the eukaryotic cell, however, the nucleus does not grow along a linear axis, as in a (rod-shaped) bacterium. During S phase it grows isometrically rather than linearly. Moreover, replicated chromosomes have to be transported within the cell over distances that largely exceed the diameter of the nucleus. Even in the small yeast cell, chromosomes are transported over more than 6 μm (7), whereas this distance is less than 2 μm in a bacterium (Fig. 1B).

In eukaryotes, partitioning of the chromosomes over the daughter cells occurs in two steps: first, the separation of replicated daughter strands, involving both disentanglement and decatenation; and second, the transport or movement of separated chromatids to opposite poles by the cytoskeletal spindle apparatus in the abrupt process of mitotic segregation. In bacteria, segregation takes place in one step, in which physical strand separation occurs continuously and gradually along with replication and cell growth. The hierarchical mode of segregation (12) implies that soon after termination of a round of replication, the daughter nucleoids are structurally separated, while they can have started new rounds of replication (Fig. 2B).

In the bacterial cell, the process of strand separation is mediated by transcription in combination with supercoiling and phase separation between nucleoid and cytoplasm. Together with cell elongation, transcriptional activities represent an expansion force which is sufficient to fully disentangle the replicated chromosomes and to position them in the prospective daughter cells. In the eukaryotic cell, a similar phase separation between chromatin and hnRNP particles may help in the process of chromatin condensation and may cause an initial disentanglement of the sister chromatids and their partial separation over a small distance perpendicular to the long axes of the chromatids (Fig. 5A). The distance of this separation is comparable to that in the bacterium (Fig. 1B). However, because eukaryotic cells are large relative to bacterial cells, they require additional mitotic transport mechanisms to fully dissolve the chromatids and to bring them to the daughter cells.

While many microbiologists compare DNA segregation in bacteria with the second step of eukaryotic segregation, i.e., mitosis, it should rather be compared with the disentanglement or separation of sister chromatids occurring during the first step. As in bacteria, this separation could start soon after replication initiation (Fig. 5A) and continue throughout G_2 phase until prophase, when it is sustained and enhanced by chromosome condensation (Fig. 5B). This suggests that a transcription-mediated process could be the driving force behind chromosome separation in both prokaryotes (Fig. 3B and 4B) and eukaryotes (Fig. 5). In both organisms, therefore, the genomic organization may help in chromosome segregation, the expression of genes fulfilling a role in the fundamental process of daughter strand separation.

ACKNOWLEDGMENTS

We thank N. Nanninga and J. Veuskens for critical reading of the manuscript and H. Oud, A. Houtsmuller, J.-L. Sikorav, and A. C. Fijnvandraat for helpful discussions.

REFERENCES

1. **Alberts, B., D. Bray, J. Lewis, M. Raff, K. Roberts, and J. Watson.** 1994. *Molecular Biology of the Cell*, 3rd ed. Garland Publishers, New York, N.Y.

2. **Begg, K., and W. D. Donachie.** 1991. Experiments on chromosome separation and positioning in *Escherichia coli*. *New Biol.* **3:**475–486.

3. **Belmont, A. S., M. B. Braunfeld, J. W. Sedat, and D. A. Agard.** 1989. Large-scale chromatin structural domains within mitotic and interphase chromosomes in vivo and in vitro. *Chromosoma* **98:**129–143.

4. **Belmont, A. S., and K. Bruce.** 1994. Visualization of G1 chromosomes: a folded, twisted, supercoiled chromonema model of interphase chromatid structure. *J. Cell Biol.* **127:**287–302.

5. **Bloomfield, V. A.** 1996. DNA condensation. *Curr. Opin. Struct. Biol.* **6:**334–341.

6. **Bohrmann, B., M. Haider, and E. Kellenberger.** 1993. Concentration evaluation of chromatin in unstained resin-embedded sections by means of low-dose ratio-contrast imaging in STEM. *Ultramicroscopy* **49:**235–251.

7. **Byers, B., and L. Goetsch.** 1975. Duplication of spindle plaques in the cell cycle and conjugation of *Saccharomyces cerevisiae*. *J. Bacteriol.* **124:**511–523.

8. **Cooper, S., and C. E. Helmstetter.** 1968. Chromosome replication and the division cycle of Escherichia coli B/r. *J. Mol. Biol.* **31:**519–540.

9. **Cremer, T., A. Kurz, R. Zirbel, S. Dietzel, B. Rinke, E. Schrock, M. R. Speicher, U. Mathieu, A. Jauch, P. Emmerich, H. Scherthan, T. Ried, C. Cremer, and P. Lichter.** 1993. Role of chromosome territories in the functional compartmentalization of the cell nucleus. *Cold Spring Harbor Symp. Quant. Biol.* **58:**777–792.

10. **Daneholt, B.** 1997. A look at messenger RNP moving through the nuclear pore. *Cell* **88:**585–588.

11. **De Jong, L., M. A. Grande, K. A. Mattern, W. Schul, and R. Van Driel.** 1996. Nuclear domains involved in RNA synthesis, RNA processing and replication. *Crit. Rev. Eukaryot. Gene Expr.* **6:**215–246.

12. **Donachie, W. D., S. Addinall, and K. Begg.** 1995. Cell shape and chromosome partition in prokaryotes or, why *E. coli* is rod-shaped and haploid. *Bioessays* **17:**569–576.

13. **Duplantier, B., G. Jannink, and J.-L. Sikorav.** 1995. Anaphase chromatid motion: involvement of type II DNA topoisomerases. *Biophys. J.* **69:**1596–1605.

14. **Eils, R., S. Dietzel, E. Bertin, E. Schrock, M. R. Speicher, T. Ried, M. RobertNicoud, C. Cremer, and T. Cremer.** 1996. Three-dimensional reconstruction of painted human interphase chromosomes: active and inactive X chromosome territories have similar volumes but differ in shape and surface structure. *J. Cell Biol.* **135:**1427–1440.

15. **Glaser, P., M. E. Sharpe, B. Raether, M. Perego, K. Ohlsen, and J. Errington.** 1997. Dynamic, mitotic-like behavior of a bacterial protein required for accurate chromosome partitioning. *Genes Dev.* **11:**1160–1168.

16. **Gordon, G. S., D. Sitnikov, C. D. Webb, A. Teleman, A. Straight, R. Losick, A. W. Murray, and A. Wright.** 1997. Chromosome and low copy plasmid segregation in E. coli: visual evidence for distinct mechanisms. *Cell* **90:**1113–1121.

17. **Graumann, P. L., R. Losick, and A. V. Strunnikov.** 1998. Subcellular localization of *Bacillus subtilis* SMC, a protein involved in chromosome condensation and segregation. *J. Bacteriol.* **180:**5749–5755.

18. **Guacci, V., E. Hogan, and D. Koshland.** 1994. Chromosome condensation and sister chromatid pairing in budding yeast. *J. Cell Biol.* **125:**517–530.

19. **Guacci, V., D. Koshland, and A. Strunnikov.** 1997. A direct link between sister chromatid cohesion and chromosome condensation revealed through the analysis of *MCD1* in S. cerevisiae. *Cell* **91:**47–57.

20. **Haaf, T., and D. C. Ward.** 1996. Inhibition of RNA polymerase II transcription causes chromatin decondensation, loss of nucleolar structure, and dispersion of chromosomal domains. *Exp. Cell Res.* **224:**163–173.

21. **Helmstetter, C. E.** 1996. Timing of synthetic activities in the cell cycle, p. 1627–1639. *In* F. C. Neidhardt, R. Curtiss III, J. L. Ingraham, E. C. C. Lin, K. B. Low, B. Magasanik, W. S. Reznikoff, M. Riley, M. Schaechter, and H. E. Umbarger (ed.), Escherichia coli *and* Salmonella: *Cellular and Molecular Biology*, 2nd ed., vol. 2. American Society for Microbiology, Washington, D.C.

22. **Hiraga, S.** 1992. Chromosome and plasmid partition in *Escherichia coli*. *Annu. Rev. Biochem.* **61:**283–306.

23. **Hirano, T., and T. J. Mitchison.** 1993. Topoisomerase II does not play a scaffolding role in the organization of mitotic chromosomes assembled in *Xenopus* egg extracts. *J. Cell Biol.* **120:**601–612.

24. **Hirano, T.** 1995. Biochemical and genetic dissection of mitotic chromosome condensation. *Trends Biol. Sci.* **20:**357–361.

25. **Jackson, D. A., A. B. Hassan, R. J. Errington, and P. R. Cook.** 1993. Visualization of focal sites of transcription within human nuclei. *EMBO J.* **12:**1059–1065.

26. **Jackson, D. A., and P. R. Cook.** 1995. The structural basis of nuclear function. *Int. Rev. Cytol.* **162A:**125–143.

27. **Jannink, G., B. Duplantier, and J.-L. Sikorav.** 1996. Forces on chromosomal DNA during anaphase. *Biophys. J.* **71:**451–465.

28. **Kellenberger, E.** 1991. Functional consequences of improved structural information on bacterial nucleoids. *Res. Microbiol.* **142:**229–238.

29. **Kimura, K., and T. Hirano.** 1997. ATP-dependent positive supercoiling of DNA by 13S condensin: a biochemical implication for chromosome condensation. *Cell* **90:**625–634.

30. **Kiseleva, E., T. Wurtz, N. Visa, and B. Daneholt.** 1994. Assembly and disassembly of spliceosomes along a specific pre-messenger RNP fiber. *EMBO J.* **13:**6052–6061.

31. **Kornberg, A., and T. A. Baker.** 1992. *DNA Replication*, 2nd ed. W. H. Freeman & Co., New York, N.Y.

32. **Koshland, D., and A. Strunnikov.** 1996. Mitotic chromosome condensation. *Annu. Rev. Cell Dev. Biol.* **12:**305–333.

33. **Koski, P., H. Saarilahti, S. Sukupolvi, S. Taira, P. Riikonen, K. Österlund, R. Hurme, and M. Rhen.** 1992. A new a-helical coiled-coil protein encoded by the *Salmonella typhimurium* virulence plasmid. *J. Biol. Chem.* **267:**12258–12265.

34. **Kurz, A., S. Lampel, J. E. Nickolenko, J. Bradl, A. Benner, R. M. Zirbel, T. Cremer, and P. Lichter.** 1996. Active and inactive genes localize preferentially in the periphery of chromosome territories. *J. Cell Biol.* **135:**1195–1205.

35. **Lawrence, J. B., C. A. Vilinave, and R. H. Singer.** 1988. Sensitive, high-resolution chromatin and chromosome mapping in situ: presence and orientation of two closely integrated copies of EBV in a lymphoma line. *Cell* **52:**51–61.

36. **Lerman, L. S.** 1971. A transition to a compact form of DNA in polymer solutions. *Proc. Natl. Acad. Sci. USA* **68:**1886–1890.

37. **Lewis, P. J., and J. Errington.** 1997. Direct evidence for active segregation of *oriC* regions of the *Bacillus subtilis* chromosome and co-localization with the Spo0J partition protein. *Mol. Microbiol.* **25:**945–954.

38. **Lin, D. C.-H., P. A. Levin, and A. Grossman.** 1997. Bipolar localization of a chromosome partition protein in *Bacillus subtilis*. *Proc. Natl. Acad. Sci. USA* **94:**4721–4726.

39. **Luirink, J., and B. Dobberstein.** 1994. Mammalian and Escherichia coli signal recognition particles. *Mol. Microbiol.* **11:**9–13.

40. **Lutkenhaus, J.** 1993. FtsZ ring in bacterial cytokinesis. *Mol. Microbiol.* **9:**404–409.

41. **Lynch, S., and J. C. Wang.** 1993. Anchoring of DNA to the bacterial cytoplasmic membrane through cotranscriptional synthesis of polypeptides encoding membrane proteins or proteins for export: a mechanism of plasmid hypernegative supercoiling in mutants deficient in DNA topoisomerase I. *J. Bacteriol.* **175:**1645–1655.

42. **Manuelidis, L.** 1990. A view of interphase chromosomes. *Science* **250:**1533–1540.

43. **Manuelidis, L., and T. L. Chen.** 1990. A unified model of eukaryotic chromosomes. *Cytometry* **11:**8–25.

44. **Marko, J. F., and E. D. Siggia.** 1995. Statistical mechanics of supercoiled DNA. *Phys. Rev. E* **52:**2912–2938.

45. **Marsden, M. P. F., and U. K. Laemmli.** 1982. Metaphase chromosome structure: evidence for a radial loop model. *Cell* **17:**849–858.

46. **Mason, D. J., and D. M. Powelson.** 1958. Nuclear division as observed in live bacteria by a new technique. *J. Bacteriol.* **71:**474–479.

47. **Mehlin, H., and B. Daneholt.** 1993. The Balbiani ring particle: a model for the assembly and export of RNPs from the nucleus? *Trends Cell Biol.* **3:**443–447.

48. **Minsky, A., R. Ghirlando, and Z. Reich.** 1997. Nucleosomes: a solution to a crowded intracellular environment? *J. Theor. Biol.* **188:**379–385.

49. **Miyazaki, W. Y., and T. L. Orr-Weaver.** 1994. Sister-chromatid cohesion in mitosis and meiosis. *Annu. Rev. Genet.* **28:**167–187.

50. **Mohl, D. A., and J. W. Gober.** 1997. Cell cycle-dependent polar localization of chromosome partitioning proteins in Caulobacter crescentus. *Cell* **88:**675–684.

51. **Murphy, L. D., and S. B. Zimmerman.** 1995. Condensation and cohesion of λ DNA in cell extracts and other media: implications for the structure and function of DNA in prokaryotes. *Biophys. Chem.* **57:**71–92.

52. **Nakayasu, H., and R. Berezney.** 1989. Mapping replication sites in the eucaryotic nucleus. *J. Cell Biol.* **108:**1–11.

53. **Nanninga, N.** 1998. Morphogenesis of *Escherichia coli*. *Microbiol. Mol. Biol. Rev.* **62:**1–20.

54. **Neidhardt, F. C., and H. E. Umbarger.** 1996. Chemical composition of *Escherichia coli*, p. 13–17. *In* F. C. Neidhardt, R. Curtiss III, J. L. Ingraham, E. C. C. Lin, K. B. Low, B. Magasanik, W. S. Reznikoff, M. Riley, M. Schaechter, and H. E. Umbarger (ed.), Escherichia coli *and* Salmonella: *Cellular and Molecular Biology*, 2nd ed., vol. 1. American Society for Microbiology, Washington, D.C.

55. **Newlon, C. S.** 1988. Yeast chromosome replication and segregation. *Microbiol. Rev.* **52:**568–601.

56. **Niki, H., R. Imamura, M. Kitaoka, K. Yamanaka, T. Ogura, and S. Hiraga.** 1992. *E. coli* MukB protein involved in chromosome partition forms a homodimer with a rod-and-hinge structure having DNA binding and ATP/GTP binding activities. *EMBO J.* **11:**5101–5109.

57. **Norris, V.** 1995. Hypothesis: chromosome separation in *E. coli* involves autocatalytic gene expression, transertion and membrane domain formation. *Mol. Microbiol.* **16:**1051–1057.

58. **Notarnicola, S. M., M. A. McIntosh, and K. S. Wise.** 1991. A *Mycoplasma hyorhinis* protein with sequence similarities to nucleotide binding enzymes. *Gene* **97:**77–85.

59. **Odijk, T.** 1998. Osmotic compaction of supercoiled DNA into a bacterial nucleoid. *Biophys. Chem.* **73:**23–30.

60. **Philpott, A., and G. H. Leno.** 1992. Nucleoplasmin remodels sperm chromatin in Xenopus egg extracts. *Cell* **69:**759–767.

61. **PinolRoma, S.** 1997. hnRNP proteins and nuclear export of mRNA. *Semin. Cell Dev. Biol.* **8:**57–63.

62. **Reich, Z., E. J. Wachtel, and A. Minsky.** 1994. Liquid-crystalline mesophases of plasmid DNA in bacteria. *Science* **264:**1460–1463.

63. **Schaechter, M., and U. von Freiesleben.** 1993. The equivalent of mitosis in bacteria, p. 61–73. *In* J. Heslop-Harrison and R. B. Flavell (ed.), *The Chromosome.* BIOS Scientific Publishers, Ltd., Oxford, England.

64. **Sharpe, M. E., and J. Errington.** 1996. The *Bacillus subtilis soj-spo0J* locus is required for a centromere-like function involved in prespore chromosome partitioning. *Mol. Microbiol.* **21:**501–509.

65. **Sharpe, M. E., and J. Errington.** 1999. Upheaval in the bacterial nucleoid: an active chromosome segregation mechanism. *Trends Genet.* **15:**70–74.

66. **Ulbrandt, N. D., J. A. Newitt, and H. D. Bernstein.** 1997. The E. coli signal recognition particle is required for the insertion of a subset of inner membrane proteins. *Cell* **88:**187–196.

67. **Van Helvoort, J. M. L. M., and C. L. Woldringh.** 1994. Nucleoid partitioning in *Escherichia coli* during steady state growth and upon recovery from chloramphenicol treatment. *Mol. Microbiol.* **13:**577–583.

68. **Van Helvoort, J. M. L. M., J. Kool, and C. L. Woldringh.** 1996. Chloramphenicol causes fusion of separated nucleoids in *Escherichia coli* K-12 cells and filaments. *J. Bacteriol.* **178:**4289–4293.

69. **Van Helvoort, J. M. L. M., P. G. Huls, N. O. E. Vischer, and C. L. Woldringh.** 1998. Fused nucleoids resegregate faster than cell elongation in *Escherichia coli pbpB*(Ts) filaments after release from chloramphenicol inhibition. *Microbiology* **44:**1309–1317.

70. **Wake, R. G., and J. Errington.** 1995. Chromosome partitioning in bacteria. *Annu. Rev. Genet.* **29:**41–67.

71. **Walter, H., and D. E. Brooks.** 1995. Phase separation in cytoplasm, due to macromolecular crowding, is the basis for microcompartmentation. *FEBS Lett.* **361:**135–139.

72. **Wansink, D. G., W. Schul, I. Van der Kraan, B. Van Steensel, R. Van Driel, and L. De Jong.** 1993. Fluorescent labeling of nascent RNA reveals transcription by RNA polymerase-II in domains scattered throughout the nucleus. *J. Cell Biol.* **122:**283–293.

73. **Wansink, D. G., E. E. M. Manders, I. van der Kraan, J. A. Aten, R. van Driel, and L. de Jong.** 1994. RNA polymerase II transcription is concentrated outside replication domains throughout S-phase. *J. Cell Sci.* **107:**1449–1456.

74. **Webb, C. D., A. Teleman, S. Gordon, A. Straight, A. Belmont, D. C. H. Lin, A. D. Grossman, A. Wright, and R. Losick.** 1997. Bipolar localization of the replication origin regions of chromosomes in vegetative and sporulating cells of B. subtilis. *Cell* **88:**667–674.

75. **Wei, X., J. Samarabandu, R. S. Devdhar, A. J. Siegel, R. Acharya, and R. Berezney.** 1998. Segregation of transcription and replication sites into higher order domains. *Science* **281:**1502–1505.

76. **Wheeler, R. T., and L. Shapiro.** 1997. Bacterial chromosome segregation: is there a mitotic apparatus? *Cell* **88:**577–579.

77. **Woldringh, C. L.** 1976. Morphological analysis of nuclear separation and cell division during the life cycle of *Escherichia coli. J. Bacteriol.* **125:**248–257.

78. **Woldringh C. L., and N. Nanninga.** 1985. Structure of nucleoid and cytoplasm in the intact cell, p. 161–197. *In* N. Nanninga (ed.), *Molecular Cytology of Escherichia coli.* Academic Press, London, England.

79. **Woldringh, C. L., P. G. Huls, and N. O. E. Vischer.** 1993. Volume growth of daughter and parent cells during the cell cycle of *Saccharomyces cerevisiae* a/a as determined by image cytometry. *J. Bacteriol.* **175:**3174–3181.

80. **Woldringh, C. L., A. Zaritsky, and N. B. Grover.** 1994. Nucleoid partitioning and the division plane in *Escherichia coli. J. Bacteriol.* **176:**6030–6038.

81. **Woldringh, C. L., P. R. Jensen, and H. V. Westerhoff.** 1995. Structure and partitioning of bacterial DNA: determined by a balance of compaction and expansion forces? *FEMS Microbiol. Lett.* **131:**235–242.

82. **Zirbel, R. M., U. R. Mathieu, A. Kurz, T. Cremer, and P. Lichter.** 1993. Evidence for a nuclear compartment of transcription and splicing located at chromosome domain boundaries. *Chromosome Res.* **1:**92–106.

COMPARING MICROBIAL GENOMES: HOW THE GENE SET DETERMINES THE LIFESTYLE

Michael Y. Galperin, Roman L. Tatusov, and Eugene V. Koonin

6

Sequencing of complete microbial genomes, pioneered in 1995 by J. C. Venter and colleagues (21), continues at an ever-increasing pace. At the time of this writing, complete sequences of one eukaryotic, four archaeal, and thirteen bacterial genomes were available (1, 5, 10, 12, 15, 21–23, 28, 29, 33, 35, 36, 43, 64, 65, 69). Many other genome projects are well under way, and by the year 2000 more than 50 microbial genomes are expected to be finished. Importantly, these genome projects span most of the major lineages in the prokaryotic phylogenetic tree as we know it today (54, 56) and include both free-living and parasitic organisms. This makes genome comparisons a particularly promising avenue of research that will profoundly affect every branch of biology.

The availability of complete genome sequences has had a major impact on our view of microbial evolution. Not so long ago, microorganisms were classified based on a number of easily measurable criteria, such as G+C content, that had little to do with the bio-chemical, physiological, or ecological properties of the organisms (73, 74). Introduction of the 16S rRNA-based grouping has put microbial phylogeny on a solid molecular basis (54, 72). On the other hand, phylogenetic trees based upon sequences of various proteins often contradicted each other and poorly correlated with 16S rRNA-based trees (see reference 9 for a recent review). Knowledge of the complete genomic sequences has redefined the task of assessing the evolutionary relationships between different groups of microorganisms. The sequences of all of the protein- and RNA-encoding genes, as well as all the intrinsic properties of the genome organization, such as codon usage, open reading frame (ORF) density, operon structure, sequences of promoters, and ribosome-binding sites, became amenable to exact numerical analysis, and the problem has shifted to deciding which set of parameters to rely on for a particular task. It has also become possible to compare organisms on the basis of redundancy of the complete protein sets encoded in their genomes. In addition, as shown below, organisms can be compared by the functions (biosynthetic pathways, DNA repair mechanisms, etc.) that are not encoded in their genomes and whose absence must be compensated for by adaptive changes in the lifestyles of these

Michael Y. Galperin, Roman L. Tatusov, and Eugene V. Koonin, National Center for Biotechnology Information, National Library of Medicine, National Institutes of Health, Bethesda, MD 20894.

Organization of the Prokaryotic Genome, Edited by Robert L. Charlebois,
© 1999 American Society for Microbiology, Washington, D.C.

organisms. Here, we review the ways one can compare microbial genomes and discuss the first results gained by using these approaches.

GENERAL ORGANIZATION OF MICROBIAL GENOMES

Comparative analysis of the complete genomes of several diverse microorganisms on the basis of such properties as codon usage (see, e.g., www.genome.ad.jp/kegg/codon_table/index.html), ORF density, and the lengths of coding regions shows many common trends in their organization. The information density is usually very high, with ~90 to 95% of the DNA coding for proteins or stable RNA. The distances between genes vary, but there seems to be no evidence for any regulatory function of larger intergenic regions. The ORF length distributions (Fig. 1) are generally similar in all prokaryotes, particularly the ones that belong to the same phylogenetic lineages, such as gram-positive bacteria (Fig. 1B) and archaea (Fig. 1C). *Saccharomyces cerevisiae*, the only eukaryote with a completely sequenced genome at this time, contains a larger number of genes encoding long, multidomain proteins than ORFs encoding 150 to 300 amino acid residues (Fig. 1C). Yeast also appears to contain a significant number of ORFs encoding ca. 100 amino acids (14), which have not been experimentally characterized or even shown to be expressed. Whether most of these predicted ORFs are artifactual or code for genuine proteins remains to be seen (4, 14).

Cross-genome comparisons on the DNA level have been mostly aimed at assessing the degree of nonrandomness of the genome structure by statistical analysis of compositional biases in chromosomal DNAs. Karlin et al. (34) compared the distributions of short oligonucleotides in prokaryotic genomes and revealed significant biases in the relative abundances of certain di- and tetranucleotides. These data were used to develop the concept of the "genome signature" and to infer possible evolutionary relationships among various groups of prokaryotes. Palindromic tetranucleotides (most notably CTAG) were found to be underrepresented in bacterial genomes, particularly in *Haemophilus influenzae*. Gelfand and Koonin (26) showed that short palindromic sequences (4, 5, or 6 bp) are also avoided in *Methanococcus jannaschii* and *Synechocystis* sp. and, to a lesser extent, in *Mycoplasma genitalium*, while they are normally represented in the genomes of chloroplasts and mitochondria. This effect was correlated with the presence in these organisms, but not in organelles, of restriction-modification systems with the respective specificities. While the reasons for over- or underrepresentation of nonpalindromic oligonucleotides, identified by Karlin and coworkers (34), still remain unclear, it probably has to do with the fact that ~90% of bacterial DNA comprises RNA- or protein-coding regions (5, 21, 23), which imposes certain constraints on their composition. This underscores the importance of comparing the RNAs and proteins encoded in each complete genome. While comparisons of microorganisms on the basis of their RNA sequences form the core of the current understanding of organismic evolution (74), they are beyond the scope of this chapter. The recent developments in this area have been covered in the classical, comprehensive treatise by Woese (72) and several recent reviews (49, 56).

LACK OF GENE ORDER CONSERVATION IN EVOLUTION

Comparison of the complete genomes of closely related bacterial species, *M. genitalium* and *Mycoplasma pneumoniae*, showed a significant degree of synteny between these organisms (30). At larger phylogenetic distances, however, the conservation of the gene order progressively decreases. The *Escherichia coli* and *H. influenzae* genomes, for example, contain many conserved gene strings, corresponding to *E. coli* operons, but there appears to be no detectable conservation of the gene order at the genome scale (37, 68, 71). The genome organization is even less conserved in more distant bacterial genomes and involves only a relatively small number of essential operons

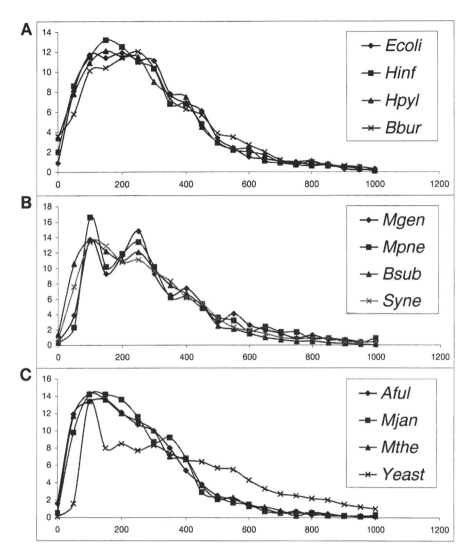

FIGURE 1 Lengths of proteins encoded in completely sequenced genomes. The lengths of predicted ORFs were averaged over 50-amino-acid intervals and plotted as fractions of the total number of ORFs predicted in each particular organism. The plots were normalized so that the total area under each curve is the same. Data for *B. burgdorferi*, *Synechocystis* sp., and *S. cerevisiae* were added to the plots of length distributions in proteobacteria (A), low-G+C gram-positive bacteria (B), and archaea (C), respectively, for illustrative purposes. *Hinf, H. influenzae; Hpyl, H. pylori; Bbur, B. burgdorferi; Mgen, M. genitalium; Mpne, M. pneumoniae; Bsub, B. subtilis; Syne, Synechocystis* sp.; *Aful, A. fulgidus; Mjan, M. jannaschii; Mthe, M. thermoautotrophicum.*

(13, 40, 52). Gene order conservation between bacteria and archaea is limited mostly to ribosomal protein operons, genes for two subunits of the DNA-dependent RNA polymerase, and some ion transport operons (Table 1). In yeast, the only eukaryotic genome available at this time, there appears to be no conservation of prokaryotic

ble 1). Even in these few conserved operons, the alignment of genes is typically distorted by insertions and deletions. In yeast, the only eukaryotic genome available at this time, there appears to be no conservation of prokaryotic

TABLE 1 Universally conserved gene strings (operons) in bacteria and archaea

Conserved gene string	Gene organization									
	E. coli	*H. influenzae*	*H. pylori*	*B. subtilis*	*M. genitalium, M. pneumoniae*	*B. burgdorferi*	*Synechocystis* sp.	*M. thermoautotrophicum*	*A. fulgidus*	*M. jannaschii*
Ribosomal proteins										
spc operon	*rplNXE-rpsNH-rplFR-rpsE-rpmD-rplO-secY-rpmJ*	Same as in *E. coli*	*rpmD* missing; insertion of *map, infA*	Insertion of *adk, map, infA*	*rpmD* missing; insertion of *adk, map, infA*	Same as in *E. coli*	*rpmD* missing; *rpsN* translocated	*rpmJ* missing; two insertions	*rpmJ* missing; two insertions	*rpmJ* missing; two insertions
S10 operon	*rpsJ-rplCDWB-rpsS-rplV-rpsC-rplP-rpmC-rpsQ*	Same as in *E. coli*	Same as in *E. coli*	Same as in *E. coli*	Same as in *E. coli*	Same as in *E. coli*	*rpsJ* translocated	*rpsJ, rplP* translocated; one insertion	*rpsJ, rplP* translocated; one insertion	*rpsJ, rplP* translocated; one insertion, operon split into two
α operon	*rpsMKD-rpoA-rplQ*	Same as in *E. coli*	Same as in *E. coli*	*rpsD* translocated	*rpsD* translocated	*rpsD* translocated	*rpsD* translocated	*rplQ* missing; *rpsD* and *rpsK* in inverse order	*rplQ* missing; *rpsD* and *rpsK* in inverse order	*rplQ* missing; *rpsD* and *rpsK* in inverse order
str operon	*rpsLG-fusA-tufA*	Same as in *E. coli*	*tufA* translocated	Same as in *E. coli*	*tufA* translocated	*fusA, tufA* translocated	Same as in *E. coli*	Same as in *E. coli*	*tufA* translocated	*tufA* translocated
L13 operon	*rplM-rpsI*	Same as in *E. coli*	Same as in *E. coli*	Same as in *E. coli*	Same as in *E. coli*	Same as in *E. coli*	Same as in *E. coli*	Same as in *E. coli*	Same as in *E. coli*	Same as in *E. coli*
RNA polymerase *β-* and *β'*-subunits	*rpoBC*	Same as in *E. coli*	Same as in *E. coli*	Same as in *E. coli*	Same as in *E. coli*	Same as in *E. coli*	*rpoC* is split into two, *rpoC1* translocated	*rpoB* and *rpoC* each split into two subunits	*rpoB* and *rpoC* each split into two subunits	*rpoB* and *rpoC* each split into two subunits
Phosphate transport operon	*pstSCAB-phoU*	*phoU* missing	Operon missing	*phoU* missing; *pstB* duplicated	*pstC* missing	*phoU* translocated	Two copies of the operon; single *phoU* translocated	Same as in *E. coli*; *psiS* and *phoU* duplicated	*phoU* truncated	*phoU* truncated

gene order. Even the genes coding for mito-chondrial (bacterial-type) ribosomal proteins that were apparently transferred to the yeast nucleus from a promitochondrial symbiont relatively recently are found on different chromosomes in no particular order. The yeast genes YDR012w (*rplD* homolog) and YDR025w (*rpsQ* homolog), located on chromosome IV not far from each other, appear to be the only remnants of the S10 operon that is fairly well conserved in bacteria and archaea (Table 1).

In some cases, however, archaea and bacteria display a very high level of gene order conservation. For example, the operons coding for vacuolar (archaeal)-type H$^+$ (or Na$^+$)-transporting ATP synthase (V-ATPase) in the bacteria *Enterococcus hirae* and *Thermus aquaticus* are almost identical to the archaeal operons (Table 2). *Borrelia burgdorferi*, *Treponema pallidum*, and *Chlamydia trachomatis* have a slightly different order of the V-ATPase genes, most of which are still organized in an operon. Such cases are likely to reflect relatively recent lateral gene transfer between bacterial and archaeal genomes. The operon coding for F$_0$F$_1$-type ATP synthase is also well conserved in most bacteria (Table 2). It is noteworthy that this operon is least conserved in *Aquifex aeolicus*, which is believed to belong to one of the earliest branches of the bacterial phylogenetic tree (6, 58). This observation leads to an interesting question regarding the origin of the the F$_0$F$_1$-type ATP synthase operon. On one hand, it could have been present in the last common ancestor of all bacteria and have been disrupted only in the evolution of the particular branch leading to *A. aeolicus*; on the other hand, genes coding for different subunits of the F$_0$F$_1$-type ATP synthase could have evolved independently and been organized in an operon after the branching of *Aquifecales*. Genes coding for several subunits of F$_0$F$_1$-type ATPase and for the V-type ATPase are present in the yeast nucleus (Table 2), but they are spread over 11 different chromosomes and are never adjacent.

An unusual case of gene order conservation involves two highly conserved genes that are induced under conditions of nutritional stress (7, 25, 57). These genes form operons in the bacteria *Bacillus subtilis* and *H. influenzae* and in the archaea *Archaeoglobus fulgidus* and *Pyrococcus horikoshii* but are unlinked in *M. jannaschii* and *Methanobacterium thermoautotrophicum*. Remarkably, in yeast these genes are adjacent but are transcribed in opposite directions and are likely to be coregulated (57).

The observation that genes that form an operon in one organism may be dispersed in another organism points to the crucial problem in analyzing operon evolution: is operon organization of genes ancestral or derived? The presence of homologous ribosomal operons in bacteria and archaea (Table 1) (2, 3, 32, 47, 50, 60) suggests that the last common ancestor of all modern organisms possessed at least some operon organization. On the other hand, many operons might have been disseminated by horizontal transfer and, in particular, could have entered the archaeal world by transfer from bacteria (see references 31, 40, and 46 for a discussion).

UNIVERSALLY CONSERVED PROTEIN DOMAINS

The original estimates of the fraction of gene products in the sequenced genomes that had detectable homologs in protein sequence databases ranged from ~78% for *H. influenzae* to ~44% for *M. jannaschii* (10, 21). Subsequent, more detailed analyses, however, revealed very similar degrees of protein sequence conservation among bacterial and archaeal genomes. Statistically significant sequence similarities to other proteins and/or conserved sequence motifs were detected for 73% of the *M. jannaschii* gene products and for 75 to 90% of bacterial gene products (40). The fraction of proteins that contain ancient conserved regions (regions conserved over large evolutionary distances) appeared to be nearly the same at ~70%. This ratio between highly conserved and more variable proteins may reflect the balance between the proteins with

TABLE 2 Conservation of ATP synthase operons in microbial genomes

Organism	F_0F_1-type ATP synthase genes[a]	Operon organization	Archaeal/vacuolar type ATP synthase genes[a]	Operon organization
Bacteria				
E. coli	atpBEFHAGDC	Single operon	None	None
H. influenzae	HI0485-HI0478[b]	Same as in E. coli	None	None
H. pylori	HP0828, HP1212, HP1137-HP1131	atpB and atpE translocated	None	None
B. subtilis	atpBEFHAGDC	Same as in E. coli	None	None
M. genitalium	MG405-MG399	Same as in E. coli	None	None
M. pneumoniae	MP238-MP244	Same as in E. coli	None	None
Synechocystis sp.	sll1322-sll1327; slr1329-slr1330	Operon split into two; atpD, atpC inverted	None	None
A. aeolicus	aq_179; aq_177; aq_1586-aq_1588; aq_679; aq_2041-aq_2038; aq_673	Operon split into separate genes; only atpFH and atpGD form operons	None	None
E. hirae	atpBEFHAGDC	Same as in E. coli	ntpIKECGABD	Single operon
B. burgdorferi	None	None	BB0091-BB0090; BB0096; BB00094-BB0092	ntpIK and ntpE translocated; ntpC and ntpG missing
C. trachomatis	None	None	CT0158-CT0157; CT0163; CT0161-CT0159	ntpIK and ntpE translocated; ntpC and ntpG missing
Thermus thermophilus	Not known[c]	Not known	vtpDLEGHABI	Same as in E. hirae

Archaea				
M. jannaschii	None	MJ0222-MJ0216; MJ0615		*ntpD* translocated
M. thermoautotrophicum	None	MTH960-MTH953		Same as in *E. hirae*
A. fulgidus	None	AF1159-AF1160; AF1162-AF1168		Interrupted by an insertion
P. horikoshii	None	PHBD017-PHBD028		Same as in *E. hirae*
Methanosarcina mazei	Not known[c]	*ahaIKECFABD*		Same as in *E. hirae*
Desulfurococcus sp.	Not known[c]	*ntpKECCGABD*		Same as in *E. hirae* (*ntpI* not sequenced)
Eukaryotes				
S. cerevisiae	YKL016c; YDR298c; YBL099w; YBR039w; YJR121w; YDL004w	YOR270c; YMR054w; YEL027w; YPL234c; YHR026w; YNL091w; YLR447c; YGR020c; YDL185w; YBR127c; YEL051w	Mitochondrial enzyme; separate genes on different chromosomes; *apB*, *apE*, and *apF* are mitochondrial genes; YKL016c is not homologous to *atpD*	Vacuolar enzyme; separate genes on different chromosomes

[a] The ATP synthase gene sequences were compared with the operons from *E. coli* (70) and *E. hirae* (66). The leading genes of both operons (*atpI* and *ntpF*) were omitted, as it was not known whether their products are actual ATP synthase subunits.

[b] Genes are listed under their original authors' designations.

[c] As complete genomic sequences of these organisms are not available, the presence of F_0F_1-type ATP synthase has not been ruled out.

conserved, housekeeping roles and those that are specific for each particular genus or species.

HOMOLOGOUS PROTEINS IN DIFFERENT MICROORGANISMS

Homologous genes and their products can be classified into orthologs, related by vertical descent (e.g., speciation), and paralogs, related by duplication (19). Typically, in the course of evolution, orthologs retain the same function while paralogs acquire novel, usually similar or related functions. Paralogous genes constitute a considerable fraction, from ∼30 to ∼55%, of all genes in microbial genomes (8, 40, 42, 44, 45, 61, 68). The fraction of paralogs generally correlates with the genome size, i.e., it is greater in free-living bacteria that have larger genomes than their relatives with the parasitic lifestyle. Remarkably, in each of the sequenced prokaryotic genomes, the largest classes of paralogs are the same. These include ATPases and GTPases with the "Walker-type" NTP-binding motifs, NAD- or FAD (flavin adenine dinucleotide)-utilizing enzymes, and helix-turn-helix DNA-binding proteins (40). The few exceptions to this rule, such as the low number of helix-turn-helix DNA-binding proteins in *M. genitalium* and the abundance of FeS oxidoreductases in *M. jannaschii*, are clearly due to the ideosyncrasies in the lifestyles of these organisms. It is noteworthy, however, that these large classes of paralogs may contain unique protein families (e.g., putative ATPases in *M. jannaschii* [38]).

Identification of orthologous genes in different organisms is complicated by ostensibly vastly different rates of evolution in different protein families and in different taxa. However, only by comparing orthologs can one construct phylogenetic trees that reflect vertical evolutionary descent (20). In addition, identification of a group of orthologs allows one to extend the functional assignment made for one member to all the members of the group, thus greatly simplifying the functional annotation of sequenced genomes. Several classifications of orthologous proteins encoded in complete genomes have been constructed, based on all-against-all protein sequence comparisons (11, 55, 67). These systems usually rely on the high level of sequence similarity between orthologous proteins in different organisms (11, 55). In contrast, the system of clusters of orthologous groups (COGs) (41, 67) was designed to accommodate the vastly different rates of evolution observed for different genes and to identify the closest homologs in each of the sequenced genomes for each protein, even if the similarity is low and not statistically significant by itself. This approach relies upon the transitivity of orthologous relationships, i.e., the notion that any group of three or more genes from distant genomes which are more similar to each other than they are to any other genes from the same genomes is likely to belong to an orthologous family. The COG database (www.ncbi.nlm.nih.gov/COG) currently contains proteins from 12 complete genomes. The functional distributions of proteins in all of these genomes (Fig. 2) are similar, which probably reflects some general balance among various cellular functions in a living cell.

FIGURE 2 Functional roles of conserved proteins in completely sequenced genomes. Each pie chart represents the numbers of proteins in each genome that are not included in the current set of COGs (the largest sector in every genome), and clockwise, the numbers of proteins that are responsible for (i) information storage and processing, including transcription, translation, DNA replication, recombination, and repair, and ribosome biogenesis; (ii) cellular processes, such as membrane and cell wall biogenesis, protein folding, and secretion; (iii) cellular metabolism, including carbohydrate, amino acid, lipid, and nucleotide metabolism and energy production and conversion. The last sector indicates poorly characterized conserved proteins. The total number is larger than the number of proteins in each particular genome, as different domains of the same protein may be included in different COGs.

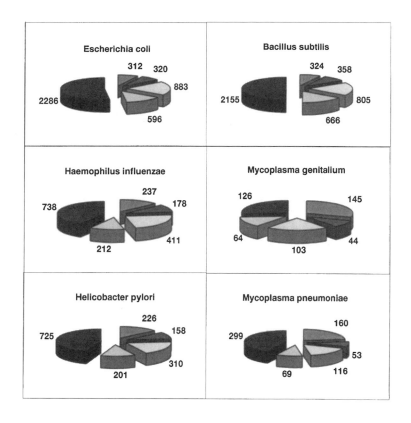

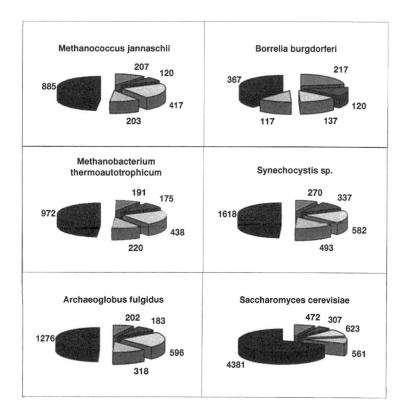

MISSING GENE FAMILIES AND EVOLUTION OF METABOLIC PATHWAYS

Comparison of the available complete genomes shows that metabolic diversity generally correlates with genome size. Pathogenic bacteria import a variety of metabolites from their hosts, which allows them to shed genes encoding enzymes for some of the metabolic pathways (21, 23, 29, 30, 68, 69). Superimposing the phylogenetic patterns in COGs onto the chart of metabolic pathways provides a convenient way to visualize the pathways that are likely to be operative or not operative in each particular organism and allows a systematic identification of functional systems that are missing in a particular phylogenetic lineage (Fig. 3 and 4) (41). Thus, using the COG approach for the analysis of carbohydrate metabolism in the gastric pathogen *Helicobacter pylori* revealed the absence of several key enzymes, such as phosphofructokinase, pyruvate kinase, and 6-phosphogluconate dehydrogenase (41). This meant that glycolysis and the pentose phosphate shunt are not functional in *H. pylori*, leaving the Entner-Doudoroff pathway as the only venue of sugar catabolism. On the other hand, phosphoenolpyruvate synthase and fructose bisphosphatase are both present in *H. pylori*, suggesting that it can perform gluconeogenesis, producing sugars which are required for nucleic acids and peptidoglycan biosynthesis. Such an organization of metabolism makes perfect sense for this bacterium, given the challenge of maintaining near-neutral intracellular pH in the highly acidic (pH 2 to 3) gastric environment. Sugar fermentation, which results in intracellular production of acid, would place an additional burden on the pH maintenance mechanism, while gluconeogenesis converts organic acids into sugars and thus removes H^+ from the cytoplasm. For the purposes of energy production, *H. pylori* apparently depends on fermentation of amino acids and oligopeptides that are produced by gastric proteolysis and transported into the cells by ABC-type transporters. Amino acid fermentation results

in alkalinization of the cytoplasm and could relieve part of the problem of pH maintenance in *H. pylori*.

Examining the phylogenetic distribution of the enzymes participating in pyrimidine (Fig. 3) and purine (Fig. 4) biosynthesis pathways brings even more interesting observations. The complete pyrimidine biosynthesis pathway is apparently missing in both *M. genitalium* and *M. pneumoniae*. As many other grampositive bacteria do possess the complete set of genes for pyrimidine biosynthesis (18, 27, 48, 59), it appears that the complete set of *pyr* genes, with the sole exception of *pyrH*, coding for uridylate kinase, has been precisely excised in the course of mycoplasmal evolution. Another interesting trend revealed by such a comparison is the presence of multiple cases of nonorthologous gene displacement (39). Indeed, while *pyrH*-type uridylate kinase is not encoded in the yeast genome (Fig. 3), UMP phosphorylation in yeast is catalyzed by one of the three paralogous members of the adenylate/cytidylate kinase family (63). Similarly, in addition to the dihydroorotase encoded by the *pyrC* gene, *E. coli, H. pylori,* and *Synechocystis* sp. also possess genes for a second type of this enzyme, which is found in gram-positive bacteria (59, 62) and is homologous to yeast allantoinase and human dihydropyrimidinase. This second form of dihydroorotase is the only one encoded in archaeal genomes, while neither form is encoded in the *H. influenzae* genome (Fig. 3). Remarkably, the *H. influenzae* genome contains the genes coding for all the reactions that lead from dihydroorotate to CTP. Thus, while this organism evidently is not capable of de novo pyrimidine biosynthesis, it has preserved sufficient metabolic plasticity to accommodate whatever pyrimidine it can get from its host. A similar trend can be seen in the even-smaller genomes of *B. burgdorferi, T. pallidum,* and *C. trachomatis,* which still contain genes coding for the downstream reactions of this pathway (Fig. 3).

The same trends can be seen in the purine biosynthetic pathway (Fig. 4). *M. genitalium,*

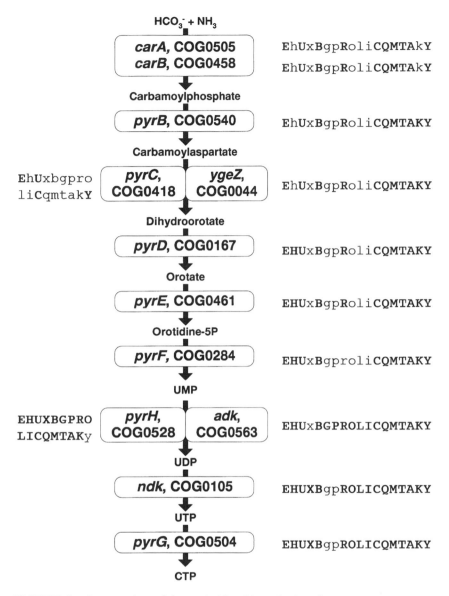

FIGURE 3 Conservation of the pyrimidine biosynthesis pathway in various microorganisms. The enzymes are listed under *E. coli* gene names. The COG numbers are from the COG database (www.ncbi.nlm.nih.gov/COG [67]). The designations of species in the phylogenetic patterns are as follows: E, *E. coli*; H, *H. influenzae*; U, *H. pylori*; X, *Rickettsia prowazekii*; B, *B. subtilis*; G, *M. genitalium*; P, *M. pneumoniae*; R, *Mycobacterium tuberculosis*; O, *B. burgdorferi*; L, *T. pallidum*; I, *C. trachomatis*; Q, *A. aeolicus*; C, *Synechocystis* sp.; M, *M. jannaschii*; T, *M. thermoautotrophicum*; A, *A. fulgidus*; K, *P. horikoshii*; Y, *S. cerevisiae*. The uppercase letters indicate organisms that have proteins in a corresponding COG; the lowercase letters indicate the organisms that are not represented in a given COG. Nonorthologous enzymes catalyzing the same biochemical reaction are shown side by side, where known. Different subunits or conserved domains of the same enzyme are shown in the same frame, one under the other.

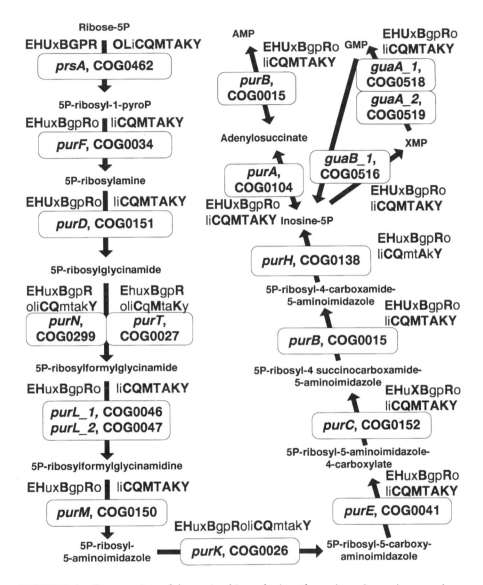

FIGURE 4 Conservation of the purine biosynthesis pathway in various microorganisms. For an explanation of the symbols, see the legend to Fig. 3.

M. pneumoniae, and *B. burgdorferi* do not have the genes that code for enzymes catalyzing the committed steps of this pathway. In contrast to pyrimidine biosynthesis, the pathway of de novo synthesis of purines is seen in *H. influenzae* but is missing in *H. pylori*. While *H. pylori* is capable of interconverting AMP and GMP, and of making them from IMP, most of the other *pur* genes are missing. In fact, the only

genes of the IMP biosynthesis part of the pathway found in *H. pylori* are *purD*, whose product is likely nonfunctional due to mutations in its active center (24), and *purB*, which is also required for purine interconversion. Thus, in the course of its evolution toward the parasitic lifestyle, *H. pylori* appears to have shed some of the *pur* genes while still preserving some degree of metabolic plasticity. Analysis of the

purine biosynthesis pathway in *M. jannaschii,
M. thermoautotrophicum*, and *A. fulgidus* points
out several still-unresolved problems in the reconstruction of the metabolisms of poorly
studied organisms, particularly extremophiles.
None of these three genomes, for example,
contains an ortholog of *purH*. While methanogenic archaea have a complex array of donors and acceptors of the C-1 residues and can
be expected to use a nonorthologous displacement for PurH, such an enzyme has not been
found by computer searches and its identification will probably require direct experimental analysis. The absence of the *purK*
ortholog in these archaea, however, does not
mean that this activity has to be assigned to
other members of the ATP-grasp superfamily
(24) that are encoded in their genomes. In
fact, carboxylation of 5-phosphoribosyl-5-
carboxyaminoimidazole can occur spontaneously (51), especially under the conditions
(high concentration of HCO_3^- and high temperature) that are optimal for the growth of
these organisms. These examples illustrate
how adaptation of a microorganism to certain
living conditions leads to changes in its genome and at the same time determines the set
of genes that the organism cannot afford to
lose. On the other hand, loss of certain metabolic functions makes the organism dependent on external sources for the missing compounds, mandating the symbiotic, parasitic, or
saprophytic lifestyle.

WHAT CAN WE EXPECT FROM
GENOME ANALYSIS?

Comparative analysis of the available microbial genomes reveals both conservation of protein families and diversity of gene repertoires
and gene organization among organisms that
belong to diverse phylogenetic lineages. While
surprises certainly lie ahead, there is little
doubt that the major protein families are already known or can be deduced from the
available sequences. On the other hand, exact
cellular functions of at least 200 conserved
protein families are still unknown, and for
some of them even a general prediction of bi-

ochemical activity does not seem possible at
this time (16, 67). An extensive experimental
effort will be needed to ascertain which of
these proteins participate, e.g., in cell division
or in still-unresolved steps of pyridoxin, bio-
tin, and thiamin biosynthetic pathways.

Genome comparisons suggest that in the
evolution of prokaryotes, horizontal gene
transfer has been common and intense. Analysis of available archaeal genomes shows that
while protein components of replication, transcription, and translation machinery resemble
their eukaryotic counterparts (17, 53), most of
the other proteins, including metabolic enzymes, proteins involved in cell wall biogenesis, and many uncharacterized proteins, are
much closer to their bacterial homologs (40,
64). It appears that multiple instances of horizontal gene transfer and perhaps even genome
fusion events were a major factor in evolution,
leading to the emergence of modern archaea
and eukaryotes. In many cases, bacterial genes
seem to have substituted for the original genes
of the archaeal-eukaryotic lineage, making
phylogenetic reconstructions extremely complicated. For the next breakthrough in understanding the early evolution of life, we might
have to wait for complete genomic sequences
of some deep-branching prokaryotes (e.g., ko-
rarchaeotes) and/or eukaryotes.

Barring completely unexpected discoveries
as new complete microbial genomes continue
to become available, the major focus will
probably shift from the identification of common, conserved protein families to the analysis
of the peculiarities of each organism related to
its adaptation to particular ecological conditions. Such analysis can be expected to produce a better insight into the unique metabolic properties of each type of human
pathogen, providing data for the design of
new antimicrobial agents, e.g., those that will
specifically target narrow groups of pathogenic
microorganisms, thus limiting the spread of
resistance to these agents.

Last but not least, the data obtained in the
course of sequencing of microbial genomes
provide invaluable information for the ongo-

ing eukaryotic genome projects. In the near-future, feedback from these projects will further increase the impact of comparative genomics.

REFERENCES

1. **Andersson, S. G., A. Zomorodipour, J. O. Andersson, T. Sicheritz-Ponten, U. C. Alsmark, R. M. Podowski, A. K. Naslund, A. S. Eriksson, H. H. Winkler, and C. G. Kurland.** 1998. The genome sequence of *Rickettsia prowazekii* and the origin of mitochondria. *Nature* **396:**133–140.

2. **Arndt, E.** 1990. Nucleotide sequence of four genes encoding ribosomal proteins from the 'S10 and spectinomycin' operon equivalent region in the archaebacterium *Halobacterium marismortui*. *FEBS Lett.* **267:**193–198.

3. **Auer, J., G. Spicker, and A. Bock.** 1989. Organization and structure of the Methanococcus transcriptional unit homologous to the *Escherichia coli* "spectinomycin operon." Implications for the evolutionary relationship of 70 S and 80 S ribosomes. *J. Mol. Biol.* **209:**21–36.

4. **Basrai, M. A., P. Hieter, and J. D. Boeke.** 1997. Small open reading frames: beautiful needles in the haystack. *Genome Res.* **7:**768–771.

5. **Blattner, F. R., G. Plunkett III, C. A. Bloch, N. T. Perna, V. Burland, M. Riley, J. Collado-Vides, J. D. Glasner, C. K. Rode, G. F. Mayhew, J. Gregor, N. W. Davis, H. A. Kirkpatrick, M. A. Goeden, D. J. Rose, B. Mau, and Y. Shao.** 1997. The complete genome sequence of *Escherichia coli* K-12. *Science* **277:**1453–1474.

6. **Bocchetta, M., E. Ceccarelli, R. Creti, A. M. Sanangelantoni, O. Tiboni, and P. Cammarano.** 1995. Arrangement and nucleotide sequence of the gene (*fus*) encoding elongation factor G (EF-G) from the hyperthermophilic bacterium *Aquifex pyrophilus*: phylogenetic depth of hyperthermophilic bacteria inferred from analysis of the EF-G/fus sequences. *J. Mol. Evol.* **41:**803–812.

7. **Braun, E. L., E. K. Fuge, P. A. Padilla, and M. Werner-Washburne.** 1996. A stationary-phase gene in *Saccharomyces cerevisiae* is a member of a novel, highly conserved gene family. *J. Bacteriol.* **178:**6865–6872.

8. **Brenner, S. E., T. Hubbard, A. Murzin, and C. Chothia.** 1995. Gene duplications in *H. influenzae*. *Nature* **378:**140.

9. **Brown, J. R., and W. F. Doolittle.** 1997. Archaea and the prokaryote-to-eukaryote transition. *Microbiol. Mol. Biol. Rev.* **61:**456–502.

10. **Bult, C. J., O. White, G. J. Olsen, L. Zhou, R. D. Fleischmann, G. G. Sutton, J. A. Blake, L. M. FitzGerald, R. A. Clayton, J. D. Gocayne, A. R. Kerlavage, B. A. Dougherty, J.-F. Tomb, M. D. Adams, C. I. Reich, R. Overbeek, E. F. Kirkness, K. G. Weinstock, J. M. Merrick, A. Glodek, J. L. Scott, N. S. M. Geoghagen, J. F. Weidman, J. L. Fuhrmann, D. Nguyen, T. R. Utterback, J. M. Kelley, J. D. Peterson, P. W. Sadow, M. C. Hanna, M. D. Cotton, K. M. Roberts, M. A. Hurst, B. P. Kaine, M. Borodovsky, H.-P. Klenk, C. M. Fraser, H. O. Smith, C. R. Woese, and J. C. Venter.** 1996. Complete genome sequence of the methanogenic archaeon, *Methanococcus jannaschii*. *Science* **273:** 1058–1073.

11. **Clayton, R. A., O. White, K. A. Ketchum, and J. C. Venter.** 1997. The first genome from the third domain of life. *Nature* **387:**459–462.

12. **Cole, S. T., R. Brosch, J. Parkhill, T. Garnier, C. Churcher, D. Harris, S. V. Gordon, K. Eiglmeier, S. Gas, C. E. Barry III, F. Tekaia, K. Badcock, D. Basham, D. Brown, T. Chillingworth, R. Connor, R. Davies, K. Devlin, T. Feltwell, S. Gentles, N. Hamlin, S. Holroyd, T. Hornsby, K. Jagels, and B. G. Barrell.** 1998. Deciphering the biology of *Mycobacterium tuberculosis* from the complete genome sequence. *Nature* **393:**537–544.

13. **Dandekar, T., B. Snel, M. Huynen, and P. Bork.** 1998. Conservation of gene order: a fingerprint of proteins that physically interact. *Trends Biochem. Sci.* **23:**324–328.

14. **Das, S., L. Yu, C. Gaitatzes, R. Rogers, J. Freeman, J. Bienkowska, R. M. Adams, T. F. Smith, and J. Lindelien.** 1997. Biology's new Rosetta stone. *Nature* **385:**29–30.

15. **Deckert, G., P. V. Warren, T. Gaasterland, W. G. Young, A. L. Lenox, D. E. Graham, R. Overbeek, M. A. Snead, M. Keller, M. Aujay, R. Huber, R. A. Feldman, J. M. Short, G. J. Olsen, and R. V. Swanson.** 1998. The complete genome of the hyperthermophilic bacterium *Aquifex aeolicus*. *Nature* **392:**353–358.

16. **Doerks, T., A. Bairoch, and P. Bork.** 1998. Protein annotation: detective work for function prediction. *Trends Genet.* **14:**248–250.

17. **Edgell, D. R., and W. F. Doolittle.** 1997. Archaea and the origin(s) of DNA replication proteins. *Cell* **89:**995–998.

18. **Elagoz, A., A. Abdi, J. C. Hubert, and B. Kammerer.** 1996. Structure and organisation of the pyrimidine biosynthesis pathway genes in *Lactobacillus plantarum*: a PCR strategy for sequencing without cloning. *Gene* **182:**37–43.

19. **Fitch, W. M.** 1970. Distinguishing homologous from analogous proteins. *Syst. Zool.* **19**:99–113.

20. **Fitch, W. M.** 1995. Uses for evolutionary trees. *Phil. Trans. R. Soc. Lond. B* **349**:93–102.

21. **Fleischmann, R. D., M. D. Adams, O. White, R. A. Clayton, E. F. Kirkness, A. R. Kerlavage, C. J. Bult, J.-F. Tomb, B. A. Dougherty, J. M. Merrick, K. McKenney, G. G. Sutton, W. FitzHugh, C. Fields, J. D. Gocayne, J. Scott, R. Shirley, L.-I. Liu, A. Glodek, J. M. Kelley, J. F. Weidman, C. A. Phillips, T. Spriggs, E. Hedblom, M. D. Cotton, T. R. Utterback, M. C. Hanna, D. Nguyen, D. M. Saudek, R. C. Brandon, L. D. Fine, J. L. Frichtman, J. L. Fuhrmann, N. S. M. Geoghagen, C. L. Gnehm, L. A. McDonald, K. V. Small, C. M. Fraser, H. O. Smith, and J. C. Venter.** 1995. Whole-genome random sequencing and assembly of *Haemophilus influenzae* Rd. *Science* **269**:496–512.

22. **Fraser, C. M., S. Casjens, W. M. Huang, G. G. Sutton, R. Clayton, R. Lathigra, O. White, K. A. Ketchum, R. Dodson, E. K. Hickey, M. Gwinn, B. Dougherty, J.-F. Tomb, R. D. Fleischmann, D. Richardson, J. Peterson, A. R. Kerlavage, J. Quackenbush, S. Salzberg, M. Hanson, R. van Vugt, N. Palmer, M. D. Adams, J. Gocayne, J. Weidman, T. Utterback, L. Watthey, L. McDonald, P. Artiach, C. Bowman, S. Garland, C. Fujii, M. D. Cotton, K. Horst, K. Roberts, B. Hatch, H. O. Smith, and J. C. Venter.** 1997. Genomic sequence of a Lyme disease spirochaete, *Borrelia burgdorferi*. *Nature* **390**: 580–586.

23. **Fraser, C. M., J. D. Gocayne, O. White, M. D. Adams, R. A. Clayton, R. D. Fleischmann, C. J. Bult, A. R. Kerlavage, G. Sutton, J. M. Kelley, J. L. Fritchman, J. F. Weidman, K. V. Small, M. Sandusky, J. Fuhrmann, D. Nguyen, T. R. Utterback, D. M. Saudek, C. A. Phillips, J. M. Merrick, J.-F. Tomb, B. A. Dougherty, K. F. Bott, P.-C. Hu, T. S. Lucier, S. N. Peterson, H. O. Smith, C. A. Hutchinson III, and J. C. Venter.** 1995. The minimal gene complement of *Mycoplasma genitalium*. *Science* **270**:397–403.

24. **Galperin, M. Y., and E. V. Koonin.** 1997. A diverse superfamily of enzymes with ATP-dependent carboxylate-amine/thiol ligase activity. *Protein Sci.* **6**:2639–2643.

25. **Galperin, M. Y., and E. V. Koonin.** 1997. Sequence analysis of an exceptionally conserved operon suggests enzymes for a new link between histidine and purine biosynthesis. *Mol. Microbiol.* **24**:443–445.

26. **Gelfand, M. S., and E. V. Koonin.** 1997. Avoidance of palindromic words in bacterial and archaeal genomes: a close connection with restriction enzymes. *Nucleic Acids Res.* **25**:2430–2439.

27. **Ghim, S. Y., and J. Neuhard.** 1994. The pyrimidine biosynthesis operon of the thermophile *Bacillus caldolyticus* includes genes for uracil phosphoribosyltransferase and uracil permease. *J. Bacteriol.* **176**:3698–3707.

28. **Goffeau, A., B. G. Barrell, H. Bussey, R. W. Davis, B. Dujon, H. Feldmann, F. Galibert, J. D. Hoheisel, C. Jacq, M. Johnston, E. J. Louis, H. W. Mewes, Y. Murakami, P. Philippsen, H. Tettelin, and S. G. Oliver.** 1996. Life with 6000 genes. *Science* **274**:546–567.

29. **Himmelreich, R., H. Hilbert, H. Plagens, E. Pirkl, B. C. Li, and R. Herrmann.** 1996. Complete sequence analysis of the genome of the bacterium *Mycoplasma pneumoniae*. *Nucleic Acids Res.* **24**:4420–4449.

30. **Himmelreich, R., H. Plagens, H. Hilbert, B. Reiner, and R. Herrmann.** 1997. Comparative analysis of the genomes of the bacteria *Mycoplasma pneumoniae* and *Mycoplasma genitalium*. *Nucleic Acids Res.* **25**:701–712.

31. **Huynen, M. A., and P. Bork.** 1998. Measuring genome evolution. *Proc. Natl. Acad. Sci. USA.* **95**:5849–5856.

32. **Itoh, T.** 1988. Complete nucleotide sequence of the ribosomal 'A' protein operon from the archaebacterium, *Halobacterium halobium*. *Eur. J. Biochem.* **176**:297–303.

33. **Kaneko, T., S. Sato, H. Kotani, A. Tanaka, E. Asamizu, Y. Nakamura, N. Miyajima, M. Hirosawa, M. Sugiura, S. Sasamoto, T. Kimura, T. Hosouchi, A. Matsuno, A. Muraki, N. Nakazaki, K. Naruo, S. Okumura, S. Shimpo, C. Takeuchi, T. Wada, A. Watanabe, M. Yamada, M. Yasuda, and S. Tabata.** 1996. Sequence analysis of the genome of the unicellular cyanobacterium Synechocystis sp. strain PCC6803. II. Sequence determination of the entire genome and assignment of potential protein-coding regions. *DNA Res.* **3**:109–136.

34. **Karlin, S., J. Mrazek, and A. M. Campbell.** 1997. Compositional biases of bacterial genomes and evolutionary implications. *J. Bacteriol.* **179**: 3899–3913.

35. **Kawarabayasi, Y., M. Sawada, H. Horikawa, Y. Haikawa, Y. Hino, S. Yamamoto, M. Sekine, S. Baba, H. Kosugi, A. Hosoyama, Y. Nagai, M. Sakai, K. Ogura, R. Otsuka, H. Nakazawa, M. Takamiya, Y. Ohfuku, T. Funahashi, T. Tanaka, Y. Kudoh, J. Yamazaki, N. Kushida, A. Oguchi, K. Aoki, and H. Kikuchi.** 1998. Complete se-

quence and gene organization of the genome of a hyper-thermophilic archaebacterium, *Pyrococcus horikoshii* OT3. *DNA Res.* **5:**147–155.

36. **Klenk, H.-P., R. A. Clayton, J.-F. Tomb, O. White, K. E. Nelson, K. A. Ketchum, R. J. Dodson, M. Gwinn, E. K. Hickey, J. D. Peterson, D. L. Richardson, A. R. Kerlavage, D. E. Graham, N. C. Kyrpides, R. D. Fleischmann, J. Quackenbush, N. H. Lee, G. G. Sutton, S. Gill, E. F. Kirkness, B. A. Dougherty, K. McKenney, M. D. Adams, B. Loftus, J. D. Peterson, C. I. Reich, L. K. McNeil, J. H. Badger, A. Glodek, L. Zhou, R. Overbeek, J. D. Gocayne, J. F. Weidman, L. McDonald, T. R. Utterback, M. D. Cotton, T. Spriggs, P. Artiach, B. P. Kaine, S. M. Sykes, P. W. Sadow, K. P. D'Andrea, C. Bowman, C. Fujii, S. A. Garland, T. M. Mason, G. J. Olsen, C. M. Fraser, H. O. Smith, C. R. Woese, and J. C. Venter.** 1997. The complete genome sequence of the hyperthermophilic, sulphate-reducing archaeon *Archaeoglobus fulgidus. Nature* **390:**364–370.

37. **Kolsto, A. B.** 1997. Dynamic bacterial genome organization. *Mol. Microbiol.* **24:**241–248.

38. **Koonin, E. V.** 1997. Evidence for a family of archaeal ATPases. *Science* **275:**1489–1490.

39. **Koonin, E. V., A. R. Mushegian, and P. Bork.** 1996. Non-orthologous gene displacement. *Trends Genet.* **12:**334–336.

40. **Koonin, E. V., A. R. Mushegian, M. Y. Galperin, and D. R. Walker.** 1997. Comparison of archaeal and bacterial genomes: computer analysis of protein sequences predicts novel functions and suggests a chimeric origin for the archaea. *Mol. Microbiol.* **25:**619–637.

41. **Koonin, E. V., R. L. Tatusov, and M. Y. Galperin.** 1998. Beyond the complete genomes: from sequences to structure and function. *Curr. Opin. Struct. Biol.* **8:**355–363.

42. **Koonin, E. V., R. L. Tatusov, and K. E. Rudd.** 1995. Sequence similarity analysis of Escherichia coli proteins: functional and evolutionary implications. *Proc. Natl. Acad. Sci. USA* **92:** 11921–11925.

43. **Kunst, F., N. Ogasawara, I. Moszer, A. M. Albertini, G. Alloni, V. Azevedo, M. G. Bertero, P. Bessières, A. Bolotin, S. Borchert, R. Borriss, L. Boursier, A. Brans, M. Braun, S. C. Brignell, S. Bron, S. Brouillet, C. V. Bruschi, B. Caldwell, V. Capuano, N. M. Carter, S.-K. Choi, J.-J. Codani, I. F. Connerton, N. J. Cummings, R. A. Daniel, F. Denizot, K. M. Devine, A. Düsterhoft, S. D. Ehrlich, P. T. Emmerson, K. D. Entian, J. Errington, C. Fabret, E. Ferrari, D.** Foulger, C. Fritz, M. Fujita, Y. Fujita, S. Fuma, A. Galizzi, N. Galleron, S.-Y. Ghim, P. Glaser, A. Goffeau, E. J. Golightly, G. Grandi, G. Guiseppi, B. J. Guy, K. Haga, J. Haiech, C. R. Harwood, A. Hénaut, H. Hilbert, S. Holsappel, S. Hosono, M.-F. Hullo, M. Itaya, L. Jones, B. Joris, D. Karamata, Y. Kasahara, M. Klaerr-Blanchard, C. Klein, Y. Kobayashi, P. Koetter, G. Koningstein, S. Krogh, M. Kumano, K. Kurita, A. Lapidus, S. Lardinois, J. Lauber, V. Lazarevic, S.-M. Lee, A. Levine, H. Liu, S. Masuda, C. Mauël, C. Médigue, N. Medina, R. P. Mellado, M. Mizuno, D. Moestl, S. Nakai, M. Noback, D. Noone, M. O'Reilly, K. Ogawa, A. Ogiwara, B. Oudega, S.-H. Park, V. Parro, T. M. Pohl, D. Portetelle, S. Porwollik, A. M. Prescott, E. Presecan, P. Pujic, B. Purnelle, G. Rapoport, M. Rey, S. Reynolds, M. Rieger, C. Rivolta, E. Rocha, B. Roche, M. Rose, Y. Sadaie, T. Sato, E. Scanlan, S. Schleich, R. Schroeter, F. Scoffone, J. Sekiguchi, A. Sekowska, S. J. Seror, P. Serror, B.-S. Shin, B. Soldo, A. Sorokin, E. Tacconi, T. Takagi, H. Takahashi, K. Takemaru, M. Takeuchi, A. Tamakoshi, T. Tanaka, P. Perpstra, A. Tognoni, V. Tosato, S. Uchiyama, M. Vandenbol, F. Vannier, A. Vassarotti, A. Viari, R. Wambutt, E. Wedler, H. Wedler, T. Weitzenegger, P. Winters, A. Wipat, H. Yamamoto, K. Yamane, K. Yasumoto, K. Yata, K. Yoshida, H.-F. Yoshikawa, E. Zumstein, H. Yoshikawa, and A. Danchin.** 1997. The complete genome sequence of the Gram-positive bacterium *Bacillus subtilis. Nature* **390:**249–256.

44. **Labedan, B., and M. Riley.** 1995. Gene products of Escherichia coli: sequence comparisons and common ancestries. *Mol. Biol. Evol.* **12:** 980–987.

45. **Labedan, B., and M. Riley.** 1995. Widespread protein sequence similarities: origins of *Escherichia coli* genes. *J. Bacteriol.* **177:**1585–1588.

46. **Lawrence, J. G., and J. R. Roth.** 1996. Selfish operons: horizontal transfer may drive the evolution of gene clusters. *Genetics* **143:**1843–1860.

47. **Lechner, K., G. Heller, and A. Bock.** 1989. Organization and nucleotide sequence of a transcriptional unit of *Methanococcus vannielii* comprising genes for protein synthesis elongation factors and ribosomal proteins. *J. Mol. Evol.* **29:**20–27.

48. **Li, X., G. M. Weinstock, and B. E. Murray.** 1995. Generation of auxotrophic mutants of *Enterococcus faecalis. J. Bacteriol.* **177:**6866–6873.

49. **Maidak, B. L., G. J. Olsen, N. Larsen, R. Overbeek, M. J. McCaughey, and C. R.**

Woese. 1997. The RDP (Ribosomal Database Project). *Nucleic Acids Res.* **25**:109–111.

50. **Mankin, A. S.** 1989. The nucleotide sequence of the genes coding for the S19 and L22 equivalent ribosomal proteins from *Halobacterium halobium*. *FEBS Lett.* **246**:13–16.

51. **Meyer, E., N. J. Leonard, B. Bhat, J. Stubbe, and J. M. Smith.** 1992. Purification and characterization of the *purE*, *purK*, and *purC* gene products: identification of a previously unrecognized energy requirement in the purine biosynthetic pathway. *Biochemistry* **31**:5022–5032.

52. **Mushegian, A. R., and E. V. Koonin.** 1996. Gene order is not conserved in bacterial evolution. *Trends Genet.* **12**:289–290.

53. **Olsen, G. J., and C. R. Woese.** 1997. Archaeal genomics: an overview. *Cell* **89**:991–994.

54. **Olsen, G. J., C. R. Woese, and R. Overbeek.** 1994. The winds of (evolutionary) change: breathing new life into microbiology. *J. Bacteriol.* **176**:1–6.

55. **Overbeck, R., N. Larsen, W. Smith, N. Maltsev, and E. Selkov.** 1997. Representation of function: the next step. *Gene* **191**:GC1–GC9.

56. **Pace, N. R.** 1997. A molecular view of microbial diversity and the biosphere. *Science* **276**:734–740.

57. **Padilla, P. A., E. K. Fuge, M. E. Crawford, A. Errett, and M. Werner-Washburne.** 1998. The highly conserved, coregulated SNO and SNZ gene families in *Saccharomyces cerevisiae* respond to nutrient limitation. *J. Bacteriol.* **180**:5718–5726.

58. **Pitulle, C., Y. Yang, M. Marchiani, E. R. Moore, J. L. Siefert, M. Aragno, P. Jurtshuk, Jr., and G. E. Fox.** 1994. Phylogenetic position of the genus *Hydrogenobacter*. *Int. J. Syst. Bacteriol.* **44**:620–626.

59. **Quinn, C. L., B. T. Stephenson, and R. L. Switzer.** 1991. Functional organization and nucleotide sequence of the *Bacillus subtilis* pyrimidine biosynthetic operon. *J. Biol. Chem.* **266**:9113–9127.

60. **Ramirez, C., L. C. Shimmin, P. Leggatt, and A. T. Matheson.** 1994. Structure and transcription of the L11-L1-L10-L12 ribosomal protein gene operon from the extreme thermophilic archaeon *Sulfolobus acidocaldarius*. *J. Mol. Biol.* **244**:242–249.

61. **Riley, M., and B. Labedan.** 1997. Protein evolution viewed through *Escherichia coli* protein sequences: introducing the notion of a structural segment of homology, the module. *J. Mol. Biol.* **268**:857–868.

62. **Schenk-Groninger, R., J. Becker, and M. Brendel.** 1995. Cloning, sequencing, and char-

acterizing the *Lactobacillus leichmannii pyrC* gene encoding dihydroorotase. *Biochimie* **77**:265–272.

63. **Schricker, R., V. Magdolen, A. Kaniak, K. Wolf, and W. Bandlow.** 1992. The adenylate kinase family in yeast: identification of URA6 as a multicopy suppressor of deficiency in major AMP kinase. *Gene* **122**:111–118.

64. **Smith, D. R., L. A. Doucette-Stamm, C. Deloughery, H. Lee, J. Dubois, T. Aldredge, R. Bashirzadeh, D. Blakely, R. Cook, K. Gilbert, D. Harrison, L. Hoang, P. Keagle, W. Lumm, B. Pothier, D. Qiu, R. Spadafora, R. Vicaire, Y. Wang, J. Wierzbowski, R. Gibson, N. Jiwani, A. Caruso, D. Bush, H. Safer, D. Patwell, S. Prabhakar, S. McDougall, G. Shimer, A. Goyal, S. Pietrokovsky, G. M. Church, C. J. Daniels, J.-I. Mao, P. Rice, J. Nolling, and J. N. Reeve.** 1997. Complete genome sequence of *Methanobacterium thermoautotrophicum* ΔH: functional analysis and comparative genomics. *J. Bacteriol.* **179**:7135–7155.

65. **Stephens, R. S., S. Kalman, C. Lammel, J. Fan, R. Marathe, L. Aravind, W. Mitchell, L. Olinger, R. L. Tatusov, Q. Zhao, E. V. Koonin, and R. W. Davis.** 1998. Genome sequence of an obligate intracellular pathogen of humans: *Chlamydia trachomatis*. *Science* **282**:754–759.

66. **Takase, K., S. Kakinuma, I. Yamato, K. Konishi, K. Igarashi, and Y. Kakinuma.** 1994. Sequencing and characterization of the *ntp* gene cluster for vacuolar-type Na(+)-translocating ATPase of *Enterococcus hirae*. *J. Biol. Chem.* **269**:11037–11044.

67. **Tatusov, R. L., E. V. Koonin, and D. J. Lipman.** 1997. A genomic perspective on protein families. *Science* **278**:631–637.

68. **Tatusov, R. L., A. R. Mushegian, P. Bork, N. P. Brown, W. S. Hayes, M. Borodovsky, K. E. Rudd, and E. V. Koonin.** 1996. Metabolism and evolution of *Haemophilus influenzae* deduced from a whole-genome comparison with *Escherichia coli*. *Curr. Biol.* **6**:279–291.

69. **Tomb, J.-F., O. White, A. R. Kerlavage, R. A. Clayton, G. G. Sutton, R. F. Fleishmann, K. A. Ketchum, H. P. Klenk, S. Gill, B. A. Dougherty, K. A. Nelson, J. Quackenbush, L. Zhou, E. F. Kirkness, S. Peterson, B. Loftus, D. Richardson, R. Dodson, H. G. Khalak, A. Glodek, K. McKenney, L. M. Fitzgerald, N. Lee, M. D. Adams, E. K. Hickey, D. E. Berg, J. D. Gocayne, T. R. Utterback, J. D. Peterson, J. M. Kelley, M. D. Cotton, J. M. Weldman, C. Fujii, C. Bowman, L. Watthey, E. Wallin, W. S. Hayes, M. Borodovsky, P. D. Karp, H. O.**

Smith, C. M. Fraser, and J. C. Venter. 1997. The complete genome sequence of the gastric pathogen *Helicobacter pylori*. *Nature* **388:**539–547.

70. **Walker, J. E., and N. J. Gay.** 1983. Analysis of *Escherichia coli* ATP synthase subunits by DNA and protein sequencing. *Methods Enzymol.* **97:** 195–218.

71. **Watanabe, H., H. Mori, T. Itoh, and T. Gojobori.** 1997. Genome plasticity as a paradigm of eubacteria evolution. *J. Mol. Evol.* **44:**57–64.

72. **Woese, C. R.** 1987. Bacterial evolution. *Microbiol. Rev.* **51:**221–271.

73. **Woese, C. R.** 1994. There must be a prokaryote somewhere: microbiology's search for itself. *Microbiol. Rev.* **58:**1–9.

74. **Woese, C. R., O. Kandler, and M. L. Wheelis.** 1990. Towards a natural system of organisms: proposal for the domains Archaea, Bacteria, and Eucarya. *Proc. Natl. Acad. Sci. USA* **87:**4576–4579.

IMPACT OF HOMOLOGOUS RECOMBINATION ON GENOME ORGANIZATION AND STABILITY

Diarmaid Hughes

7

The central question in this review is whether it is reasonable to view a bacterial chromosome simply as a collection of linked genes and operons or whether the relative positioning and orientation of genes on a chromosome are themselves selected parameters. To address this question, I shall ask (i) to what degree chromosomal organization is preserved in evolution; (ii) what the rates of the recombinogenic mechanisms that shuffle the order of genes are; and (iii) whether the degree of preservation is consistent with the rates of recombination and, if not, whether one can identify selective forces which oppose the shuffling of genomes. This review concerns bacterial genomes, primarily those of *Escherichia coli* and *Salmonella typhimurium* (proper name, *Salmonella enterica* serovar Typhimurium). For these bacteria, abundant information is available on evolutionary relationships, high-quality genetic maps exist (*E. coli* is completely sequenced [22]), and there is extensive knowledge of mechanisms of recombination. Comparing the genomes of *E. coli* and *S. typhimurium* is therefore a natural starting point

for discussing the forces which determine genome organization and stability in general.

STABILIZING CHROMOSOME MAPS ON EVOLUTIONARY TIMESCALES

E. coli and *S. typhimurium* probably diverged from a common ancestor about 140 million years ago (99, 100). Estimating 200 generations per year (111), this corresponds to over 50 billion generations of separate evolution. In spite of the impressive number of genome replications in their separate evolutions, the genetic maps of these two species are extensively superimposable (16, 106, 109). At the nucleotide level their homologous protein-coding sequences differ by 1 to 25% (122). The majority of phenotypic characters which distinguish these species come from genome segments present in only one of them (95, 99, 106). These include regions coding for proteins involved in virulence and invasion of host cells, and there is considerable evidence that the regions have been acquired by horizontal transfer from distantly related species (15, 99). However, these accretions are sufficiently limited so as not to obliterate the overall genome similarity. Therefore, the genomes must have evolved over 50 billion generations, accumulating nucleotide substitutions and acquiring novel sequences through horizontal

Diarmaid Hughes, Department of Molecular Biology, Box 590, The Biomedical Center, Uppsala University, S-751 24 Uppsala, Sweden.

Organization of the Prokaryotic Genome, Edited by Robert L. Charlebois, transfers (and losing other sequences by deletion), without significantly rearranging their ancestral chromosomal gene order in all that time.

Why, then, is genome organization preserved? In principle, low rates of recombination and/or selection against the fixation in the population of bacteria with shuffled genomes could explain why the genomes are not rearranged. In the following sections I shall review the available data on rates of recombination, in particular that relating to transpositions and inversions, the events which shuffle gene order on a chromosome. However, a problem with the data is that recombination rates as such have not been measured. The available data is typically an expression of the frequency of a particular recombinant found in a population (for example, an overnight liquid culture) after selection for a phenotype on solid medium (for example, Lac$^+$, Tc^r, or Tc^r and prototrophy). For the purpose of comparison with mutation rate data, throughout this review I shall make a crude conversion of the recombination frequency data into a rate per cell per generation. The conversion assumes that the published frequency (F) of a recombination event (the fraction of cells with a particular phenotype in a population) equals the rate of the event per cell per generation (μ), multiplied by time (t) expressed as the log of the final population size over the initial population size. Thus, $F = \mu t$. For simplicity, it is assumed that the initial population size is 10^3 cells and that the final population size is 10^9 cells, giving a t value of 13.8. It is also assumed that selected recombinants have no significant growth advantage or disadvantage and that recombination occurs during growth of the liquid culture and not on the selective plate. With all of these assumptions, and the formula given above, the outcome is that recombination rates are approximately 1 order of magnitude lower than the published frequencies of recombination, and these are the rates I use for comparison with nucleotide substitution rates. In addition, in the following sections I shall discuss, where

available, the data on forces potentially selecting against the fixation of shuffled genomes. Selective forces fall into two groups: (i) those affecting gene expression and (ii) those affecting chromosome replication. The first group includes alterations in gene dosage, alterations in the direction of DNA transcription relative to DNA replication, and alterations affecting the prior adaptation of gene expression to local superhelical context (26). The second group includes the biased orientation of chi sites relative to the origin of DNA replication (24, 92) and the polarized organization of the DNA replication terminus region of the chromosome (46, 105).

The first section of the review considers data on recombination between directly oriented repeat sequences. This recombination has important consequences on a physiological timescale for reversibly adapting the physiology of cell growth to novel environments, but its primary importance for genome shuffling in evolution is its generation of fragments of DNA which may translocate to novel regions of the genome. Next I discuss the phenomenon of gene conversion, whose importance in genome organization is to prevent repeat sequences from diverging, thus maintaining them as substrates for homologous recombination throughout evolution. Thirdly, I discuss the data on recombination between inversely oriented repeat sequences, an event which can cause inversion of the region between the repeat sequences. Finally, at the end of this review I try to draw some reasonable conclusions from the data and, in addition, suggest areas where additional experimental data is required if this issue is to be resolved.

DIRECTLY ORIENTED REPEATS

Homologous recombination between directly oriented repeat sequences present on one chromosome can (i) duplicate all of the genetic material between the repeats (by a sister chromosome exchange) or (ii) delete the intervening material (by both interchromosomal and intrachromosomal exchanges) and (iii) create a novel join sequence in each case. De letion produces a DNA fragment which might integrate at other positions carrying the same repeat sequence (translocation).

Naturally occurring, directly oriented repeat sequences known to give rise to tandem duplications (and presumed to give rise to deletions) are found in both *E. coli* and *S. typhimurium*. These include the seven *rrn* operons (6, 69), which in both *S. typhimurium* and *E. coli* are all transcribed away from the origin of replication (in direct orientation on each half of the chromosome [Fig. 1]). In addition, IS*200* in *S. typhimurium* (40) and IS*1* (55), REP (125), and Rhs (71) in *E. coli* have each been implicated in duplication formation. There are also short repeat sequences, of 15 nucleotides and less, known to generate DNA amplifications (33, 134).

The Roles of RecA, RecBCD, and RecF

The degree to which ectopic exchanges between directly repeated sequences are RecA dependent varies with size and distance. Large chromosomal duplications are genetically unstable but are stabilized by *recA* mutations (4, 39, 40, 45, 71), implicating homologous recombination in their formation and loss. However, in plasmid model systems deletion between repeats often occurs efficiently in *recA* strains (82, 89). This is not specifically a plasmid-based phenomenon, as it is also seen for deletions between the same repeated sequences engineered onto the *E. coli* chromosome (82). In general, if repeats are small (less than 1 kb in length) and in tandem (or separated by only a few hundred nucleotides), then a proportion of the deletions can occur in a RecA-independent manner (17). As the distance between repeats increases to several kilobase pairs, deletions occur exclusively in a RecA-dependent manner. The lower-threshold length for Rec-dependent recombination is about 20 to 40 bp (123, 136). Smaller sequence repeats (8 to 17 bp) are observed to recombine specifically (2, 32, 134,

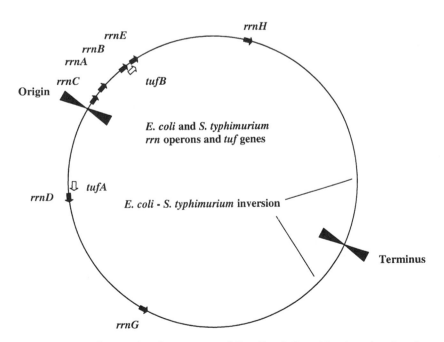

FIGURE 1 The circular chromosome of *E. coli* and *S. typhimurium*, showing the orientations and positions of the *rrn* operons and the *tuf* genes relative to the origin and terminus of DNA replication. The region around the terminus which is inverted in *E. coli* relative to that in *S. typhimurium* is also indicated.

137), but often this is *recA* independent (2, 56, 82, 83, 87, 90). Models to account for RecA independent deletion and duplication formation include strand slippage during DNA replication or repair, similar to proposals for the generation of frameshift mutations (131, 135), and the pairing of single-stranded regions of sister chromosome sequences immediately after passage of the replication fork (82). RecA-independent recombination provides a simple explanation for the generation of small duplications and deletions, which may have locally important phenotypes, in duplicating individual genes or domains within genes or in generating deletions which fuse novel sequences together (within or between genes).

Long repeat sequences have a greater probability of containing or being close to chi, the sequence which stimulates RecBCD recombinase activity (98), suggesting processing by RecBCD as another reason why ectopic recombination frequencies increase with repeat length. However, duplications and deletions between long sequence repeats, although reduced in frequency, are not abolished by *recB* mutations (39, 84, 85, 117, 118). The explanation is that recombination between long repeats on the chromosome is equally dependent on RecBCD and RecF functions (39), probably reflecting the availability of different initial substrates for recombination, with either double-strand breaks (RecBCD dependent) or single-strand nicks or gaps (RecF dependent). Increased mutation rates or defects in the DNA replication and repair enhance recombination between repeat sequences, probably because the increased number of lesions in the DNA provides additional substrates for recombination enzymes (9, 48, 49, 60, 86). Thus, conditions under which the genome is damaged correlate with extensive duplications. This coupling between damage and duplication may increase the chances that cells will survive extensive DNA damage. It may therefore be a selected rather than an accidental property.

Tandem Duplication and Gene Dosage

Tandem duplications occur spontaneously and may be selectively retained but are lost in Rec$^+$ strains when the selection pressure is removed, restoring the bacterial cell to its original haploid state. Genetic methods for selecting tandem duplications have been described (6, 7), and they show that spontaneous duplications of all regions of the *Salmonella* chromosome (except the terminus, where they are not found) are found at frequencies (F) of 10^{-2} to 10^{-4} per cell (equivalent to rates [μ] of 10^{-3} to 10^{-5}). The highest frequencies are for regions bounded by directly repeated *rrn* operons, with the region between *rrnB* and *rrnE* duplicated in 3% of fast-growing cells (Fig. 1). The short distance separating *rrnB* from *rrnE* (1 min; about 40 kbp) may contribute to a high intrinsic rate of duplication, but it is interesting to speculate that selection for increased levels of the duplicated gene products at high growth rates (they include EF-Tu and the β and β' subunits of RNA polymerase) may also select for maintenance of this duplication.

Genes expressed at high levels are generally located in the origin-proximal half of the chromosome, presumably because closeness to the origin of DNA replication gives a gene dosage effect (115, 128). Tandem duplications in bacteria have been selected on the basis of the extra enzyme activity associated with the duplicated region (3, 5, 34, 40, 50, 127, 129, 130). Many of these duplications are large— 10 to 30% of the chromosome—and are unstable in Rec$^+$ cells. If the ability to duplicate a set of interacting genes is of advantage to bacteria, it might select for the clustering of these genes to one chromosomal segment which is readily duplicated. Strains with chromosomal rearrangements, such as inversions or translocations, which alter gene order might then be counterselected in nature because they disrupt the gene sets whose duplication is occasionally advantageous (127). Genetic duplications, then, can be viewed as a regulatory device, contributing to the fitness of a genome by facilitating a flexible and reversible response to a variety of selections.

Amplification beyond Duplication

In some cases, selection for increased gene dosage results in amplification of short chromosomal segments up to 100 times. Examples are ampicillin resistance, encoded by the β-lactamase-producing *ampC* gene in *E. coli*, with up to 10 tandemly repeated 10-kbp regions containing *ampC* (32, 33); β-galactosidase activity from a poorly expressed *lacIZ* fusion gene in *E. coli*, with up to 100 tandemly repeated chromosomal segments of 5 to 37 kbp, placing *lacIZ* under the control of a foreign promoter (134, 137); a poorly expressed *lacIZ* fusion on an F′ plasmid in *S. typhimurium*, where a 20- to 40-kbp region is amplified up to 50-fold during selections for Lac$^+$ "adaptive mutations" (8); and the *argF* gene of *E. coli*, which is flanked by IS*1* sequences and is amplified up to 45-fold in mutants selected for overproduction of the enzyme (55). The common characteristics of these amplifications are that the regions involved are relatively short (less than 40 kbp) and the direct-repeat sequences involved in the first two examples are also short (12 to 15 bp). Because of the number of generations it would take to generate tandem repeats of 10 to 100 copies by successive RecA-mediated sister chromosome exchanges (approximately 6 to 11 generations if the choice of recombination partner is random at each generation), an alternative "do-loop" model has been proposed (103, 107). The mechanism requires double-strand break repair by RecA to lead to the formation of a replication fork and subsequent rolling-circle replication to generate a long tandem array. This eventually has to be resolved, either by repairing the broken end or by integrating a circle of repeats into the chromosome. It is interesting to note that the amplification of the *argF* gene requires an F factor in *cis* to the gene (27), suggestive of the proposed role F plasmid transfer replica-

tion plays in gene amplification during some adaptive-mutation selections (8, 38). The ability to amplify these 10- to 40-kbp regions, which contributes to fitness, should not be compromised by larger genome order rearrangements which leave these short regions intact.

Novel Join Points (Operon Fusions)

Tandem duplications, and their associated deletions and translocations, create novel sequence join points which potentially have selective value for the cell. This possibility has been demonstrated experimentally in selections based on the fusion of promoterless *trp* or *his* genes to foreign promoter elements (4, 37). The *his* duplications are facilitated by a short repeated sequence, the 40-bp REP element, located immediately upstream of *hisD*, with another REP element near *argA* providing the second recombining site (125), placing the *his* operon under the control of the *argA* promoter. These selected duplications are frequently large (up to 25% of the *Salmonella* genome) and are unstable in Rec$^+$ strains unless selection for prototrophy is maintained.

A deletion creating a novel join point with selective value occurs naturally in the *hisM* gene of *S. typhimurium*. *hisM* codes for a histidine transport protein and contains internal duplicated sequences which frequently lead to an in-frame deletion within the gene. Each copy of the protein is active but has different specificities, the wild type transporting histidine and the deletion mutants transporting histidinol (102).

Translocations

Recombination between directly oriented repeat sequences can create a DNA fragment (linear or circular, depending on the mechanism of recombination) which can recombine with the chromosome. This might, for example, result from recombination between two *rrn* operons. The excised fragment should be capable of recombining with any one of the other *rrn* operons, thus translocating the excised region to a novel location. There is a paucity of data in the literature on how frequently these translocation events occur. Experiments with *S. typhimurium* demonstrate that a partial *his* operon placed between two directly oriented Tn*10* elements at minute 2 is excised as a circle by intrachromosomal recombination between the Tn*10* elements and that this circle can reintegrate into the chromosome by recombination with the normal *his* operon located at minute 42 (84). The efficiency of this circle capture (translocation) is 1%, with 1 kbp of *his* operon homology between the circle and the chromosome, but it increases to 40% when the region of homology is extended to 4 kbp (84). *rrn* operons are in the 4-kbp size range, suggesting that capture of excised inter-*rrn* regions should also be efficient. Taking 10^{-3} to 10^{-5} as the rate of circle generation (based on average duplication frequencies) and 10^{-2} to 10^{-3} as the rate of circle capture (a frequency of 10^{-1} to 10^{-2}) gives a predicted translocation rate of 10^{-5} to 10^{-8} per cell per generation for inter-*rrn* regions. However, such translocations have not been observed to rearrange the chromosomes of either *E. coli* or *S. typhimurium*, suggesting either that the rates are insufficient or that there is selection against rearranged chromosomes. Strains constructed in *E. coli* with the *rrnB-rrnE* segment translocated to *rrnC*, *rrnD*, *rrnG*, and *rrnH* are viable but suffer a reduction of up to 5% in growth rate in rich medium (44). However, in *S. typhi* natural translocations involving regions bounded by *rrn* operons appear to be common (74, 77).

GENE CONVERSION

Homologous recombination can occur between repeated sequences which are not completely identical. The frequency of recombination is very sensitive to the degree of sequence divergence (123, 124). One explanation for this is that the mismatch repair system eliminates intermediates containing a mismatched heteroduplex (104). Recombination between dispersed near-identical repeat sequences can result in duplications, deletions, and inversions, depending on the relative orientation of the repeats. However, there are two other interesting recombination events between near-identical repeats, namely, reciprocal recombination and gene conversion. Reciprocal recombination here means that the repeat sequences undergo a double recombination event within the repeat sequence, resulting in an exchange of segments between the two repeats without any associated outside chromosomal rearrangement. If this occurs between two nonidentical genes on the same chromosome, the result is a reciprocal translocation of genetic information between the two genes. As a consequence, sequence A will now be present at location B, while the equivalent sequence B will be present at location A. This can result in the simultaneous appearance of A-B-A and B-A-B hybrid genes at each location. When reciprocal recombination occurs between nonallelic genes on sister chromosomes, it results in apparent gene conversion after completed chromosome replication (94, 118, 119). Gene conversion, on the other hand, means that a recombination event between two near-identical sequences results in the transfer of a stretch of sequence from one gene to the other without any reciprocal transfer, leaving one gene sequence unchanged. Gene conversion was initially described in fungal systems where all the products of meiosis within an ascus could be recovered and examined (72, 96, 97, 101). These observations showed that genetic information was sometimes converted from that of one parent into that of the other. This means that the recombinational process was found to be non-reciprocal. It has been suggested that conversion results from repair of recombinational heteroduplexes with one or the other of the parental types as a template (51). Conversion could also occur by a recombination mechanism, such as that described for double-strand break repair (132).

Gene conversion is often invoked to explain the presence of families of dispersed sequences evolving in concert within one genome. A good example of concerted evolution involving a bacterial gene family is provided by the *tuf* genes. The two *tuf* genes of *S. typhimurium* are located in inverse orientation, about 700 kb apart, on opposite sides of the origin of replication (Fig. 1). Tracks of sequence information are transferred between these genes at a rate of about 10^{-8} per generation (1). This transfer is RecA and RecB dependent (about 1,000-fold effects) and is strongly inhibited by the MutSLH mismatch repair enzymes (about 1,000-fold), indicating that it depends on formation of an intermediate *tuf* gene heteroduplex. Similar exchanges might also help to preserve sequence similarity within other gene families, for example, the *rrn* operons, but currently there are no experimental data. Gene conversion-type mechanisms prevent repeat sequences in a gene family from diverging, which means that they remain in the chromosome as tools for continued genomic rearrangements over long periods. Gene conversion mechanisms thus contribute not only to coevolution within a gene family but also to the preservation of a mechanism for continued rearrangements within the chromosome.

INVERSIONS

Homologous recombination between repeated sequences in inverse orientation on the same chromosome can generate inversions of the sequence between the repeats. Inversions do not increase the amount of repetitive DNA in the chromosome; therefore, the rate of reversal due to homologous recombination between the repeats is expected to equal the rate of initial occurrence. Inversions can occur between homologous short and long sequences. Experiments demonstrating naturally occurring as well as artifically selected inversions in *E. coli* and *S. typhimurium* are examined below.

Natural Inversions

A major difference in genome order between *E. coli* K-12 and *S. typhimurium* LT2 is the large inversion of the replication terminus region of the chromosome (25, 108). Inversions of different lengths of this region are also found in *Salmonella enteritidis* (78) and the related species *Klebsiella aerogenes* (13). Another difference is inversion of the region between *rrn*D and *rrn*E for the commonly used W3110 strain of *E. coli* K-12 (43). All of these inversions are symmetrical about the origin-terminus axis of chromosome replication, thus maintaining gene orientation and gene distance relative to the origin of replication (Fig. 1). The higher-order genome organizations of several *Salmonella* species have recently been analyzed by restriction digestion targeted to their repetitive *rrn* operons (74–77). A variety of rearrangements relative to *S. typhimurium* LT2 were found, explainable by recombination among the *rrn* operons, including inversions and translocations of fragments to different locations. In almost all cases these rearrangements do not alter distance from, or orientation relative to, the origin of replication. These data suggest that alternative gene orders are evolutionarily viable provided that an equal division of the chromosome into two arms and the relative distances of genes from the origin are maintained. In principle, orientation may also be selected for, but it results mechanistically in this case from recombination involving *rrn* operons which are all oriented away from the origin (Fig. 1). It is not obvious whether these rearrangements occur because they are conservative or because the *rrn* operons, the longest repeat sequences in the chromosome, are the most likely source of rearrangements. It is tempting to propose that less conservative rearrangements might reduce the growth rate. An *E. coli* strain that has been in the laboratory for several decades (42) has a remarkably unbalanced inversion between *rrn*E and *rrn*G (Fig. 1). It also suffers from a severe growth defect, with strong selection for compensatory rearrangements restoring the normal *E. coli* gene order.

Selected Inversions

Early experiments by Konrad (60) indicated that inversions might be problematic. Extensive selection for recombination between homologous *lac* operon sequences in inverse orientation placed at *lac* and at Φ80*att* gave no simple inversions. A plausible explanation for failure in this case is the large-scale reversal of gene orientation it would occasion in one arm of the chromosome. Later, a selection in *S. typhimurium* for activation of an unexpressed *hisD* gene by fusion to a new promoter did result in inversions, but 20 times less frequently than either duplications or deletions in the same selection (114). Thus, these inversions occur at a rate of 10^{-7} or greater, assuming a duplication rate of 10^{-5}. The site of recombination in these mutants is a REP sequence (125) present in the *hisG-hisD* intercistronic region and many other locations around the chromosome (41). Despite the many potential locations for the other end point of the inversion, two particular large inversions were isolated repeatedly. Furthermore, most large inversions isolated included the origin (or terminus) of replication. These results suggested that not only are inversions rare, but their locations on the chromosome are nonrandom.

An example of inversion involving short sequence homologies is the inversion of an 800-bp fragment in a constructed *lac* region of *E. coli* placed between inverted repeat sequences of 12 or 23 bp occurring with the frequency 10^{-8} to 10^{-7} (116), equivalent to a rate of approximately 10^{-9} to 10^{-8}. There is no evidence that short inversely oriented repeats can facilitate inversions over large distances.

One factor which might influence inversion frequencies is the orientation of genes relative to the direction of replication (22, 24). In yeast, replication fork barriers block replication moving in the direction opposite to transcription (23). In vitro experiments with *E. coli* components show that the replication fork pauses when it encounters RNA polymerase moving in the opposite direction (73). Such effects might select against major changes in gene orientation within the chromosome. However, at least some small inversions within one replication arm are viable (57, 114), suggesting that the chromosome arms

are not saturated with orientation-sensitive sites. Also, *E. coli* strains can replicate any chromosome segment in either direction from a plasmid origin of replication (30).

There is a striking bias in the orientation of chi sites on the chromosome of *E. coli*: 80% of the sites are oriented in such a way that they will activate a RecBCD complex moving toward the origin of replication (24, 92, 98). This arrangement may facilitate the repair of double-strand breaks associated with the passage of replication forks (66). If this arrangement is important for the repair of chromosome damage, then rearrangements that reverse the bias will reduce fitness.

Another possible problem with inversion viability is the polar relationship between the origin and terminus of DNA replication on the circular chromosome. Normally, the terminus of DNA replication (63, 79) lies halfway across the chromosome from the origin, so that each chromosome arm is equally sized. However, some of the large inversions selected by Schmid and Roth (114) grossly alter the relative lengths of the replication arms so that one is less than 1 Mbp while the other is almost 4 Mbp in length. These strains are viable and apparently do not suffer major growth defects (114). Two studies which try to systematically address the nature of the inversion problem are described below.

Nondivisible Zones in *E. coli*

Evidence that large-scale aspects of chromosome organization may be under selection and may limit the possibilities for generating inversions comes from studies of the terminus region of DNA replication in *E. coli*. Pairs of defective Tn*10* transposons placed in inverse orientation at many different positions on the *E. coli* chromosome were used to select for specific inversions (105). Although inversions covering all regions of the chromosome were isolated, for many pairs of Tn*10* transposons it proved impossible to select inversions. Strikingly, each of the failed inversions had one endpoint within either of two regions close to (and on either side of) the terminus of DNA replication. These regions extend from 17 to 29 min and from 44 to 33 min and are called nondivisible zones (105) (Fig. 2). These regions of the chromosome contain the physiologically important replication terminator loci (35, 46), *TerA, -B, -C*, and *-D* (Fig. 2). The terminator loci unidirectionally inhibit replication forks moving in the unnatural direction (30), which are particularly slow in the one-third of the chromosome around the terminators. This suggested that there might also be features of nucleoid organization which are symmetrical on the two *oriC-TerA* or *-TerC* arms and polarized with respect to the direction of replication (30). The nondivisible zones may represent the polarized organization of the chromosome in these regions, and these zones would define the two chromosome arms. The *Ter* loci are 23-bp sequences that are bound by the Tus protein, and this binding is both necessary and sufficient to form a polar DNA replication barrier by inhibiting helicase action (47, 68). However, in a strain with a mutation in the *tus* gene in which replication does not pause at termina-

tors, there is no change in the distribution of invertible and noninvertible segments (36), eliminating replication pausing itself as the problem in generating these inversions. It is interesting to note that some inversions selected in *S. typhimurium* which seem to be similar in structure to those causing deleterious growth in *E. coli* (105) do not cause any obvious growth defect in *Salmonella* (61).

Forbidden Inversions in *S. typhimurium*

In *S. typhimurium*, specific inversions were selected by placing a part of the *his* operon in inverse orientation at many sites distant from the normal *his* locus, which was also mutant (120). His prototrophs were selected and could be created if the inverted *his* sequences recombined to create a functional operon. The successful selection of an inversion was dependent on the position of the *his* sequences. In general, inversions crossing the terminus region of the chromosome were found, whereas those involving the region from the normal *his* locus in the opposite direction, so-called nonpermissive intervals (Fig.

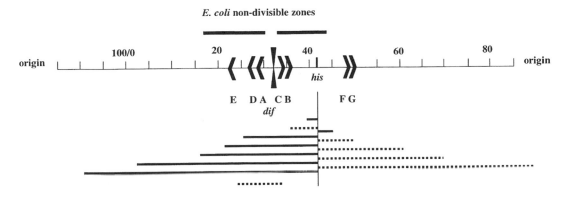

FIGURE 2 Linear map of the *E. coli* chromosome opened at the origin of DNA replication. *E. coli* map positions (approximately equivalent in *S. typhimurium*) are shown above the line, and the positions of *Ter* sites (E, D, A, C, B, F, and G) and *dif* are shown on the line and labeled below. The bars above the line indicate the nondivisible zones for inversion endpoints based on the data in reference 105. The *his* operon position is shown on the line to facilitate comparison with the results from *S. typhimurium* shown below the line. The solid bars indicate permissive inversion intervals, and the dashed lines indicate nonpermissive inversion intervals for *S. typhimurium*. This is a representative sample of the data from reference 120. Note that there is currently no experimental evidence that *S. typhimurium* has the same organization of *dif* and *Ter* sites as *E. coli*.

2), were not found. Similar selections made with Mud *lac* sequences present in inverse orientation at various locations confirmed that most large permissive inversions include the terminus region, but small and some large inversions distant from the terminus are also permissive. Most nonpermissive inversions have one endpoint in the terminal one-third of the chromosome but do not invert the terminus. Of the inversions which are permissive, only one causes a marked reduction in growth rate (120). One notable feature of these results is that the sequences within some of the nonpermissive intervals, and the slow-growing inversion, are themselves contained within other permissive inversions without obvious growth defects (Fig. 2). This suggests that the problems are not due to the orientations of the sequences contained within the inverted segment but may be related to the specific end points of the inversions.

Forbidden Inversions Are Not Lethal

An important question regarding nonpermissive intervals for inversions (and nondivisible zones) is whether the inversion is lethal or whether there is a mechanistic problem in making the inversion. With regard to the latter, it should be noted that inversions are generated by recombination between two sequences present in the same circular chromosome. Inversion frequencies are often strongly reduced in *recB* strains (118), possibly reflecting the need to make a fully reciprocal intrachromosomal exchange. Other recombinant types are not limited in this way and can occur by recombination between repeated sequences on sister chromosomes after replication. In principle, inversions can also be generated by sister chromosome exchanges, but this requires two rare independent events. Extrapolating from the average rate of duplication (10^{-5}), one would expect the two coincident exchanges required for the generation of an inversion by sister chromosome exchanges to be below the detection level of the experiments. To address this question, several of these inversions were constructed by genetic means, using two-fragment transductions

(113). These constructions showed that for three nonpermissive intervals tested the inversion itself was not lethal or growth impaired (94, 118, 120). This shows that the failure to select inversions in *Salmonella* (120) may not be due to lethal phenotypes but rather to problems associated with creating the inversions. What then is the nature of the mechanistic problem? In experiments which revealed nonpermissive intervals for inversion selection, the same chromosomal repeat sequences yielded double recombinants and apparent gene convertants (94, 118). This shows that the inverse repeat sequences can interact with each other to produce other types of recombinants. However, both double recombinants and apparent gene convertants, as well as duplications or deletions, can be generated by either full- or half-reciprocal events between sister chromosomes. The formation of permissive inversions, on the other hand, involves fully reciprocal exchanges that rejoin both sets of flanking sequences (84) and is strongly dependent on the RecBCD recombination system (84, 119). Thus, nonpermissive intervals for inversion formation can be explained if sequences at certain pairs of sites are barred from intrachromosomal exchanges (118), whereas recombination between sister chromosomes for the same pairs of sites is not subject to these constraints. Nonpermissive sites for inversion formation could reflect a paucity of sites for initiating or resolving the required full exchange. The transductional method of forming these inversions may overcome these problems by providing extrachromosomal fragments with long homologous sequences and highly recombinogenic double-stranded ends. However, there are other possibilities (94), including that recombination events that lead to inversion formation in certain regions generate "toxic" intermediates. We will return to this possibility later in the context of replication and recombination.

RECOMBINATION EFFECTS IN THE TERMINUS REGION

Xer and *dif*

The catenated chromosomes that form at the end of the DNA replication cycle can be physically separated by topoisomerase IV as long as the sister chromosomes undergo an even number of recombination events during replication and the chromosomes exist as monomers. The *E. coli* chromosome harbors a recombination site, *dif*, that maps to where DNA replication normally terminates (Fig. 3). Recombination between two *dif* sites is catalyzed by the site-specific recombinases XerC and XerD (21). It has been proposed that the function of this recombination is to convert chromosome dimers back to monomers so that they can be segregated prior to cell division, for example, by resolving chromosome dimers that result from an odd number of exchanges between sister chromosomes (20, 64). The only sequences required for *dif* function are contained within the 33 bp which bind XerC and XerD (70, 133). The Xer-*dif* site-specific recombination of normal chromosome segregation occurs only within the DNA replication terminus region DAZ (for Dif acting zone). It does not occur when *dif* is inserted at other distant chromosomal locations (70, 133), which, however, remain recombinogenic with other chromosomal *dif* sites and stimulate inversions and deletions (133). Thus, the location of *dif* in a narrow zone of about 10 kb on either side of its natural position (28, 62), the terminal domain of the nucleoid (the last domain of the chromosome to be pulled apart during segregation), is required for its proper function. This suggests that distinct domains of the chromosome become separated soon after passage of the replication fork (12, 20).

Terminus Hyperrecombination Zone

The frequency of RecA- and RecBCD-dependent recombination in the terminus region of *E. coli* is very significantly higher than in the rest of the chromosome (81). Prophage excision occurs at a uniform rate of about 10^{-5} per generation at most chromosomal posi-

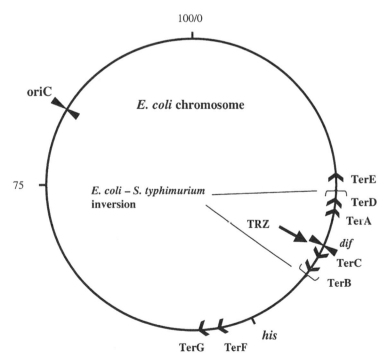

FIGURE 3 Chromosome map of *E. coli* showing the relative positions of the TRZ, the *dif* site-specific recombination site, and the *Ter* sites. The region that is inverted relative to that in *S. typhimurium* is indicated and bracketed.

tions, but the rates are 10- to 100-fold higher at the edge of the terminus region, reaching a maximum of 10^{-2} close to *dif*. This region of maximum recombination is about 40 kbp in length and has been named the terminus recombination zone (TRZ) (Fig. 3). The TRZ and DAZ (see above) are coincident. Deletion of the TRZ region does not eliminate hyperrecombination but simply shifts it in the direction of the deletion. This suggests that hyperrecombination in the TRZ is related to its position in the nucleoid rather than to its sequence (80). Altering the position of the *Ter* DNA replication trap (by moving *Ter* sites to another position on the chromosome) showed that hyperrecombination in the TRZ occurs independently of DNA replication termination (80). This suggests that the TRZ is defined by the structure of the terminal domain of the nucleoid. Hyperrecombination in the TRZ may be a consequence of the circularity of the *E. coli* chromosome, with catenated daughter chromosomes that are joined after the completion of replication, during the final stages of nucleoid assembly and decatenation, providing a good substrate for recombination (46, 80). A current model proposes that, after each replication cycle, a sequence-dependent structure centered around *dif* must be assembled on each side of the nucleoid terminus (29). According to this model, the nucleoid assembly process pushes catenation links between sister chromosomes toward this center, creating a synapsis between the chromosomes. Cuts introduced by topoisomerases initiate the terminal recombination events. Recent experiments show that the hyperrecombination events have a polarity from the middle of the TRZ (*dif*), suggesting that the orientation of sequences within the TRZ influences recombination (29).

The experiments described above indicate that the nucleoid terminus of the *E. coli* chromosome is organized to facilitate chromosome segregation. The mechanism involves sequence polarity, polarity in the direction of nucleoid organization, and a specialized recombination system. The primary function of the polarized *Ter* sites may therefore be to confine the terminal events of DNA replication to the terminus region, which would facilitate chromosome segregation. These features suggest that the organization of the terminal region of the chromosome is preserved in evolution to allow orderly cell division. The functional importance of the structural organization of this region may in fact account for many of the problems noted in selecting inversions with end points in this region (105). However, there may still be significant differences between *E. coli* and *S. typhimurium* (120) in this respect, since the terminus region of *S. typhimurium* is still not as well defined as that in *E. coli*.

REPLICATION AND RECOMBINATION

History of the Idea

Gerald Smith proposed on theoretical grounds that conjugational recombination in *E. coli* might involve chromosome replication (126). A problem with earlier models is that reciprocal recombination between a fragment of DNA and a chromosome generates another fragment and, unless this fragment is degraded, exchanges with the chromosome would continue ad infinitum. The solution, he proposed, is to view recombination as a nonreciprocal process. This means that during recombination, the D loop in the recipient chromosome containing the invading 3' DNA end is converted into a replication fork. This is proposed to create two replication forks, one at each end of the fragment, which proceed around the chromosome in opposite directions, eventually resulting in two complete chromosomes. One chromosome is identical to the parental chromosome, and the other is a recombinant incorporating the foreign fragment of DNA. The model of Smith (126) can be applied to transduction, transformation, and double-strand break repair and predicts that there is a strong link between DNA recombination and DNA replication.

Experimental evidence for a connection between recombination and replication comes from a number of sources. One is the finding that DNA damage induces the SOS response, particularly activation of so-called damage-inducible stable DNA replication (58). This mode of replication does not require *oriC* but does require RecA and RecBCD activities. This suggests that, during the SOS response, double-strand breaks in the chromosome are processed by RecBCD and RecA, which generate D loops in a homologous region of a sister chromosome from which DNA replication can be initiated (10, 11). This raises the question of how D loops are converted into replication forks. Recent evidence suggests that the PriA protein catalyzes assembly of a replication primosome at the D loop, thereby creating a replication fork (91). Mutations in *priA* which block the first step in primosome assembly also block recombination-dependent DNA replication (59, 88). *priA* null mutations which block damage- and recombination-dependent replication also reduce (but do not eliminate) P1 transduction efficiency and Hfr conjugation (59, 110). Thus, *priA* null mutations are recombination deficient. Ironically, Hfr conjugation, for which Smith originally proposed his model, is only moderately dependent on PriA, being reduced threefold. Extragenic suppressors of the UV sensitivity phenotype of *priA* mutants map in *dnaC* (110), the gene for one of the other components of the primosome assembly reaction, which is also required for recombination-dependent replication. PriA function in primosome assembly is also required for double-strand break repair (59). Thus, the proposal that recombination might proceed through replication (126) is supported by experimental evidence. This type of double-strand break repair can also be used to repair collapsed replication forks (11, 65, 67). In fact, the latter function may be the primary reason why the orientation of chi sites is biased on the chromosome (66). Collapsed replication forks may occur fortuitously when gaps in one DNA strand are encountered (66), but there is also evidence that blockage of replication forks is associated with double-strand breaks, resulting in SOS

induction and recombination (52–54, 93, 121). Related to this is the observation that Rec-dependent deletion of tandem duplications is enhanced by a variety of different mutations affecting DNA polymerase III (112). The above-mentioned evidence suggests that much, but not all, homologous recombination involves extensive DNA replication. This leads to the expectation that features of the chromosome which influence DNA replication should have pronounced effects on homologous recombination.

Consequences for Rearrangments of Long-Range Replication

No systematic study has yet been made of the influence of *priA* mutations and replication on the formation of duplications, deletions, translocations, and inversions. Most cases where there are difficulties in selecting inversions relate to the region close to the normal terminus of replication (105, 120), which is highly organized with regard to replication. A possible explanation for some of these difficulties is that an inverse-order sequence that includes a *Ter* site and excludes *dif* will block recombinational replication (Fig. 2). This would be scored as a nonpermissive interval for inversion. From this argument follows the prediction that it should become permissive in a *tus* mutant strain which allows replication to proceed through *Ter* sites to the terminus. However, in *E. coli* the distribution of invertible and noninvertible segments was unchanged in a *tus* mutant (36), suggesting that other factors, for example, polarized nucleoid organizing sequences (29), may be involved.

CONCLUSIONS

It seems reasonable to conclude (despite the additional detail that a complete genome sequence of *S. typhimurium* might reveal) that at least for *E. coli* and *S. typhimurium*, genome order is preserved on an evolutionary timescale. Is this genetic stability a consequence of a low rate of rearrangement, or is it a consequence of selection against bacteria with rearranged genomes? The events most likely to scramble genome order are the translocation of fragments to novel locations and the inversion of chromosome segments. Are the rates of these rearrangements sufficiently high to have scrambled the genomes of *E. coli* and *S. typhimurium* in the time since their separation? In fact, the literature contains no quantitative data on the rates of either of these rearrangements involving natural repeat sequences in the chromosome. However, using the available data on the frequencies of similar events, one can crudely calculate that translocations and inversions involving long repeat sequences (such as *rrn*, *tuf*, and IS elements) occur at a rate of at least 10^{-8} per target per generation. This calculation assumes that frequencies can be converted to rates as described in the introduction and that the quantitative data on frequencies mentioned in the sections above are similar to rates of rearrangement involving natural sequences. This calculated rate of rearrangement can be compared with the rate of less than 10^{-9} for specific base substitution mutations in these genomes (1, 31). Furthermore, it is calculated that most synonymous nucleotide changes between *E. coli* and *S. typhimurium* are saturated (14). If translocation and inversion rates are significantly greater than substitution rates, then selectively neutral rearrangements occurring during the separate evolution of *E. coli* and *S. typhimurium* should also be saturated. This discrepancy in rates of occurrence versus rates of fixation of genetic rearrangements is a strong argument in favor of selective pressures to maintain the current genome order.

What might the selective forces maintaining genome order be? The interconnections between DNA replication and homologous recombination could explain selection against the occurrence and maintenance of rearrangements disrupting the normal terminus of DNA replication. Throughout the rest of the genome, one possible selective force is the interrelationship between regulation of gene expression and position on the chromosome (e.g., the adaptation of gene expression to gene dosage effects close to the origin of replication, to transcription direction relative to replication direction, or to local superhelical contexts). If changing these situations by large genetic rearrangements causes a fitness problem, it might be difficult to solve by compensatory mutations because potentially dozens to hundreds of genes would be simultaneously affected, leading instead to the extinction of the rearranged cell line. Another selection pressure which could act to maintain genome order outside of the terminus region is the biased orientation of chi sites, which might be important for double-strand break repair during replication. However, this selection pressure would only counteract a subset of rearrangements, not, for example, inversions involving the *rrn* operons. What is the evidence in the literature that large genetic rearrangements have negative effects on fitness? Most inversions isolated in *E. coli* and *S. typhimurium* are reported as having no significant effects on growth rate, but the few translocations made and tested in *E. coli* are associated with decreases in growth rate of up to a few percent. One interesting possibility regarding fitness and selection is that rearrangements which have little or no detectable growth rate effect in laboratory media might affect bacterial infection or growth in eukaryotic hosts. In this context it is worth noting that for a variety of point mutations in *S. typhimurium* affecting the gene expression machinery (transcription and translation), relative fitness defects of a few percent in laboratory media are magnified to fitness differences of several orders of magnitude in competition in eukaryotic hosts (18, 19). More quantitative data is required regarding both the rates with which rearrangements of each type occur (in particular those involving natural sequences) and their associated fitness effects in relevant environments before it can be determined to what extent growth selection rather than rates of recombination are responsible for the lack of genome rearrangement between *E. coli* and *S. typhimurium*.

ACKNOWLEDGMENTS

My research is funded by the Swedish Natural Sciences Research Council.

I thank John Roth for suggestions on the outline of the chapter, Otto Berg for discussions on mutation rates, and Måns Ehrenberg, Dan Andersson, and the two reviewers for helpful and critical comments on the manuscript.

REFERENCES

1. **Abdulkarim, F., and D. Hughes.** 1996. Homologous recombination between the *tuf* genes of *Salmonella typhimurium. J. Mol. Biol.* **260:** 506–522.

2. **Albertini, A. M., M. Hofer, M. P. Calos, and J. H. Miller.** 1982. On the formation of spontaneous deletions: the importance of short sequence homologies in the generation of large deletions. *Cell* **29:**319–328.

3. **Ames, G. F., D. P. Biek, and E. N. Spudich.** 1978. Duplications of histidine transport genes in *Salmonella typhimurium* and their use for the selection of deletion mutants. *J. Bacteriol.* **136:** 1094–1108.

4. **Anderson, R. P., and J. R. Roth.** 1978. Tandem chromosomal duplications in *Salmonella typhimurium*: fusion of histidine genes to novel promoters. *J. Mol. Biol.* **119:**147–166.

5. **Anderson, R. P., and J. R. Roth.** 1978. Tandem genetic duplications in *Salmonella typhimurium*: amplification of the histidine operon. *J. Mol. Biol.* **126:**53–71.

6. **Anderson, R. P., and J. R. Roth.** 1981. Spontaneous tandem genetic duplications in *Salmonella typhimurium* arise by unequal recombination between rRNA (*rrn*) cistrons. *Proc. Natl. Acad. Sci. USA* **78:**3113–3117.

7. **Anderson, R. P., C. G. Miller, and J. R. Roth.** 1976. Tandem duplications of the histidine operon observed following generalized transduction in *Salmonella typhimurium. J. Mol. Biol.* **105:**201–218.

8. **Andersson, D. I., E. S. Schlecta, and J. R. Roth.** 1998. Evidence that gene amplification underlies adaptive mutability of the bacterial *lac* operon. *Science* **282:**1133–1135.

9. **Arthur, H. M., and R. G. Lloyd.** 1980. Hyper-recombination in *uvrD* mutants of *Escherichia coli* K-12. *Mol. Gen. Genet.* **180:**185–191.

10. **Asai, T., S. Sommer, A. Bailone, and T. Kogoma.** 1993. Homologous recombination-dependent initiation of DNA replication from DNA damage-inducible origins in *Escherichia coli. EMBO J.* **12:**3287–3295.

11. **Asai, T., D. B. Bates, and T. Kogoma.** 1994. DNA replication triggered by double-stranded breaks in *E. coli*: dependence on homologous recombination functions. *Cell* **78:**1051–1061.

12. **Baker, T. A.** 1991. ... and then there were two. *Nature* **353:**794–795.

13. **Bender, R. A., A. Macaluso, and B. Magasanik.** 1976. Glutamate dehydrogenase: genetic mapping and isolation of regulatory mutants of *Klebsiella aerogenes. J. Bacteriol.* **127:**141–148.

14. **Berg, O. G.** 1999. Synonymous nucleotide divergence and saturation: effects of site-specific variations in codon bias and mutation rates. *J. Mol. Evol.* **48:**398–407.

15. **Bergthorsson, U., and H. Ochman.** 1995. Heterogeneity of genome sizes among natural isolates of *Escherichia coli. J. Bacteriol.* **177:**5784–5789.

16. **Berlyn, M. K. B., K. B. Low, K. E. Rudd, and M. Singer.** 1996. Linkage map of *Escherichia coli* K-12, edition 9, p. 1715–1902. *In* F. C. Neidhardt, R. Curtiss III, J. L. Ingraham, E. C. C. Lin, K. B. Low, B. Magasanik, W. S. Reznikoff, M. Riley, M. Schaechter, and H. E. Umbarger (ed.), *Escherichia coli and Salmonella: Cellular and Molecular Biology,* 2nd ed., vol. 2. American Society for Microbiology, Washington, D.C.

17. **Bi, X., and L. F. Liu.** 1994. *recA*-independent and *recA*-dependent intramolecular plasmid recombination. Differential homology requirement and distance effect. *J. Mol. Biol.* **235:**414–423.

18. **Björkman, J., D. Hughes, and D. I. Andersson.** 1998. Virulence of antibiotic-resistant *Salmonella typhimurium. Proc. Natl. Acad. Sci. USA* **95:**3949–3953.

19. **Björkman, J., P. Samuelsson, D. I. Andersson, and D. Hughes.** 1999. Novel mutations affecting translational accuracy, antibiotic resistance and virulence of *Salmonella typhimurium. Mol. Microbiol.* **31:**53–58.

20. **Blakely, G., S. Colloms, G. May, M. Burke, and D. Sherratt.** 1991. *Escherichia coli* XerC recombinase is required for chromosomal segregation at cell division. *New Biol.* **3:**789–798.

21. **Blakely, G., G. May, R. McCulloch, L. K. Arciszewska, M. Burke, S. T. Lovett, and D. J. Sherratt.** 1993. Two related recombinases are required for site-specific recombination at *dif* and *cer* in *E. coli. Cell* **75:**351–361.

22. **Blattner, F. R., G. Plunkett III, C. A. Bloch, N. T. Perna, V. Burland, M. Riley, J. Collado-Vides, J. D. Glasner, C. K. Rode, G. F. Mayhew, J. Gregor, N. W. Davis, H. A. Kirkpatrick, M. A. Goeden, D. J. Rose, B. Mau, and Y. Shao.** 1997. The complete genome sequence of Escherichia coli K-12. *Science* **277:**1453–1474.

23. **Brewer, B. J., and W. L. Fangman.** 1988. A replication fork barrier at the 3′ end of yeast ribosomal RNA genes. *Cell* **55:**637–643.

24. **Burland, V., G. Plunkett III, D. L. Daniels, and F. R. Blattner.** 1993. DNA sequence and

analysis of 136 kilobases of the *Escherichia coli* genome: organizational symmetry around the origin of replication. *Genomics* **16**:551–561.

25. **Casse, F., M.-C. Pascal, and M. Chippaux.** 1973. Comparison between the chromosomal maps of *Escherichia coli* and *Salmonella typhimurium.* Length of the inverted segment in the *trp* region. *Mol. Gen. Genet.* **124**:253–257.

26. **Charlebois, R. L., and A. St. Jean.** 1995. Supercoiling and map stability in the bacterial chromosome. *J. Mol. Evol.* **41**:15–23.

27. **Clugston, C. K., and A. P. Jessop.** 1991. A bacterial position effect: when the F factor in *E. coli* K-12 is integrated *in cis* to a chromosomal gene that is flanked by IS1 repeats the elements are activated so that amplification and other regulatory changes that affect the gene can occur. *Mutat. Res.* **248**:1–15.

28. **Cornet, F., J. Louarn, J. Patte, and J.-M. Louarn.** 1996. Restriction of the activity of the recombination site *dif* to a small zone of the *Escherichia coli* chromosome. *Genes Dev.* **10**:1152–1161.

29. **Corre, J., F. Cornet, J. Patte, and J.-M. Louarn.** 1997. Unraveling a region-specific hyper-recombination phenomenon: genetic control and modalities of terminal recombination in *Escherichia coli.* *Genetics* **147**:979–989.

30. **de Massey, B., S. Bejar, J. Louarn, J.-M. Louarn, and J. P. Bouche.** 1987. Inhibition of replication forks exiting the terminus region of the *Escherichia coli* chromosome occurs at two loci separated by 5 min. *Proc. Natl. Acad. Sci. USA* **84**:1759–1763.

31. **Drake, J. W.** 1991. A constant rate of spontaneous mutation in DNA-based microbes. *Proc. Natl. Acad. Sci. USA* **88**:7160–7164.

32. **Edlund, T., and S. Normark.** 1981. Recombination between short DNA homologies causes tandem duplication. *Nature* **292**:269–271.

33. **Edlund, T., T. Grundström, and S. Normark.** 1979. Isolation and characterization of DNA repetitions carrying the chromosomal beta-lactamase gene of *Escherichia coli* K-12. *Mol. Gen. Genet.* **173**:115–125.

34. **Folk, W. R., and P. Berg.** 1971. Duplication of the structural gene for glycyl-transfer RNA synthetase in *Escherichia coli.* *J. Mol. Biol.* **58**:595–610.

35. **François, V., J. Louarn, and J.-M. Louarn.** 1989. The terminus of the *Escherichia coli* chromosome is flanked by several polar replication pause sites. *Mol. Microbiol.* **3**:995–1002.

36. **François, V., J. Louarn, J. Patte, J. E. Rebollo, and J.-M. Louarn.** 1990. Constraints in chromosomal inversions in *Escherichia coli* are not explained by replication pausing at inverted ter-

minator-like sequences. *Mol. Microbiol.* **4**:537–542.

37. **Fulcher, C. A., and R. Bauerle.** 1978. Reinitiation of tryptophan operon expression in a promoter deletion strain of *Salmonella typhimurium.* *Mol. Gen. Genet.* **158**:239–250.

38. **Galitski, T., and J. R. Roth.** 1995. Evidence that F plasmid transfer replication underlies apparent adaptive mutations. *Science* **268**:421–423.

39. **Galitski, T., and J. R. Roth.** 1997. Pathways for homologous recombination between chromosomal direct repeats in *Salmonella typhimurium.* *Genetics* **146**:751–767.

40. **Haack, K. R., and J. R. Roth.** 1995. Recombination between chromosomal IS*200* elements supports frequent duplication formation in *Salmonella typhimurium.* *Genetics* **141**:1245–1252.

41. **Higgins, C. F., G. F.-L. Ames, W. M. Barnes, J. M. Clement, and M. Hofnung.** 1982. A novel intercistronic regulatory element of prokaryotic operons. *Nature* **298**:760–762.

42. **Hill, C. W., and J. A. Gray.** 1988. Effects of chromosomal inversion on cell fitness in *Escherichia coli* K-12. *Genetics* **119**:771–778.

43. **Hill, C. W., and B. W. Harnish.** 1981. Inversions between ribosomal RNA genes of *Escherichia coli.* *Proc. Natl. Acad. Sci. USA* **78**:7069–7072.

44. **Hill, C. W., and B. W. Harnish.** 1982. Transposition of a chromosomal segment bounded by redundant rRNA genes into other rRNA genes in *Escherichia coli.* *J. Bacteriol.* **149**:449–457.

45. **Hill, C. W., J. Foulds, L. Soll, and P. Berg.** 1969. Instability of a missense suppressor resulting from a duplication of genetic material. *J. Mol. Biol.* **39**:563–581.

46. **Hill, T. M.** 1996. Features of the chromosome terminus region, p. 1602–1614. *In* F. C. Neidhardt, R. Curtiss III, J. L. Ingraham, E. C. C. Lin, K. B. Low, B. Magasanik, W. S. Reznikoff, M. Riley, M. Schaechter, and H. E. Umbarger (ed.), Escherichia coli *and* Salmonella: *Cellular and Molecular Biology,* 2nd ed., vol. 2. American Society for Microbiology, Washington, D.C.

47. **Hill, T. M., and K. J. Marinus.** 1990. *Escherichia coli* Tus protein acts to arrest the progression of DNA replication forks *in vitro.* *Proc. Natl. Acad. Sci. USA* **87**:2481–2485.

48. **Hoffmann, G. R., R. W. Morgan, and R. C. Harvey.** 1978. Effects of chemical and physical mutagens on the frequency of a large genetic duplication in *Salmonella typhimurium.* I. Induction of duplications. *Mutat. Res.* **52**:73–80.

49. **Hoffmann, G. R., R. W. Morgan, and R. Kirven.** 1978. Effects of chemical and physical mutagens on the frequency of a large genetic duplication in *Salmonella typhimurium.* II. Stimula-

tion of duplication-loss from merodiploids. *Mutat. Res.* **52**:81–86.

50. **Hoffmann, G. R., M. J. Walkowicz, J. M. Mason, and J. F. Atkins.** 1983. Genetic instability associated with the *aroC321* allele in *Salmonella typhimurium* involves genetic duplication. *Mol. Gen. Genet.* **190**:183–188.

51. **Holliday, R.** 1964. A mechanism for gene conversion in fungi. *Genet. Res.* **5**:282–304.

52. **Horiuchi, T., and Y. Fujimura.** 1995. Recombinational rescue of the stalled DNA replication fork: a model based on analysis of an *Escherichia coli* strain with a chromosome region difficult to replicate. *J. Bacteriol.* **177**:783–791.

53. **Horiuchi, T., Y. Fujimura, H. Nishitani, Y. Kobayashi, and M. Hidaka.** 1994. The DNA replication fork blocked at the Ter site may be an entrance for the RecBCD enzyme into duplex DNA. *J. Bacteriol.* **176**:4656–4663.

54. **Horiuchi, T., H. Nishitani, and T. Kobayashi.** 1995. A new type of *E. coli* recombinational hotspot which requires for activity both DNA replication termination events and the Chi sequence. *Adv. Biophys.* **31**:133–147.

55. **Jessop, A. P., and C. Clugston.** 1985. Amplification of the ArgF region in strain HfrP4X of *E. coli* K-12. *Mol. Gen. Genet.* **201**:347–350.

56. **Jones, I. M., S. B. Primrose, and S. D. Erlich.** 1982. Recombination between short direct repeats in a RecA host. *Mol. Gen. Genet.* **188**:486–489.

57. **Kleckner, N., K. Reichardt, and D. Botstein.** 1979. Inversions and deletions of the *Salmonella* chromosome generated by the translocatable tetracycline resistance element Tn*10. J. Mol. Biol.* **127**:89–115.

58. **Kogoma, T.** 1997. Stable DNA replication: the interplay between DNA replication, homologous recombination, and transcription. *Microbiol. Mol. Biol. Rev.* **61**:212–238.

59. **Kogoma, T., G. W. Cadwell, K. G. Barnard, and T. Asai.** 1996. The DNA replication priming protein, PriA, is required for homologous recombination and double-strand break repair. *J. Bacteriol.* **178**:1258–1264.

60. **Konrad, E. B.** 1977. Method for the isolation of *Escherichia coli* mutants with enhanced recombination between chromosomal duplications. *J. Bacteriol.* **130**:167–172.

61. **Krug, P. J., A. Z. Gileski, R. J. Code, A. Torjussen, and M. B. Schmid.** 1994. Endpoint bias in large Tn*10*-catalyzed inversions in *Salmonella typhimurium. Genetics* **136**:747–756.

62. **Kuempel, P., A. Hogaard, M. Nielsen, O. Nagappan, and M. Tecklenburg.** 1996. Use of a transposon (Tn*dif*) to obtain suppressing and nonsuppressing insertions of the *dif* resolvase site of *Escherichia coli. Genes Dev.* **10**:1162–1171.

63. **Kuempel, P. L., S. A. Duerr, and N. R. Seeley.** 1977. Terminus region of the chromosome in *Escherichia coli* inhibits replication forks. *Proc. Natl. Acad. Sci. USA* **74**:3927–3931.

64. **Kuempel, P. L., J. M. Henson, L. Dircks, M. Tecklenburg, and D. F. Lim.** 1991. *dif*, a *recA*-independent recombination site in the terminus region of *Escherichia coli. New Biol.* **3**:799–811.

65. **Kuzminov, A.** 1995. Instability of inhibited replication forks in *E. coli. Bioessays* **17**:733–741.

66. **Kuzminov, A.** 1995. Collapse and repair of replication forks in *Escherichia coli. Mol. Microbiol.* **16**:373–384.

67. **Kuzminov, A., E. Schabtach, and F. W. Stahl.** 1994. χ sites in combination with RecA protein increase the survival of linear DNA in *Escherichia coli* by inactivating exoV activity of RecBCD nuclease. *EMBO J.* **13**:2764–2776.

68. **Lee, E. H., A. Kornberg, M. Hikada, T. Kobayashi, and T. Horiuchi.** 1989. Escherichia coli replication termination protein impedes the action of helicases. *Proc. Natl. Acad. Sci. USA* **86**:9104–9108.

69. **Lehner, A. F., and C. W. Hill.** 1980. Involvement of ribosomal ribonucleic acid operons in *Salmonella typhimurium* chromosomal rearrangements. *J. Bacteriol.* **143**:492–498.

70. **Leslie, N. R., and D. J. Sherratt.** 1995. Site-specific recombination in the replication terminus of *Escherichia coli*: functional replacement of *dif. EMBO J.* **14**:1561–1570.

71. **Lin, R. J., M. Capage, and C. W. Hill.** 1984. A repetitive DNA sequence, *rhs*, responsible for duplications within the *Escherichia coli* K-12 chromosome. *J. Mol. Biol.* **177**:1–18.

72. **Lindegren, C. C.** 1953. Gene conversion in *Saccharomyces. J. Genet.* **51**:625–637.

73. **Liu, B., and B. M. Alberts.** 1995. Head-on collision between a DNA replication apparatus and RNA polymerase transcription complex. *Science* **267**:1131–1136.

74. **Liu, S.-L., and K. E. Sanderson.** 1995. Rearrangements in the genome of the bacterium *Salmonella typhi. Proc. Natl. Acad. Sci. USA* **92**:1018–1022.

75. **Liu, S.-L., and K. E. Sanderson.** 1995. I-*Ceu*I reveals conservation of the genome of independent strains of *Salmonella typhimurium. J. Bacteriol.* **177**:3355–3357.

76. **Liu, S.-L., and K. E. Sanderson.** 1995. The chromosome of *Salmonella paratyphi* A is inverted by recombination between *rrn*H and *rrn*G. *J. Bacteriol.* **177**:6585–6592.

77. **Liu, S.-L., and K. E. Sanderson.** 1996. Highly plastic chromosomal organization in *Salmonella typhi. Proc. Natl. Acad. Sci. USA* **93:**10303–10308.

78. **Liu, S.-L., A. Hessel, and K. E. Sanderson.** 1993. The *Xba*I-*Bln*I-*Ceu*I genomic cleavage map of *Salmonella enteritidis* shows an inversion relative to *Salmonella typhimurium* LT2. *Mol. Microbiol.* **10:**655–664.

79. **Louarn, J., J. Patte, and J. M. Louarn.** 1977. Evidence for a fixed termination site of chromosome replication in *Escherichia coli* K12. *J. Mol. Biol.* **115:**295–314.

80. **Louarn, J., F. Cornet, V. François, J. Patte, and J.-M. Louarn.** 1994. Hyperrecombination in the terminus region of the *E. coli* chromosome: possible relation to nucleoid organization. *J. Bacteriol.* **176:**7524–7531.

81. **Louarn, J.-M., J. Louarn, V. François, and J. Patte.** 1991. Analysis and possible role of hyperrecombination in the termination region of the *Escherichia coli* chromosome. *J. Bacteriol.* **173:** 5097–5104.

82. **Lovett, S. T., P. T. Drapkin, V. A. Sutura, Jr., and T. J. Gluckman.** 1993. A sister-strand exchange mechanism for *recA*-independent deletion of repeated DNA sequences in *Escherichia coli. Genetics* **135:**631–642.

83. **Lovett, S. T., T. J. Gluckman, P. J. Simon, V. A. Sutera, Jr., and P. T. Drapkin.** 1994. Recombination between repeats in *Escherichia coli* by a *recA*-independent, proximity-sensitive mechanism. *Mol. Gen. Genet.* **245:**294–300.

84. **Mahan, M. J., and J. R. Roth.** 1988. Reciprocality of recombination events that rearrange the chromosome. *Genetics* **120:**23–35.

85. **Mahan, M. J., and J. R. Roth.** 1989. Role of *recBC* function in formation of chromosomal rearrangements: a two-step model for recombination. *Genetics* **121:**433–443.

86. **Marinus, M. G., and E. B. Konrad.** 1976. Hyper-recombination in *dam* mutants of *Escherichia coli* K-12. *Mol. Gen. Genet.* **149:**273–277.

87. **Marvo, S. L., R. S. King, and S. R. Jaskunas.** 1983. Role of short regions of homology in intermolecular illegitimate recombination events. *Proc. Natl. Acad. Sci. USA* **80:**2452–2456.

88. **Masai, H., T. Asai, Y. Kubota, K. Arai, and T. Kogoma.** 1994. *Escherichia coli* PriA protein is essential for inducible and constitutive stable DNA replication. *EMBO J.* **13:**5338–5346.

89. **Matfield, M., R. Badawi, and W. J. Brammar.** 1985. Rec-dependent and rec-independent recombination of plasmid-borne duplications in *Escherichia coli* K12. *Mol. Gen. Genet.* **199:**518–523.

90. **Mazin, A. V., A. V. Kuzminov, G. L. Dianov, and R. I. Salganik.** 1991. Molecular mechanisms of deletion formation in *Escherichia coli* plasmids. II. Deletion formation mediated by short direct repeats. *Mol. Gen. Genet.* **228:** 209–214.

91. **McGlynn, P., A. A. Al-Deib, J. Lui, K. J. Marians, and R. G. Lloyd.** 1997. The DNA replication protein PriA and the recombination protein RecG bind D-loops. *J. Mol. Biol.* **270:** 212–221.

92. **Médigue, C., A. Viari, A. Henaut, and A. Danchin.** 1993. Colibri: a functional database for the *Escherichia coli* genome. *Microbiol. Rev.* **57:**623–654.

93. **Michel, B., S. D. Erlich, and M. Uzest.** 1997. DNA double-strand breaks caused by replication arrest. *EMBO J.* **16:**430–438.

94. **Miesel, L., A. Segall, and J. R. Roth.** 1994. Construction of chromosomal rearrangements in *Salmonella* by transduction: inversions of nonpermissive segments are not lethal. *Genetics* **137:** 919–932.

95. **Mills, D. M., V. Bajaj, and C. A. Lee.** 1995. A 40 kb chromosomal fragment encoding *Salmonella typhimurium* invasion genes is absent from the corresponding region of the *Escherichia coli* K-12 genome. *Mol. Microbiol.* **15:**749–759.

96. **Mitchell, M. B.** 1955. Aberrant recombination of pyroxidine mutants of *Neurospora. Proc. Natl. Acad. Sci. USA* **41:**215–220.

97. **Mitchell, M. B.** 1955. Further evidence of aberrant recombination in *Neurospora. Proc. Natl. Acad. Sci. USA* **41:**935–937.

98. **Myers, R. S., and F. W. Stahl.** 1994. χ and the RecBCD enzyme of *Escherichia coli. Annu. Rev. Genet.* **28:**49–70.

99. **Ochman, H., and E. A. Groisman.** 1994. The origin and evolution of species differences in *Escherichia coli* and *Salmonella typhimurium. EXS* **69:**479–493.

100. **Ochman, H., and A. C. Wilson.** 1987. Evolution in bacteria: evidence for a universal substitution rate in cellular genomes. *J. Mol. Evol.* **26:**74–86.

101. **Olive, S.** 1959. Aberrant tetrads in *S. fimicola. Proc. Natl. Acad. Sci. USA* **45:**727–732.

102. **Payne, G. M., E. N. Spudich, and G. F. Ames.** 1985. A mutational hot-spot in the *hisM* gene of the histidine transport operon in *Salmonella typhimurium* is due to deletion of repeated sequences and results in an altered specificity of transport. *Mol. Gen. Genet.* **200:**493–496.

103. **Petit, M.-A., J. M. Mesas, P. Noiret, F. Morel-Deville, and S. D. Erlich.** 1992. Induction of DNA amplification in the *Bacillus subtilis* chromosome. *EMBO J.* **11:**1317–1326.

104. **Rayssiguier, C., D. S. Thaler, and M. Radman.** 1989. The barrier to recombination between *Escherichia coli* and *Salmonella typhimurium* is disrupted in mismatch-repair mutants. *Nature* **342:**396–401.

105. **Rebello, J.-E., V. François, and J.-M. Louarn.** 1988. Detection and possible role of two large nondivisible zones on the *Escherichia coli* chromosome. *Proc. Natl. Acad. Sci. USA* **85:**9391–9395.

106. **Riley, M., and S. Krawiec.** 1987. Genome organization, p. 967–981. *In* F. C. Neidhardt, J. L. Ingraham, K. B. Low, B. Magasanik, M. Schaechter, and H. E. Umbarger (ed.), Escherichia coli *and* Salmonella typhimurium: *Cellular and Molecular Biology*. American Society for Microbiology, Washington, D.C.

107. **Roth, J. R., N. Benson, T. Galitski, K. Haack, J. G. Lawrence, and L. Miesel.** 1996. Rearrangements of the bacterial chromosome: formation and applications, p. 2256–2276. *In* F. C. Neidhardt, R. Curtiss III, J. L. Ingraham, E. C. C. Lin, K. B. Low, B. Magasanik, W. S. Reznikoff, M. Riley, M. Schaechter, and H. E. Umbarger (ed.), Escherichia coli *and* Salmonella: *Cellular and Molecular Biology*, 2nd ed., vol. 2. American Society for Microbiology, Washington, D.C.

108. **Sanderson, K. E., and C. A. Hall.** 1968. F-prime factors of *Salmonella typhimurium* and an inversion between *S. typhimurium* and *E. coli. Genetics* **64:**215–228.

109. **Sanderson, K. E., A. Hessel, S.-L. Liu, and K. E. Rudd.** 1996. The genetic map of *Salmonella typhimurium*, edition VIII, p. 1903–1999. *In* F. C. Neidhardt, R. Curtiss III, J. L. Ingraham, E. C. C. Lin, K. B. Low, B. Magasanik, W. S. Reznikoff, M. Riley, M. Schaechter, and H. E. Umbarger (ed.), Escherichia coli *and* Salmonella: *Cellular and Molecular Biology*, 2nd ed., vol. 2. American Society for Microbiology, Washington, D.C.

110. **Sandler, S. J., S. Samra, and A. J. Clark.** 1996. Differential suppression of *priA2::kan* phenotypes in *Escherichia coli* K-12 by mutations in *priA*, *lexA*, and *dnaC. Genetics* **143:**5–13.

111. **Savageau, M. A.** 1983. *Escherichia coli*: habitats, cell types, and molecular mechanisms of gene control. *Am. Nat.* **122:**732–744.

112. **Saveson, C. J., and S. T. Lovett.** 1997. Enhanced deletion formation by aberrant DNA replication in *Escherichia coli. Genetics* **146:**457–470.

113. **Schmid, M. B., and J. R. Roth.** 1983. Genetic methods for analysis and manipulation of inversion mutations in bacteria. *Genetics* **105:**517–537.

114. **Schmid, M. B., and J. R. Roth.** 1983. Selection and endpoint distribution of bacterial inversion mutations. *Genetics* **105:**539–557.

115. **Schmid, M. B., and J. R. Roth.** 1987. Gene location affects expression level in *Salmonella typhimurium. J. Bacteriol.* **169:**2872–2875.

116. **Schofield, M. A., R. Agbunag, and J. H. Miller.** 1992. DNA inversions between short inverted repeats in *Escherichia coli. Genetics* **132:**295–302.

117. **Sclafani, R. A., and J. A. Wechsler.** 1981. High frequency of genetic duplications in the *dnaB* region of the *Escherichia coli* K12 chromosome. *Genetics* **98:**677–690.

118. **Segall A. M., and J. R. Roth.** 1989. Recombination between homologies in direct and inverse orientation in the chromosome of *Salmonella*: intervals which are nonpermissive for inversion formation. *Genetics* **122:**737–747.

119. **Segall, A. M., and J. R. Roth.** 1994. Approaches to half-tetrad analysis in bacteria: recombination between repeated, inverse-order chromosomal sequences. *Genetics* **136:**27–39.

120. **Segall, A. M., M. J. Mahan, and J. R. Roth.** 1988. Rearrangement of the bacterial chromosome: forbidden inversions. *Science* **241:**1314–1318.

121. **Seigneur, M., V. Bidnenko, S. D. Erlich, and B. Michel.** 1998. RuvAB acts at arrested replication forks. *Cell* **95:**419–430.

122. **Sharp, P. M.** 1991. Determinants of DNA sequence divergence between *Escherichia coli* and *Salmonella typhimurium*: codon usage, map position, and concerted evolution. *J. Mol. Evol.* **33:**23–33.

123. **Shen, P., and H. V. Huang.** 1986. Homologous recombination in *Escherichia coli*: dependence on substrate length and homology. *Genetics* **112:**441–457.

124. **Shen, P., and H. V. Huang.** 1989. Effect of base pair mismatches on recombination via the RecBCD pathway. *Mol. Gen. Genet.* **218:**358–360.

125. **Shyamala, V., E. Schneider, and G. F. Ames.** 1990. Tandem chromosomal duplications: role of REP sequences in the recombination event at the joint-point. *EMBO J.* **9:**939–946.

126. **Smith, G. R.** 1991. Conjugational recombination in *E. coli*: myths and mechanisms. *Cell* **64:**19–27.

127. **Sonti, R. V., and J. R. Roth.** 1989. Role of gene duplications in the adaptation of *Salmonella typhimurium* to growth on limiting carbon sources. *Genetics* **123:**19–28.

128. **Sousa, C., V. de Lorenza, and A. Cebolla.** 1997. Modulation of gene expression through

chromosomal positioning in *Escherichia coli*. *Microbiology* **143:**2071–2078.

129. **Straus, D. S.** 1975. Selection for a large genetic duplication in *Salmonella typhimurium*. *Genetics* **80:**227–237.

130. **Straus, D. S., and L. D. Straus.** 1976. Large overlapping duplications in *Salmonella typhimurium*. *J. Mol. Biol.* **103:**143–153.

131. **Streissinger, G.** 1985. Mechanisms of spontaneous and induced frameshift mutations. *Genetics* **109:**633–659.

132. **Szostak, J. W., T. Orr-Weaver, R. J. Rothstein, and F. W. Stahl.** 1983. The double-strand-break repair model for recombination. *Cell* **33:**25–35.

133. **Tecklenburg, M., A. Naumer, A. Nagappan, and P. Kuempel.** 1995. The *dif* resolvase locus of the *Escherichia coli* chromosome can be replaced by a 33-bp sequence, but function depends on location. *Proc. Natl. Acad. Sci. USA* **92:**1352–1356.

134. **Tlsty, T. D., A. M. Albertini, and J. H. Miller.** 1984. Gene amplification in the *lac* region of *E. coli*. *Cell* **37:**217–224.

135. **Trinh, T. Q., and R. R. Sinden.** 1993. The influence of primary and secondary DNA structure in deletion and duplication between direct repeats in *Escherichia coli*. *Genetics* **134:**409–422.

136. **Watt, V. M., C. J. Ingles, M. S. Urdea, and W. J. Rutter.** 1985. Homology requirements for recombination in *Escherichia coli*. *Proc. Natl. Acad. Sci. USA* **82:**4768–4772.

137. **Whoriskey, S. K., V. H. Nghiem, P. M. Leong, J. M. Masson, and J. H. Miller.** 1987. Genetic rearrangements and gene amplification in *Escherichia coli*: DNA sequences at the junctures of amplified gene fusions. *Genes Dev.* **1:**227–237.

ILLEGITIMATE RECOMBINATION IN BACTERIA

Bénédicte Michel

8

Illegitimate recombination is a generic term used to designate the formation of new DNA molecules by the junction of nonhomologous or short homologous DNA sequences. Rearrangements caused by illegitimate recombination include deletions, inversions, amplifications, and translocations. Since there is no requirement for the presence of homologous or specific sequences, they may occur anywhere on a genome and have dramatic consequences. Illegitimate recombination is a ubiquitous phenomenon and includes three types of events. In the first class, rearrangements occur by recombination between short homologous sequences. A second class is associated with site-specific elements (a lysogenic bacteriophage, a transposable element, or an inversion system). A last class groups all rearrangements in which the newly linked sequences share less than 3 bp of homology and have no homology with known specific sites. There is increasing evidence that these three classes of illegitimate-recombination events proceed by different pathways. The present review will focus mainly on the progress made

in the field in the last decade (for recent reviews, see references 38, 39, 53, 63, and 70).

RECOMBINATION BETWEEN SHORT HOMOLOGOUS SEQUENCES

The size of the minimal sequence sufficient to initiate homologous recombination in vivo varies with the bacterial species. In *Escherichia coli*, the minimal homology required for RecA-mediated homologous recombination, or the MEPS (minimum efficient processing segment), is 20 to 30 bp (62, 120, 139). It may be slightly longer in *Bacillus subtilis* (70 bp [61]) and is 10-fold longer in *Campylobacter coli* (270 bp [104]). Recombination between sequences shorter than 20 bp will be considered here as illegitimate. Spontaneous rearrangements in bacterial chromosomes occur in short homologous sequences, 3 to 20 bp long, with an efficiency significantly higher than would be expected from purely random events (55, 90, 111, 144). This has led to systematic studies of the effects of short homologous sequences on genome stability.

Bénédicte Michel, Génétique Microbienne, Institut National de la Recherche Agronomique, Domaine de Vilvert, 78352 Jouy en Josas Cedex France.

Organization of the Prokaryotic Genome, Edited by Robert L. Charlebois,
© 1999 American Society for Microbiology, Washington, D.C.

Structural Factors Influencing Recombination between Short Repeats

RECOMBINATION FREQUENCY INCREASES WITH THE LENGTH AND THE GC CONTENT OF REPEATED SEQUENCES

The frequency of recombination between two short repeated sequences increases with their lengths. Deletion systems that allow a systematic study of this parameter were constructed in *E. coli* and *B. subtilis* (99, 100). Plasmids were used in *B. subtilis*, and the T7 phage was used in *E. coli*. In both organisms, the increase is exponential with the length of the repeats. Compilation of the results allowed the conclusion that the frequency of recombination doubled for each base pair added for homologies between 8 and 21 bp. For longer homologies, the increase was more gradual, becoming linear as expected for events in which RecA promoted the homology search (11, 82, 99). Consistent with the idea that increasing the lengths of short homologies promoted rearrangements by stabilization of a paired intermediate, GC-rich sequences were reported to be preferentially used for deletion formation both in T4 (124) and in *E. coli* plasmids (14).

RECOMBINATION FREQUENCY DECREASES WITH THE LENGTH OF THE INTERVENING SEQUENCE

The effects of the distances between the short homologous sequences were also systematically tested in *E. coli* and *B. subtilis* plasmids (25, 78) and in the *B. subtilis* chromosome (25). In all cases, the frequency of intramolecular recombination decreased exponentially with the distance between the homologous sequences, and a biphasic distribution of the frequencies was observed. Although sequences in close proximity recombine at a relatively high frequency (5×10^{-6} per generation for 18-bp direct repeats separated by 51 bp), the recombination efficiency decreases very rapidly with increasing distances (a decrease of two log units for distances between 30 and 300 bp [25]). It has been proposed that the biphasic distribution reflects the existence of two different mechanisms for short and long distances. For a distance of less than 400 bp, contacts may occur during replication when the DNA is single stranded (see below for recombination mechanisms), since for these short distances, contacts of the double-stranded repeats may be limited by the lack of flexibility of short sequences. For longer distances, the ability of the double-stranded DNA to bend may allow contacts between the repeats. In T7 bacteriophage, recombination frequency also decreases with the distance between short homologous sequences (110).

THE PRESENCE OF SECONDARY STRUCTURES STIMULATES REARRANGEMENTS

When a DNA sequence flanked by short direct repeats is palindromic, recombination between the repeats occurs at a higher frequency. This was first recognized by the analysis of the spectrum of spontaneous deletions occurring in the *lac* operon (45) and during studies of precise and imprecise transposon excision (42, 143). Composite transposons were often used for systematic studies of the effects of palindromes. Tn*5* and Tn*10* are flanked by long inverted repeats of about 1.5 kb, and they create a 9-bp duplication of the target DNA upon insertion. Precise transposon excision occurs independently of the transposase, by recombination between the short duplicated sequences adjacent to the inverted repeats. When short homologous sequences are separated by a palindrome, the frequency of deletion depends on the length of this palindrome (99, 141, 142). It also depends on the palindrome's capacity to adopt a hairpin structure, itself determined by the composition of the loop and the stem and by the topology of the molecule (22, 32, 33, 101a, 123). The formation of a hairpin structure has two consequences: it brings the recombining sequences closer together, which will increase their probability of interaction (see above), and it induces replication pauses (9, 20, 67), which favor recombination between repeated sequences (see below). The stimulation of illegitimate recombination by palindromic sequences is due to a combination of these two effects, and whether one or the other is limiting may depend on the particular system considered. For example, the magnitude of the effect in multicopy plasmids (several log units of increase in recombination frequency for a twofold increase in the length of the palindromic sequence, from 40 to 50 to 90 to 100 bp) likely results from the additive effects of an increase in the efficiency of deletion formation and an increase in the level of enrichment in the recombinant plasmids that have lost a sequence which impedes replication (13, 26). For recent reviews of the instability of palindrome sequences, see references 69 and 71.

STABILITY OF NATURALLY REPEATED SEQUENCES IN BACTERIAL CHROMOSOMES

Several enterobacteria contain a number of highly repetitive sequences in noncoding regions, some of which are composed of one or several palindromic sequences (reviewed in reference 5). BIMEs (bacterial interspersed mosaic elements) are mosaic combinations of several short sequence motifs (44). One of these motifs is a palindromic sequence, called PU (palindromic unit) or REP (repetitive extragenic sequence), about 40 bp long separated by short (up to 40-bp) extra-PU sequences. Comparison of six intergenic regions in 51 *E. coli* and *Shigella* natural isolates showed multiple sequence variations within or near BIME elements (6). The rearrangements include excision of a whole BIME and expansion or deletion within BIMEs, ending in a variable number of PUs. They were caused by recombination between short repeated sequences. For example, excision of a whole BIME was observed when the BIME sequences were bracketed by direct repeats of 28 to 37 bp and was presumably facilitated by the palindromic structure of the deleted sequence.

A second family of repeated sequences, called IRUs (intergenic repeat units [119]) or ERIC (enterobacterial repetitive intergenic consensus [51]), are imperfect palindromes, about 127 bp in length, found in multiple copies in enterobacteria. Comparison of ERIC sequences from different enterobacteria revealed that they differ by the presence or absence of palindromic sequences bracketed by short homologies, which suggests that these sequences have evolved by palindrome-induced deletions between direct-repeat sequences (118). Palindrome-induced deletion may thus control the number and the lengths of these repeated elements, which were proposed to play a role in the functional organization of bacterial chromatin (5, 118).

STABILITY OF EUKARYOTIC REPEATED SEQUENCES INTRODUCED IN *E. COLI*: MICROSATELLITE SEQUENCES

In view of the availability of bacterial genetic tools and the ease of manipulation of small genomes, *E. coli* represents the recipient of choice for studies of the stability of eukaryotic sequences of interest. Therefore, although microsatellites have not been found so far in sequenced prokaryotic genomes, the discovery that several human disorders result from the expansion of repeated nucleotide triplets (reviewed in references 4 and 84) and the observation that certain cancers are associated with microsatellite instability (reviewed in reference 64) led to the study of microsatellite stability in *E. coli*, used as a model organism.

Microsatellites were introduced into the *E. coli* chromosome (91). As expected for repeated sequences, they are unstable in *E. coli*, with a tendency to be deleted rather than to expand, which probably accounts for their absence from prokaryotic genomes. The frequency of deletion and expansion of $(AC)_n$ repeats varies from 10^{-6} per generation for repeats of 18 bp to 10^{-4} for repeats of 51 bp, with 5 to 10 times more deletions than expansions and variation of 1 or 2 dinucleotides representing 80 to 90% of total events (91).

Some microsatellite sequences can adopt a palindromic structure in vitro, or complex structures that also impede replication (60, 96, 132). As discussed below, elements that slow down or block replication progression are in general factors of instability. Detailed studies of lambda bacteriophage revealed that the propensity of microsatellite sequences to adopt a hairpin structure in vivo is an important determinant of their stability (32). AC repeats also tend to be deleted when inserted in M13 bacteriophage (74). In contrast, in T7 bacteriophage, insertions are more frequent than deletions, since addition of a dinucleotide occurs at a frequency of 10^{-3} and loss occurs at a frequency of 10^{-4} (149). This may reflect some specific properties of the T7 replication machinery.

Repeats also tend to shrink when inserted in plasmids (12, 18, 114). At least for long repeats of triplet sequences, this can be explained by the observation that they prevent plasmid propagation (18). Consequently, molecules that have lost repeats have a replicative advantage, which may increase the apparent recombination frequency, as observed for other sequences that impede plasmid replication (13, 26). In addition, some of these plasmids prevent cell growth, allowing enrichment in cells containing deletant plasmids (18). This renders difficult the interpretation of quantitative studies of the stability of long microsatellite sequences in multicopy plasmid systems. In contrast with the bias for the loss of triplet repeats at 37°C, propagation of plasmids carrying CTG repeats at 25°C in cells deficient for several recombination proteins allowed the selection of tract expansions (109a). It would be interesting to test whether the same phenomenon can also be observed when the triplet repeats are in the *E. coli* chromosome.

Influence of Functional Parameters on Direct-Repeat Stability

In addition to particular features of DNA sequences, the frequency of illegitimate recombination depends on the mode of replication of the molecule and on the presence of sequences that induce replication arrest. It is also influenced by the presence of DNA lesions and by the transcription status of the region.

MODE OF REPLICATION

In *E. coli*, transposons are highly unstable when present on single-stranded bacteriophages (48) or when a single-stranded replication origin is activated on a hybrid molecule (31). Transposon excision from an F' episome is strongly stimulated by conjugation (10, 125). In *B. subtilis*, transposon excision is increased up to 1,000-fold when the transposon is carried by a rolling-circle plasmid (56). Altogether, these results point to a protective role against illegitimate recombination of a coupled leading-lagging-strand DNA synthesis. Differences in the stabilities of repeats on leading and lagging strands of the *E. coli* chromosome were also reported: for example, TG-AC repeats are about twofold more unstable when the TG repeats are on the leading strand (91). This difference could be explained either by different properties of the polymerase molecules that replicate the two strands or by the single-stranded nature of the lagging-strand template. In bacteriophage λ, deletions of palindromic DNA sequences occur preferentially on the lagging strand of the replication fork (101a), consistent with the idea that formation of a hairpin structure is facilitated on this strand.

REPLICATION PAUSES

Replication pauses cause instability in a variety of systems (reviewed in reference 16). Rearrangements may occur by different mechanisms, depending on the nature of the replication arrest. Replication blockage is caused by elements such as nucleoprotein complexes bound to DNA, which block the progression of helicase and consequently that of the entire replication machinery, resulting in the formation of a Y structure. The effects of such replication blockage were studied with monocopy plasmids that carry specific *E. coli* replication terminators (*Ter*). *Ter* sites form a complex with the Tus protein, which mediates the arrest of DNA replication through interaction with the replicative helicase (reviewed in reference 50). Replication arrest at *Ter* induces the formation of deletions between repeated sequences (14). This suggests that other nucleoprotein complexes capable of impeding the progression of replication forks may be a source of illegitimate recombination.

Encounters with sequences such as palindromes or DNA lesions also cause replication pause. The arrest of DNA polymerase by a UV lesion may be followed by the opening of the helix downstream of the arrest site by the helicase (reference 109; reviewed in reference 108). This type of blockage may favor polymerase errors, possibly because of the single-stranded nature of the DNA on which replication restarts, as discussed below.

DNA LESIONS
UV irradiation induces recombination between short direct repeats. It increases transposon excision in *E. coli*, and this induction is SOS dependent (24, 95). It also increases the formation of lambda transducing particles (147). Such particles can be formed via (i) the erroneous action of the *int-xis*-specific system (reviewed in reference 38), (ii) recombination between short homologous sequences, or (iii) recombination between nonhomologous sequences (121). Of these three events, only recombination between short homologous sequences is stimulated by UV irradiation (reference 147 and references therein).

Oxidative lesions are known to induce rearrangements. Deletions by illegitimate recombination between short homologous sequences were reported in *E. coli fur* mutants, in which a defect in iron metabolism regulation results in increased oxidative damage (126, 127). Exposure to hydrogen peroxide induces instability of microsatellite sequences carried by *E. coli* plasmids (55a). Finally, deletions by recombination between short repeated sequences also increase in carotenoidless mutants of *Rubivivax gelatinosus*, defective

in protection against photooxidative stress (97).

TRANSCRIPTION
In bacteria, transcription inhibits deletion between tandemly repeated sequences 10-fold (26). In contrast, transcription was shown to increase recombination between nonhomologous sequences (see below).

Molecular Models Proposed for Recombination between Short Repeated Sequences
Two models have been proposed for illegitimate recombination by the junction of short homologous sequences. The first model, referred to as the "slipped-mispairing model," "replication slippage," or "copy choice recombination," derives from the frameshift mispairing model originally proposed to account for the high incidence of frameshift mutations in short regions of reiterated base pairs (40). In this model (Fig. 1), a replication pause leads to dissociation of the newly synthesized strand from its template and pairing with the other repeated sequence. Upon continuation of replication, a DNA strand containing either a deletion or a duplication is synthesized. Deletions occur when the short homologous sequence used for reannealing is located downstream of the pause site relative to the direction of replication. Conversely, an upstream location will result in duplication. Intermolecular-recombination events of this type will lead to translocations. Switching from a leading-strand to a lagging-strand template (or vice versa) produces complex rearrangements associated with deletions or duplications.

The single-strand annealing (SSA) model was originally proposed to account for homologous-recombination events in higher eukaryotes (Fig. 2) (75). The initiating event is a DNA double-strand break (DSB). Exonucleolytic erosion of the double-stranded ends allows the pairing of exposed complementary single-stranded sequences. Subsequent exonucleolytic degradation of the

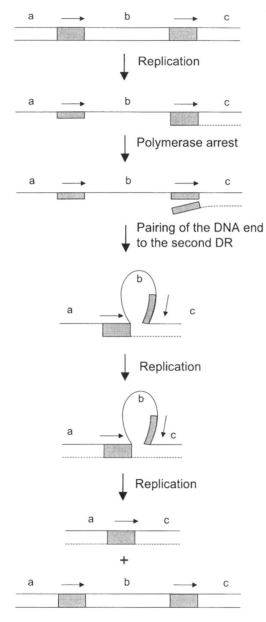

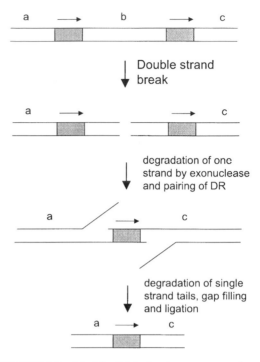

FIGURE 2 Model of deletion between two short homologous sequences by SSA. The directly repeated (DR) sequences are indicated by grey boxes and the arrows above the boxes. The initiating event is a DNA DSB occurring between the repeated sequences. Exonucleolytic degradation of one of the strands renders short homologous sequences single stranded, which allows them to pair. Degradation by exonucleases of the single-stranded tails, gap filling by a polymerase, and ligation lead to the deletion of the b region (defined in the legend to Fig. 1) and of one DR.

FIGURE 1 Model of deletion by replication slippage between two short homologous sequences. The directly repeated (DR) sequences are indicated by shaded boxes and the arrows above the boxes. The hatched lines represent newly synthesized DNA strands, and a, b, and c are the DNA regions flanking (a and c) and between (b) the repeated sequences. A pause of the DNA polymerase during the synthesis of the first DR encountered allows the opening of the DNA. Erroneous pairing with the other DR leads to the loss of the b region and of one of the repeated sequences. The deletion is stabilized by replication

single-stranded tails, filling of eventual gaps by a DNA polymerase, and ligation of the remaining nicks lead to a deleted molecule. This model can explain deletions between repeated sequences.

Several lines of evidence argue in favor of the existence of both the replication slippage and SSA pathways.

continuation. Duplication of the b sequence and of one of the DR occurs when DNA synthesis pauses at the second DR encountered and the newly synthesized DNA folds back, allowing erroneous pairing with the first DR and a second replication of the b region (not shown).

ILLEGITIMATE RECOMBINATION BY REPLICATION SLIPPAGE

A prediction of the replication slippage mechanism is that recombinant molecules are synthesized de novo and do not carry parental DNA. Thus, a radioactively labelled parental molecule should give rise to unlabelled recombinant molecules. This prediction was tested for the excision of a transposon from molecules replicated by the M13 replication origin (31). No transfer of parental DNA to the recombinant molecules could be detected during transposon excision, which strongly supports a replication slippage mechanism. The stimulation of transposon excision by rolling-circle replication adds to the long list of indirect evidence that supports the occurrence of the replication slippage events in vivo (reviewed in reference 16). More complex rearrangements may also result from template switching of a DNA polymerase. For example, a replication switch from the leading-strand to the lagging-strand template was proposed to explain the occurrence of deletion-inversions in long palindromes and in quasipalindromic sequences (101, 105).

In vitro experiments also support the existence of the replication slippage pathway: the three *E. coli* polymerases, *E. coli* polymerase I and polymerase II, involved in DNA repair, and *E. coli* polymerase III, which replicates the bacterial chromosome, were all shown to mediate rearrangements by replication slippage (19, 20, 98). This may be a general property of DNA polymerases, since it was also observed for several bacteriophage DNA polymerases (19). Interestingly, the efficiency of slippage is inversely related to the ability of the polymerases to catalyze strand displacement (19).

ILLEGITIMATE RECOMBINATION BY SSA

The initiating event in the formation of deletions via an SSA pathway is a DSB (Fig. 2). Studies of deletions between directly repeated sequences formed upon transformation with linear plasmid DNA (29), or upon linearization of T7 phage molecules (65, 148) have provided direct evidence that SSA takes place.

Although the SSA model explains deletions by an error in the repair of a DSB, the question of the origin of spontaneous DSBs remains. DSBs may be due to spontaneous oxidative lesions (references 52 and references therein), and as expected, deletions are increased in cells defective for oxidative protection (97, 126, 127). They also occur upon UV irradiation, as proposed to explain the occurrence of λ transducing particles induced by UV irradiation (147). DSBs may also result from a defect in the sealing of lagging strands during chromosomal replication (21), from a defect in postreplicative mismatch repair (137), and from replication fork blockage (88). Replication terminators of *E. coli* induce recombination between short directly repeated sequences and were therefore used to test the mechanism of replication pause-induced deletions (14). The occurrence of deletions increased in a strain devoid of exonuclease V, in which linear intermediates are stabilized by the inactivation of the major double-strand exonuclease. This indicated that deletions between microhomologies induced by replication arrest at *Ter* sites required a linear intermediate. Hence, these deletions resulted from spontaneous DSBs at stalled replication forks (Fig. 3) (14).

REPLICATION SLIPPAGE VERSUS SSA

In summary, the nature of the replicating element, the bacterial growth conditions, and the DNA sequences surrounding the short homologies will determine the initial event leading to replication slippage or SSA. Both pathways depend on the lengths and the GC contents of the repeats, since they imply a pairing step. Both can be induced by replication pauses, and the nature of the pause may dictate the pathway. A nucleoprotein complex that blocks the progression of helicases, and hence of the whole replisome, may induce breakage and initiate an SSA pathway (14). The transient pause of a polymerase caused by a small palindrome will be more likely to in-

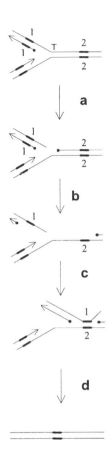

FIGURE 3 Repair of a broken replication fork by SSA (reproduced from reference 14). A replication fork arrested at a *Ter* (T) site is represented. Short repeated sequences (1 and 2) are shown as bold lines. (a) Breakage of the lagging-strand template in the vicinity of *Ter*, generating a DSB. (b) Nucleolytic degradation of the exposed 5' ends, generating a 3'-tailed single-stranded region. (c) Annealing of complementary sequences 1 and 2. (d) The intermediate is repaired by removal of the 3' tail, gap filling, and ligation. A round of replication produces the deletant plasmid molecule. Arrows indicate the 3' ends of leading and lagging strands.

duce polymerase slippage. Both mechanisms may be involved in UV-induced deletions. On the one hand, the helicase creates a single-stranded region downstream of the polymerase arrest site on which slippage may be facilitated (reviewed in reference 108). On the other hand, UV lesions induce DNA breakage (138).

Genetic Studies of Illegitimate Recombination between Short Repeated Sequences

Most of the genetic studies of illegitimate recombination were performed in *E. coli*, either on the chromosome or with bacteriophages. Experiments in which multicopy plasmids were allowed to propagate for several generations in mutant strains may be less reliable, since the mutation may modify the replication and segregation properties of parental and recombinant plasmids. Experiments with high-copy-number plasmids are therefore not systematically reported here.

DOWN MUTANTS

Most of the recombination events between short homologous sequences occur independently from the action of RecA, since the length considered is far below the *E. coli* MEPS. It should be noted, however, that on the F' episome, deletions between short homologous sequences were reduced 25-fold by a *recA* mutation, suggesting that RecA-dependent recombination between sequences of less than 15 bp occurs on this element (1). Similarly, the inversion of an 800-bp sequence flanked by short direct repeats close to the MEPS (21 bp) depends on RecA and RecBCD (113). This observation is not surprising, since inversions are reciprocal events which cannot be accounted for by either of the two main pathways of RecA-independent recombination.

Mutants reported to decrease the frequency of recombination between short homologous sequences include *recJ* mutants that reduce 5- to 10-fold the formation of lambda transducing phages (131), *xth* mutants deficient in exonuclease III, and *lig* (ligase) mutants that decrease the formation of certain plasmid deletions (28, 29, 86). In addition, ligase-over-producing mutations increase the frequency of deletions formed upon transformation with linear-plasmid DNA (27). The three enzymes, RecJ, exonuclease III, and ligase, have been proposed to promote deletion formation through their participation in an SSA reaction.

RecJ or exonuclease III exonucleolytic action would expose short repeated sequences after a DSB (29, 131), and ligase would seal the molecule (27, 28, 86). Fis mutations decrease the formation of λ transducing phages, possibly by modification of the local bending (116).

UP MUTANTS

Most of the mutations that modify illegitimate-recombination frequencies have a stimulatory effect, which suggests that bacteria have evolved protective systems against this type of event. Three classes of up mutations can be distinguished. A null mutation may confer a hyperrecombination phenotype when it inactivates a protein that would destroy a recombination intermediate. A second class of hyperrecombination phenotypes results from the synthesis of a mutated protein that directly favors rearrangements. Finally, several hyperrecombination mutations are pleiotropic and may have an indirect effect.

Mutations that increase the frequency of transposon excision were searched for and found in the genes responsible for mismatch repair (mutHLS [79]). It has been proposed that excision involves the formation of a hairpin structure comprising several mismatches. This intermediate would be recognized and destabilized by the mismatch repair enzymes. mutS mutations also increase the instability of microsatellite sequences inserted into the E. coli chromosome (91) and in plasmids (114). This corroborates the observations made in several other organisms and reflects the capacity of the MutHLS system to excise 1- or 2-nucleotide loops formed during replication slippage on repeated sequences.

Deletions at Ter sites are increased by recD mutations that inactivate exonuclease V, the main double-strand exonuclease of E. coli (14). Inactivation of exonuclease V may stabilize linear molecules by allowing resection by single-strand exonucleases, thereby promoting SSA (Fig. 3). Similarly, an increase of the deletion frequency by addAB mutations in B. subtilis was reported, albeit in a multicopy plasmid system (85). The formation of trans-

ducing λ particles via recombination between short homologies was also proposed to occur by an SSA mechanism. These recombination events are stimulated by UV irradiation (147) and are also increased in a recQ mutant (47). RecQ is a 3'-to-5' helicase that may destroy the intermediate produced by annealing of complementary single-stranded ends in an SSA reaction. The synergetic effect of UV irradiation and recQ mutations supports a model in which UV lesions stimulate illegitimate recombination by causing the initial DSBs.

In wild-type strains, λ phage molecules carrying palindromes of a few hundred base pairs cannot propagate (43, 68, 122). During replication, these long palindromes are rendered single stranded and may adopt a hairpin structure, thereby preventing replication progression. However, these phages can propagate in mutants defective for SbcCD, a nuclease that specifically cleaves hairpin structures (30, 43, 68, 122). A model has been proposed to account for these observations, based on a study with cells that carry λ integrated into the chromosome (72) (Fig. 4). In wild-type cells, SbcCD introduces a double-strand break in the stem-loop structures that may either kill the phage or be repaired by homologous recombination with the sister chromosome. This restores the original sequence but still prevents replication, creating a strong selective pressure for the loss of the palindrome. On the other hand, persistence of the hairpin structure in the absence of SbcCD may favor deletion by polymerase slippage. In support of this hypothesis, the propagation of palindromic sequences of several kilobases was only possible in sbcCD mutants and was accompanied by shrinkage of the palindrome (122). In contrast, the sbcCD mutation was essential for the expansion of CTG tracts on multicopy plasmids at low temperature, presumably because it allowed the stabilization of secondary structure (109a). The role of the other mutations carried by the strain in which expansion took place (recB, recJ, uvrD, and umuC) was not tested.

A second class of hyperrecombination mutations corresponds to the synthesis of an al-

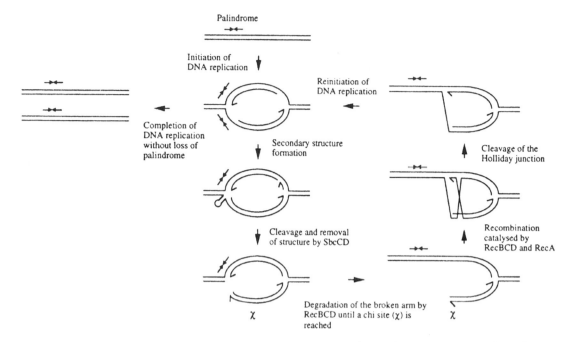

FIGURE 4 Model of SbcCD action (reproduced, with permission, from reference 72). During the replication of a palindromic sequence, intrastrand base pairing can cause pausing of DNA replication and is a potential precursor for deletion. To reinitiate replication, the SbcCD protein removes the secondary structure and generates a DSB. This DSB is repaired by homologous recombination with the sister chromosome to allow the reconstitution of a replication fork. This model specifically predicts that the intact copy of the palindromic sequence is replicated again after reconstitution of the replication fork.

tered protein, which promotes rearrangements via its modified activity. *xonA* mutations are alleles of the *sbcB* gene encoding exonuclease I, a 3′-to-5′ single-stranded DNA exonuclease. Some *xonA* mutations increase deletion frequency in multicopy plasmids up to 100-fold (2). These dominant mutations have been proposed to direct the synthesis of an altered SbcB protein with residual or modified activity. Mutations in *ssb*, encoding the *E. coli* single-stranded DNA binding protein, stimulate excision of transposons from the chromosome (103). They were also reported to stimulate deletions in mini-F plasmids (92). In *ssb* mutants, leading- and lagging-strand synthesis may be out of balance or *ssb* mutations may cause replication pauses that are deletion prone. Alternatively, wild-type SSB protein may directly prevent erroneous annealing between short homologous sequences (92).

The last class of hyperrecombination mutations may have an indirect effect. In a screen for mutants that increase the frequency of deletion between short homologies in the *E. coli* chromosome, a mutation was isolated in the *topB* gene, encoding topoisomerase III (112, 145). Topoisomerase III of *E. coli* is a type I topoisomerase that catalyzes the relaxation of negatively supercoiled DNA and possesses a potent decatenase activity (36, 37). The protein is present at 1 to 10 copies per cell, and the mutant strain has no known phenotype. It has been proposed that local modification of supercoiling was responsible for the 5- to 10-fold increase of deletion frequency in the *topB* mutant (112). Effects of *topB* mutations on plasmid deletions have also been reported in a multicopy plasmid system (130). Mutations in the *bglY* gene increase 10- to 100-fold the frequency of deletions in the *E. coli* chromosome (73). This gene encodes the histone-like protein H1. *bglY* mutations are pleiotropic and also affect supercoiling (49). Finally, mutations in the *uup* gene of unknown function stimulate transposon excision (103).

Inactivation of the SOS repressor LexA [*lexA*(Def) mutations] increases microsatellite instability in the *E. coli* chromosome (91). This supports the idea that illegitimate recombination may be stimulated under stress conditions. Effects of SOS on plasmid deletions have also been reported in a multicopy plasmid system (7).

ILLEGITIMATE RECOMBINATION ASSOCIATED WITH SITE-SPECIFIC ELEMENTS

Rearrangements have been found associated with lysogenic bacteriophages, with various kinds of transposons and insertion sequences and with site-specific DNA inversion systems. All these elements encode their own recombination enzymes. Site-specific recombinases catalyze break and join reactions at precise target sequences. Transposases recognize the ends of transposons and random or specific target sites. Rearrangements associated with transposable elements are generally less frequent than actual transposition. However, a high number of genes presumably of bacteriophage origin, defective prophages, and defective insertion sequences could be identified in the *E. coli* and *B. subtilis* genomes (17, 66). They are the remnants of intense rearrangements associated with mobile elements during evolution. At least one type of rearrangement due to a mobile element, *araB-lacZ* fusions mediated by Mu, depends on selection conditions and arises only in nongrowing cells (41, 81). This observation, combined with the observation of intense transposition activity in resting cells (93, 94), suggests that under stress conditions, a controlled induction of rearrangements could occur, generating a population of new bacteria. This process may account for the participation of illegitimate recombination in the natural selection of bacteria with increased fitness. The evolutionary relevance of such "adaptive" illegitimate recombination is discussed by Shapiro (117).

Products of Intramolecular Transposition Reactions

Certain rearrangements associated with transposons are the physiological products of intramolecular or intermolecular reactions. They will not be discussed here in detail, since they result from "legitimate" transposase activity. Adjacent deletions or insertion-inversions are the expected products of intramolecular transposition of replicative transposons like Mu (117, 140). Intermolecular reactions lead to the formation of cointegrates and hence to replicon fusions. Intramolecular transposition of IS*1* also leads to adjacent deletions, possibly by a similar mechanism (129). Finally, nonreplicative elements, such as IS*10*, can also mediate adjacent deletions, insertion-inversions, or the formation of cointegrates. This has been reported to occur when the element is duplicated by the passage of the replication fork, the ends of two different replicated elements being used for transposition (23).

Errors of Site-Specific Enzymes

Some rearrangements associated with site-specific elements are indeed illegitimate. They may result from the erroneous action of a site-specific recombinase on a sequence that resembles a canonical site of action (termed a pseudosite or quasisite). Lambda transducing phages can be formed by the action of Xis on pseudo-attachment sites (reviewed in reference 38). The Cin recombinase mediates DNA inversions between two wild-type *cix* sites flanking genes that determine the host range of bacteriophage P1. Cin can also act with low frequency at secondary (or quasi-) sites, designated *cix*Q. These events were originally documented in plasmids. More recent studies present evidence for inversion of 100 to 200 kb of the *E. coli* chromosome, mediated by the erroneous action of Cin on chromosomal *cix*Q sites (107). In vivo, a wild-type IS*911* end can utilize a fortuitously occurring pseudo-end sequence as a partner (102). Binding of Tn5 transposase to pseudo-Tn5 outside ends (one-ended transposition) will catalyze the formation of a deletion by excision of the pseudotransposon without actual transposition (Fig. 5). Adjacent deletions thus result from unsuccessful attempts at transposition (58). The action of Tn10 transposase in vitro on pseudo-end sequences confirmed the existence of transposon-associated rearrangements mediated by such a mechanism (23). IS10-promoted adjacent deletions occur at 2% of the frequency of IS10 transposition and are catalyzed by a variety of processes. References 23 and 63 provide a detailed discussion on rearrangements associated with Tn10-IS10 and Tn5-IS5.

Erroneous Processing of Transposition Intermediates

The erroneous processing of a Mu transposition intermediate by cellular enzymes induces

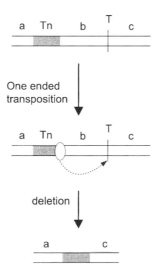

FIGURE 5 Schematic model of deletion formation by one-ended transposition. The transposon is shown as a shaded box; the bound transposase is shown as an oval. One end of the transposon and the target (T) are acted upon by the transposase. The region (b) between the transposon end and the target is excised by the transposase and deleted. Regions a and c are the regions flanking the deleted sequences.

deletion formation. The strand transfer complex is a branched structure formed when the nucleoprotein complex (or "transposasome") containing Mu termini, MuA protein, and the host HU protein attacks a target sequence and attaches one strand of the target to each end of Mu (89). In the absence of the MuB protein, transposition cannot proceed further, and the intermediate with a palindrome-like structure likely favors deletions of Mu and adjacent DNA. Analysis of Mu-promoted ara-lacZ fusion revealed that deletions may occur either by direct religation of a DNA DSB, leaving part of a Mu end, or by replication slippage of the leading strand onto the template of the lagging strand, leaving two partial Mu ends in opposite orientations. In both cases, direct repeats are found at the junction (80).

ILLEGITIMATE RECOMBINATION BETWEEN NONHOMOLOGOUS SEQUENCES

As mentioned above, we will consider a nonhomologous junction to be any recombination event joining sequences that share less than 2 bp of homology that is not recognized by site-specific enzymes. In studies of spontaneous deletion junctions in E. coli, about equal numbers of rearrangements occurring between short homologous and nonhomologous sequences were found (111). This ratio, as well as the global efficiency of illegitimate recombination, varies with the bacterial species. Illegitimate integration of plasmids into the chromosome occurred in B. subtilis mostly via stretches of short homologies (6 to 14 nucleotides [8, 34]). Shorter microhomologies were found at junctions upon plasmid integration into the chromosome of Rhodococcus formans (2 to 4 nucleotides [35]), whereas no significant microhomology was found in C. coli (104). Mycobacterium tuberculosis undergoes illegitimate incorporation of nonreplicative elements at a high efficiency (reference 59; reviewed in reference 83); however, the nature of the junctions was not analyzed.

End joining, which is an efficient process in eukaryotes (reviewed in reference 46), is dependent on the properties of ligases (146). Since, in contrast with eukaryotic ligases, *E. coli* ligase cannot join blunt ends, but only cohesive ends, microhomologies in end-joining reactions are used via an SSA pathway and do not occur between strictly nonhomologous sequences. Therefore, at least in this model bacterium, rearrangements via sequences devoid of any homology point to specific illegitimate recombination pathways (14, 121).

Gyrase Errors

Topoisomerases are enzymes that modify the supercoiling of molecules through transient breakage and ligation of DNA strands. The first evidence that topoisomerases may promote rearrangements in bacteria came from the work of Ikeda and collaborators (54). Gyrase was shown to promote the fusion of lambda and pBR322 molecules by joining of nonhomologous sequences. Gyrase is a type II topoisomerase that introduces negative superhelicity through breakage of the two DNA strands and ligation (reviewed in reference 136). It was proposed that the exchange of gyrase subunits bound to two different molecules leads to covalent linkage of these molecules (Fig. 6). The bacteriophage T4 topoisomerase II, and two eukaryotic topoisomerases II, may promote similar events (reviewed in reference 53). Additional evidence for the formation of nonhomologous deletions via the erroneous action of gyrase came from the isolation of temperature-sensitive *gyrA* mutants that cause a 30-fold increase in the frequency of spontaneous illegitimate-recombination events when shifted to a restrictive temperature (121). Selection of transducing particles of lambda was used to measure illegitimate recombination in this study, and only nonhomologous recombination was promoted by the gyrase mutation, while short homologous recombination was not affected. Gyrase-mediated rearrangements are therefore clearly independent of the presence of short homologies.

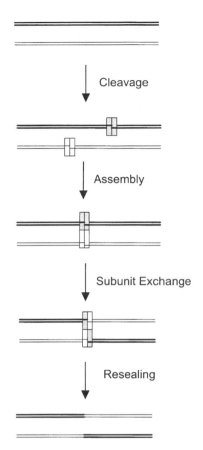

FIGURE 6 Model for deletion formation by erroneous action of gyrase. Gyrase is shown as two rectangles, corresponding to the two subunits of the enzyme. Gyrase binds to double-stranded DNA and introduces a double-strand cut. Exchange of subunits between two gyrase molecules, acting at different places on a DNA molecule or on two different DNA molecules, and resealing by gyrase lead to rearrangement. Bold and thin lines are two different DNA molecules. White and gray rectangles are two different gyrase molecules.

Topoisomerase I Errors

The second main topoisomerase of *E. coli*, topoisomerase I, was also shown to promote rearrangements. In contrast with gyrase, this enzyme relaxes DNA through transient breakage and ligation of one strand. In bacteria, topoisomerase I is linked transiently to the 5′ end of the DNA (128). The evidence for the involvement of this enzyme in illegitimate recombination was obtained in a study of replication blockage by the *Ter*–Tus complex, which causes recombination not only between short homologous sequences (see above) but also between nonhomologous sequences (14). The latter deletions occurred by joining of the sequences immediately preceding the two replication arrest sites present on a monocopy plasmid. They did not occur in a *topA10* mutant of *E. coli*, deficient in topoisomerase I activity. It was proposed that a topoisomerase acting at one blocked replication fork could directly join the 5′ end to which it is attached to a free 3′ end at the other blocked replication fork (Fig. 7). This emphasizes the role of replication blockage in illegitimate recombination, since arrest of replication can promote the three main pathways of illegitimate recombination: replication slippage, SSA, and topoisomerase errors.

Deletions at the M13 Replication Origin

When carried on plasmids derived from pBR322, the M13 replication origin is a deletion hot spot. The nick site at the origin is linked to a nonrelated sequence, in a process that depends on the nicking action of the phage replication protein but does not depend on phage replication (87). The second deletion endpoint is not random; some sequences are preferentially linked to the phage origin regardless of DNA sequence or structural homology (86). This system was used, under conditions where M13 does not replicate, to search for sequences that may induce illegitimate recombination. Introduction of a replication terminator on pBR322-M13 hybrids provokes the formation of deletions that join the phage origin to the sequence preceding the replication terminator (15). This was the first indication that replication arrest could induce nonhomologous rearrangements, which was later confirmed in a model system that does not carry the M13 origin (14). The pBR322-oriM13 hybrid system was used to test whether other structures that may impede pBR322 replication would also induce deletions. Another nucleoprotein complex, the Lac repressor bound to the *lac* operator, was

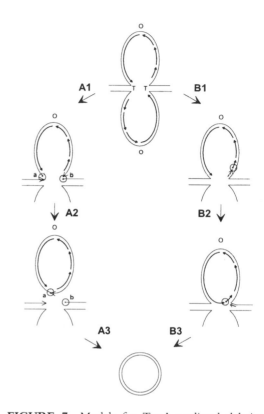

FIGURE 7 Models for TopA-mediated deletion formation between divergent replication forks blocked at *Ter* sites (reproduced from reference 14). The *Ter1-Ter2* replication intermediate is partially represented; template strands are shown as thin lines, and newly synthesized leading and lagging strands are shown as continuous and interrupted thick arrows, respectively. O and T represent *oriC* and replication terminators, respectively. The 5′ and 3′ ends generated by topoisomerase cleavage are represented by a point and a thin arrow, respectively. Deletion can result either from the junction of the template strands (model A) or from the junction of the newly synthesized strands (model B). In model A, topoisomerase-mediated cleavage occurs at each of the two replication forks, in the vicinity of *Ter* sites (A1); a topoisomerase molecule (a) covalently linked to the 5′ end generated on the leading-strand template catalyzes by error ligation with the 3′ end created by another topoisomerase molecule (b) at the other replication fork (A2); this leads to the excision of a gap-containing molecule, which is converted into a circular double-stranded plasmid by continuation of leading-strand synthesis (A3). In model B, a topoisomerase molecule cleaves the lagging strand in the vicinity of a *Ter* site (B1); this molecule, bound to the 5′ end, catalyzes the joining to the 3′ end of the leading strand at the other blocked replication fork (B2); a circular double-stranded deletant plasmid is generated by another round of replication (B3).

also a deletion hot spot in this system. The deletion endpoints were located in the operator sequence. They occurred regardless of the operator orientation, provided that the repressor–operator complex was in the path of pBR322 replication (135). This suggests that the deletion-prone properties of the Ter-Tus complex may be extended to other nucleoprotein complexes. Transcription from the strong P_{Tac} promoter also induced deletions, provided that transcription progressed toward pBR322 replication (133, 135). Deletion endpoints were located in and downstream of the transcribed region, and it was proposed that collision between the replication fork and the transcription apparatus induces a pause long enough to cause deletions. Deletions ending downstream of the transcription terminator may be induced by replication pauses caused by the lower superhelicity of DNA due to active transcription (76, 133). When conditions permissive for M13 replication were used, the instability caused by concomitant replication and transcription in opposite orientations was increased further (134). This high instability may explain the structure of the phage, in which transcription and rolling-circle replication progress in the same direction. Interestingly, replication and transcription tend to be codirectional in bacterial genomes (17, 66), possibly to minimize instability due to illegitimate recombination.

Nonhomologous Rearrangements Induced by DNA Lesions

The effect of UV irradiation on nonhomologous deletions was also tested in pBR322-oriM13 hybrids. UV irradiation of a plasmid region creates a deletion hot spot (115). However, in contrast with deletion hot spots created in this system by nucleoprotein complexes, deletions caused by UV lesions were dependent on M13 replication and were not observed upon blockage of pBR322 replication. pBR322 and M13 use the DnaB and Rep helicases, respectively, for replication. The encounter of the pBR322 replication fork with a UV lesion blocks the progression of the polymerase but presumably not the

DnaB helicase (108), whereas arresting M13 replication may also block the Rep helicase, due to its tight association with the replisome. This structure, akin to that resulting from helicase blocking by a nucleoprotein complex (Tus-Ter or LacI-lac_{op}) could be deletion prone. Deletion formation at UV lesions would thus depend on the structure and composition of the replisome.

γ-Irradiation can also induce nonhomologous rearrangements (reviewed in reference 52). After γ-irradiation, a B. subtilis strain was isolated that possesses two extensive adjacent chromosomal rearrangements: a translocation of the trpE-ilvA segment comprising 4% of the genome and an inversion involving 36% of the chromosome. These rearrangements were transferred to the wild-type strain and extensively studied (reviewed in reference 3). Two of the junctions were sequenced, and no homology was found (57). They could result from topoisomerase errors at replication forks blocked by DNA damage. γ-Irradiation induces the formation of deleterious oxygen species. Oxidative damages may therefore induce illegitimate recombination between short homologous sequences as well as between nonhomologous sequences.

CONCLUSION

The multiple pathways for illegitimate recombination suggest that prokaryotic genomes have a high potential for apparently random rearrangements. Errors of enzymes which cut and join DNA either at specific sites or anywhere on a genome, repair of DNA breakage by linkage of DNA ends, and polymerase errors caused by perturbation of the smooth progression of DNA synthesis all participate in rearrangements. Presumably, the remodeling of genomes via illegitimate recombination is limited by the presence of a powerful homologous-recombination machinery that reduces the lifetime of hyperrecombinogenic structures and, by ensuring faithful repair, maintains a precise organization of genomic structures (reference 77; reviewed in reference 106). However, it would be important to determine how the processes involved in the preservation of the genome structure may be perturbed by mutations, exogenous stress, or suboptimal growth conditions.

Illegitimate recombination is a major issue in eukaryotes, because it is at the origin of numerous pathological disorders. Since the various processes involved make use of universally conserved functions, prokaryotes remain a tool of choice for the study of illegitimate recombination mechanisms and the determination of the role of various internal and external stimuli in the occurrence of rearrangements that are apparently fortuitous.

ACKNOWLEDGMENTS

I am very grateful to Era Cassuto, Dusko Ehrlich, Philippe Noirot, and Marie-Agnes Petit for helpful reading of the manuscript and to David Leach for providing Fig. 4 and comments on the model.

REFERENCES

1. **Albertini, A. M., M. Hofer, M. P. Calos, and J. H. Miller**. 1982. On the formation of spontaneous deletions: the importance of short sequence homologies in the generation of large deletions. *Cell* **29:**319–328.

2. **Allgood, N. D., and T. J. Silhavy**. 1991. *Escherichia coli xonA* (*sbcB*) mutants enhance illegitimate recombination. *Genetics* **127:**671–680.

3. **Anagnostopoulos, C.** 1990. Genetic rearrangements in *Bacillus subtilis*, p. 361–371. *In* K. Drlika and M. Riley (ed.), *The Bacterial Chromosome*. American Society for Microbiology, Washington, D.C.

4. **Ashley, C. T., Jr., and S. T. Warren**. 1995. Trinucleotide repeat expansion and human disease. *Annu. Rev. Genet.* **29:**703–728.

5. **Bachellier, S., E. Gilson, M. Hoffnung, and C. W. Hill**. 1996. Repeated sequences, p. 2012–2040. *In* F. C. Neidhardt, R. Curtiss III, J. L. Ingraham, E. C. C. Lin, K. B. Low, B. Magasanik, W. S. Reznikoff, M. Riley, M. Schaechter, and H. E. Umbarger (ed.), Escherichia coli *and* Salmonella typhimurium: *Cellular and Molecular Biology,* 2nd ed. American Society for Microbiology, Washington, D.C.

6. **Bachellier, S., J. M. Clement, M. Hofnung, and E. Gilson**. 1997. Bacterial interspersed mosaic elements (BIMEs) are a major source of sequence polymorphism in *Escherichia coli* intergenic regions including specific associations with a new insertion sequence. *Genetics* **145:**551–562.

7. **Balbinder, E., B. Coll, J. Hutchinson, A. S. Bianchi, T. Groman, K. A. Wheeler, and M. Meyer**. 1993. Participation of the SOS system in producing deletions in *E. coli* plasmids. *Mutat. Res.* **286:**253–265.

8. **Bashkirov, V. I., F. K. Khasanov, and A. A. Prozorov**. 1987. Illegitimate recombination in *Bacillus subtilis*: nucleotide sequences at recombinant DNA junctions. *Mol. Gen. Genet.* **210:**578–580.

9. **Bedinger, P., M. Munn, and B. M. Alberts**. 1989. Sequence-specific pausing during *in vitro* DNA replication on double-stranded DNA templates. *J. Biol. Chem.* **264:**16880–16886.

10. **Berg, D. E., C. Egner, and J. B. Lowe**. 1983. Mechanism of F factor-enhanced excision of transposon Tn*5*. *Gene* **22:**1–7.

11. **Bi, X., and L. F. Liu**. 1994. RecA-independent and RecA-dependent intramolecular plasmid recombination—differential homology requirement and distance effect. *J. Mol. Biol.* **235:**414–423.

12. **Bichara, M., S. Schumacher, and R. P. P. Fuchs**. 1995. Genetic instability within monotonous runs of CpG sequences in *Escherichia coli*. *Genetics* **140:**897–907.

13. **Bierne, H., S. D. Ehrlich, and B. Michel**. 1995. Competition between parental and recombinant plasmids affects the measure of recombination frequencies. *Plasmid* **33:**101–112.

14. **Bierne, H., S. D. Ehrlich, and B. Michel**. 1997. Deletions at stalled replication forks occur by two different pathways. *EMBO J.* **16:**3332–3340.

15. **Bierne, H., S. D. Ehrlich, and B. Michel**. 1991. The replication termination signal *TerB* of the *Escherichia coli* chromosome is a deletion hot spot. *EMBO J.* **10:**2699–2705.

16. **Bierne, H., and B. Michel**. 1994. When replication forks stop. *Mol. Microbiol.* **13:**17–23.

17. **Blattner, F. R., G. Plunkett III, C. A. Bloch, N. T. Perna, V. Burland, M. Riley, J. Collado-Vides, J. D. Glasner, C. K. Rode, G. F. Mayhew, J. Gregor, N. W. Davis, H. A. Kirkpatrick, M. A. Goeden, D. J. Rose, B. Mau, and Y. Shao**. 1997. The complete genome sequence of *Escherichia coli* K-12. *Science* **277:**1453–1474.

18. **Bowater, R. P., W. A. Rosche, A. Jaworski, R. R. Sinden, and R. D. Wells**. 1996. Relationship between *Escherichia coli* growth and deletions of CTG · CAG triplet repeats in plasmids. *J. Mol. Biol.* **264:**82–96.

19. **Canceill, D., E. Viguera, and S. D. Ehrlich**. Replication slippage of different DNA polymerases is inversely related to their strand displacement efficiency. Submitted for publication.

20. **Canceill, D. and S. D. Ehrlich.** 1996. Copy-choice recombination mediated by DNA polymerase III holoenzyme from *Escherichia coli*. *Proc. Natl. Acad. Sci. USA* **93**:6647–6652.

21. **Cao, Y., and T. Kogoma.** 1995. The mechanism of *recA polA* lethality: suppression by RecA-independent recombination repair activated by the *lexA*(Def) mutation in *Escherichia coli*. *Genetics* **139**:1483–1494.

22. **Chalker, A. F., E. A. Okely, A. Davison, and D. R. F. Leach.** 1993. The effects of central asymmetry on the propagation of palindromic DNA in bacteriophage lambda are consistent with cruciform extrusion *in vivo*. *Genetics* **133**: 143–148.

23. **Chalmers, R. M., and N. Kleckner.** 1996. IS*10*/Tn*10* transposition efficiently accommodates diverse transposon end configurations. *EMBO J.* **15**:5112–5122.

24. **Chan, A., M. S. Levy, and R. Nagel.** 1994. RecN SOS gene and induced precise excision of Tn*10* in *Escherichia coli*. *Mutat. Res. Lett.* **325**: 75–79.

25. **Chedin, F., E. Dervyn, R. Dervyn, S. D. Ehrlich, and P. Noirot.** 1994. Frequency of deletion formation decreases exponentially with distance between short direct repeats. *Mol. Microbiol.* **12**:561–569.

26. **Chedin, F., R. Dervyn, S. D. Ehrlich, and P. Noirot.** 1997. Apparent and real recombination frequencies in multicopy plasmids: the need for a novel approach in frequency determination. *J. Bacteriol.* **179**:754–761.

27. **Conley, E. C., and J. R. Saunders.** 1984. Recombination-dependent recircularization of linearized pBR322 plasmid DNA following transformation of *Escherichia coli*. *Mol. Gen. Genet.* **194**:211–218.

28. **Conley, E. C., V. A. Saunders, V. Jackson, and J. R. Saunders.** 1986. Mechanism of intramolecular recyclization and deletion formation following transformation of *Escherichia coli* with linearized plasmid DNA. *Nucleic Acids Res.* **14**: 8919–8932.

29. **Conley, E. C., V. A. Saunders, and J. R. Saunders.** 1986. Deletion and rearrangement of plasmid DNA during transformation of *Escherichia coli* with linear plasmid molecules. *Nucleic Acids Res.* **14**:8905–8917.

30. **Connelly, J. C., E. S. de Leau, E. A. Okely, and D. R. Leach.** 1997. Overexpression, purification and characterisation of the SbcCD protein from *Escherichia coli*. *J. Biol. Chem.* **272**: 19819–19826.

31. **D' Alencon, E., M. Petranovic, B. Michel, P. Noirot, A. Aucouturier, M. Uzest, and S. D. Ehrlich.** 1994. Copy-choice illegitimate DNA recombination revisited. *EMBO J.* **13**: 2725–2734.

32. **Darlow, J. M., and D. R. F. Leach.** 1995. The effects of trinucleotide repeats found in human inherited disorders on palindrome inviability in *Escherichia coli* suggest hairpin folding preferences *in vivo*. *Genetics* **141**:825–832.

33. **Davison, A., and D. R. Leach.** 1994. The effects of nucleotide sequence changes on DNA secondary structure formation in *Escherichia coli* are consistent with cruciform extrusion *in vivo*. *Genetics* **137**:361–368.

34. **Dempsey, L. A., and D. A. Dubnau.** 1989. Identification of plasmid and *Bacillus subtilis* chromosomal recombination sites used for pE194 integration. *J. Bacteriol.* **171**:2856–2865.

35. **Desomer, J., M. Crespi, and M. Van Montagu.** 1991. Illegitimate integration of non-replicative vectors in the genome of *Rhodococcus fascians* upon electrotransformation as an insertional mutagenesis system. *Mol. Microbiol.* **5**: 2115–2124.

36. **DiGate, R. J., and K. J. Marians.** 1988. Identification of a potent decatenating enzyme from *Escherichia coli*. *J. Biol. Chem.* **263**:13366–13373.

37. **DiGate, R. J., and K. J. Marians.** 1989. Molecular cloning and DNA sequence analysis of *Escherichia coli topB*, the gene encoding topoisomerase III. *J. Biol. Chem.* **264**:17924–17930.

38. **Ehrlich, S. D.** 1989. Illegitimate recombination in bacteria, p. 799–832. *In* D. E. Berg and M. M. Howe (ed.), *Mobile DNA*. American Society for Microbiology, Washington, D.C.

39. **Ehrlich, S. D., H. Bierne, E. d'Alencon, D. Vilette, M. Petranovic, P. Noirot, and B. Michel.** 1993. Mechanisms of illegitimate recombination. *Gene* **135**:161–166.

40. **Farabaugh, P. J., U. Schmeissner, M. Hofer, and J. H. Miller.** 1978. Genetic studies of the lac repressor. VII. On the molecular nature of spontaneous hotspots in the *lacI* gene of *Escherichia coli*. *J. Mol. Biol.* **126**:847–857.

41. **Foster, P. L., and J. Cairns.** 1994. The occurrence of heritable Mu excisions in starving cells of *Escherichia coli*. *EMBO J.* **13**:5240–5244.

42. **Foster, T. J., V. Lundblad, S. Hanley-Way, S. M. Halling, and N. Kleckner.** 1981. Three Tn10-associated excision events: relationship to transposition and role of direct and inverted repeats. *Cell* **23**:215–227.

43. **Gibson, F. P., D. R. F. Leach, and R. G. Lloyd.** 1992. Identification of *sbcD* mutations as cosuppressors of *recBC* that allow propagation of DNA palindromes in *Escherichia coli* K-12. *J. Bacteriol.* **174**:1222–1228.

44. **Gilson, E., W. Saurin, D. Perrin, S. Bachellier, and M. Hofnung.** 1991. Palindromic

units are part of a new bacterial interspersed mosaic element (BIME). *Nucleic Acids Res.* **19:** 1375–1383.

45. **Glickman, B. W., and L. S. Ripley.** 1984. Structural intermediates of deletion mutagenesis: a role for palindromic DNA. *Proc. Natl. Acad. Sci. USA* **81:**512–516.

46. **Haber, J. E.** 1995. *In vivo* biochemistry: physical monitoring of recombination induced by site-specific endonucleases. *Bioessays* **17:**609–620.

47. **Hanada, K., T. Ukita, Y. Kohno, K. Saito, J. Kato, and H. Ikeda.** 1997. RecQ DNA helicase is a suppressor of illegitimate recombination in *Escherichia coli*. *Proc. Natl. Acad. Sci. USA* **94:** 3860–3865.

48. **Herrmann, R., K. Neugebauer, H. Zentgraf, and H. Schaller.** 1978. Transposition of a DNA sequence determining kanamycin resistance into the single-stranded genome of bacteriophage fd. *Mol. Gen. Genet.* **159:**171–178.

49. **Higgins, C. F., C. J. Dorman, D. A. Stirling, L. Waddell, I. R. Booth, G. May, and E. Bremer.** 1988. A physiological role for DNA supercoiling in the osmotic regulation of gene expression in *S. typhimurium* and *E. coli. Cell* **52:** 569–584.

50. **Hill, T. M.** 1996. Features of the chromosomal terminus region, p. 1602–1614. *In* F. C. Neidhardt, R. Curtiss III, J. L. Ingraham, E. C. C. Lin, K. B. Low, B. Magasanik, W. S. Reznikoff, M. Riley, M. Schaechter, and H. E. Umbarger (ed.), Escherichia coli *and* Salmonella typhimurium: *Cellular and Molecular Biology,* 2nd ed. American Society for Microbiology, Washington, D.C.

51. **Hulton, C. S., C. F. Higgins, and P. M. Sharp.** 1991. ERIC sequences: a novel family of repetitive elements in the genomes of *Escherichia coli, Salmonella typhimurium* and other enterobacteria. *Mol. Microbiol.* **5:**825–834.

52. **Hutchinson, F.** 1993. Induction of large DNA deletions by persistent nicks—a new hypothesis. *Mutat. Res.* **299:**211–218.

53. **Ikeda, H.** 1990. DNA topoisomerase-mediated illegitimate recombination, p. 341–356. *In* N. R. Cozarelli and J. C. Wang (ed.) *DNA Topology and Its Biological Effects.* Cold Spring Harbor Laboratory Press, Cold Spring Harbor, N.Y.

54. **Ikeda, H., K. Moriya, and T. Matsumoto.** 1981. In vitro study of illegitimate recombination: involvement of DNA gyrase. *Cold Spring Harbor Symp. Quant. Biol.* **45:**399–408

55. **Ishiura, M., N. Hazumi, H. Shinagawa, A. Nakata, T. Uchida, and Y. Okada.** 1990. RecA-independent high-frequency deletion of recombinant cosmid DNA in *Escherichia coli. J. Gen. Microbiol.* **136:**69–79.

55a.**Jackson, A., R. Chen, and L. A. Loeb.** 1998. Induction of microsatellite instability by oxidative DNA damage. *Proc. Natl. Acad. Sci. USA* **95:** 12468–12473.

56. **Janniere, L., C. Bruand, and S. D. Ehrlich.** 1990. Structurally stable *Bacillus subtilis* cloning vectors. *Gene* **87:**53–61.

57. **Jarvis, E. D., S. Cheng, and R. Rudner.** 1990. Genetic structure and DNA sequences at junctions involved in the rearrangements of *Bacillus subtilis* strains carrying the *Trp26* mutation. *Genetics* **126:**785–797.

58. **Jilk, R. A., J. C. Makris, L. Borchardt, and W. S. Reznikoff.** 1993. Implications of Tn*5*-associated adjacent deletions. *J. Bacteriol.* **175:** 1264–1271.

59. **Kalpana, G. V., B. R. Bloom, and W. R. Jacobs.** 1991. Insertional mutagenesis and illegitimate recombination in mycobacteria. *Proc. Natl. Acad. Sci. USA* **88:**5433–5437.

60. **Kang, S. M., K. Ohshima, M. Shimizu, S. Amirhaeri, and R. D. Wells.** 1995. Pausing of DNA synthesis *in vitro* at specific loci in CTG and CGG triplet repeats from human hereditary disease genes. *J. Biol. Chem.* **270:**27014–27021.

61. **Khasanov, F. K., D. J. Zvingila, A. A. Zainullin, A. A. Prozorov, and V. I. Bashkirov.** 1992. Homologous recombination between plasmid and chromosomal DNA in *Bacillus subtilis* requires approximately 70 bp of homology. *Mol. Gen. Genet.* **234:**494–497.

62. **King, S. R., and J. P. Richardson.** 1986. Role of homology and pathway specificity for recombination between plasmids and bacteriophage lambda. *Mol. Gen. Genet.* **204:**141–147.

63. **Kleckner, N., R. M., Chalmers, D. Kwon, J. Sakai, and S. Bolland.** 1996. Tn*10* and IS*10* transposition and chromosome rearrangements: mechanisms and regulation *in vivo* and *in vitro*, p. 49–84. *In* H. Saedler and A. Gierl (ed.), *Transposable Elements*, vol. 204. Springer, Berlin, Germany.

64. **Kolodner, R.** 1996. Biochemistry and genetics of eukaryotic mismatch repair. *Genes Dev.* **10:** 1433–1442.

65. **Kong, D. C., and W. Masker.** 1994. Deletion between direct repeats in T7 DNA stimulated by double-strand breaks. *J. Bacteriol.* **176:**5904–5911.

66. **Kunst, F., N. Ogasawara, I. Moszer, A. M. Albertini, G. Alloni, V. Azevedo, M. G. Bertero, P. Bessières, A. Bolotin, S. Borchert, R. Borriss, L. Boursier, A. Brans, M. Braun, S. C. Brignell, S. Bron, S. Brouillet, C. V. Bruschi, B. Caldwell, V. Capuano, N. M. Carter, S.-K. Choi, J.-J. Codani, I. F. Connerton, N. J. Cummings, R. A. Daniel,**

F. Denizot, K. M. Devine, A. Düsterhöft, S. D. Ehrlich, P. T. Emmerson, K. D. Entian, J. Errington, C. Fabret, E. Ferrari, D. Foulger, C. Fritz, M. Fujita, Y. Fujita, S. Fuma, A. Galizzi, N. Galleron, S.-Y. Ghim, P. Glaser, A. Goffeau, E. J. Golightly, G. Grandi, G. Giuseppi, B. J. Guy, K. Haga, J. Haiech, C. R. Harwood, A. Hénaut, H. Hilbert, S. Holsappel, S. Hosono, M.-F. Hullo, M. Itaya, L. Jones, B. Joris, D. Karamata, Y. Kasahara, M. Klaerr-Blanchard, C. Klein, Y. Kobayashi, P. Koetter, G. Koningstein, S. Krogh, M. Kumano, K. Kurita, A. Lapidus, S. Lardinois, J. Lauber, V. Lazarevic, S.-M. Lee, A. Levine, H. Liu, S. Masuda, C. Mauël, C. Médigue, N. Medina, R. P. Mellado, M. Mizuno, D. Moestl, S. Nakai, M. Noback, D. Noone, M. O'Reilly, K. Ogawa, A. Ogiwara, B. Oudega, S.-H. Park, V. Parro, T. M. Pohl, D. Portetelle, S. Porwollik, A. M. Prescott, E. Presecan, P. Pujic, B. Purnelle, G. Rapoport, M. Rey, S. Reynolds, M. Rieger, C. Rivolta, E. Rocha, B. Roche, M. Rose, Y. Sadaie, T. Sato, E. Scanlan, S. Schleich, R. Schroeter, F. Scoffone, J. Sekiguchi, A. Sekowska, S. J. Seror, P. Serror, B.-S. Shin, B. Soldo, A. Sorokin, E. Tacconi, T. Takagi, H. Takahashi, K. Takemaru, M. Takeuchi, A. Tamakoshi, T. Tanaka, P. Terpstra, A. Tognoni, V. Tosato, S. Uchiyama, M. Vandenbol, F. Vannier, A. Vassarotti, A. Viari, R. Wambutt, E. Wedler, H. Wedler, T. Weitzenegger, P. Winters, A. Wipat, H. Yamamoto, K. Yamane, K. Yasumoto, K. Yata, K. Yoshida, H.-F. Yoshikawa, E. Zumstein, H. Yoshikawa, and A. Danchin. 1997. The complete genome sequence of the Gram-positive bacterium *Bacillus subtilis*. *Nature* **390:**249–256.

67. LaDuca, R. J., P. J. Fay, C. Chuang, C. S. McHenry, and R. A. Bambara. 1983. Site-specific pausing of deoxyribonucleic acid synthesis catalyzed by four forms of *Escherichia coli* DNA polymerase III. *Biochemistry* **22:**5177–5188.

68. Leach, D. R., and F. W. Stahl. 1983. Viability of lambda phages carrying a perfect palindrome in the absence of recombination nucleases. *Nature* (London) **305:**448–451.

69. Leach, D. R. F. 1996. Cloning and characterisation of DNAs with palindromic sequences, p. 1–11. *In* J. K. Setlow (ed.), *Genetic Engineering*, vol. 18. Plenum Press, New York, N.Y.

70. Leach, D.R.F. 1996. *Genetic Recombination*. Blackwell Science, Oxford, England.

71. Leach, D. R. F. 1994. Long DNA palindromes, cruciform structures, genetic instability and secondary structure repair. *Bioessays* **16:** 893–900.

72. Leach, D. R. L., E. A. Okely, and D. J. Pinder. 1997. Repair by recombination of DNA containing a palindromic sequence. *Mol. Microbiol.* **26:**597–606.

73. Lejeune, P., and A. Danchin. 1990. Mutations in the *bglY* gene increase the frequency of spontaneous deletions in *Escherichia coli* K-12. *Proc. Natl. Acad. Sci. USA* **87:**360–363.

74. Levinson, G., and G. A. Gutman. 1987. High frequencies of short frameshifts in poly-CA/TG tandem repeats borne by bacteriophage M13 in *Escherichia coli* K-12. *Nucleic Acids Res.* **15:**5323–5338.

75. Lin, F. L., K. Sperle, and N. Sternberg. 1984. Model for homologous recombination during transfer of DNA into mouse L cells: role for DNA ends in the recombination process. *Mol. Cell. Biol.* **4:**1020–1034.

76. Liu, L. F., and J. C. Wang. 1987. Supercoiling of the DNA template during transcription. *Proc. Natl. Acad. Sci. USA* **84:**7024–7027.

77. Louarn, J., F. Cornet, V. Francois, J. Patte, and J. M. Louarn. 1994. Hyperrecombination in the terminus region of the *Escherichia coli* chromosome: possible relation to nucleoid organization. *J. Bacteriol.* **176:**7524–7531.

78. Lovett, S. T., T. J. Gluckman, P. J. Simon, V. A. Sutera, and P. T. Drapkin. 1994. Recombination between repeats in *Escherichia coli* by a *recA*-independent, proximity-sensitive mechanism. *Mol. Gen. Genet.* **245:**294–300.

79. Lundblad, V., and N. Kleckner. 1984. Mismatch repair mutations of *Escherichia coli* K12 enhance transposon excision. *Genetics* **109:**3–19.

80. Maenhaut-Michel, G., C. E. Blake, D. R. F. Leach, and J. A. Shapiro. 1997. Different structures of selected and unselected *araB-lacZ* fusions. *Mol. Microbiol.* **23:**1133–1145.

81. Maenhaut-Michel, G., and J. A. Shapiro. 1994. The roles of starvation and selective substrates in the emergence of *araB-lacZ* fusion clones. *EMBO J.* **13:**5229–5239.

82. Mazin, A. V., A. V. Kuzminov, G. L. Dianov, and R. I. Salganik. 1991. Mechanisms of deletion formation in *Escherichia coli* plasmids. II. Deletions mediated by short direct repeats. *Mol. Gen. Genet.* **228:**209–214.

83. McFadden, J. 1996. Recombination in mycobacteria. *Mol. Microbiol.* **21:**205–211.

84. Mcmurray, C. T. 1995. Mechanisms of DNA expansion. *Chromosoma* **104:**2–13.

85. Meima, R., B. J. Haijema, H. Dijkstra, G. J. Haan, G. Venema, and S. Bron. 1997. Role of enzymes of homologous recombination in illegitimate plasmid recombination in *Bacillus subtilis*. *J. Bacteriol.* **179:**1219–1229.

86. **Michel, B., E. D'Alencon, and S. D. Ehrlich.** 1989. Deletion hot spots in chimeric *Escherichia coli* plasmids. *J. Bacteriol.* **171:**1846–1853.

87. **Michel, B., and S. D. Ehrlich.** 1986. Illegitimate recombination at the replication origin of bacteriophage M13. *Proc. Natl. Acad. Sci. USA* **83:**386–390.

88. **Michel, B., S. D. Ehrlich, and M. Uzest.** 1997. DNA double-strand breaks caused by replication arrest. *EMBO J.* **16:**430–438.

89. **Mizuuchi, K.** 1992. Transpositional recombination—mechanistic insights from studies of Mu and other elements. *Annu. Rev. Biochem.* **61:**1011–1051.

90. **Mollet, B., and M. Delley.** 1991. A betagalactosidase deletion mutant of *Lactobacillus bulgaricus* reverts to generate an active enzyme by internal DNA sequence duplication. *Mol. Gen. Genet.* **227:**17–21.

91. **Morel P., C. Reverdy, B. Michel, S. D. Ehrlich, and E. Cassuto.** The role of SOS and flap processing in microsatellite instability in *E. coli. Proc. Natl. Acad. Sci. USA* **95:**10003–10008.

92. **Mukaihara, T., and M. Enomoto.** 1997. Deletion formation between the two *Salmonella typhimurium* flagellin genes encoded on the mini F plasmid: *Escherichia coli ssb* alleles enhance deletion rates and change hot-spot preference for deletion endpoints. *Genetics* **145:**563–572.

93. **Naas, T., M. Blot, W. M. Fitch, and W. Arber.** 1995. Dynamics of IS-related genetic rearrangements in resting *Escherichia coli* K-12. *Mol. Biol. Evol.* **12:**198–207.

94. **Naas, T., M. Blot, W. M. Fitch, and W. Arber.** 1994. Insertion sequence-related genetic variation in resting *Escherichia coli* K-12. *Genetics* **136:**721–730.

95. **Nagel, R., A. Chan, and E. Rosen.** 1994. *ruv* and *recG* genes and the induced precise excision of Tn*10* in *Escherichia coli. Mutat. Res.* **311:**103–109.

96. **Ohshima, K., and R. D. Wells.** 1997. Hairpin formation during DNA synthesis primer realignment in vitro in triplet repeat sequences from human hereditary disease genes. *J. Biol. Chem.* **272:**16798–16806.

97. **Ouchane, S., M. Picaud, C. Vernotte, and C. Astier.** 1997. Photooxidative stress stimulates illegitimate recombination and mutability in carotenoid-less mutants of *Rubrivivax gelatinosus. EMBO J.* **16:**4777–4787.

98. **Papanicolaou, C., and L. S. Ripley.** 1989. Polymerase-specific differences in the DNA intermediates of frameshift mutagenesis. *In vitro* synthesis errors of *Escherichia coli* DNA polymerase I and its large fragment derivative. *J. Mol. Biol.* **207:**335–353.

99. **Peeters, B. P., J. H. de Boer, S. Bron, and G. Venema.** 1988. Structural plasmid instability in *Bacillus subtilis*: effect of direct and inverted repeats. *Mol. Gen. Genet.* **212:**450–458.

100. **Pierce, J. C., D. C. Kong, and W. Masker.** 1991. The effect of the length of direct repeats and the presence of palindromes on deletion between directly repeated DNA sequences in bacteriophage T7. *Nucleic Acids Res.* **19:**3901–3905.

101. **Pinder, D. J., C. E. Blake, and D. R. F. Leach.** 1997. DIR: a novel DNA rearrangement associated with inverted repeats. *Nucleic Acids Res.* **25:**523–529.

101a. **Pinder, D. J., C. E. Blake, J. C. Lindsey, and D. R. F. Leach.** 1998. Replication strand preference for deletions associated with palindromes. *Mol. Microbiol.* **28:**719–727.

102. **Polard, P., L. Seroude, O. Fayet, M. F. Prere, and M. Chandler.** 1994. One-ended insertion of IS*911. J. Bacteriol.* **176:**1192–1196.

103. **Reddy, M., and J. Gowrishankar.** 1997. Identification and characterization of *ssb* and *uup* mutants with increased frequency of precise excision of transposon Tn*10* derivatives: nucleotide sequence of *uup* in *Escherichia coli. J. Bacteriol.* **179:**2892–2899.

104. **Richardson, P. T., and S. F. Park.** 1997. Integration of heterologous plasmid DNA into multiple sites on the genome of *Campylobacter coli* following natural transformation. *J. Bacteriol.* **179:**1809–1812.

105. **Rosche, W. A., T. Q. Trinh, and R. R. Sinden.** 1997. Leading strand specific spontaneous mutation corrects a quasipalindrome by an intermolecular strand switch mechanism. *J. Mol. Biol.* **269:**176–187.

106. **Roth, J. R., N. Benson, T. Galinski, K. Haack, J. G. Lawrence, and L. Miesel.** 1996. Rearrangements of the bacterial chromosome: formation and applications, p. 2256–2276. *In* F. C. Neidhardt, R. Curtiss III, J. L. Ingraham, E. C. C. Lin, K. B. Low, B. Magasanik, W. S. Reznikoff, M. Riley, M. Schaechter, and H. E. Umbarger (ed.), *Escherichia coli and Salmonella typhimurium: Cellular and Molecular Biology,* 2nd ed. American Society for Microbiology, Washington, D.C.

107. **Rozsa, F. W., P. Viollier, M. Fussenegger, R. Hiestand-Nauer, and W. Arber.** 1995. Gin-mediated recombination at secondary crossover sites on the *Escherichia coli* chromosome. *J. Bacteriol.* **177:**1159–1168.

108. **Rupp, W. D.** 1996. DNA repair mechanisms, p. 2277–2294. *In* F. C. Neidhardt, R. Curtiss III, J. L. Ingraham, E. C. C. Lin, K. B. Low, B. Magasanik, W. S. Reznikoff, M. Riley, M. Schaechter, and H. E. Umbarger (ed.), Escher-

ichia coli *and* Salmonella typhimurium: *Cellular and Molecular Biology*, 2nd ed. American Society for Microbiology, Washington, D.C.

109. **Rupp, W. D., and P. Howard-Flanders.** 1968. Discontinuities in the DNA synthesized in an excision-defective strain of *Escherichia coli* following ultraviolet irradiation. *J. Mol. Biol.* **31:** 291–304.

109a.**Sarkar, P. S., H. C. Chang, F. B. Boudi, and S. Reddy.** 1998. CTG repeats show bimodal amplification in *E. coli*. *Cell* **95:**531–540.

110. **Scearce, L. M., J. C. Pierce, B. Mcinroy, and W. Masker.** 1991. Deletion mutagenesis independent of recombination in bacteriophage T7. *J. Bacteriol.* **173:**869–878.

111. **Schaaper, R. M., and R. L. Dunn.** 1991. Spontaneous mutation in the *Escherichia coli* lacI gene. *Genetics* **129:**317–326.

112. **Schofield, M. A., R. Agbunag, M. L. Michaels, and J. H. Miller.** 1992. Cloning and sequencing of *Escherichia coli* mutR shows its identity to *topB*, encoding topoisomerase III. *J. Bacteriol.* **174:**5168–5170.

113. **Schofield, M. A., R. Agbunag, and J. H. Miller.** 1992. DNA inversions between short inverted repeats in *Escherichia coli*. *Genetics* **132:** 295–302.

114. **Schumacher, S., R. P. P. Fuchs, and M. Bichara.** 1997. Two distinct models account for short and long deletions within sequence repeats in *Escherichia coli*. *J. Bacteriol.* **179:**6512–6517.

115. **Seigneur, M., S. D. Ehrlich, and B. Michel.** 1997. Blocking rolling circle replication with a UV lesion creates a deletion hotspot. *Mol. Microbiol.* **26:**569–580.

116. **Shanado, Y., J. Kato, and H. Ikeda.** 1997. Fis is required for illegitimate recombination during formation of lambda *bio* transducing phage. *J. Bacteriol.* **179:**4239–4245.

117. **Shapiro, J. A.** 1997. Genome organization, natural genetic engineering and adaptive mutation. *Trends Genet.* **13:**98–104.

118. **Sharp, P. M., and D. R. F. Leach.** 1996. Palindrome-induced deletion in enterobacterial repetitive sequences. *Mol. Microbiol.* **22:**1055–1056.

119. **Sharples, G. J., and R. G. Lloyd.** 1990. A novel repeated DNA sequence located in the intergenic regions of bacterial chromosomes. *Nucleic Acids Res.* **18:**6503–6508.

120. **Shen, P., and H. V. Huang.** 1986. Homologous recombination in *Escherichia coli*: dependence on substrate length and homology. *Genetics* **112:**441–457.

121. **Shimizu, H., H. Yamaguchi, Y. Ashizawa, Y. Kohno, M. Asami, J. Kato, and H.**

Ikeda. 1997. Short-homology-independent illegitimate recombination in *Escherichia coli*: distinct mechanism from short-homology-dependent illegitimate recombination. *J. Mol. Biol.* **266:**297–305.

122. **Shurvinton, C. E., M. M. Stahl, and F. W. Stahl.** 1987. Large palindromes in the lambda phage genome are preserved in a *rec+* host by inhibiting lambda DNA replication. *Proc. Natl. Acad. Sci. USA* **84:**1624–1628.

123. **Sinden, R. R., G. X. Zheng, R. G. Brankamp, and K. N. Allen.** 1991. On the deletion of inverted repeated DNA in *Escherichia coli*: effects of length, thermal stability, and cruciform formation in vivo. *Genetics* **129:**991–1005.

124. **Singer, B. S., and J. Westlye.** 1988. Deletion formation in bacteriophage T4. *J. Mol. Biol.* **202:**233–243.

125. **Syvanen, M., J. D. Hopkins, T. J. Griffin IV, T. Y. Liang, K. Ippen-Ihler, and R. Kolodner.** 1986. Stimulation of precise excision and recombination by conjugal proficient F' plasmids. *Mol. Gen. Genet.* **203:**1–7.

126. **Touati, D.** Personal communication.

127. **Touati, D., M. Jacques, B. Tardat, L. Bouchard, and S. Despied.** 1995. Lethal oxidative damage and mutagenesis are generated by iron in delta *fur* mutants of *Escherichia coli*: protective role of superoxide dismutase. *J. Bacteriol.* **177:** 2305–2314.

128. **Tse-Dinh, Y. C., B. G. McCarron, R. Arentzen, and V. Chowdhry.** 1983. Mechanistic study of *E. coli* DNA topoisomerase I: cleavage of oligonucleotides. *Nucleic Acids Res.* **11:** 8691–8701.

129. **Turlan, C., and M. Chandler.** 1995. IS1-mediated intramolecular rearrangements: formation of excised transposon circles and replicative deletions. *EMBO J.* **14:**5410–5421.

130. **Uematsu, N., S. Eda, and K. Yamamoto.** 1997. An *Escherichia coli* topB mutant increases deletion and frameshift mutations in the *supF* target gene. *Mutat. Res.* **383:**223–230.

131. **Ukita, T., and H. Ikeda.** 1996. Role of the *recJ* gene product in UV-induced illegitimate recombination at the hotspot. *J. Bacteriol.* **178:** 2362–2367.

132. **Usdin, K., and K. J. Woodford.** 1995. CGG repeats associated with DNA instability and chromosome fragility form structures that block DNA synthesis *in vitro*. *Nucleic Acids Res.* **23:** 4202–4209.

133. **Vilette, D., S. D. Ehrlich, and B. Michel.** 1995. Transcription-induced deletions in *Escherichia coli* plasmids. *Mol. Microbiol.* **17:**493–504.

134. **Vilette, D., S. D. Ehrlich, and B. Michel.** 1996. Transcription-induced deletions in plasmid vectors: M13 DNA replication as a source of instability. *Mol. Gen. Genet.* **252**:398–403.

135. **Vilette, D., M. Uzest, S. D. Ehrlich, and B. Michel.** 1992. DNA transcription and repressor binding affect deletion formation in *Escherichia coli* plasmids. *EMBO J.* **11**: 3629–3634.

136. **Wang, J. C.** 1996. DNA topoisomerases. *Annu. Rev. Biochem.* **65**:635–692.

137. **Wang, T. C., and K. C. Smith.** 1986. Inviability of *dam recA* and *dam recB* cells of *Escherichia coli* is correlated with their inability to repair DNA double-strand breaks produced by mismatch repair. *J. Bacteriol.* **165**:1023–1025.

138. **Wang, T. C., and K. C. Smith.** 1983. Mechanisms for *recF*-dependent and *recB*-dependent pathways of postreplication repair in UV-irradiated *Escherichia coli uvrB. J. Bacteriol.* **156**: 1093–1098.

139. **Watt, V. M., C. J. Ingles, M. S. Urdea, and W. J. Rutter.** 1985. Homology requirements for recombination in *Escherichia coli. Proc. Natl. Acad. Sci. USA* **82**:4768–4772.

140. **Weinert, T. A., N. A. Schaus, and N. D. Grindley.** 1983. Insertion sequence duplication in transpositional recombination. *Science* **222**: 755–765.

141. **Weston-Hafer, K., and D. E. Berg.** 1991. Deletions in plasmid pBR322: replication slippage involving leading and lagging strands. *Genetics* **127**:649–655.

142. **Weston-Hafer, K., and D. E. Berg.** 1991. Limits to the role of palindromy in deletion formation. *J. Bacteriol.* **173**:315–318.

143. **Weston-Hafer, K., and D. E. Berg.** 1989. Palindromy and the location of deletion endpoints in *Escherichia coli. Genetics* **121**:651–658.

144. **Whoriskey, S. K., V. H. Nghiem, P. M. Leong, J. M. Masson, and J. H. Miller.** 1987. Genetic rearrangements and gene amplification in *Escherichia coli*: DNA sequences at the junctures of amplified gene fusions. *Genes Dev.* **1**:227–237.

145. **Whoriskey, S. K., M. A. Schofield, and J. H. Miller.** 1991. Isolation and characterization of *Escherichia coli* mutants with altered rates of deletion formation. *Genetics* **127**:21–30.

146. **Wilson, T. E., U. Grawunder, and M. R. Lieber.** 1997. Yeast DNA ligase IV mediates non-homologous DNA end joining. *Nature* (London) **388**:495–498.

147. **Yamaguchi, H., T. Yamashita, H. Shimizu, and H. Ikeda.** 1995. A hotspot of spontaneous and UV-induced illegitimate recombination during formation of lambda *bio* transducing phage. *Mol. Gen. Genet.* **248**:637–643.

148. **Yang, Y., and W. Masker.** 1997. Double-strand breaks increase the incidence of genetic deletion associated with intermolecular recombination in bacteriophage T7. *Mol. Gen. Genet.* **255**:277–284.

149. **Yang, Y., and W. Masker.** 1996. Instability of repeated dinucleotides in bacteriophage T7 genomes. *Mutat. Res.* **354**:113–122.

INSERTION SEQUENCES
AND TRANSPOSONS

Ronald Chalmers and Michel Blot

9

DEFINITIONS

Transposable elements (TEs) have been defined as DNA sequences able to insert at many sites in the genome (25). Within prokaryotes, two groups of TEs were initially distinguished: insertion sequences, or IS elements, were defined as short TEs (0.7 to 2 kb) which carry no genes other than those related to transposition, while transposons were defined as larger TEs (>2 kb) which also contain genes unrelated to transposition, such as antibiotic resistance genes. Later, two other groups of TEs were discovered: retrons, which are short TEs transposing via an RNA intermediate (59), and mobile introns, which are found inserted at highly specific positions within prokaryotic genes and are capable of homing transposition into the wild-type allele if it does not already contain a copy of the intron (38). Although they are able to transpose their sequences to new sites, retrons and prokaryotic introns do not have the capacity to disseminate rapidly in the genome and are not thought to make a significant contribution to genome rearrangement.

With very few exceptions, the IS elements have very similar structures and organizations (77). A protein–coding region of about 1.2 kb is flanked by two small inverted repeats of 9 to 40 bp which are specific to a given IS element. In the central region there is usually a single open reading frame coding for a single transposase, but occasionally there may be more than one protein produced, sometimes by the mechanism of translational frame shifting. Within the inverted repeat there are sequences which are recognized and bound by the transposase as the first step of the transposition reaction. These recognition sequences are located close to the terminal base pairs, where the chemical steps of the transposition reaction occur.

A new classification for TEs has been proposed based on these structural features of IS elements. Class I includes most IS elements which have the structural features mentioned above, as well as composite transposons, which are composed of a pair of IS elements that cooperate to mobilize the DNA lying between them. Class II includes the noncomposite transposons and a limited number of IS elements (e.g., IS*91*, IS*110*, and IS*200*) with unusual structural features (79). In class III are

Ronald Chalmers, Department of Biochemistry, University of Oxford, South Parks Road, Oxford OX1 3QU, United Kingdom. *Michel Blot*, Genomique Bactérienne et Evolution, Université Joseph Fourier-CNRS EP 2029-CEA LRC12, Grenoble 38041 Cedex, France.

Organization of the Prokaryotic Genome, Edited by Robert L. Charlebois,
© 1999 American Society for Microbiology, Washington, D.C.

the mutator phages, such as Mu, which replicate their sequences in the host genome by a transposition mechanism when they enter the lytic phase. Finally, conjugative transposons do not fit well into this classification, as they are functionally linked to some plasmids (79).

At present there are about 500 TEs identified in many different bacterial species (77). Most IS elements and transposons were discovered after their transposition into genes of interest. However, sometimes transposons were identified from the phenotypes of their genes for antibiotic or heavy metal resistance and catabolic pathways for xenobiotics. More recently, sequencing projects have also revealed IS-like sequences which have not yet been shown to transpose. It is also clear that our knowledge of the genomes of many species of soil and oceanic microorganisms is so sparse that a large number of TEs certainly remain to be discovered. There are several technical tools available to search for TEs based on activation or mutation of appropriately designed reporter genes (88).

Although TEs are a highly diverse group with respect to both structural organization and DNA or protein sequences, they appear to have evolved from a common ancestor. Sequence alignments can be used to group the TEs into 17 families, although due to the extreme divergence of the amino acid sequences, alignments are often little better than would be expected by chance (77). However, the common ancestry of the TEs has become more clear from the comparison of a limited number of crystal structures and the identification of a conserved DDE catalytic triad. These characteristics show that many of the TEs share an ancient common ancestor with each other as well as with the genes encoding retroviral integrases, the Holliday junction-resolving enzyme RuvC, and RNaseH (48, 90a, 91). Interestingly, with few exceptions, the 17 families of elements are represented within many taxa of eubacteria and archaebacteria, suggesting that horizontal transfer is easy. This is perhaps not surprising, since

many TEs are on plasmids and phages. A statistical analysis of codon usage classified *Escherichia coli* DNA into three groups: highly expressed genes; moderately or rarely expressed genes; and a group of genes described as inherited from horizontal transfer, which includes IS elements, genes for restriction endonucleases, pili, and plasmid transfer (78). This again suggests that TEs are involved in genetic exchange among and within taxa.

TEs AS THE CAUSE OF REARRANGEMENTS

In bacteria the relative juxtapositions of genes are not necessarily important because most will be involved in producing *trans*-acting factors, which are able to fulfill their function irrespective of their location or arrangement in the genome. However, there are a number of other reasons why the orientation and juxtaposition of genes may be vitally important. For example, in bacterial species which are conjugative, or highly transformable, natural selection will favor the linkage of genes with related functions. Also, genes with related functions are often grouped together into operons to enhance regulation. Cotranscription of the genes in an operon requires that the genes all be oriented in the same direction, but other factors are important as well. For example, in *E. coli*, neighboring operons are often transcribed in the same direction, and this is most likely to be the same as the direction of replication (16). Also, the orientation of Chi sites, which are involved in homologous recombination, correlates with the direction of replication in large areas of the genome (16). In these examples, the orientation and arrangement of large stretches of sequences could come under selection because of the interplay among replication, transcription, and repair.

The extent to which TEs are responsible for the differences among the genetic maps of related strains of bacteria is not yet clear. At present, there is only one whole genome sequence comparison available, between strains of the human gastric pathogen *Helicobacter pylori* (4a). In order to align the genome sequence of each strain, it was necessary to artificially invert and/or transpose 10 regions, ranging in size from 1 to 83 kb. Three of these 10 regions were associated with TEs; most of the others were associated with other types of repeated sequence and the insertion of restriction modification systems (4a).

The rearrangements promoted by transposons are deletions, inversions, and replicon fusions. In combination, these are sufficient to bring about any conceivable change in the arrangement of a group of genes. Although the forms of these rearrangements are very simple, the underlying mechanisms can be quite complex and may involve a cascade of host-promoted recombination events following transposition.

In bacteria, the most frequent recombination is that by the *recA*-dependent homologous pathway. Illegitimate recombination between unrelated sequences or microhomologies is much less frequent and may, for example, follow a failure to repair a collapsed replication fork by homologous recombination. Although transposons provide mechanisms for generating rearrangements independently of homologous or illegitimate recombination, they interact with these mechanisms as well. Transposon insertions themselves represent regions of "portable homology," and the cleavage of DNA during the transposition reaction may trigger host-promoted illegitimate recombination.

Mechanisms of Transposition

The well-studied transposons of bacteria can be divided into two large classes according to whether they become duplicated during transposition. Nonreplicative (or "cut-and-paste") transposons are excised from the donor site by double-strand breaks and inserted at the target site without the duplication of the transposon sequences (e.g., IS10 and IS50) (6, 7, 9, 29, 52, 60, 65, 66). In contrast, replicative transposons are duplicated during the transposition reaction by the passage of a replication fork

specifically through the transposon sequences (e.g., Tn3 and Mu) (5, 54, 86, 105, 107).

The most common types of nonreplicative transposons in bacteria are found among the IS elements, such as the members of the large IS4 family, which includes the thoroughly studied IS10 and IS50 (77). IS elements have a minimal structure and consist very simply of short terminal inverted repeats on either side of a gene for the transposase protein. When IS elements are associated with specific phenotypes, such as drug resistance, it is because a pair of IS elements cooperate to form a composite transposon in which they mobilize the DNA lying between them. Examples include Tn10 and Tn5, which contain copies of IS10 and IS50, respectively (9, 65).

Replicative transposons are much larger than IS elements. In addition to the usual transposon-encoded phenotypes, such as antibiotic resistance, they have a transposase and a resolvase which is required to separate the donor from the target replicon as the last step of the transposition reaction (107). Some bacteriophages are also considered to group with the replicative transposons because they use transposition to replicate during the lytic phase of their life cycles. The best known example is phage Mu, which is over 40 kb long (80, 86).

Nonreplicative Transposition

The mechanism of nonreplicative transposition, based on our knowledge of IS10 (66), is illustrated in Fig. 1. The first chemical step of the transposition reaction is cleavage of the transposon from the donor DNA by double-strand breaks, which generate 5′ phosphate and 3′ hydroxyl groups precisely at the ends of the transposon. Next, the transposon interacts with a potential target site and the 3′ hydroxyl groups at each end of the transposon undergo strand transfer to 5′ phosphate groups at the target site. The 5′ phosphate groups, which accept the transposon at the target site, overhang each other on opposite strands of the DNA; in Fig. 1, they are illustrated with a 5-bp stagger, but this distance varies with dif-

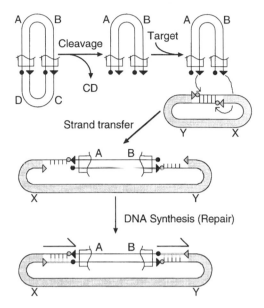

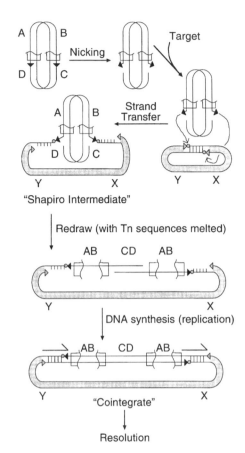

FIGURE 1 Mechanism of nonreplicative transposition. The DNA components of the reactions are all shown, but supercoiling and the protein components have been omitted for clarity. Markers A and B are in the transposon; C and D are in the flanking donor DNA, which is lost; X and Y are in the target DNA. Symbols: half boxes, transposon ends; solid and shaded circles, 5′ phosphate groups; solid and shaded triangles, 3′ hydroxyl groups; half arrows, direct repeats of target sequences.

FIGURE 2 Mechanism of replicative transposition. The transposon and donor (AB and CD, respectively) are fused with the target molecule, XY, to form a Shapiro intermediate, which is converted to a cointegrate by replication. Symbols: half boxes, transposon ends; shaded circles, 5′ phosphate groups; solid and shaded triangles, 3′ hydroxyl groups; half arrows, direct repeats of target sequences.

ferent transposons. Note that it is the staggered nature of the accepting phosphates which produces the short direct repeats of target sequence which always flank transposon insertions and which are the hallmarks of the transposition process.

Strand transfer is the last transposase-promoted step of the transposition reaction. The host cell completes the process with a small amount of replication to repair the short single-stranded sequences flanking the transposon. DNA synthesis is confined to the flanking sequences and does not extend through the transposon as in replicative transposition (see below).

Replicative Transposition

The mechanism of replicative transposition, illustrated in Fig. 2, has many features in com-

mon with nonreplicative transposition (80, 107). The first chemical step of the reaction cleaves the 3′ strand of the DNA precisely at the end of the transposon. The opposite strand at the transposon end is never cleaved, and the donor and the transposon remain joined. Next, the free 3′ hydroxyl groups at the ends of the transposon are strand transferred to the 5′ phosphate groups at the target site. Strand transfer joins the donor and the transposon to the target molecule in a complicated structure called a Shapiro intermediate (105). As was the case for nonreplicative transposition, the free 3' hydroxyl groups in the target molecule serve as primers for the DNA replication which completes the transposition reaction. In this case, however, replication is not confined to the single-stranded gaps at the target site but continues through the transposon sequences from both directions. This can be difficult to visualize, and in Fig. 2 the Shapiro intermediate has been redrawn with the two DNA strands of the transposon sequences melted out in readiness for replication.

The product of replicative transposition is a cointegrate, which is a fusion between the donor and the target molecules with a copy of the transposon at each of the two junction points. The difference between a cointegrate and the simple insertion generated by nonreplicative transposition is due to differences in the first step of the reaction. In replicative transposition, there are no double-strand breaks at the ends of the transposon. Instead, the ends are only nicked, leaving the donor molecule attached to the transposon segment by the other strands. This is the only mechanistic difference between the two types of transposition.

To complete the replicative transposition process, the cointegrate has to be resolved back into a donor molecule and a target molecule with a new insertion. This is achieved by a transposon-encoded resolvase, which catalyzes a site-specific reciprocal recombination between resolvase sites in the transposon (107).

Regions of Portable Homology

The existence of more than one copy of a transposon in a cell gives rise to the possibility of generating rearrangements by homologous recombination (33, 65, 67, 68). One function of homologous recombination is to repair double-strand breaks by using the information on a sister replicon. This process does not promote rearrangements and in fact represents a force for genome stabilization. This is true because homologous sequences are normally found in homologous locations and the products of a reciprocal cross are identical to the substrates. Although there are many examples of duplicated (and therefore homologous) genes in bacteria, these have usually diverged sufficiently to preclude homologous recombination between them. In contrast, however, a new transposon insertion is identical to the original element and may be located in a random position.

Homologous recombination between transposons promotes inversions, deletions, and replicon fusions (Fig. 3). If the new insertion is in the same orientation as the original (i.e., a direct repeat), a reciprocal cross will delete the markers between the elements. If the new insertion is in the inverted-repeat configuration, the markers between the elements will be inverted by a reciprocal cross.

Precise and Nearly Precise Excision

Deletions, inversions, and replicon fusions produced by transposons acting as regions of portable homology have transposons at the junction points of the rearrangements (Fig. 3). For the new markers to become closely and stably linked, the transposon sequences have to be lost.

Transposon sequences can be lost by the mechanism of precise and nearly precise excision (9, 51, 63–65, 76, 106). In precise excision, the original target site is regenerated exactly and no trace of the insertion or duplication of target sequences remains (Fig. 4). In nearly precise excision, the target duplications and remnants of the inverted repeats remain at the site of the insertion. The frequencies of these events vary for different chromosomal loci, genetic backgrounds, and transposons. For Tn10 in E. coli, the frequencies of transposition, precise excision, and nearly precise excision are approximately 10^{-7}, 10^{-9}, and 10^{-6}, respectively (65, 76). Precise excision is not dependent on transposon functions, but mutations in many different host genes can increase the frequency. For example, uup3 increases precise excision of Tn5 from 10^{-7} to almost 10^{-5} (57, 89). For transposons located on F episomes or multicopy

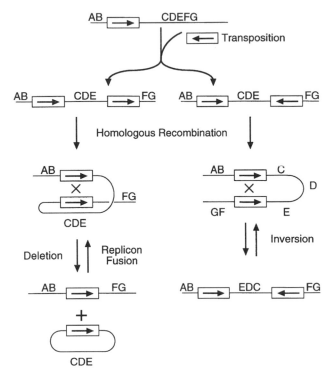

FIGURE 3 Transposons as regions of portable homology. Transposons are represented as rectangles, with arrows indicating the relative orientations of the insertions. Homologous recombination between insertions in the direct-repeat configuration deletes one copy of the transposon and the markers in between. Recombination between inverted repeats causes inversion of one copy of the transposon and the markers in between.

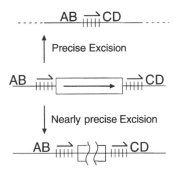

FIGURE 4 Precise and nearly precise excision. (Middle) A transposon insertion is shown as a rectangle flanked by 5-bp direct repeats of target site duplications. (Top) Precise excision restores the locus to wild type by the deletion of the transposon and one of the 5-bp target site repeats. (Bottom) Nearly precise excision leaves behind remnants of the transposon and both of the target duplications. Nearly precise excision can go on to give precise excisions.

plasmids, the frequency of precise excision can be even higher, reaching 10^{-4} for Tn5 on a pBR-based plasmid (34, 110).

The mechanism of precise and nearly precise excision involves replication slippage between the short direct repeats at or near the transposon ends (26, 40, 51). During replication, when the DNA is transiently single stranded, slippage is promoted by base pairing between the inverted repeats at either end of the transposon. This explains why the process is relatively efficient for the composite transposons Tn10 and Tn5, in which the flanking IS elements provide large inverted repeats. However, precise excision can be detected at a very low frequency for IS elements themselves, which have much smaller flanking inverted repeats (63, 65, 106).

It should be noted that the transposons in the portable-homology rearrangements (Fig. 3) are no longer flanked by the same direct repeats of target sequences as they were before the rearrangements. Instead, each insertion has swapped ends and is now flanked by nonidentical target sequences. This should preclude precise excision but still allow for nearly precise excision, which uses direct repeats associated with palindromes embedded within the ends of the transposon (51).

Precise and nearly precise excision by the slippage mechanism is not dependent on host homologous recombination. However, there is probably another *recA*-dependent pathway for transposon precise excision which is induced by UV and other types of DNA damage (30).

Intramolecular Replicative Transposition

The standard steps of the replicative transposition reaction (Fig. 2) produce deletions and inversions if the target site is located in the same DNA molecule as the transposon (5, 15) (Fig. 5). When the ends of the transposon interact with a potential target site, there are two possible orientations, which correspond simply to a 180° rotation of the target site. In standard intermolecular transposition, the orientation of the target will only affect the orientation of the transposon insertion. However, in the case of intramolecular target sites, the products will be either deletions or inversions, depending on whether the orientation of the transposon ends with respect to the target site is in *cis* or in *trans* (Fig. 5). In the orientation producing *cis* attack, the 5′ phosphate group at the target site is located on the same strand of the DNA as the 3′ hydroxyl group at the transposon end. In the *trans* attack orientation, the groups are on opposite strands of the DNA. In either orientation of the target site, replication duplicates the transposon sequences. However, markers between one end of the transposon and the target site become deleted or inverted in the case of *cis* attack and *trans* attack, respectively. This type of rearrangement can be very frequent, and in the

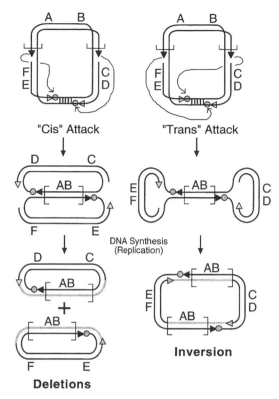

FIGURE 5 Intramolecular replicative transposition. The ends of the transposon are represented as interrupted half boxes on either side of transposon markers AB. An arbitrary target site is located between CD and EF. Following strand transfer, replication of the transposon sequences results in deletions and inversions. Symbols: half boxes, transposon ends; circles, 5′ phosphate groups; triangles, 3′ hydroxyl groups.

case of Tn*1*, a close relative of Tn*3*, the rate equals transposition at 10^{-4} (15).

Inside-Out Transposition

In nonreplicative composite transposons, deletions and inversions are generated by inside-out transposition (Fig. 6A). A composite transposon contains two copies of an IS element and a total of four transposon ends. In principle, the transposase protein can engage in an interaction with any pair of ends, and this promiscuity results in a number of different outcomes to the reaction (28) (Fig. 6A).

Inside-out transposition occurs when transposase interacts with the two innermost ends of the IS*10* elements (65, 106). In this situation, the entire replicon behaves as a transposon and intermolecular transposition will result in a replicon fusion, whereas intramolecular transposition will result in rearrangements. Intramolecular strand transfer to an arbitrary target site is illustrated in each of the two possible orientations (Fig. 6A). The orientation shown on the left results in a deletion extending from the inside end of one IS element to the target site. This brings markers C and G into juxtaposition, although they remain separated by one of the IS elements. In the opposite orientation of the target site, the result is an inversion extending through one IS element to the target site. This rearrangement is referred to as a deletion-inversion because the transpo-

son sequences XYZ are deleted during the cleavage step of the reaction.

In the case of Tn*10*, the rate of inside-out transposition is approximately 10^{-5}, or about 100-fold greater than the rate of normal transposition (65). In contrast, for Tn*5* the rate of inside-out transposition is at least 100-fold lower than the rate of transposition (9).

The deletion-inversion product resembles a new composite transposon which now carries the markers FED instead of XYZ (Fig. 6A). However, with the IS elements in the direct-repeat configuration there is a likelihood that homologous recombination will delete the composite structure, leaving only a single IS element. Usually composite transposons have their IS elements in the inverted-repeat

FIGURE 6 Inside-out transposition, dimer donors, and cryptic ends. (A) The IS elements on the left (L) and right (R) of a composite transposon are represented as rectangles, with arrows indicating their relative orientations. The transposon carries the unique sequences XYZ, and an arbitrary target site is located between markers DEF and GHI. During inside-out transposition, the innermost pair of ends from the IS elements are cleaved and transferred to the target site. The cleavage step always deletes the transposon sequences XYZ. Strand transfer produces either deletions or inversions, depending on the orientation of the target. A target site on a different DNA molecule would produce a replicon fusion (not shown). (B) A replicon with a single IS element will have multiple ends available if it exists as a dimer. If opposite ends of the sister IS elements are used for transposition, cleavage will delete half of the dimer and yield an intermediate, almost identical to inside-out transposition. Strand transfer will likewise produce deletions, inversions, and replicon fusions (not shown). (C) A cryptic transposon end (half box) may be present in the correct location and orientation with respect to one of the transposon ends.

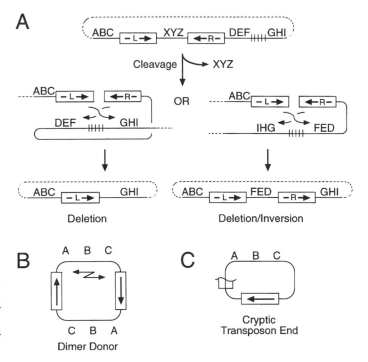

configuration so that homologous recombination only causes the inversion of the markers within the transposon.

The essential feature of inside-out transposition is that a large portion of the host replicon behaves as a transposon, resulting in deletions and inversions if the target sites are intramolecular. However, this process does not require a complete composite transposon but can be achieved by a single IS element accompanied by a third transposon end in the appropriate location (Fig. 6B and C). The first possibility is that a replicon with a single IS element can exist as a dimer, which will provide the additional transposon end required (8, 9). This mechanism can also explain the curious observation that under certain circumstances, supposedly nonreplicative IS elements promote the formation of cointegrates just like the replicative transposons (4, 8, 9, 55, 75). The second possibility is the presence of a cryptic transposon end (Fig. 6C), which is a DNA sequence with sufficient fortuitous resemblance to a transposon end that it is recognized by transposase (28, 61, 87).

Bimolecular Synapsis

Bimolecular synapsis is the interaction of transposase with a pair of transposon ends located on different molecules. It provides a mechanism for the nonreplicative IS elements to produce cointegrates, adjacent deletions, and replicative inversions (4, 75, 93) (Fig. 7).

IS*10* transposition is regulated by a *dam* methylation site at the inside end of the element (65, 92). The fully methylated transposon end is not a good substrate for transposition, but after the passage of a replication fork through the element, one of the two transiently hemimethylated species becomes activated. This suggested the possibility that transposase interacts with the opposite ends of sister elements following the passage of a replication fork (Fig. 7). If the target site is found on a different molecule, the product will be a cointegrate following degradation of the collapsed replication fork. If the target site is intramolecular, the products will be deletions and inversions, depending on the relative orientation of the transposon ends and the target site (Fig. 7).

It should be noted that support for the bimolecular synapsis hypothesis is based largely on observations in vitro (28) and that there may be constraints on the process in vivo. Even so, the frequency at which nonreplicative IS elements promote cointegrates, adjacent deletions, and replicative inversions is very low, and it is not possible to rule out the bimolecular synapsis hypothesis with the data available at present.

INTERACTIONS BETWEEN TEs AND THE HOST

In this section we will focus on the impact of IS elements and transposons on the fitness of their host cells. The main differences between these two types of TEs is that the IS elements are phenotypically cryptic while the transposons carry genes which directly improve the host physiology. It is easy to see how the genes for antibiotic resistance or the catabolism of xenobiotic compounds can increase the fitness of the host, but the IS elements might be expected to be largely detrimental except on the very rare occasions when they cause beneficial genetic rearrangements.

Mutation Rates and Distribution of TEs

Systematic studies of the spectrum of spontaneous mutagenesis in *E. coli* have shown native IS elements to be responsible for the largest number of mutations in a neutral marker (62, 94, 100, 101). Transposition accounts for approximately 60% of all mutations, while deletions and point mutations account for only 30 and 10%, respectively. In specific studies, TEs show a wide range of transposition rates, from 10^{-3} per genome per generation for Tn*9* when introduced into naive cells to nearly a null rate for IS*200* (27). A conservative estimate of the transposition rate of all IS elements present in *E. coli* is approximately 10^{-5} per genome per generation (ca. 10^{-8} per gene per

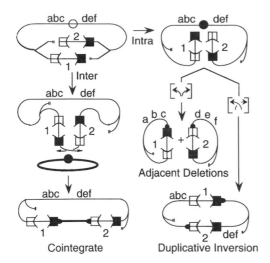

Intra

Inter

Adjacent Deletions

Cointegrate

Duplicative Inversion

FIGURE 7 Bimolecular synapsis. Transposon ends are represented as open and solid half boxes to distinguish the left and right ends, respectively. An arbitrary intramolecular target site is illustrated, flanked by markers ABC and DEF. After the transposon sequences have been duplicated by the passage of a replication fork, transposase cleaves at the opposite ends of each of the sister elements. If the target site is on a different DNA molecule, the strand transfer product is a cointegrate (left side of figure). If the target site is intramolecular, the products are replicative deletions or replicative inversions, depending on the relative orientations of the ends and the target site. Open circle, potential target site; solid circle, actual target site used.

generation). In *E. coli*, the most frequent insertions are made by IS5 and IS1, and to a lesser extent IS2, IS4, IS30, and IS186, while IS3 and IS150 are relatively inactive (43, 77). Recent data suggest, however, that IS150 can promote a burst of transposition (85), as was previously observed for IS30 (83).

Another way to measure the activity of IS elements is to study their distribution across the chromosome by restriction fragment length polymorphism (RFLP) after hybridization with internal sequences of IS elements (RFLP-IS). The IS elements are mapped in several *E. coli* strains (11, 12, 44, 112, 113), and their insertion sites are precisely known in the full sequence of *E. coli* (16). In K-12, there are ca. 50 copies of eight different IS elements,

corresponding to 1% of the genome. These provided about 20 changes in the history of eight major strains of K-12, from which a phylogenetic tree can be constructed (43). Their distribution on the circular chromosome reveals some regions with a larger number of IS elements that might be due to historical contingency, although it has been suggested recently that IS1 displays a pattern with local preferences for transposition (22).

In *E. coli* (the ECOR collection), the different IS elements could be grouped by the level of contribution to genetic diversity as follows: IS5 and IS1, high; IS2, IS4, and IS30, moderate; IS3, null (98, 99). This is in agreement with the insertion rates obtained with different reporter genes cited above. A similar ranking was observed in studies of the IS element dynamics within the *E. coli* K-12 strain W3110 during storage as stabs during a period of up to 30 years (82, 83). In these studies, the different copies of a given IS element did not behave in the same fashion. Some of them seem very stable and can be used as phylogenetic markers (13, 71), while others are unstable and drive important chromosomal polymorphisms. However, the sequence comparison of these different copies has not shown significant changes that might explain a difference in their transposition behavior (72). Similar data were obtained with *Salmonella typhimurium* except that, based on RFLP-IS, these strains were not very divergent, indicating a smaller contribution of IS elements to genomic variation (14).

When TEs are present more than once on a chromosome, the different copies can serve as homologous sequences for site-specific recombination (45). Some known rearrangements within the commonly used K-12 strains are due to inversions between IS elements (see reference 43 for a review). Also, IS-mediated rearrangements are responsible for Hfr and F′ formation (114). Moreover, the end points of some small inversions (97), as well as very large inversions (42), correspond exactly to the location of IS elements. Using the complete *E. coli* K-12 sequence, it is now possible to have a dynamic view of IS-related rearrangements within cultured clones (85). It is important at this point to distinguish the rate at which mutation events arise in populations from the frequency at which these mutants are found later in the population. In the latter case, natural selection, as well as genetic drift, may make the most important contribution to the persistence of a new mutation.

Negative Selection

The negative impact of TEs on host abilities has never been extensively tested by experiment. However, data banks report many IS element insertions during the cloning steps in bacteria, which is a measure of their toxicity as mutators. The fact that almost any gene can be mapped by transposon mutagenesis (41) is also a sign that TEs can insert at many sites in the genome and disrupt genes, so that their effect is deleterious to the host bacterium. For this reason, TEs were thought of as "selfish" DNA (47, 84).

The view of TEs as purely selfish DNA was quickly contradicted. First, on a theoretical basis, it was argued that these elements might be, or become, involved in euchromosomal evolution (reference 23; for a discussion, see reference 24). Further, it was proposed that TEs might not be maintained unless they had a positive role in bacterial evolution (35, 37). Second, competition experiments showed that strains carrying a TE were more successful than their isogenic TE-free counterparts (10, 31, 32, 56). It was proposed that TEs can promote and disseminate genomic plasticity and thus serve, in hindsight, the adaptation of their hosts. In the following paragraphs we will evaluate the nature of TE-related rearrangements and their impact on the long-term fitness of bacteria and show that (i) TEs have adapted to provide a tolerable amount of mutation which is not a great burden to their hosts and (ii) TEs are involved in beneficial mutations.

TEs were initially identified because they insert at high rates into phenotypically visible genes (50, 103). Depending on which gene has been disrupted and the prevailing environmental conditions, insertions can either kill or decrease fitness over a large range. Nonviable mutants are directly eliminated from the population; however, bacteria that have a slight fitness reduction are counterselected at a rate which depends on their loss of fitness and their environment. In a more general view, it has been proposed that the chromosomal distribution of IS elements in the ECOR collection reflects the fact that these elements are "mildly" deleterious to their hosts (53, 99).

If TEs are indeed deleterious, we might expect in the very long term that this would lead to the elimination of TEs from bacteria. Exactly the contrary is observed. TEs are widespread across bacterial genera, and there is no sign that their existence is compromised, except that some IS element copies seem to have been mostly immobile during the evolution of enterobacteria. It is thus tempting to believe that the current transposition rate reflects a coadaptation of the host and the passenger. A good example is given by Tn5, which has one of the highest rates of transposition (ca. 10^{-5} per cell per division). The transposition frequency of Tn5 can be increased even further by point mutations in transposase (115) or in the transposon ends, which are apparently maintained at a suboptimal level to depress the transposition rate (117). However, these mutants, which have many chances to arise spontaneously in nature, are not the wild type, perhaps because they would promote "too much" transposition. This situation is analogous to the way in which a virus can adapt to a new host by becoming less pathogenic.

TEs must have evolved mechanisms to nullify the cost of their presence in bacterial genomes. We can see three main possible strategies that TEs may use to balance the negative fitness inevitably provoked by their insertions.

1. As discussed above, TEs might reduce their transposition rates to a level which is not too harmful (i.e., become less transposable).

2. TEs might become "infectious." The proportion of TEs lost after deleterious mutations would be replaced by TEs transmitted horizontally on plasmids, phages, or conjugative transposons.

3. TEs might provide some benefits and thus "pay" for their presence in the host.

We will now discuss examples of TEs which use these strategies to reduce the impact of the mutation load on the genomes.

Control of Transposition

TEs use a number of mechanisms to minimize the rate of transposition (reviewed in reference 77). These are (i) transcriptional control of the transposase gene, (ii) efficiency of translation of the mRNA into a functional transposase, (iii) access of the transposase to the inverted repeats flanking the TEs, (iv) a *trans* acting inhibitor, and (v) transposition immunity (mostly transposons).

Promoters within IS elements are often very weak, although usually strong enough to produce mRNA and obtain a transposase. However, access to the promoter sequence can be restricted to limit transcription. Quite often the transposase promoter is located within one of the inverted repeats so that occupancy by transposase blocks further transcription of the transposase gene (69). It seems that the active form of transposase is a multimer, and thus at low concentration it acts as a repressor of transposition.

More complex is the programmed translational frameshift required by IS1 and IS150. The IS1 element has two recognizable encoding regions, *insA* and *insB*. After a normal mRNA transcription, the translation of *insA* produces a protein which binds the inverted repeats but is not active for transposition. A ribosomal shift is required to create the fusion protein InsAB, which is the active transposase. The rate of transposition is therefore always less than the rate of ribosomal shifting, especially if the competition between InsA and InsAB is taken into account (49).

The access of the transposase to the inverted repeat is controlled by *dam* methylation in IS10 (65). The inside end of the element is

inactive for transposition when fully methylated. One of the two hemimethylated forms of the inverted repeat, present after the passage of a replication fork, is active, thus coupling transposition to replication. In contrast, for the related element Tn5, transposition is repressed by the competition between the two Tsp proteins encoded by IS50L and IS50R. Only TspR is an active transposase, while TspL is truncated and blocks access to the transposon ends (90).

IS10 also produces an antisense RNA which blocks expression of transposase (65). In contrast to the transposase, which is *cis* acting, the inhibitor is diffusible (or *trans* acting). The combination of a *cis*-acting transposase and a *trans*-acting inhibitor serves to limit the number of copies of the element in a cell because the abundance of the inhibitor increases with copy number whereas that of the transposase does not.

Finally, many transposons display target immunity, in which the presence of an insertion provides immunity to large regions of surrounding sequences. For phage Mu this phenomenon is an active mechanism involving MuB protein (2, 3). For Tn3 the basis of the effect is still not clear but depends on both the transposase and the presence of the inverted repeat (107).

At this point it is interesting to look at the strategies used by some viruses and to compare them with those of the IS elements. Phage λ inserts into the bacterial genome at a single attachment site, BOB′, while the phage P1 can insert into any site of the genome. One might imagine that a TE with a strong specificity of insertion would therefore not create a large mutation burden. There is in fact one such example. Tn7 normally transposes to a single conserved site in the *E. coli* genome, which by analogy with phage lambda is also referred to as the attachment site. Tn7 does not appear to be closely related to the IS elements, and this is reflected in differences in the mechanism of transposition (39). One important difference between the IS elements and Tn7 transposition is that whereas the IS elements transpose only inefficiently to random sites, Tn7 transposes efficiently to its single attachment site.

Another example of how site specificity prevents the buildup of a transposition load can be found in the very few prokaryotic introns discovered in the last decade (70). The mobile introns of phage T4 are a special case that might belong to this category, with a very conserved sequence for homing which does not seem to affect the expression of the genes in which the introns are found. These TEs are found only once on the genome, at a very precise location controlled by their integrase, and they do not seem to be able to spread elsewhere except by recombination (81).

Transmission of TEs

TEs are known to be present on many phages and plasmids. The F fertility plasmid has copies of IS2, IS3, and Tn1000 (γδ). These may recombine with their homologs on the chromosome to create Hfr strains (45). When TEs participate in genetic transmission through cells, they can be thought of as beneficial to their hosts. They can increase the genetic plasticity of the cells and thus promote rapid adaptation, perhaps through a transfer and integration of foreign genes, such as those for xenobiotic resistance, that become essential for survival in nature. Conjugative transposons are the most extreme case of this strategy, since they harbor within the same sequence the cellular (transposition) and the intercellular (conjugation) mobilities. The amount of data on their active role in microbial ecology is now increasing (58, 96).

Although the exchange of genetic material, and the presence of TEs across bacterial taxa, is an important point, it probably does not have a tremendous effect on the dynamics of a population. There are no specific data to measure the role of TEs on gene transmission, only proof that they are able to do it occasionally. We believe that the consequence of the occurrence of a given TE is more qualitative (presence versus absence) than quantitative (ratio of bacteria carrying the TE). After

a TE has entered a taxon, it is the direct contribution to fitness (negative or positive) which will dictate whether it is able to spread.

Positive Selection

Transposons differ from IS elements because they carry genes unrelated to transposition. These include a large range of resistance against antibiotics and heavy metals and catabolic genes allowing degradation of toxic compounds. It is obvious that the presence of these kinds of transposons within wild bacteria provides for a selective advantage in many natural environments, and therefore it can account for the wide distribution of transposons (17). It is believed that the capture of such genes within TEs has been important in the dissemination of these resistances, especially when the TEs could combine with plasmids or become conjugative transposons.

One example of an increase of fitness imparted by a transposon has been studied at the mutational, the population, and the molecular levels. Transposon Tn5 has one of the highest transposition rates (46, 90), and it might be expected to be highly detrimental to bacterial populations in the long term. Competitions between strains with and without Tn5 showed a positive fitness for Tn5-harboring strains (10), initially attributed to IS50 (56) but in fact due to the constitutive expression of the bleomycin resistance gene in the absence of the drug (20, 21). It appears that the expression of bleomycin resistance from Tn5 protects the cell against oxidative damage (1) and provides a growth advantage sufficient for invasion of an experimental population. It has been suggested that the continuous positive fitness of Tn5 might allow for the higher-than-normal rate of transposition (19).

Physiological Regulation and Experimental Evolution

IS elements can promote changes in the host cell physiology by affecting the pattern of gene expression. In the simplest case, an insertion can directly inactivate a regulator. Insertion into an inducer or a repressor will cause silencing or constitutive expression, respectively. However, IS elements also contain promoters which can drive changes in the expression of genes near an insertion. For example, the pOUT promoter of IS10 drives constitutive transcription outwards from the end of the element (108, 109). On the other hand, IS5 has a −35 box close to one end, as well as a partial −10 box which overlaps the inverted repeat. Upon insertion the −10 box can be completed, thus creating a new transcriptional unit (102).

To explain the establishment and maintenance of IS elements within a genome, some theoretical models require a positive contribution to fitness by the elements (35, 36). One possibility for IS elements is that rarely they promote a very beneficial mutation, which then allows the full set of TEs carried on the chromosome to hitchhike into future generations (18). Since these mutations are certainly rare, but have potentially drastic effects, special experimental designs must be used to study the long-term evolution of populations.

In chemostat experiments, where strong selection can be applied for many generations, the successful mutations are not usually found in structural genes but in their regulatory regions. In many cases these changes are due to IS element insertions. It appears that these allow for more rapid adaptation than is possible for changes in structural genes. If structural genes are already highly adapted and occupy local maxima on the fitness landscape, mutations which increase fitness are likely to be much rarer than regulatory changes which achieve the same effect. In a recent example, cross-feeding on acetate for 1,750 generations produced mutants with semiconstitutive overexpression of acetyl coenzymeA synthetase. These were due to IS element insertions or a point mutation in the 5′-regulatory region (95, 111).

In another competition study, using isogenic strains either with or without IS10, the bacteria with IS10 rapidly took over the culture, suggesting that the mutator function of IS10 was promoting beneficial mutations (31,

32). Unfortunately, the exact locations of these insertions were not mapped. Similar experiments were performed with the eukaryotic element Ty in *Saccharomyces cerevisiae* (116), and these also suggested that TEs aid adaptation to new environments.

In an experiment lasting almost 10 years, 12 isogenic cultures of *E. coli* B were propagated at a rate of 6.6 generations per day. The fitness increase for growth on a glucose-limited broth was significant and was accompanied by changes in cell shape (73, 74). RFLP-IS on two of these cultures detected more than 200 IS-related rearrangements. The pedigrees of these cultures revealed six pivotal mutations which became fixed (85), suggesting that the corresponding IS-related rearrangements had either hitchhiked together with a beneficial mutation or were themselves beneficial. The cloning and sequencing of the regions around these IS elements showed that regulatory genes had been hit, leading to overexpression of genes involved in metabolism (47a). In another of these cultures, a polymorphism was discovered at generation 6,000 in which large cells and small cells were able to coexist (95a). This was correlated with an RFLP-IS polymorphism which was maintained for 14,000 generations. It was shown that small colonies had an IS element inserted in a gene for menaquinone synthesis, and this suggested the possibility of a cross-feeding relationship with the large cells (101a). The ratio of small to large cells was maintained by a frequency-dependent selection. However, if menaquinone was added to the growth medium, the small cells were able to take over the culture (95a). Both examples show that IS elements are able to promote mutations that may favor their carriers through changes in physiology.

CONCLUDING REMARKS

In *E. coli*, the genetic burden imposed by TEs is relatively low. In the sequenced isolate of K12 they account for only 1% of the genome, and the total rate of transposition is about 10^{-5} per cell per generation. Even so, TEs make the largest contribution to mutations in a neutral marker, followed by deletions and base substitutions (94).

TEs are the leading causes of genomic instability in *E. coli* and most probably in other species as well. The question naturally arises whether TEs are purely selfish or if they somehow pay their way into future generations by providing the host with some benefit. For transposons which carry selectable traits, such as antibiotic resistance, the benefits are obvious. Such traits as antibiotic resistance are not essential for survival except on rare occasions in nature. In such situations, transposons are able to mobilize resistances and make them available to a wider section of the microbial community. For the IS elements, however, the arguments are not so clear.

In the few experiments where selection has been maintained for many generations, IS elements have been found to have a major effect on the genetic structure of the population. Successful mutations are usually rearrangements which alter patterns of gene expression rather than changes which improve the activity of structural genes. In this context, it is tempting to view TEs as natural engineers (104), providing the host with a supply of genetic variation as a substrate for selection and adaptation. This, however, is a very difficult argument which can be undermined by considering the long-standing observation that almost all mutations are either neutral or detrimental to the individual in which they take place. It would therefore seem to us that IS elements are mostly selfish but that they contribute to evolution nevertheless.

ACKNOWLEDGMENTS

M.B. thanks J. Geiselmann and D. Schneider for valuable criticism of the manuscript.

Research by M.B. is supported by an ATIPE from CNRS. R.C. is a Royal Society University Research Fellow, and he thanks The Wellcome Trust and The Biotechnology and Biological Sciences Research Council for funding work on transposition.

REFERENCES

1. **Adam, E., M. R. Volkert, and M. Blot.** 1998. Cytochrome C biogenesis is involved in the transposon Tn*5*-mediated bleomycin resistance and associated fitness effect in *E. coli. Mol. Microbiol.* **28**:15–24.

2. **Adzuma, K., and K. Mizuuchi.** 1989. Interaction of proteins located at a distance along DNA: mechanism of target immunity in the Mu DNA strand-transfer reaction. *Cell* **57**:41–47.

3. **Adzuma, K., and K. Mizuuchi.** 1988. Target immunity of Mu transposition reflects a differential distribution of Mu B protein. *Cell* **53**:257–266.

4. **Ahmed, A.** 1991. A comparison of intramolecular rearrangements promoted by transposons Tn*5* and Tn*10. Proc. R. Soc. Lond. B* **244**:1–9.

4a. **Alm, R. A., L. S. Ling, D. T. Moir, B. L. King, E. D. Brown, P. C. Doig, D. R. Smith, B. Noonan, B. C. Guild, B. L. deJonge, G. Carmel, P. J. Tummino, A. Caruso, M. Uria-Nickelsen, D. M. Mills, C. Ives, R. Gibson, D. Merberg, S. D. Mills, Q. Jiang, D. E. Taylor, G. F. Vovis, and T. J. Trust.** 1999. Genomic-sequence comparison of two unrelated isolates of the human gastric pathogen *Helicobacter pylori. Nature* **397**:176–180.

5. **Arthur, A., and D. Sherratt.** 1979. Dissection of the transposition process: a transposon-encoded site-specific recombination system. *Mol. Gen. Genet.* **175**:267–274.

6. **Bender, J., and N. Kleckner.** 1986. Genetic evidence that Tn10 transposes by a nonreplicative mechanism. *Cell* **45**:801–815.

7. **Benjamin, H. W., and N. Kleckner.** 1992. Excision of Tn*10* from the donor site during transposition occurs by flush double-strand cleavages at the transposon termini. *Proc. Natl. Acad. Sci. USA* **89**:4648–4652.

8. **Berg, D. E.** 1983. Structural requirement for IS*50*-mediated gene transposition. *Proc. Natl. Acad. Sci. USA* **80**:792–796.

9. **Berg, D. E.** 1989. Transposon Tn*5*, p. 185–210. *In* D. E. Berg and M. M. Howe (ed.), *Mobile DNA.* American Society for Microbiology, Washington, D.C.

10. **Biel, S. W., and D. L. Hartl.** 1983. Evolution of transposons: natural selection for Tn*5* in *E. coli* K12. *Genetics* **103**:581–592.

11. **Birkenbihl, R. P., and W. Vielmetter.** 1989. Complete maps of IS*1*, IS*2*, IS*3*, IS*4*, IS*5*, IS*30*, and IS*150* locations in *E. coli* K12. *Mol. Gen. Genet.* **220**:147–153.

12. **Birkenbihl, R. P., and W. Vielmetter.** 1991. Completion of the IS map in *E. coli*: IS*186* positions on the *E. coli* K12 chromosome. *Mol. Gen. Genet.* **226**:318–320.

13. **Bisercic, M., and H. Ochman.** 1993. The ancestry of insertion sequences common to *Escher-*

ichia coli and *Salmonella typhimurium. J. Bacteriol.* **175:**7863–7868.

14. **Bisercic, M., and H. Ochman.** 1993. Natural populations of Escherichia coli and Salmonella typhimurium harbor the same classes of insertion sequences. *Genetics* **133:**449–454.

15. **Bishop, R., and D. Sherratt.** 1984. Transposon Tn*1* intra-molecular transposition. *Mol. Gen. Genet.* **196:**117–122.

16. **Blattner, F. R., I. G. Plunkett, C. A. Bloch, N. T. Perna, V. Burland, M. Riley, J. Collado-Vides, J. D. Glasner, C. K. Rode, G. F. Mayhew, J. Gregor, N. W. Davis, H. A. Kirkpatrick, M. A. Goeden, D. J. Rose, B. Mau, and Y. Shao.** 1997. The complete genome sequence of *E. coli* K-12. *Science* **277:**1453–1461.

17. **Blazquez, J., A. Navas, P. Gonzalo, J. L. Martinez, and F. Baquero.** 1996. Spread and evolution of natural plasmids harboring transposon Tn*5. FEMS Microbiol. Ecol.* **19:**63–71.

18. **Blot, M.** 1994. Transposable elements and adaptation of host bacteria. *Genetica* **93:**5–12.

19. **Blot, M., B. Hauer, and G. Monnet.** 1994. Growth advantage, better survival and the bleomycin resistance gene of Tn*5. Genetics* **242:**595–601.

20. **Blot, M., J. Heitman, and W. Arber.** 1993. Tn*5*-mediated bleomycin resistance in *Escherichia coli* requires the expression of host genes. *Mol. Microbiol.* **8:**1017–1024.

21. **Blot, M., J. Meyer, and W. Arber.** 1991. Bleomycin-resistance gene derived from the transposon Tn*5* confers selective advantage to *Escherichia coli* K-12. *Proc. Natl. Acad. Sci. USA* **88:**9112–9116.

22. **Boyd, E. F., and D. L. Hartl.** 1997. Nonrandom location of IS*1* elements in the genomes of natural isolates of *Escherichia coli. Mol. Biol. Evol.* **14:**725–732.

23. **Campbell, A.** 1981. Evolutionary significance of accessory DNA elements in bacteria. *Annu. Rev. Microbiol.* **35:**55–83.

24. **Campbell, A.** 1981. Some questions about movable elements and their implications. *Cold Spring Harbor Symp. Quant. Biol.* **45:**1–9.

25. **Campbell, A. M., D. Berg, D. Botstein, E. Lederberg, R. Novick, P. Starlinger, and W. Szybalski.** 1977. Nomenclature of transposable elements in Prokaryotes, p. 15–22. *In* A. I. Bukhari, J. A. Shapiro, and S. L. Adhya (ed.), *DNA Insertion Elements, Plasmids and Episomes.* Cold Spring Harbor Laboratory, Cold Spring Harbor, N.Y.

26. **Canceill, D., and S. D. Ehrlich.** 1996. Copy-choice recombination mediated by DNA

polymerase III holoenzyme from *Escherichia coli. Proc. Natl. Acad. Sci. USA* **93:**6647–6652.

27. **Casadesus, J., and J. R. Roth.** 1989. Absence of insertions among spontaneous mutants of *Salmonella typhimurium. Mol. Gen. Genet.* **216:**210–216.

28. **Chalmers, R. M., and N. Kleckner.** 1996. IS*10*/Tn*10* transposition efficiently accommodates diverse transposon end configurations. *EMBO J.* **15:**5112–5122.

29. **Chalmers, R. M., and N. Kleckner.** 1994. Tn*10*/IS*10* transposase purification, activation, and in vitro reaction. *J. Biol. Chem.* **269:**8029–8035.

30. **Chan, A., and R. Nagel.** 1997. Involvement of *recA* and *recF* in the induced precise excision of Tn*10* in *Escherichia coli. Mutat. Res.* **381:**111–115.

31. **Chao, L., and S. M. McBroom.** 1985. Evolution of transposable elements: an IS*10* insertion increases fitness in *E. coli. Mol. Biol. Evol.* **2:**359–369.

32. **Chao, L., C. Vargas, B. B. Spear, and E. C. Cox.** 1983. Transposable elements as mutator genes in evolution. *Nature* **303:**633–635.

33. **Chumley, F. G., and J. R. Roth.** 1980. Rearrangements of the bacterial chromosome using Tn*10* as a region of homology. *Genetics* **94:**1–14.

34. **Collins, J., G. Volckaert, and P. Nevers.** 1982. Precise and nearly-precise excision of the symmetrical inverted repeats of Tn*5*; common features of *recA*-independent deletion events in *Escherichia coli. Gene* **19:**139–146.

35. **Condit, R.** 1990. The evolution of transposable elements: conditions for establishment in bacterial populations. *Evolution* **44:**347–359.

36. **Condit, R., and B. R. Levin.** 1990. The evolution of plasmid carrying multiple resistance genes: the role of segregation, transposition and homologous recombination. *Am. Nat.* **135:**573–596.

37. **Condit, R., F. Stewart, and B. Levin.** 1988. The population biology of bacterial transposons: a priori conditions for maintenance as parasitic DNA. *Am. Nat.* **132:**129–147.

38. **Cousineau, B., D. Smith, S. Lawrence-Cavanagh, J. Mueller, J. Yang, D. Mills, D. Manias, G. Dunny, A. Lambowitz, and M. Belfort.** 1998. Retrohoming of a bacterial group II intron: mobility via complete reverse splicing, independent of homologous DNA recombination. *Cell* **21:**4.

39. **Craig, N. L.** 1996. Transposon Tn*7. Curr. Top. Microbiol. Immunol.* **204:**27–48.

40. **d'Alencon, E., M. Petranovic, B. Michel, P. Noirot, A. Aucouturier, M. Uzest, and S. D. Ehrlich.** 1994. Copy-choice illegitimate DNA

recombination revisited. *EMBO J.* **13:**2725–2734.

41. **de Bruijn, F. J.** 1987. Transposon Tn5 mutagenesis to map genes. *Methods Enzymol.* **154:** 175–196.

42. **de-Massy, B., J. Patte, J. M. Louarn, and J. P. Bouche.** 1984. oriX: a new replication origin in E. coli. *Cell* **36:**221–227.

43. **Deonier, R. C.** 1996. Native insertion sequence elements: locations, distributions and sequence relationships, p. 2000–2012. *In* F. C. Neidhardt, R. Curtiss III, J. L. Ingraham, E. C. C. Lin, K. B. Low, B. Magasanik, W. S. Reznikoff, M. Riley, M. Schaechter, and H. E. Umbarger (ed.), *Escherichia coli and Salmonella: Cellular and Molecular Biology*, 2nd ed., vol. 2. ASM Press, Washington D.C.

44. **Deonier, R. C., and R. C. Hadley.** 1976. Distribution of inverted IS-length sequences in the *E. coli* K12 genome. *Nature* **264:**191–193.

45. **Deonier, R. C., and R. G. Hadley.** 1980. IS2-IS2 and IS3-IS3 recombination frequencies in F integration. *Plasmid* **3:**48–64.

46. **Dodson, K. W., and D. E. Berg.** 1989. Factors affecting transposition activity of IS50 and Tn5 ends. *Gene* **76:**207–213.

47. **Doolittle, W. F., and C. Sapienza.** 1980. Selfish genes, the phenotype paradigm and genome evolution. *Nature* **284:**601–603.

47a. **Duperchy, E., D. Schneider, and M. Blot.** Unpublished data.

48. **Dyda, F., A. B. Hickman, T. M. Jenkins, A. Engelman, R. Craigie, and D. R. Davies.** 1994. Crystal structure of the catalytic domain of HIV-1 integrase: similarity to other polynucleotidyl transferases. *Science* **266:**1981–1986.

49. **Escoubas, J. M., M. F. Prère, O. Fayet, I. Salvignol, D. Galas, D. Zerbib, and M. Chandler.** 1991. Translational control of transposition activity of the bacterial insertion sequence IS1. *EMBO J.* **10:**705–712.

50. **Fiandt, M., W. Szybalski, and M. Malamy.** 1972. Polar insertions in *lac, gal* and phage lambda consist of a few IS-DNA sequences inserted with either orientation. *Mol. Gen. Genet.* **119:**223–231.

51. **Foster, T. J., V. Lundblad, S. Hanley-Way, S. M. Halling, and N. Kleckner.** 1981. Three Tn10-associated excision events: relationship to transposition and role of direct and inverted repeats. *Cell* **23:**215–227.

52. **Goryshin, I. Y., and W. S. Reznikoff.** 1998. Tn5 in vitro transposition. *J. Biol. Chem.* **273:** 7367–7374.

53. **Green, L., R. Miller, D. E. Dykhuizen, and D. L. Hartl.** 1984. Distribution of DNA inser-tion elements IS5 in natural isolates of *E. coli.* *Proc. Natl. Acad. Sci. USA* **81:**4500–4504.

54. **Grindley, N. D., and D. J. Sherratt.** 1979. Sequence analysis at IS1 insertion sites: models for transposition. *Cold Spring Harbor Symp. Quant. Biol.* **2:**1257–1261.

55. **Harayama, S., T. Oguchi, and T. Iino.** 1984. Does Tn10 transpose via the cointegrate molecule? *Mol. Gen. Genet.* **194:**444–450.

56. **Hartl, D. L., D. E. Dykhuizen, R. D. Miller, L. Green, and J. De Framond.** 1983. Transposable element IS50 improves growth rate of E. coli cells without transposition. *Cell* **35:**503–510.

57. **Hopkins, J. D., M. Clements, and M. Syvanen.** 1983. New class of mutations in *Escherichia coli* (*uup*) that affect precise excision of insertion elements and bacteriophage Mu growth. *J. Bacteriol.* **153:**384–389.

58. **Hughes, V. M., and N. Datta.** 1983. Conjugative plasmids in bacteria of the 'pre-antibiotic' era. *Nature* **302:**725–726.

59. **Inouye, S., and M. Inouye.** 1993. The retron: a bacterial retroelement required for the synthesis of msDNA. *Curr. Opin. Genet. Dev.* **3:**713–718.

60. **Isberg, R. R., and M. Syvanen.** 1985. Tn5 transposes independently of cointegrate resolution. Evidence for an alternative model for transposition. *J. Mol. Biol.* **182:**69–78.

61. **Jilk, R. A., J. C. Makris, L. Borchardt, and W. S. Reznikoff.** 1993. Implications of Tn5-associated adjacent deletions. *J. Bacteriol.* **175:** 1264–1271.

62. **Kitamura, K., Y. Torii, C. Matsuoka, and K. Yamamoto.** 1995. DNA sequence changes in mutations in the *tonB* gene on the chromosome of *Escherichia coli* K12: insertion elements dominate the spontaneous spectra. *Jpn. J. Genet.* **70:**35–46.

63. **Kleckner, N.** 1977. Translocatable elements in procaryotes. *Cell* **11:**11–23.

64. **Kleckner, N.** 1981. Transposable elements in prokaryotes. *Annu. Rev. Genet.* **15:**341–404.

65. **Kleckner, N.** 1989. Transposon Tn10, p. 227–268. *In* D. E. Berg and M. M. Howe (ed.), *Mobile DNA*. American Society for Microbiology, Washington, D.C.

66. **Kleckner, N., R. M. Chalmers, D. Kwon, J. Sakai, and S. Bolland.** 1996. Tn10 and IS10 transposition and chromosome rearrangements: mechanisms and regulation in vivo and in vitro. *Curr. Top. Microbiol. Immunol.* **204:**49–84.

67. **Kleckner, N., and D. G. Ross.** 1980. *recA-*dependent genetic switch generated by transposon Tn10. *J. Mol. Biol.* **144:**215–221.

68. **Kleckner, N., J. Roth, and D. Botstein.** 1977. Genetic engineering in vivo using translo-

catable drug-resistance elements. New methods in bacterial genetics. *J. Mol. Biol.* **116**:125–159.

69. **Krebs, M. P., and W. S. Reznikoff.** 1986. Transcriptional and translational initiation sites of IS*50*. Control of transposase gene and inhibitor expression. *J. Mol. Biol.* **192**:781–791.

70. **Lambowitz, A. M., and M. Belfort.** 1993. Introns as mobile genetic elements. *Annu. Rev. Biochem.* **62**:587–622.

71. **Lawrence, J. G., D. E. Dykhuizen, R. F. DuBose, and D. L. Hartl.** 1989. Phylogenetic analysis using insertion sequence fingerprinting in *Escherichia coli*. *Mol. Biol. Evol.* **6**:1–14.

72. **Lawrence, J. G., H. Ochman, and D. L. Hartl.** 1992. The evolution of insertion sequences within enteric bacteria. *Genetics* **131**:9–20.

73. **Lenski, R. E., M. R. Rose, S. C. Simpson, and S. C. Tadler.** 1991. Long-term experimental evolution in *E. coli*. I. Adaptation and divergence during 2,000 generations. *Am. Nat.* **138**:1315–1341.

74. **Lenski, R. E., and M. Travisano.** 1994. Dynamics of adaptation and diversification: a 10,000-generation experiment with bacterial populations. *Proc. Natl. Acad. Sci. USA* **91**:6808–6814.

75. **Lichens-Park, A., and M. Syvanen.** 1988. Cointegrate formation by IS*50* requires multiple donor molecules. *Mol. Gen. Genet.* **211**:244–251.

76. **Lundblad, V., A. F. Taylor, G. R. Smith, and N. Kleckner.** 1984. Unusual alleles of *recB* and *recC* stimulate excision of inverted repeat transposons Tn*10* and Tn*5*. *Proc. Natl. Acad. Sci. USA* **81**:824–288.

77. **Mahillon, J., and M. Chandler.** 1998. Insertion sequences. *Microbiol. Mol. Biol. Rev.* **62**:725–774.

78. **Médigue, C., J. P. Bouche, A. Hénaut, and A. Danchin.** 1990. Mapping of sequenced genes (700 kbp) in the restriction map of the *Escherichia coli* chromosome. *Mol. Microbiol.* **4**:169–187.

79. **Merlin, C., J. Mahillon, J. Nesvera, and A. Toussaint.** 1998. Gene recruiters and transporters: the modular structure of bacterial mobile elements. *In* C. M. Thoma (ed.), *Plasmid Ecology and Biology*. Harwood Academic Publishers, gmbh, Amsterdam, The Netherlands.

80. **Mizuuchi, K.** 1992. Transpositional recombination: mechanistic insights from studies of mu and other elements. *Annu. Rev. Biochem.* **61**:1011–1051.

81. **Mueller, J. E., T. Clyman, Y. J. Huang, M. M. Parker, and M. Belfort.** 1996. Intron mobility in phage t4 occurs in the context of recombination-dependent DNA replication by way of multiple pathways. *Genes Dev.* **10**:351–364.

82. **Naas, T., M. Blot, W. M. Fitch, and W. Arber.** 1995. Dynamics of IS-related genetic rearrangements in resting *E. coli* K-12. *Mol. Biol. Evol.* **12**:198–207.

83. **Naas, T., M. Blot, W. M. Fitch, and W. Arber.** 1994. Insertion sequence-related genetic rearrangements in resting *E. coli* K-12. *Genetics* **136**:721–730.

84. **Orgel, L. E., and F. H. C. Crick.** 1980. Selfish DNA: the ultimate parasite. *Nature* **284**:604–607.

85. **Papadopoulos, D., D. Schneider, J. Meier-Eiss, W. Arber, R. E. Lenski, and M. Blot.** 1998. Genomic evolution during a 10,000-generation experiment with bacteria. *Proc. Nat. Acad. Sci. USA* **96**:3807–3812.

86. **Pato, M. L.** 1989. Bacteriophage Mu, p. 23–52. *In* D. E. Berg and M. M. Howe (ed.), *Mobile DNA*. American Society for Microbiology, Washington, D.C.

87. **Polard, P., M. F. Prere, O. Fayet, and M. Chandler.** 1992. Transposase-induced excision and circularization of the bacterial insertion sequence IS*911*. *EMBO J.* **11**:5079–5090.

88. **Raabe, T., E. Jenny, and J. Meyer.** 1988. A selection cartridge for rapid detection and analysis of spontaneous mutations including insertions of transposable elements in Enterobacteriaceae. *Mol. Gen. Genet.* **215**:176–180.

89. **Reddy, M., and J. Gowrishankar.** 1997. Identification and characterization of *ssb* and *uup* mutants with increased frequency of precise excision of transposon Tn*10* derivatives: nucleotide sequence of *uup* in *Escherichia coli*. *J. Bacteriol.* **179**:2892–2899.

90. **Reznikoff, W. S.** 1993. The TN*5* transposon. *Annu. Rev. Microbiol.* **47**:945–963.

90a. **Reznikoff, W. S.** Personal communication.

91. **Rice, P., and K. Mizuuchi.** 1995. Structure of the bacteriophage Mu transposase core: a common structural motif for DNA transposition and retroviral integration. *Cell* **82**:209–220.

92. **Roberts, D., B. C. Hoopes, W. R. McClure, and N. Kleckner.** 1985. IS10 transposition is regulated by DNA adenine methylation. *Cell* **43**:117–130.

93. **Roberts, D. E., D. Ascherman, and N. Kleckner.** 1991. IS*10* promotes adjacent deletions at low frequency. *Genetics* **128**:37–43.

94. **Rodriguez, H., E. T. Snow, U. Bhat, and E. L. Loechler.** 1992. An *Escherichia coli* plasmid-based, mutational system in which *supF* mutants are selectable—insertion elements dominate the spontaneous spectra. *Mutat. Res.* **270**:219–231.

95. **Rosenweig, R. F., R. R. Sharp, D. S. Treves, and J. Adams.** 1994. Microbial evolution in a simple unstructured environment: ge-

netic differentiation in Escherichia coli. *Genetics* **137**:903–917.

95a. **Rozen, D.** Personal communication.

96. **Salyers, A. A., N. B. Shoemaker, A. M. Stevens, and L. Y. Li.** 1995. Conjugative transposons: an unusual and diverse set of integrated gene transfer elements. *Microbiol. Rev.* **59:** 579.

97. **Savic, D., S. Romac, and D. Ehrlich.** 1983. Inversion in the lactose region of *E. coli* K-12. Inversion termini map with IS*3* elements a3b3 and b5a5. *J. Bacteriol.* **155**:943–946.

98. **Sawyer, S., and D. Hartl.** 1986. Distribution of transposable elements in Prokaryotes. *Theor. Popul. Biol.* **30**:1–16.

99. **Sawyer, S. A., D. E. Dykhuizen, R. F. DuBose, L. Green, T. Mutangadura-Mhlanga, D. F. Wolczyk, and D. L. Hartl.** 1987. Distribution and abundance of insertion sequences among natural isolates of *Escherichia coli. Genetics* **115**:51–63.

100. **Schaaper, R. M., B. N. Danforth, and B. W. Glickman.** 1986. Mechanisms of spontaneous mutagenesis: an analysis of spontaneous mutation in the *E. coli lacI* gene. *J. Mol. Biol.* **189**:273–284.

101. **Schaaper, R. M., and R. L. Dunn.** 1991. Spontaneous mutation in the *E. coli lacI* gene. *Genetics* **129**:317–326.

101a. **Schneider, D.** Personal communication.

102. **Schnetz, K., and B. Rak.** 1992. IS*5*—a mobile enhancer of transcription in *Escherichia coli. Proc. Natl. Acad. Sci. USA* **89**:1244–1248.

103. **Shapiro, J.** 1969. Mutations caused the insertion of genetic material into the *gal* operon of *E. coli. J. Mol. Biol.* **40**:93–99.

104. **Shapiro, J. A.** 1997. Genome organization, natural genetic engineering and adaptive mutation. *Trends Genet.* **13**:98–104.

105. **Shapiro, J. A.** 1979. Molecular model for the transposition and replication of bacteriophage Mu and other transposable elements. *Proc. Natl. Acad. Sci. USA* **76**:1933–1937.

106. **Shen, M. M., E. A. Raleigh, and N. Kleckner.** 1987. Physical analysis of Tn*10-* and IS*10-* promoted transpositions and rearrangements. *Genetics* **116**:359–369.

107. **Sherratt, D.** 1989. Tn3 and related transposable elements: site-specific recombination and transposition, p. 163–184. *In* D. E. Berg and M. M. Howe (ed.), *Mobile DNA.* American Society for Microbiology, Washington, D.C.

108. **Simons, R. W., B. C. Hoopes, W. R. McClure, and N. Kleckner.** 1983. Three promoters near the termini of IS10: pIN, pOUT, and pIII. *Cell* **34**:673–682.

109. **Simons, R. W., and N. Kleckner.** 1983. Translational control of IS10 transposition. *Cell* **34**:683–691.

110. **Syvanen, M., J. D. Hopkins, T. T. Griffin, T. Y. Liang, K. Ippen-Ihler, and R. Kolodner.** 1986. Stimulation of precise excision and recombination by conjugal proficient F′ plasmids. *Mol. Gen. Genet.* **203**:1–7.

111. **Treves, D. S., S. Manning, and J. Adams.** 1998. Repeated evolution of an acetate-crossfeeding polymorphism in long-term populations of *Escherichia coli. Mol. Biol. Evol.* **15:** 789–797.

112. **Umeda, M., and E. Ohtsubo.** 1990. Mapping of insertion element IS*5* in the *Escherichia coli* K-12 chromosome. Chromosomal rearrangements mediated by IS*5. J. Mol. Biol.* **213**: 229–237.

113. **Umeda, M., and E. Ohtsubo.** 1990. Mapping of insertion element IS*30* in the *Escherichia coli* K12 chromosome. *Mol. Gen. Genet.* **222**: 317–322.

114. **Umeda, M., and E. Ohtsubo.** 1989. Mapping of insertion elements IS*1*, IS*2* and IS*3* on the *Escherichia coli* K-12 chromosome. Role of the insertion elements in formation of Hfrs and F′ factors and in rearrangement of bacterial chromosomes. *J. Mol. Biol.* **208**:601–614.

115. **Wiegand, T. W., and W. Reznikoff.** 1992. Characterization of two hypertransposing Tn*5* mutants. *J. Bacteriol.* **174**:1229–1239.

116. **Wilke, C. M., and J. Adams.** 1992. Fitness effects of Ty transposition in *Saccharomyces cerevisiae. Genetics* **131**:31–42.

117. **Zhou, M., A. Bhasin, and W. S. Reznikoff.** 1998. Molecular genetic analysis of transposase-end DNA sequence recognition: cooperativity of three adjacent base pairs in specific interaction with a mutant Tn*5* transposase. *J. Mol. Biol.* **276**:913–925.

STRUCTURE OF DNA WITHIN THE BACTERIAL CELL: PHYSICS AND PHYSIOLOGY

Conrad L. Woldringh and Theo Odijk

10

The bacterial chromosome does not occur dispersed throughout the cell, but is organized in a separate phase, called the nucleoid. This structure duplicates and segregates along with cell growth, as has been demonstrated by phase-contrast microscopy of living cells (29). More recently, fluorescence microscopy has confirmed that nucleoid regions show characteristic shapes depending on the growth rate of the cells (22, 42, 53). Biochemical determinations suggest that the DNA in the nucleoid contains relatively few bound proteins and that it is compacted by forming supercoiled domains (11, 17, 41).

In a growing cell the DNA is involved in the processes of replication and coupled transcription-translation, but also in the simultaneously occurring process of segregation of the daughter strands. Therefore, the organization of genes maintained at fixed positions on the chromosome may not only be determined by their influence on local levels of su-

percoiling and efficiency of gene expression but may also be related to their contribution to structural compaction and to their role in the process of segregation. Knowledge of complete nucleotide sequences (2) will certainly help to find relationships between the functions of genes and their positions on the chromosome. In addition, it is now experimentally possible to determine the position of specific DNA segments within the structure of the nucleoid by fluorescence in situ hybridization (reference 33 and our unpublished observations), by immunolabeling (12, 15), or by making use of fusions between DNA binding proteins and the green fluorescent protein (12, 13). However, before we can evaluate the significance of the positions of specific genes within the nucleoid or correlate their positions to the shape and movement of the nucleoid as a whole, we need to better understand the physicochemical forces that determine the phase separation between DNA and cytoplasm during the permeabilization and dehydration steps necessary for microscopic visualization.

In living cells, the organization of the DNA is influenced by two different sets of conditions: (i) physical conditions, such as the concentrations and sizes of other macromolecules, the ionic strength, and the degree of supercoiling, and (ii) physiological conditions (31),

Conrad L. Woldringh, Institute for Molecular Cell Biology, BioCentrum Amsterdam, University of Amsterdam, Kruislaan 316, 1098 SM Amsterdam, The Netherlands. *Theo Odijk*, Faculty of Chemical Engineering and Materials Science, Delft University of Technology, P.O. Box 5045, 2600 GA Delft, The Netherlands.

Organization of the Prokaryotic Genome, Edited by Robert L. Charlebois,
© 1999 American Society for Microbiology, Washington, D.C.

which determine the involvement of the DNA in processes like replication, transcription, and segregation. Here we will first consider microscopical and biochemical observations that define the volumes of cell and nucleoid and give information about the chemical composition of cytoplasm and nucleoid and about the state of supercoiling. Next, we will discuss electron microscopic preparation methods that reveal different shapes and textures of the DNA within the nucleoid. Finally, we review current ideas on the physics of DNA and use these to describe the interactions between cytoplasmic proteins and the DNA superhelix. We investigate the occurrence of phase separation between nucleoid and cytoplasm. This review is an interdisciplinary attempt to bring together physical and biological perspectives on the organization of DNA within the bacterial cell.

LIGHT MICROSCOPY OF LIVING CELLS

A historical overview of the microscopic visualization of the bacterial nucleoid has been given by Robinow and Kellenberger (44). Some of the points important for the physicochemical properties discussed below will be summarized here.

In living bacteria the nucleoid can be observed by phase-contrast microscopy when the cells are immersed in a medium with a high refractive index, such as aqueous gelatin or polyvinylpyrrolidone, adjusted to give iso-osmotic conditions (51). The high refractive index of the background, which matches that of the cytoplasm, abolishes the bright halo around the cell and allows the visualization of the small difference in refractive index between the hydrated nucleoplasm and the denser cytoplasm. This technique was first applied in an almost inimitable way by Mason and Powelson in 1958 (29).

Previously, the confocal scanning light microscope (CSLM) has been used for live imaging of slow-growing *Escherichia coli* B/r (H266) cells at a resolution of 130 nm (Fig. 1A). These images, obtained by absorption contrast alone, have been used for the calculation of cell and nucleoid volumes (cf. Fig. A2 in reference 51) (Table 1). In addition, the refractive index values of cytoplasm and nucleoid have been estimated by immersive refractometry with conventional phase-contrast microscopy (51). These determinations suggested that the nucleoid must contain, besides the DNA, an additional amount of 8.6% (wt/vol) protein, whereas the cytoplasm consists of 21% (wt/vol) protein. Because the nucleoids in these small, slow-growing cells could be visualized with the improved resolution of the CSLM and because the pattern of chromosome replication is known, we have used these data to calculate the volume of a nonreplicating chromosome, which is the starting point for our physicochemical considerations presented below.

Alternatively, the nucleoid can be stained by growing cells in the presence of the fluorochrome DAPI (4',6-diamidino-2-phenylindole) (22, 53), which is specific for DNA and does not seem to influence the growth of the cells at concentrations below 200 ng/ml. However, irradiation with a mercury lamp, necessary to excite the fluorochrome, kills the cells rapidly (53). With both methods, phase-contrast microscopy and fluorescence microscopy, characteristic shapes of the nucleoid can be recognized. Especially in rapidly growing cells, dumbell shapes and lobular "V" and "W" shapes can be distinguished (29). In slow-growing cells, the nucleoid approximates a simple rod structure that elongates concomitantly with the lengthwise extension of the cell (Fig. 1A).

Recent fluorescence microscope studies of OsO_4-fixed and DAPI-stained *E. coli* cells and filaments (54) have shown that nucleoids segregate through the action of coupled transcription-translation and that in filaments, already segregated nucleoids coalesce if these processes are inhibited. This transcription-mediated movement is also demonstrated when cells grow into filaments because of DNA synthesis inhibition (58). In Fig. 1B and C it can be seen how the nonreplicating nu-

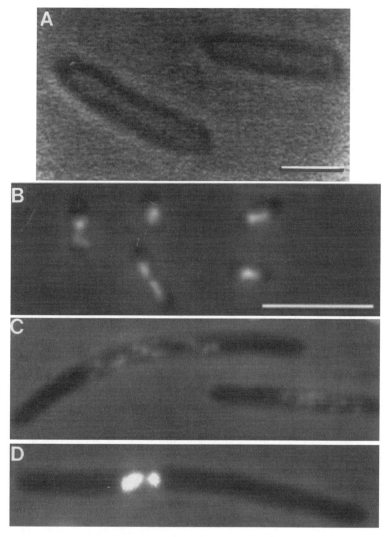

FIGURE 1 Light microscope images of the nucleoid in *E. coli*. (A) CSLM image (absorption contrast; resolution, 130 nm [51]) of *E. coli* B/r (H266) cells grown in alanine minimal medium with a doubling time of 150 min. The smaller cell can be assumed to represent a newborn cell, containing an unreplicated chromosome. (B) Fluorescence microscopy of an OsO₄-fixed and DAPI-stained *E. coli* MC4100 *dnaX*(Ts) cell, grown in glucose minimal medium at permissive temperature (30°C). (C) Cells from the same culture as in panel B but grown at restrictive temperature (42°C) for two mass doublings. The nucleoid(s) can be seen to be pulled out into small lobules (58). (D) Cell from the same population as in panel C but treated with 300 μg of chloramphenicol/ml to inhibit protein synthesis. The DNA has retracted into confined regions (54). It should be emphasized that there is no indication of an increased concentration (packing density) or a different organizational state of this DNA, as previously noted by Kellenberger (18). Bars, 1 μm (A) and 5 μm (B through D).

TABLE 1 Macromolecular compositions of cell, cytoplasm, and nucleoid in *E. coli* B/r (H266) cells grown in alanine medium ($T_d = 150$ min)[a]

Parameter	Value		
	Cell	Cytoplasm	Nucleoid
Volume (μm^3)	0.46[b]	0.29[c] (V_c)	0.08[d] (V_n)
Amt of protein (10^{-12} g)	0.086[e]	0.061 (21%)[f]	0.007 (8.6%)[f]
No. of 40-kDa proteins[g]	1.3×10^6	0.9×10^6 (m_c)	0.1×10^6 (m_n)
Total protein vol (μm^3)[h]	0.066	0.046	0.005
No. of ribosomes[i]		5,000	
No. of ribosomal proteins[i]		0.1×10^6	
Total ribosomal vol (μm^3)[j]		0.041	
Protein vol fraction	0.144	0.166[k] (v_c)	0.06 (v_n)
Ribosome vol fraction		0.14	
Amt of DNA (10^{-12} g)	0.005[l]		0.005
DNA concnt (mg/ml)			65[m]

[a] All data refer to the average cell in the population. Compare with similar canonical cells, grown in minimal glucose medium, as given by Neidhardt and Umbarger (32) and Bremer and Dennis (7). T_d, doubling time.

[b] Calculated from average cell length (2.5 μm) and diameter (0.5 μm), assuming the shape of the cell to be a right cylinder with hemispherical polar caps.

[c] Obtained by subtracting the volume of the envelope (0.09 μm^3) and the volume of the nucleoid (0.08 μm^3) from the total cell volume (cf. Table 1 in reference 57).

[d] Calculated from estimates of nucleoid length (1.9 μm) and diameter (0.24 μm) as visualized by CSLM (51) (Fig. 1A).

[e] From Table 1 in reference 57 and Churchward et al. (9).

[f] From Valkenburg and Woldringh (51), who estimated the protein concentrations in cytoplasm and nucleoid to be 21 and 8.6%, respectively, using immersive refractometry.

[g] The mean of the length distribution of *E. coli* proteins is 321 amino acids (24).

[h] Assuming the radius of a protein *a* to be 2.3 nm (47).

[i] From reference 57; cf. Neidhardt and Umbarger (32). The cumulative molecular weight of ribosomal proteins was taken to be 0.77×10^6 Da (34).

[j] Ribosomes are assumed to be spheres with diameters of 25 nm.

[k] v_c, the cytoplasmic protein volume fraction, was calculated for soluble proteins by subtracting the number (0.1×10^6) and volume (0.041 μm^3) of ribosomal proteins.

[l] Calculated as the amount of a nonreplicating chromosome.

[m] Compare with the DNA concentration of 100 mg/ml as determined by ratio contrast imaging in the scanning transmission electron microscope (5).

cleoid is pulled into small lobules that coalesce again into a confined region (Fig. 1D) upon inhibition of protein synthesis with chloramphenicol. These and similar observations (55) have led to the transcription-mediated segregation model, in which physicochemical compaction forces (see below) are counteracted by coupled transcription, translation, and translocation of membrane proteins acting as an expansion force (35, 60).

DNA TOPOLOGY

Two biochemical properties of DNA are important for understanding its organization within an inactive (i.e., under conditions of no replication or transcription) bacterial nucleoid: (i) the degree of supercoiling and (ii) the association with DNA binding proteins like HU (45).

In vivo, the (right-handed) DNA double helices of plasmids and the bacterial chromosome are untwisted or underwound in a left-handed way, causing the helices to become negatively supercoiled (1). As the untwisting requires energy, the negative superhelicity represents an excess of free energy. This supercoiling may have several functions: (i) it causes compaction of the long DNA molecule in a smaller volume (see the theoretical considerations below); (ii) it favors the binding of proteins, like those involved in replication and transcription, which require the separation (melting) of the two strands of the double helix; and (iii) it may help in the formation of daughter nucleoids during chromosome segregation (26).

To quantify the degree of supercoiling, the linking number (Lk) of the DNA molecule has to be considered (1). Lk is the number of times the two strands of the double helix turn around each other. The "standard" linking number of a relaxed closed-circular DNA molecule, Lk_0, is equal to the total number of base pairs, N, divided by the number of bases per helical turn. Under standard conditions (0.2 M NaCl; 37°C; pH 7 [1]), the helical repeat (equivalent to 10.5 bp per turn at minimum torsional stress), h_0, is 3.5 nm. Thus, for the *E. coli* chromosome with $N = 4,639,221$ bp (2), we have $Lk_0 = 441,831$. This means that during a round of bidirectional replication of 40 min, each fork has to unwind $441,831/2 \cdot 40 = 5,523$ links, or 92 links per s.

The measure of the extent of supercoiling is expressed as the linking difference between the in vivo (energy-requiring) linking numbers, Lk and Lk_0. Normalization to the size of the chromosome gives the specific linking difference, s, as follows: $s = (Lk - Lk_0)/Lk_0$. Plasmids from isolated *E. coli* cells have a typical s value of -0.06 (1). If it is assumed that the same value applies to the *E. coli* chromosome, the linking difference is as follows: $\Delta Lk = Lk - Lk_0 = 415,321 - 441,831 = -26,510$.

The negative superhelical tension in the chromosome is maintained through the combined action of topoisomerases I and IV and DNA gyrase (11). The latter enzyme actively supercoils the DNA at the expense of free energy of ATP hydrolysis. It has been estimated (1, 3) that in *E. coli* about -0.025 of the above-mentioned specific linking difference of -0.06 is unconstrained and able to adopt the conformation of a plectonemic supercoil. The remainder is stabilized by DNA binding proteins like HU, H-NS (H1), IHF, FIS, and RNA polymerase (11, 45, 50).

The introduction of some nicks into the DNA of the isolated genome causes the appearance of only a few relaxed loops, while the rest of the DNA remains supercoiled (61). From the number of nicks that have to be introduced to fully unwind the whole chromosome, it has been estimated that the chromosome could contain about 46 independently supercoiled domains of 100 kbp.

Electron microscopy of isolated nucleoids, prepared according to the Kleinschmidt spreading technique, has indeed revealed that the naked (protein-free) DNA emerges from the lysed cell as supercoiled loops (41), which is compatible with the idea of domains. These could be formed by some DNA binding proteins that cross-link two helices and prevent rotation of the helix beyond the binding sites. However, the boundaries of the loops have not been defined and it is suspected that fixed boundaries do not exist (30). Moreover, it could be established that the degree of supercoiling in all the loops is the same and that environmental signals (like osmolarity, availability of oxygen and nutrients, and pH) change supercoiling in all the loops to the same extent (40).

These observations suggest that the independently supercoiled domains originate from temporary anchoring or restraint of the DNA. Anchoring of DNA to a fixed structure like the plasma membrane could occur if envelope proteins are being cotranscriptionally and cotranslationally inserted into the membrane (27). Such anchoring has also been proposed to play a role in chromosome segregation (35, 60). On the whole, the problem of supercoiled domains warrants further investigation.

NUCLEOID STRUCTURE AS REVEALED BY ELECTRON MICROSCOPY

Due to the limited resolution of the light microscope, the organization of the DNA and the boundary of the nucleoid cannot be accurately determined. What can we learn about nucleoid structure from the high-resolution images obtained by electron microscopy? As emphasized by Robinow and Kellenberger in their review (44), bacterial DNA is not associated with as many proteins as the chromatin of eukaryotic cells. This property led Kellenberger to introduce the term "low-protein chromatin" (19) as distinct from eukaryotic, histone-containing chromatin. Low-protein chromatin also occurs in the nuclei of dinoflagellates and in viruses, mitochondria, and chloroplasts. Because of its low content of bound proteins, bacterial DNA is more sensitive than eukaryotic DNA to aggregation during the dehydration procedure necessary for electron microscopy (Fig. 2A). As a result it cannot be fixed by conventional fixatives but needs to be preserved under so-called Ryter-Kellenberger conditions. The principal ingredients are, apart from OsO_4, a mixture of amino acids, together with calcium or magnesium ions, and uranyl acetate as a secondary fixative. In this way the DNA-plasms in nucleoids become visible (Fig. 2B) as "loose networks composed of strands of various thicknesses" (44). The overall shape of the nucleoid after OsO_4 fixation resembles that observed with the light microscope (51). However, with the introduction of glutaraldehyde as a prefixative, a more dispersed appearance of the nucleoid is obtained (Fig. 2C).

It is a general experience that fixation with OsO_4 enhances the visibility of the nucleoid both under the phase-contrast microscope (51) and, after DAPI staining, under the fluorescence microscope (53). Also, in thin sections observed by electron microscopy, OsO_4 fixation shows a more confined nucleoid and a more enhanced phase separation between nucleoplasm and cytoplasm (Fig. 2B) than does fixation with glutaraldehyde (Fig. 2C). An explanation for this phenomenon is that OsO_4 causes the dissociation of DNA binding proteins and a migration of proteins from the nucleoid to the cytoplasmic phase, which is maintained during the subsequent dehydration and shrinkage of the cells (52). In contrast, glutaraldehyde may preserve the association of DNA binding proteins to the DNA by cross-linking them to other proteins throughout the nucleoid and the cytoplasm. Upon dehydration and cell shrinkage, this may lead to a more dispersed image of the nucleoplasm (Fig. 2C). Therefore, both types of fixation, with OsO_4 and with glutaraldehyde, may give misleading information about the organization of the DNA within the nucleoid and about the border between the nucleoplasmic and cytoplasmic phases.

A third image of the bacterial nucleoid is obtained by cryofixation and freeze substitution (CFS) (Fig. 2D). In this procedure, living cells are first frozen very quickly in liquid propane or helium. Subsequently, the ice in the frozen specimen is substituted by a mixture of 2.5% OsO_4 and acetone at $-90°C$ for 64 h (16). Once the ice has been replaced, the temperature of the now chemically fixed and dehydrated cells is increased to allow conventional embedding in plastic and thin sectioning (16). After such a CFS, the bacterial nucleoid consists of a phase of dispersed granules without any sign of fibrils. Kellenberger (19, 44) suggested that the fibrous aspect of DNA after chemical fixation is caused by a dissociation of the relatively few proteins and a loss of ions upon fixation, allowing lateral aggregation of the DNA into fibrillar structures.

After conventional OsO_4 fixation, the DNA fibrils sometimes condense into a liquid crystal upon dehydration, as suggested by the presence of arc-shaped structures (21) indicative of the fibrillar cholesteric organization of DNA described by Livolant and Leforestier (25). If these structures are interpreted as artificial rearrangements, then a procedure like CFS, which does not produce such fibrillar arrangements, can be considered a better method.

Kellenberger (4, 17, 18) has relied mainly on the CFS technique in proposing his model of the nucleoid, consisting of a confined interior with dispersed, ribosome-free regions filled with 4- to 6-nm-diameter globules (compactosomes) and extending at its periphery into numerous "corraline projections" (19). These excrescences are proposed to represent DNA regions engaged in transcription. Previously, the location of transcription activity at the nucleoid periphery was suggested by Kleppe et al. (23) and demonstrated by autoradiographic visualization of pulse-labeled RNA in thin sections of chemically fixed cells (46). It supports the transcription-mediated segregation model (54) discussed above. Concerning the global shape of the nucleoid, however, it should be noted that the CFS

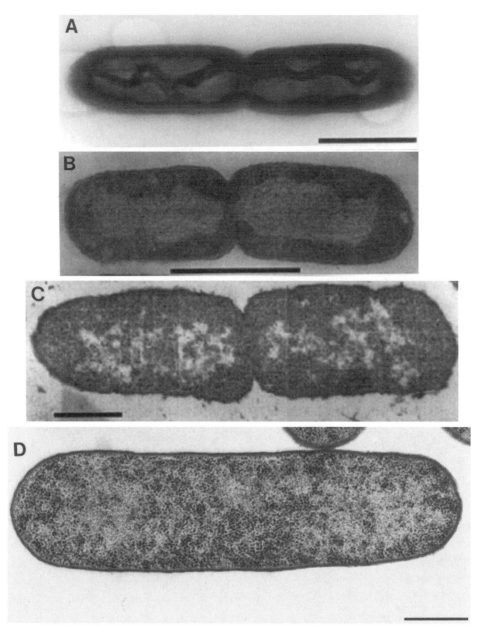

FIGURE 2 Electron microscope images of the nucleoid in *E. coli*. (A) Whole-mount transmission electron micrograph of an *E. coli* B/r A cell, prefixed only in growth medium (minimal alanine medium; doubling time, 126 min) with 0.1% OsO_4 and directly dehydrated with alcohol for critical-point drying. Note that this strain differs in size and shape from the *E. coli* B/r (H266) cell in Fig. 1A. Within the region of the nucleoplasm of the transparent cell, the DNA, containing few bound proteins and not properly fixed, can be seen to have precipitated into an undulating structure. Note also that the diameter of the cell (350 nm) has shrunk considerably compared to that of nondehydrated cells (640 nm [56]). (B) Thin section of an *E. coli* B/r A cell, grown in minimal alanine medium (doubling time, 90 min), and fixed with OsO_4 under Ryter-Kellenberger conditions (see the text). Within the region of the nucleoplasm, the DNA is now visible as a fibrillar network, which has resisted the collapse by dehydration visible in panel A. (C) Thin section of an *E. coli* K-12 cell, grown in broth (doubling time, 22 min) and prefixed with 2.5% glutaraldehyde. The nucleoid contains sparse fibrils and is dispersed throughout the cytoplasm. (D) Thin section of an *E. coli* B cell prepared by CFS. The dispersed nucleoid contains only a fine-grained plasm (from reference 16 with permission). All magnification bars represent 0.5 μm.

technique also causes structural changes. For instance, plasmolysis spaces visible in living cells are not preserved by CFS (reference 20; see the discussion in reference 59). Thus, it seems possible that the CFS procedure can also cause rearrangements of DNA and proteins and that the dispersed and granular appearance (compactosomes) of the DNA is induced by CFS.

Can the supercoils, so nicely visualized by the Kleinschmidt spreading technique, become visible in thin sections? We have to concede that none of the methods for thin sectioning gives a hint of supercoiled loops or domains. The large differences in the global shape and texture of the nucleoid must apparently be ascribed to subtle differences in permeabilization, extraction, shrinkage, and swelling that occur in the many steps of the procedure leading to thin sections. It seems to us imperative that more fundamental work should be devoted to the elucidation of the physicochemical factors involved in the preparative procedures discussed above. For instance, the influence of various fixatives and dehydration agents on the interaction of DNA fragments, mutually or with proteins, could be monitored by well-established physical techniques. If certain fixatives or dehydration agents do induce substantial attractive forces in these biopolymers, it may help to account for the fibrillar aspects sometimes seen in thin sections. However, in the remaining survey we concern ourselves solely with repulsive (i.e., electrostatic) forces between the particles. They give rise to the phenomenon of depletion and, in our view, lead to the phase separation of the nucleoid within the bacterial cell.

DNA PHYSICS AND TERMINOLOGY

We would now like to survey several current ideas on DNA physics relevant to the structure of the bacterial genome within the cell. Polymer physics has progressed tremendously in the last few decades. It leans heavily on statistical mechanics. Polymer physicists have of course developed their own jargon, which is probably not very accessible to the readers of this book. We therefore introduce some terminology before coming to DNA compaction proper. Microbiologists with an inclination for physics may wish to consult a recent book on polymer theory which also addresses certain aspects of the physics of DNA (14).

In equilibrium statistical mechanics one formulates the energy, E, of the system (or part of it) in terms of suitably chosen variables (for example, the coordinates of some molecules). The energy is often termed "Hamiltonian" for reasons that are too complex to discuss here. If these variables are too detailed, the ultimate form of the theory may become altogether unwieldy; if they are too coarse grained, the resulting model may have little bearing on the system we wish to study. We note that there is a tension here: the microbiologist often wants as much detail as possible; the physicist seeks simplified models in order to render the theory tractable.

In statistical mechanics, an energy is weighted by a probability, a so-called Boltzmann factor: $\exp(-E/k_B T)$, where k_B is Boltzmann's constant and T is the temperature, which is constant at equilibrium. Here we will not deal with steady-state equilibria, which are the domain of nonequilibrium statistical mechanics. The energy, E, is not the only quantity of interest; the entropy, S, equal to $k_B \ln W$, where W is the number of realizations of the system, is another variable we would like to compute. The system as a whole tends to minimize its energy, E_{tot}, and maximize its entropy, S_{tot}; at constant temperature, volume, and number of particles, we need to focus on the Helmholtz free energy, $F_{tot} = E_{tot} - TS_{tot}$, which is a minimum at equilibrium. This fact shows up in an operational way in various computations. Finally, averages in equilibrium statistical mechanics are precisely equivalent to time averages for a system at thermodynamic equilibrium. Accordingly, in polymer theory, a drawn figure of a DNA molecule, for example, implies that this is a typical (i.e., very probable) configuration of the DNA chain out of myriads of realizations.

A DNA model that has been of consider-able use in polymer theory is that of the wormlike chain. Chemical details, including the base pair sequence, are smoothed out completely. Linear, double-stranded DNA in solution is viewed as a homogeneous elastic rod undulating in a heat bath at temperature *T*. Solvent molecules are continually buffeting the rod, which adopts wormlike configura-tions. Time averages are statistical-mechanical averages, and it is possible to set up a theory for the latter. Brownian motion of the DNA and its elasticity, or resistance to bending, trade off at a persistence length, *P*, which is a statistical concept. Mathematically, it can be shown that the average inner product of two unit vectors tangential to the wormlike chains is given by Grosberg and Khokhlov (reference 14; note that these authors call the chain "per-sistent" instead of wormlike):

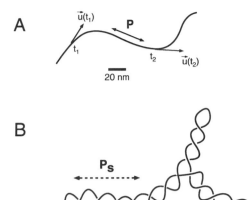

$$\langle \vec{u}(t_1)\vec{u}(t_2)\rangle = \langle \cos \Theta_{12}\rangle$$
$$= \exp - |t_1 - t_2|/P \tag{1}$$

FIGURE 3 (A) Typical configuration of linear wormlike DNA. The two unit vectors $u(t_1)$ and $u(t_2)$ are tangential to the chain at respective positions t_1 and t_2. The persistence length, *P*, is indicated as being 40 nm. (B) Typical configuration of a plectonemic DNA supercoil showing one branch. The persistence length, P_s, is unknown (see the text).

Fig. 3A shows a typical configuration of linear DNA together with the unit vectors at contour positions t_1 and t_2 along the chain. The angle between the two vectors is Θ_{12}. The average in equation 1 is over all DNA configurations. The meaning of equation 1 is as follows. A section of chain of length $|t_1 - t_2|$ that is much less than *P* is essentially rod-like, whereas two points on the DNA contour separated by distances much greater than *P* are uncorrelated, on average. One must imagine the chain to be moving incessantly, though on average a section of a size a bit less than the persistence length is straight. The persistence length of DNA is about 45 nm under physi-ological conditions (49). One often also intro-duces another scale, the Kuhn length (or seg-ment length), *A*, equal to 2*P*; a very long DNA chain may be viewed as a random walk with step length *A*.

Next, from a physical point of view, the DNA may be thought of as having a finite diameter ($d = 2$ nm) and a charge density aris-ing from the negatively charged phosphate groups. It is a charged, wormlike cylinder, which may interact with itself as it adopts var-ious configurations. Under physiological con-ditions, the solution contains $1 - 1$ electro-lyte, so the charges are screened. The Coulomb interaction between two negative charges fixed in space is shielded by small pos-itive ions drifting through the surrounding water—but not totally, in view of the inces-sant thermal motion. Accordingly, there exists a nonzero screening length, the so-called De-bye length, λ, which is inversely proportional to the square root of the ionic strength. Elec-trostatic screening is a well-known problem in statistical physics, first convincingly explained by Debye and Hückel in 1923 (48). As a re-sult, it is useful to introduce an effective di-ameter, d_{eff}, also depending on the ionic strength (48). The reader must imagine the DNA chain of Fig. 3A to fluctuate and con-tinually collide with itself. Theorists often think of two Kuhn segments of the chain interacting with each other like two colliding pencils (of length A and diameter d_{eff}). One can easily assure oneself that the excluded volume of two such slender rodlike objects should scale as $A^2 d_{eff}$ (and not as $A d_{eff}^2$). This is a basic ingredient in the theory of the excluded-volume effect of DNA, which agrees with experiments (36).

Supercoiled DNA is now often deemed to be plectonemic in solution (6). A physically convenient representation would look like that in Fig. 3, where we have included one branch. The curve represents the DNA helix. Because the superhelix may be thought of as acting like a spring, we again introduce a diameter, D_s. Recent theoretical work shows that the supercoil diameter is given in terms of the specific linking difference (39):

$$D_s = h_0/2\pi|s| \qquad (2)$$

Here, h_0 is the helical repeat of the DNA helix (3.5 nm; see above). Actually, there is a weak dependence on the superhelix pitch angle in equation 2, but we neglect it here in view of the qualitative nature of our discussion. Equation 2 is valid under conditions where the plectonemic structure is ideally springlike. Below, we remark on a case where this supposition is no longer true. Like linear DNA, the superhelix behaves like a locally linear elastic structure which is buffeted by collisions with solvent molecules. Hence, we may introduce a superhelical persistence length, P_s (Fig. 3B). This quantity has not been measured yet, but it is likely to be close to $2P$ (37).

Finally, we end this section with two remarks. First, the discussion above illustrates a principle of hierarchical conceptualization often used in polymer theory. Thus, the "chemical" DNA helix may be replaced by a wormlike chain at one level in the hierarchy and by a random walk of Kuhn length A at another level. On the other hand, in a discussion of the physics of superhelical DNA, the DNA helix itself may prove to be unimportant. The model of choice within the hierarchy depends on the physical problem we would like to address. Often, it is determined by the relevant scale in a given physical context. For instance, if this scale is larger than D_s, the molecular details of the DNA chain and even the configuration of the wormlike DNA helix within a plectonemic supercoil may be essentially irrelevant. Hence, in this case, the plectoneme may be envisaged as a superwormlike chain of persistence length P_s and diameter D_s, devoid of any other detail.

A second problem is that we often introduce free energies (like the ones discussed below) at some level of the hierarchy. It is not always illuminating to separate these into entropies and energies (as in the beginning of this section), and we will not do so below.

SUPERHELICAL DNA IN A CONFINED SUSPENSION OF PROTEINS; PHASE SEPARATION BY DEPLETION

Having set the stage, we are finally in a position to focus attention on the interactions between proteins and the chromosomal DNA in a bacterial cell. The contents of the cell are of course multifarious (7, 32) (Table 1). Nevertheless, we must not forget that a globular particle of whatever size has only about $k_B T$ of thermal energy. The majority of cytoplasmic particles consist of small globular proteins (24), and it is these that we need to account for within an approximate picture based on statistical physics. Let us then regard the cell as a compartment containing very many negatively charged proteins (32) interacting with plectonemic DNA. A $1 - 1$ electrolyte at 0.2 M (mainly K^+ [from Table 2 in reference 8]) is present and causes the electrostatic interactions to be screened by a Debye length, λ, of ≈ 0.68 nm. Such a suspension could be called congested or crowded (62) for the obvious reason that the respective particles get in each other's way. Physicists, however, like to use the term "depletion" (10). This implies something more than crowding: it also incorporates the effects of interactions and entropy. For instance, a DNA chain is depleted from a neg-

TABLE 2 Input variables used in the theoretical analysis (38)

Variable	Value
DNA	
Helix diameter, d	2 nm
Base-pair spacing	0.34 nm
Helical repeat, h_0	3.5 nm
Effective specific linking difference, s	−0.025
Helix contour length, L	1.6 mm
Superhelix contour length, L_s	0.64 mm
Protein	
Radius, a	2.3 nm
Effective number of charges	−10
Aqueous solution	
Permitivity	78.6 (at $T = 298$ K)
Ionic strength (1 − 1 electrolyte)	0.2 M[a]

[a] From reference 8.

atively charged membrane not only because it is repelled by the surface but also because it loses configurational degrees of freedom, resulting in a decrease in its entropy. Thus, depletion may depend on the temperature.

Let us start with the situation where the DNA is dispersed throughout the entire cell (Fig. 4). The problem is whether this is thermodynamically stable. In order to describe the effective interactions among the particles in the cell, a physical chemist would try to compute the so-called virial coefficients that occur in an expansion of the free energy as a function of the protein volume fraction. Here, we give a very heuristic type of argumentation for the effective interactions. A protein of radius a is excluded from the body of the DNA supercoil, whose total contour length equals L along the helical axis. The effective value of the superhelical linking difference is about −0.025, as we have seen above, so the superhelical diameter is about 20 nm, from equation 2. A small protein will swim right through the interstitial space of the plectoneme, so the excluded volume must be proportional to L. Moreover, the probability of approaching the DNA helix is governed by a Boltzmann factor $\exp(-E_p/k_BT)$: the protein cannot approach the helix more closely than a typical distance R, since otherwise the elec-

trostatic repulsion, E_p, becomes too large with respect to k_BT and the probability drops almost to zero. A rough estimate for R would be as follows: $R \approx a + 1/2d + \lambda$, the sum of the protein radius, the helix radius, and the Debye screening length. The excluded volume between one protein and the DNA superhelix is about R^2L, representing a cylindrical sheath surrounding the DNA helix.

We may now write the depletion interaction, i.e., the free energy of interaction between the DNA genome and m free proteins within the cell, as follows (38):

$$F_{dep} \approx (mR^2Lk_BT)/V \qquad (3)$$

Here, V is the volume of the cellular compartment and the "$\approx$" sign means "order of magnitude." This expression may be understood in the following way. The volume fraction excluded to one protein is R^2L/V, and there are m proteins interacting independently of one another with the superhelix, so we expect F_{dep} to be proportional to m. Furthermore, the proteins behave as free particles, to a first approximation, whose sole energy scale is k_BT. Hence, F_{dep} must be proportional to k_BT. Since an enhancement of the excluded volume implies a decrease in the freedom afforded to the proteins, the entropy must decrease: the sign in equation 3 must be positive.

A

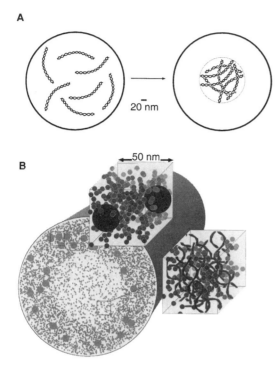

20 nm

B

50 nm

FIGURE 4 (A) Cross section of *E. coli* perpendicular to the long axis of the cell. If the superhelical DNA (here visible in the form of several sections within the left circle) were dispersed throughout the cell, the free energy would be F_n (see equation 5), but now pertaining to the total volume, $V_c + V_n$. According to the phase separation theory (38), the dispersed state is unstable because the free energy is higher than that when the DNA is confined within the nucleoid, as shown in the right circle (i.e., when the total free energy is $F_c + F_n$, as discussed in the text). (B) The nucleoid is stabilized by the osmotic pressure arising from the excess concentration of proteins in the cytoplasm. The small circles within the two cubes depict the small proteins discussed in the text. Conforming to the concentrations indicated in Table 1, the upper cube $(50 \text{ nm})^3$ contains 392 proteins and 2.2 ribosomes, reflecting the cytoplasm; the cube reflecting the nucleoid contains 152 proteins and 1 μm of supercoiled DNA. Each configuration is actually a snapshot of spheres distributed randomly via a computer program. The cross section area reflects the numbers of proteins and ribosomes of a slice of arbitrary depth, indicating the phase separation between cytoplasm and nucleoid.

There is another interaction term competing with the protein-DNA depletion energy given by equation 3. The DNA is confined within the tiny space of the cell and interacts with itself. In Fig. 3B, the superhelical Kuhn segments of length A_s collide with themselves. The excluded volume between two such segments should scale as $B_a = A_s^2 D_s$ (analogous to $A^2 d_{\text{eff}}$ for linear DNA, as we have already seen). There are $N = L_s/A_s$ segments, giving rise to $N(N - 1)/2 \approx 2N^2/2$ pairs of interactions, where L_s is the contour length of the superhelix. We express the superhelix self-energy—the interaction of the superhelix with itself—as follows (38):

$$F_{\text{self}} \approx (N^2 B_a k_B T)/V \approx (L_s^2 D_s k_B T)/V \quad \textbf{(4)}$$

which is independent of A_s. Here, $1/2N^2 B_a$ may be interpreted as an "interaction volume," which must be scaled by the system volume, V, and thermal energy, $k_B T$, in order to derive F_{self}, which must have the dimension energy. Two approximations have been made to establish equation 4: (i) the superhelical diameter, D_s, is much greater than the Debye screening length, λ, so we need not worry too much about the electrostatics; (ii) branching (Fig. 3B) does not interfere with our estimate of the excluded volume between Kuhn segments, since the density of branch points is quite small according to recent work (28).

We now review a recent theory of the bacterial nucleoid as arising from phase separation caused by the interactions given by equations 3 and 4 (38). The total free energy, F_n, of the nucleoid is expressed as the sum of the two contributions discussed above and an ideal mixing term due to the entropy of the proteins.

$$F_n = F_{\text{self},n} + F_{\text{dep},n} + F_{\text{mix},n} \quad \textbf{(5)}$$

$$F_{\text{mix},i} = m_i k_B T \ln(m_i/V_i) - m_i k_B T \quad \textbf{(6)}$$

The interactions between the proteins are simply disregarded. Next, we assume there are two phases (Fig. 4B). One phase is the cytoplasm (index $i = c$), with a total free energy, $F_{\text{mix},c}$, of $m_i = m_c$ proteins enclosed in a volume, V_i, equal to V_c. The other phase consists of the nucleoid (index $i = n$), with $m_i = m_n$ proteins, interacting with the DNA superhelix within a volume, V_i, equal to V_n. In equilibrium, we must minimize the total free energy, $F_{mix,c} + F_n$, of the total system with respect to m_c and V_c, bearing in mind that the total number of proteins, $m_c + m_n$, and the total volume, $V_c + V_n$, remain fixed. In effect, this statement is equivalent to the coexistence equations: one signifies the equality of the osmotic pressures of the two phases; the other represents the invariance of the chemical potential of the proteins throughout the cell. This program has been carried out in order to compute the predicted *E. coli* parameters (38), but here we merely summarize the main results.

We have collected various input variables employed in the theoretical analysis in Table 2. Quantities predicted by the phase-separation theory (38) are presented in Table 3. We have used the total volume ($V_c + V_n$) and the total number of proteins ($m_c + m_n$) enclosed by the plasma membrane as exhibited in Table 1. It is seen that the theory is in quite good agreement with the experimental determination of the nucleoid volume, V_n, and cytoplasmic volume fraction, v_c (Table 1 and Fig. 4B). The nucleoid volume fraction, v_n, is underestimated by a factor of two. Nevertheless, we must not forget that a certain fraction of protein is probably associated with the DNA so that the effective number of free proteins is less than that established experimentally. A further elaboration of the margins of error inherent in the theoretical analysis has been given by Odijk (38).

The picture above may be too simple, however. Our concern is that the DNA volume fraction (≈ 0.36) is quite high within the nucleoid. The superhelical Kuhn segments may want to align in one direction because of the packing constraints. We encounter the same effect when we try to pack pencils into a box. On the one hand, the orientational degrees of freedom become restricted, leading to a decrease in entropy. On the other hand, the excluded-volume effect diminishes, as one can readily see by moving two aligned pencils with respect to each other. There is an optimal state in which the superhelical segments have a fairly high degree of orientational order (38).

A second difficulty arises because the interaction of the superhelix with itself may be so intense that equation 2 is no longer legitimate. The superhelix could ripple (37). These topics have been dealt with by one of us (38). Since a precise quantitative theory is formidable, the notion of liquid-crystalline order within the nucleoid, as suggested for plasmid DNA (43), is speculative at present. The nucleoid may be so small that its cholesteric order is not seen under the microscope; the cholesteric pitch could simply be too large.

CONCLUSION

The focus of this discussion of the organization of bacterial DNA has been the volume of the nucleoid as visualized by CSLM (Fig. 1A). Because of the resolution of the CSLM (130 nm) and the size of the cell (500-nm diameter), we can obtain only a rough estimate of this volume. With the knowledge of the chemical composition of *E. coli* (7, 32, 57) and assuming a phase separation between nucleoid and cytoplasm, we can compute the concentrations and volume fractions of DNA and proteins as given in Table 1 and simulate them as shown in Fig. 4B. In doing so, we disregard the electron microscope observations of thin sections of glutaraldehyde-fixed cells (Fig. 1C) or cells prepared by CFS (Fig. 1D), which show a dispersed nucleoid and suggest an extensive mingling between DNA and cytoplasm. We thereby concede that every fixation and dehydration method necessary for electron microscopy has the potential to cause artificial rearrangements of the DNA.

Can the assumption of phase separation as suggested by light microscopy be supported by physicochemical considerations of the constraints of DNA superhelices in a concentrated suspension of proteins? If the variables as given in Table 2 are valid, the theoretical analysis developed by one of us (38) predicts a clear phase separation, as indicated by the values

TABLE 3 Variables computed from theory and derived from the coexistence equations (38)

Variable	Value
Debye screening length, λ	0.68 nm
Exclusion radius, R	4.7 nm
Superhelix diameter, D_s	22 nm
DNA-protein exclusion vol, $B_c = \pi L R^2$	0.111 μm^3
DNA total self-excluded vol, $B_s = \pi L_s^2 D_s/4$	710 μm^3
Vol of the nucleoid, V_n	0.068 μm^{3a}
Nucleoid protein vol fraction, ν_n	0.032[b]
Cytoplasmic protein vol fraction, ν_c	0.162[c]

[a] Compare with the experimental value of 0.08 μm^3 in Table 1.
[b] Compare with the experimental value of 0.06 in Table 1.
[c] Compare with the experimental value of 0.166 in Table 1.

presented in Table 3 and as depicted in Fig. 4B. The existence of such a phase separation may or may not imply an anisotropic packing of DNA superhelical loops in the form of a cholesteric liquid crystal. In order to make a definite prediction about the existence of liquid-crystalline order, we need, at the very least, a precise theory or experimental determination of the superhelical persistence length, P_s. One relevant dimensionless variable is the anisometry factor, P_s/D_s. Clearly, if it is too small, liquid-crystalline packing of the chromosomal DNA becomes impossible.

The purported coexistence of two phases in the bacterial cell—the nucleoid and the cytoplasm—further implies the existence of a well-defined surface tension at the nucleoid boundary. The tension must be positive: a mesoscopic deformation of the nucleoid would lead to an increase in the surface free energy. Accordingly, phase separation theory could play a role in understanding deformational dynamics of the nucleoid as ascribed to the transcription-mediated pulling (Fig. 1C) during the process of segregation (35, 60). Finally, we would like to stress that there are several contentious issues that will need to be addressed in future work. The cell is at best in a state of stationary dynamic equilibrium. How good the full equilibrium theory discussed here will be remains to be seen. Next, the electrolytic contents of the cytoplasm are rather complicated (8): the effects of the va-

riety of mono-, di-, and trivalent ions need to be assessed. Other causes of the compaction of DNA (e.g., RNA polymerase) also have to be studied. We have also neglected to mention the influence of tRNAs on the osmotic balance. Last but not least, we would like to know the nature of the domains alluded to in the past. We hope to come back to these issues in future work.

ACKNOWLEDGMENTS

We thank J.-L. Sikorav, H. V. Westerhoff, and J. Snoep for stimulating discussion; D. S. Cayley, E. Kellenberger, A. Minsky, V. Norris, M. T. Record, and S. Zimmerman for correspondence; P. G. Huls for providing Fig. 1B to D; J. A. Hobot and E. Kellenberger for providing Fig. 2D; and N. O. E. Vischer for calculating and drawing Fig. 4.

REFERENCES

1. **Bates, A. D., and A. Maxwell.** 1993. *DNA Topology.* Oxford University Press Inc., New York, N.Y.
2. **Blattner, F. R., G. Plunkett III, C. A. Bloch, N. T. Perna, V. Burland, M. Riley, J. Collado-Vides, J. D. Glasner, C. K. Rode, G. F. Mayhew, J. Gregor, N. W. Davis, H. A Kirkpatrick, M. A. Goeden, D. J. Rose, B. Mau, and Y. Shao.** 1997. The complete genome sequence of *Escherichia coli* K-12. *Science* **277:**1453–1462.
3. **Bliska, J. B., and N. R. Cozzarelli.** 1987. Use of site-specific recombination as a probe of DNA structure and metabolism in vivo. *J. Mol. Biol.* **194:**205–218.

4. **Bohrmann, B., W. Villiger, R. Johansen, and E. Kellenberger.** 1991. Coralline shape of the bacterial nucleoid after cryofixation. *J. Bacteriol.* **173:**3149–3158.

5. **Bohrmann, B., M. Haider, and E. Kellenberger.** 1993. Concentration evaluation of chromatin in unstained resin-embedded sections by means of low-dose ratio-contrast imaging in STEM. *Ultramicroscopy* **49:**235–251.

6. **Boles, T. C., J. H. White, and N. R. Cozzarelli.** 1990. Structure of plectonemically supercoiled DNA. *J. Mol. Biol.* **213:**931–951.

7. **Bremer, H., and P. P. Dennis.** 1996. Modulation of chemical composition and other parameters of the cell by growth rate, p. 1553–1569. *In* F. C. Neidhardt, R. Curtiss III, J. L. Ingraham, E. C. C. Lin, K. B. Low, B. Magasanik, W. S. Reznikoff, M. Riley, M. Schaechter, and H. E. Umbarger (ed.), Escherichia coli *and* Salmonella: *Cellular and Molecular Biology*, 2nd ed, vol. 2. American Society for Microbiology, Washington, D.C.

8. **Cayley, S., B. A. Lewis, H. J. Guttman, and M. T. Record, Jr.** 1991. Characterization of the cytoplasm of *Escherichia coli* K-12 as a function of external osmolarity. Implications for protein-DNA interactions *in vivo. J. Mol. Biol.* **222:**281–300.

9. **Churchward, G., H. Bremer, and R. Young.** 1982. Macromolecular composition of bacteria. *J. Theor. Biol.* **94:**651–670.

10. **de Gennes, P. G.** 1979. *Scaling Concepts in Polymer Physics.* Cornell University Press, Ithaca, N.Y.

11. **Drlica, K., and C. L. Woldringh.** 1998. Chromosomal organization: nucleoids, chromosomal folding and DNA topology, p. 12–22. *In* F. J. de Bruijn, J. R. Lupski, and G. M. Weinstock (ed.), *Bacterial Genomes: Physical Structure and Analysis.* Chapman & Hall, New York, N.Y.

12. **Glaser, P., M. E. Sharpe, B. Raether, M. Perego, K. Ohlsen, and J. Errington.** 1997. Dynamic, mitotic-like behavior of a bacterial protein required for accurate chromosome partitioning. *Genes Dev.* **11:**1160–1168.

13. **Gordon, G. S., D. Sitnikov, C. D. Webb, A. Teleman, A. Straight, R. Losick, A. W. Murray, and A. Wright.** 1997. Chromosome and low copy plasmid segregation in E. coli: visual evidence for distinct mechanisms. *Cell* **90:**1113–1121.

14. **Grosberg, A. Y., and A. R. Khokhlov.** 1994. *Statistical Physics of Macromolecules.* American Institute of Physics, New York, N.Y.

15. **Harry, E. J., K. Pogliano, and R. Losick.** 1995. Use of immunofluorescence to visualize cell-specific gene expression during sporulation in *Bacillus subtilis. J. Bacteriol.* **177:**3386–3393.

16. **Hobot, J. A., W. Villiger, J. Escaig, M. Maeder, A. Ryter, and E. Kellenberger.** 1985. Shape and fine structure of nucleoids observed on sections of ultrarapidly frozen and cryosubstituted bacteria. *J. Bacteriol.* **162:**960–971.

17. **Kellenberger, E.** 1989. Bacterial chromatin (a critical review of structure-function relationships), p. 3–25. *In* K. W. Adolph (ed.), *Chromosomes: Eukaryotic, Prokaryotic and Viral*, vol. III. CRC Press, Inc., Boca Raton, Fla.

18. **Kellenberger, E.** 1990. Intracellular organization of the bacterial genome, p. 173–186. *In* K. Drlica and M. Riley (ed.), *The Bacterial Chromosome.* American Society for Microbiology, Washington, D.C.

19. **Kellenberger, E.** 1991. Functional consequences of improved structural information on bacterial nucleoids. *Res. Microbiol.* **142:**229–238.

20. **Kellenberger, E.** 1990. The "Bayer bridges" confronted with results from improved electron microscopy methods. *Mol. Microbiol.* **4:**697–705.

21. **Kellenberger, E., and B. Arnold-Schulz-Gamen.** 1992. Chromatins of low-protein content: special features of their compaction and condensation. *FEMS Microbiol. Lett.* **100:**361–370.

22. **Kellenberger, E., and C. Kellenberger-Van der Kamp.** 1994. Unstained and *in vivo* fluorescently stained bacterial nucleoids and plasmolysis observed by a new specimen preparation method for high-power light microscopy of metabolically active cells. *J. Microsc.* **176:**132–142.

23. **Kleppe, K., S. Ovrebo, and I. Lossius.** 1979. The bacterial nucleoid. *J. Gen. Microbiol.* **112:**1–13.

24. **Koonin, E. V., R. L. Tatusov, and K. E. Rudd.** 1996. *Escherichia coli* protein sequences: functional and evolutionary implications, p. 2203–2217. *In* F. C. Neidhardt, R. Curtiss III, J. L. Ingraham, E. C. C. Lin, K. B. Low, B. Magasanik, W. S. Reznikoff, M. Riley, M. Schaechter, and H. E. Umbarger (ed.), Escherichia coli *and* Salmonella: *Cellular and Molecular Biology*, 2nd ed., vol. 2. American Society for Microbiology, Washington, D.C.

25. **Livolant, F., and A. Leforestier.** 1996. Condensed phases of DNA: structures and phase transitions. *Prog. Polym. Sci.* **21:**1115–1164.

26. **Løbner-Olesen, A., and P. L. Kuempel.** 1992. Chromosome partitioning in *Escherichia coli. J. Bacteriol.* **174:**7883–7889.

27. **Lynch, S., and J. C. Wang.** 1993. Anchoring of DNA to the bacterial cytoplasmic membrane through cotranscriptional synthesis of polypeptides encoding membrane proteins or proteins for

export: a mechanism of plasmid hypernegative supercoiling in mutants deficient in DNA topoisomerase I. *J. Bacteriol.* **175:**1645–1655.

28. **Marko, J. F., and E. D. Siggia.** 1995. Statistical mechanics of supercoiled DNA. *Phys. Rev. E* **52:**2912–2938.

29. **Mason, D. J., and D. M. Powelson.** 1958. Nuclear division as observed in live bacteria by a new technique. *J. Bacteriol.* **71:**474–479.

30. **Miller, W. G., and R. W. Simons.** 1993. Chromosomal supercoiling in *Escherichia coli. Mol. Microbiol.* **10:**675–684.

31. **Nanninga, N., and C. L. Woldringh.** 1985. Growth, genome duplication and division, p. 161–197. *In* N. Nanninga (ed.), *Molecular Cytology of Escherichia coli.* Academic Press, London, England.

32. **Neidhardt, F. C., and H. E. Umbarger.** 1996. Chemical composition of *Escherichia coli*, p. 13–17. *In* F. C. Neidhardt, R. Curtiss III, J. L. Ingraham, E. C. C. Lin, K. B. Low, B. Magasanik, W. S. Reznikoff, M. Riley, M. Schaechter, and H. E. Umbarger (ed.), Escherichia coli *and* Salmonella: *Cellular and Molecular Biology*, 2nd ed., vol. 1. American Society for Microbiology, Washington, D.C.

33. **Niki, H., and S. Hiraga.** 1997. Subcellular distribution of actively partitioning F plasmid during the cell division cycle in E. coli. *Cell* **90:**951–957.

34. **Noller, H. F., and M. Nomura.** 1996. Ribosomes, p. 167–186. *In* F. C. Neidhardt, R. Curtiss III, J. L. Ingraham, E. C. C. Lin, K. B. Low, B. Magasanik, W. S. Reznikoff, M. Riley, M. Schaechter, and H. E. Umbarger (ed.), Escherichia coli *and* Salmonella: *Cellular and Molecular Biology*, 2nd ed., vol. 1. American Society for Microbiology, Washington, D.C.

35. **Norris, V.** 1995. Hypothesis: chromosome separation in *E. coli* involves autocatalytic gene expression, transertion and membrane domain formation. *Mol. Microbiol.* **16:**1051–1057.

36. **Odijk, T.** 1979. On the ionic-strength dependence of the intrinsic viscosity of DNA. *Biopolymers* **18:**3111–3113.

37. **Odijk, T.** 1996. DNA in a liquid-crystalline environment: tight bends, rings, supercoils. *J. Chem. Phys.* **106:**1270–1286.

38. **Odijk, T.** 1998. Osmotic compaction of supercoiled DNA into a bacterial nucleoid. *Biophys. Chem.* **73:**23–30.

39. **Odijk, T., and J. Ubbink.** 1998. Dimensions of a plectonemic DNA supercoil under fairly general perturbations. *Physica A* **252:**61–66.

40. **Pavitt, G. D., and C. F. Higgins.** 1993. Chromosomal domains of supercoiling in *Salmonella typhimurium. Mol. Microbiol.* **10:**685–696.

41. **Pettijohn, D. E., and R. R. Sinden.** 1985. Structure of the isolated nucleoid, p. 199–227. *In* N. Nanninga (ed.), *Molecular Cytology of Escherichia coli.* Academic Press, London, England.

42. **Poplawski, A., and R. Bernander.** 1997. Nucleoid structure and distribution in thermophilic archaea. *J. Bacteriol.* **179:**7625–7630.

43. **Reich, Z., E. J. Wachtel, and A. Minsky.** 1994. Liquid-crystalline mesophases of plasmid DNA in bacteria. *Science* **264:**1460–1463.

44. **Robinow, C., and E. Kellenberger.** 1994. The bacterial nucleoid revisited. *Microbiol. Rev.* **58:**211–232.

45. **Rouvière-Yaniv, J., E. Kiseleva, A. Bensaid, A. Almeida, and K. Drlica.** 1992. Protein HU and bacterial supercoiling, p. 17–43. *In* S. Mohan, C. Dow, and J. Cole (ed.), *Prokaryotic Structure and Function.* Cambridge University Press, Cambridge, England.

46. **Ryter, A., and A. Chang.** 1975. Localization of transcribing genes in the bacterial cell by means of high resolution autoradiography. *J. Mol. Biol.* **98:**797–810.

47. **Srere, P.A.** 1981. Protein crystals as a model for mitochondrial matrix proteins. *Trends Biochem. Sci.* **6:**4–7.

48. **Stigter, D.** 1977. Interactions of highly charged colloidal cylinders with applications to double-stranded DNA. *Biopolymers* **16:**1435–1448.

49. **Taylor, W. H., and P. J. Hagerman.** 1990. Application of the method of T4-DNA ligase-catalyzed ring-closure to the study of DNA structure. II. NaCl-dependence of DNA flexibility and helical repeat. *J. Mol. Biol.* **212:**363–376.

50. **Ussery, D. W., J. C. D. Hinton, B. J. A. M. Jordi, P. E. Granum, A. Scirafi, R. J. Stephen, A. E. Tupper, G. Berridge, J. M. Sidebotham, and C. F. Higgins.** 1994. The chromatin-associated protein H-NS. *Biochimie* **76:**968–980.

51. **Valkenburg, J. A. C., and C. L. Woldringh.** 1984. Phase separation between nucleoid and cytoplasm in *Escherichia coli* as defined by immersive refractometry. *J. Bacteriol.* **160:**1151–1157.

52. **Valkenburg, J. A. C., C. L. Woldringh, G. J. Brakenhoff, H. T. M. van der Voort, and N. Nanninga.** 1985. Confocal scanning light microscopy of the *Escherichia coli* nucleoid: comparison with phase-contrast and electron microscope images. *J. Bacteriol.* **161:**478–483.

53. **Van Helvoort, J. M. L. M., and C. L. Woldringh.** 1994. Nucleoid partitioning in *Escherichia coli* during steady state growth and upon recovery from chloramphenicol treatment. *Mol. Microbiol.* **13:**577–583.

54. **Van Helvoort, J. M. L. M., J. Kool, and C. L. Woldringh.** 1996. Chloramphenicol

causes fusion of separated nucleoids in *Escherichia coli* K-12 cells and filaments. *J. Bacteriol.* **178:** 4289–4293.

55. **Van Helvoort, J. M. L. M., P. G. Huls, N. O. E. Vischer, and C. L. Woldringh.** 1998. Fused nucleoids resegregate faster than cell elongation in *Escherichia coli pbpB*(Ts) filaments after release from chloramphenicol inhibition. *Microbiology* **144:**1309–1317.

56. **Woldringh, C. L., M. A. de Jong, W. Van den Berg, and L. Koppes.** 1977. Morphological analysis of the division cycle of two *Escherichia coli* substrains during slow growth. *J. Bacteriol.* **131:**270–279.

57. **Woldringh, C. L., and N. Nanninga.** 1985. Structure of nucleoid and cytoplasm in the intact cell, p. 161–197. *In* N. Nanninga (ed.), *Molecular Cytology of Escherichia coli.* Academic Press, London, England.

58. **Woldringh, C. L., A. Zaritsky, and N. B. Grover.** 1994. Nucleoid partitioning and the division plane in *Escherichia coli. J. Bacteriol.* **176:** 6030–6038.

59. **Woldringh, C. L.** 1994. Significance of plasmolysis spaces as markers for periseptal annuli and adhesion sites. *Mol. Microbiol.* **14:**597–607.

60. **Woldringh, C. L., P. R. Jensen, and H. V. Westerhoff.** 1995. Structure and partitioning of bacterial DNA: determined by a balance of compaction and expansion forces? *FEMS Microbiol. Lett.* **131:**235–242.

61. **Worcel, A., and E. Burgi.** 1972. On the structure of the folded chromosome of *E. coli. J. Mol. Biol.* **71:**127–147.

62. **Zimmerman, S. B., and A. P. Minton.** 1993. Macromolecular crowding: biochemical, biophysical and physiological consequences. *Annu. Rev. Biophys. Biomol. Struct.* **22:**27–65.

DNA SUPERCOILING AND ITS CONSEQUENCES FOR CHROMOSOME STRUCTURE AND FUNCTION

N. Patrick Higgins

11

Understanding the topology of chromosomes became a serious intellectual challenge once a structural model for DNA was in hand. For example, worry about the "unwinding" problem arose immediately after Watson and Crick proposed their structure, which is a plectonemic coil with an interwound repeat length of approximately 10 bp. Since the *Escherichia coli* chromosome is very long (>4,000,000 bp) and replication is very fast (taking about 50 min from start to finish), it was recognized that the DNA molecule must (i) change its structure so that the helix straightens out to allow strand separation, (ii) spin rapidly about the long axis as strands separate (at 10,000 rpm), or (iii) be frequently cut and pasted so that tangles can pass one through another. Although solution 2 is correct, Delbruck considered this to be implausible and lobbied for a complementary paranemic helix (Fig. 1) that would have no need to untwist during replication (47). With a zero linking number, a paranemic helix would be easily paired and separated. Watson and Crick admitted the problem, but for crystallographic reasons they

could not accept a paranemic helix. Thus, they finessed. "Although it is difficult at the moment to see how these processes (chromosome replication and segregation) occur without everything getting tangled, we do not feel that this objection will be insuperable" (131).

Current understanding of supercoiling behavior in bacterial chromosomes has been shaped largely by efforts to address four types of problems. (i) How can supercoil distributions be detected and quantified in vivo? (ii) How do supercoils influence the protein machines that replicate, transcribe, and recombine cellular DNA? (iii) Which genes regulate supercoiling? (iv) How far do supercoils move inside living cells? New biochemical and genetic experiments show that three topoisomerases—gyrase, topoisomerase I (topo I), and topo IV—control supercoiling quantitatively, that chromosomes have a dynamic but more random plectonemic supercoil organization than previously thought, and that gyrase and topo IV modulate domain size in chromosomes.

FOUR TOPOISOMERASES SOLVE ALL TOPOLOGICAL PROBLEMS

As it turned out, the enzymes that solve the untwisting problem do so by breaking and re-

N. Patrick Higgins, Department of Biochemistry and Molecular Genetics, University of Alabama at Birmingham, Birmingham, AL 35294-2170.

Organization of the Prokaryotic Genome, Edited by Robert L. Charlebois,
© 1999 American Society for Microbiology, Washington, D.C.

joining DNA molecules, thereby allowing individual strands to pass through one another (as in solution 3 above). These amazing enzymes, the topoisomerases, are best understood in *E. coli*. A sketch of the four enzymes found in *E. coli* is given below. For more molecular details of the mechanisms of these proteins, see reference 130.

Topoisomerases are divided into two classes based on the number of phosphodiester bonds broken and re-formed per reaction cycle (130). Type I enzymes, which include bacterial and eukaryotic topo I and topo III, break one strand per cycle, whereas type II enzymes, which include gyrase and topo IV in prokaryotes and topo II in eukaryotes, break two strands simultaneously.

Topo I (ω protein) was the first enzyme in *E. coli* found to catalytically remove supercoils from a covalently closed DNA molecule. Topo I was found to be an essential enzyme (21, 99, 101), and with prescient insight, Wang noted in 1971 that topo I breaks and rejoins the phosphodiester backbone bonds of DNA without apparent energy input (129). Subsequently, all topoisomerases (and many site-specific recombinases) have been demonstrated to use a mechanism by which covalent phosphotyrosine linkage between protein and DNA conserves the chemical phosphodiester energy while topological transformations take place in the DNA molecule (see reference 120 for a recent model of the topological transformation of eukaryotic topo I).

Topo III is the second type I enzyme discovered in bacteria. Enigmatically, topo III is dispensable for viability, although it probably has an important role in DNA replication (19, 20). One type of topo III mutant forms deletions at positions where short, perfect repeats occur along the DNA sequence (110); this deletion phenotype is mirrored by topo III mutants in *Saccharomyces cerevisiae* (32, 56). The enzyme works in vitro as a decatenase for replication of plasmid chromosomes, requiring single-stranded regions to separate replicating molecules (40, 42).

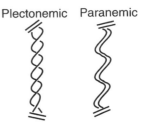

FIGURE 1 The difference between plectonemic and paranemic helices is that strand separation of a plectonemic molecule requires twisting about the long axis.

Gyrase was the first type II enzyme discovered (34), and it remains unique for its ability to introduce negative supercoils into relaxed, positively or negatively supercoiled DNA at the expense of ATP binding and hydrolysis (77, 121, 122). An essential enzyme in bacteria, gyrase is critical for nearly all complex transactions that involve DNA, including recombination, replication, transcription, and chromosome segregation (16). Gyrase is also an important drug target in bacteria. Highly potent fluoroquinolones kill bacterial cells by turning gyrase into a cytotoxin (30). Fluoroquinolones form stable ternary complexes with gyrase and DNA (18, 76) and stimulate the breakage-reunion activity of the GyrA subunit (30, 108). While poisoned with a quinolone, gyrase obstructs replication and transcription severely enough to cause double-strand breaks in DNA under a variety of circumstances (24).

Topo IV is closely related to gyrase in both primary sequence and catalytic mechanism (54, 72, 109). Topo IV requires ATP to remove either negative or positive supercoils from DNA (130), although the enzyme is passive and only changes linking in the direction of a less supercoiled state. Topo IV is essential for segregation of bacterial plasmids and chromosomes (2, 126, 139); it is the primary decatenating and unknotting activity in the cell (140), and like gyrase, topo IV is inhibited by fluoroquinolones (24).

STRUCTURAL EFFECTS OF THE SUPERCOIL DISTRIBUTION

Supercoil density is defined by the term σ, which is the average number of superhelical turns divided by the number of Watson-Crick turns of a double helix. Supercoiling is a form of energy that influences Watson-Crick structure. Like a spring, the free energy of a supercoiled DNA molecule increases exponentially with supercoil density. Enzyme machines that unwind DNA to carry out their function (RNA polymerase and DNA replisomes) can sense and utilize this energy. However, if the superhelix reaches critical density, the Watson-Crick structure will unpair locally, allowing DNA to adopt alternative structures that relieve the tension. The best-studied supercoil-dependent alternatives to the right-handed B-DNA helix include left-handed Z-DNA, cruciforms, and triple-stranded or H-form DNA. All of these structures have different sequence rules that govern their formation.

Simple sequences of alternating CG and TG base pairs adopt a stable left-handed conformation (see reference 38 for a recent review). The longer the alternating dinucleotide sequence repeat length, the lower the superhelix density at which left-handed Z-DNA forms. The base pairs at the junctions between B-DNA and Z-DNA are unpaired and sensitive to chemical probes, such as osmium tetroxide (3, 90). Sequences adopting the left-handed structure and junction bases are not substrates for the type II restriction or methylation enzymes, and this behavior serves as the basis for measuring left-handed DNA in vivo (51, 138).

Sequences that contain long stretches of polypurine-polypyrimidine can form triple-stranded (H-form) DNA (48). In an intramolecular triplex, half of either the purine- or pyrimidine-rich strand becomes unpaired, and its complement forms a triple strand by making Hoogsteen base pairs with purines in the major groove of the Watson-Crick base-pairs. H-form DNA has been detected in vivo, but only when supercoiling is driven by transcription or other unusual circumstances (85). Intermolecular triplexes involving both RNA and DNA strands have been detected in vivo (29).

Inverted-repeat sequences can also form cruciforms, with an unpaired loop at the end and a sequence resembling a Holliday junction at the base (65). Very long inverted repeats are not well tolerated in E. coli (10, 61, 62). Long runs of the simple-repeat dinucleotide AT will form a cruciform, with an unpaired AT base pair at its center serving as the loop. The sequence $(AT)_{34}$ adopts the cruciform structure in vivo in E. coli cells after salt shock (78).

THE HOMEOSTATIC CONTROL MODEL FOR REGULATING σ

In vitro, DNA gyrase is capable of supercoiling DNA to a σ value of -0.1. At this very high supercoiling level, many sequences adopt alternative structures. In vivo, the average value of σ has been measured in plasmids and the chromosome, using many different techniques (6, 9, 36, 50, 86, 97, 119). Although variation is observed with growth conditions, measurements of σ in wild-type (WT) cells fall within the range of -0.05 to -0.075. However, only half of this linking difference is carried in an energized state; half of the supercoils in bacterial DNA appear to be tied up with bound proteins, such as RNA polymerase and the histone-like proteins (9, 50, 98, 138). This makes the effective supercoiling density equivalent to a range from -0.025 to -0.038 and raises a question: how is the level of supercoiling controlled in vivo?

The homeostatic supercoil regulation model was inspired by two observations. First, many promoters sense supercoiling levels (79). Some promoters initiate transcription more frequently at high negative supercoiling levels—e.g., Mu p_E (58), topA (125), and leu-500 (103, 104, 124)—while others are more efficient when the DNA is relaxed—e.g., proU (44), his (105), and Mu p_c (58). Second, expression of gyrase increases when the chromosome becomes relaxed (1, 82), whereas the expression of topo I requires high supercoiling levels (125). Menzel and Gellert proposed a regulatory loop in which transcriptional modulation based on supercoil levels leads these competing activities to achieve equilibrium (81). Many observations of gyrase and topo I mutants and of the consequences of treating cells with various gyrase inhibitors support the model (23).

The homeostatic supercoiling hypothesis was developed before the discovery of topo III and topo IV, which, like topo I, relax negative supercoils. One hint that topo IV may contribute to supercoiling control is that the lethal effect of a *topA* deletion can be suppressed by overexpression of topo IV (54, 80). Recent results demonstrate that topo IV does contribute to regulation of σ in vivo. Both gyrase and topo IV are inhibited by the fluoroquinolones, but single-amino-acid substitutions within the active site of either enzyme will confer high-level drug resistance (24). A set of isogenic strains of *E. coli* was engineered to allow selective quinolone inhibition of gyrase, topo IV, or both in combination with different mutation states of topo I (140). The results show that three topoisomerases (gyrase, topo IV, and topo I) act together to set supercoiling levels in vivo. First, if both gyrase and topo IV were inhibited simultaneously, topo I relaxed only high-energy supercoils (σ changed from -0.075 to -0.05). This in vivo relaxation is consistent with in vitro studies, where topo I requires supercoiling to overcome an unfavorable free energy needed to unpair a segment for breaking and rejoining DNA (130). Second, if gyrase was inactivated with a fluoroquinolone in the presence of a drug-resistant topo IV (with or without topo I), supercoiling fell to very low levels ($\sigma = -0.015$). This shows that topo IV contributes to the equilibrium. Third, when topo I, gyrase, and topo IV were all inhibited simultaneously, there was no change in supercoiling, suggesting that no other topoisomerase (including topo III) contributes to average σ (140).

REPLICATION AND TRANSCRIPTION—THREE SUPERCOIL PROBLEMS

Chromosome replication has three critical stages (initiation, elongation, and segregation), each with different supercoiling problems. For initiation, the two strands of *oriC* must separate to allow assembly of the replication machinery (the replisome [57]). By providing energy to unpair the strands of the double helix, supercoiling helps determine the efficiency and temporal control of chromosome replication (4, 5, 75). In the elongation phase of DNA synthesis, two highly processive replication forks sweep leftward and rightward from *oriC* toward a rendezvous (after 50 min) at the terminus. During elongation, unwinding is the problem. Positive supercoils generated ahead of both forks are dissipated to allow the fork movement. There is an unresolved question about which enzymes contribute to this process (2, 126, 139); both gyrase and topo IV are capable of removing positive supercoils from DNA. Moreover, in an in vitro system for plasmid replication, any one of three single topoisomerases (gyrase, topo III, or topo IV) allows replication fork progression (40, 41).

Finally, after two circular chromosomes are complete, they are frequently (invariably?) interlocked (catenated). In the *E. coli* chromosome, catenane links seem to be concentrated in a 20-kb sector of DNA near the *dif* site (59). In this zone, homologous recombination between sister chromosomes is extraordinarily frequent (15, 71) and the XerC-XerD *dif* dimerization system only converts chromosomal dimers to monomers when the *dif* site lies within the zone of hyperrecombination (14, 60). The untangling of sister plasmids is blocked by quinolones, even when gyrase is quinolone resistant (140). However, if topo IV is quinolone resistant, plasmids separate efficiently. When both gyrase and topo IV are drug resistant, quinolones no longer kill bacterial cells or impair growth. These experiments demonstrate that topo IV probably performs most of the work in decatenating replicated chromosomes in vivo, which would explain its essential phenotype.

Transcription encompasses supercoiling problems similar in two respects to those of replication. First, transcription initiation requires unpairing of the DNA duplex, and supercoiling can influence this step as it does in initiation of DNA replication (see above). Second, transcription is similar to replication in that movement of RNA polymerase generates temporary positive supercoiling ahead of, and negative supercoiling behind, the DNA segment that is being transcribed (31).

The torsional effects of transcription have been studied with supercoil-sensitive promoters (11, 12, 103, 104, 114, 123) by measuring the formation of Z-DNA (3, 100) and by monitoring the extrusion of cruciforms (78). Whereas it is sometimes possible to see small transcriptional torsion effects in WT cells (12, 27, 136), chromosomes appear well buffered from the topological effects of transcription. Dramatic changes in torsional strain are only seen with mutations in topo I (23) or when gyrase is inhibited with drugs (68). Supercoiling effects are exacerbated by two conditions. When two closely linked transcription units are simultaneously transcribed in divergent or convergent orientations, DNA is twisted in opposite directions and can become hypersupercoiled (67, 135). When the transcription machinery becomes bound to a fixed structure, i.e., when an integral membrane protein becomes inserted into the membrane while it is being transcribed and translated (70, 74), positive- and negative-supercoiling fields can change DNA structure (100). Topo I plays an important role in many of these instances (its essential role?) by preventing hypernegative supercoiling. In topA mutants, hypersupercoiling seems to allow formation of an R loop involving nascent RNA and highly underwound DNA template (25, 64). Overexpression of RNase H partially corrects a topA defect (25).

SUPERCOIL DIFFUSION AND TORSIONAL DISTRIBUTION

The tension present in negative supercoils is not unrestricted or free to move throughout large chromosomes. Whole chromosomes released from cells by a variety of different cell-lysing techniques retain supercoiling, even when they carry a number of single-strand nicks (55, 73, 134). Chromosomal supercoiling is also partitioned in vivo. By treating cells with X rays, which cause repairable DNA nicks, and then measuring the supercoil-enhanced binding and photo-cross-linking of trimethyl psoralen, Sinden and Pettijohn showed that more than 100 single-strand breaks were needed to release most of the supercoil energy present in the E. coli chromosome (113). This observation has led to a search for the mechanisms and locations of barriers that constrain supercoils inside living cells.

Because numerous promoters respond to supercoiling changes, it has been proposed that supercoiling is a regulatory mechanism that allows bacteria to sense environmental stimuli (22, 43, 44). To test this hypothesis, two groups used supercoil-sensitive promoters coupled to reporter genes to look for domains with different supercoil density (84, 96). Remarkably, supercoiling behavior was uniform throughout all tested regions of the E. coli and Salmonella typhimurium chromosomes. These results demonstrate that topoisomerases can maintain an average distribution of σ throughout most (if not all) domains.

SUPERCOIL DIFFUSION AND PLECTONEMIC DISTRIBUTION

An alternative approach to locate domain barriers exploits an assay for movement of plectonemic structure rather than torsion. Many site-specific recombination systems, including λ integration and excision (34, 35, 83), bacteriophage Mu transposition (17), Tn3 and γδ resolution (7, 116), and gin-hin inversion (37, 52, 53), require a negatively supercoiled substrate. In each system, a recombinase first binds to its cognate sequences and then the recombining sites collide to produce a synapse with a precise structure that allows exchange among the four strands of DNA. In living bacteria, supercoils adopt the interwound (plectonemic) form (9), and DNA-DNA communication within the network is facilitated by at least two types of duplex plectonemic movement—slithering and supercoil branch migration (Fig. 2).

In the $\gamma\delta$/Tn3 resolvase systems, recombination only occurs if a chromosome contains two directly oriented res sites (see reference 117 for a review). Orientation specificity is achieved by movement of the plectonemic supercoils (Fig. 2). In vitro, $\gamma\delta$ resolution can be divided into three stages that proceed on distinctly different timescales. First, binding of resolvase protein to the res sites takes about a millisecond in vitro. Second, supercoil movement allows bound sites to synapse and capture precisely three supercoil nodes within the two 114-bp sites. Surprisingly, supercoil diffusion is fast enough to require stopped flow technology for its measurement (45, 91). With plasmids of several different sizes, more than 60% of the molecules form a synapse 1 s after resolvase addition. This result is stunning, as almost all direct-order-pairing possibilities in a 5-kb plasmid are scanned in this very short time. Noting that fluorescence is frequently used to visualize DNA inside living cells, this movement would be invisible to normal microscopic resolution. The slow steps in resolution are DNA cleavage, strand exchange, and religation of the four strands, which takes about a minute for completion.

To study domains in living cells, the plectonemic structure throughout a 100-kb domain of S. typhimurium has been probed by using the $\gamma\delta$ resolution system as an assay (45). Three results in this study were surprising. First, plectonemic supercoils diffuse over long distances; 100-kb deletions are readily detectable. Even over this long distance, the rule for plectonemic movement was valid because only directly oriented res sites recombined. Second, recombination became less efficient as the distance separating two recombining sites increased, but the distance penalty obeyed first-order kinetics. Recombination dropped by 50% (equivalent to the half-life term of a radioactive decay) for 20 kb of site separation in exponentially growing cells. This result indicates that barriers are random with respect to DNA sequence. Third, supercoil diffusion was more efficient when cells entered stationary phase.

DOMAIN SIZE

The analyses of the $\gamma\delta$ resolution for a 100-kb interval (45, 115) agree in several important ways with the psoralen cross-linking studies of the whole chromosome (113). Sinden and Pettijohn's estimates of 43 ± 10 domains/bacterial genome equivalent correspond to domains of $100 (\pm 20)$ kb. However, they assumed that all domains would be equal in size. A random distribution of barriers would cause an underestimate of barrier frequency because a break in a large domain relaxes more DNA than a break in a small domain. Nonetheless, even without adjustments, the two methods are within a factor of 2. Analysis by $\gamma\delta$ resolution assays indicated a median domain size of 40 kb, with some cells having small domains and others having very large domains of >100 kb (115) (Fig. 3). The psoralen and resolvase methods also agreed on two other facets of domain behavior. Slow-growing cells have larger domains than fast-growing ones. Growth on succinate as a sole carbon source caused the domain number measured with psoralen cross-linking to drop by half, a result mirrored with resolvase assays when the cells entered stationary phase (45). Moreover, no incremental increases in torsional relaxation or resolution were seen when cells were held for over an hour under nongrowing but viable conditions (115). Thus, barriers do not relax gradually when DNA replication is completed or stalled.

BARRIER STRUCTURE

Numerous theories have been proposed to explain chromosome movements in vivo. First,

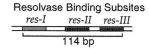

Resolvase Binding Subsites
res-I res-II res-III

114 bp

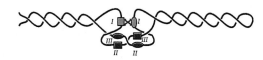

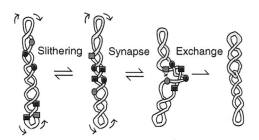

Slithering Synapse Exchange

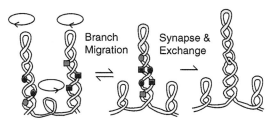

Branch Migration Synapse & Exchange

FIGURE 2 Plectonemic movement required for γδ resolution. At the top, the organization of the γδ resolvase binding site (*res*) is given along with three subsites labeled *res-I, -II,* and *-III*. Each subsite binds a dimer of resolvase, but only resolvase dimers bound at the *res-I* subsite can catalyze DNA strand exchange. Following binding, the two *res* sites must form a precise synapse, which tangles the six subsites into three interwound supercoils. Two movements of DNA allow this juxtapositioning: slithering, in which DNA moves like a conveyor belt and all points along the chain move relative to all other points, and supercoil branch migration, in which extrusion and resorption of supercoil branches cause sites to become plectonemically interwound. The recombination products include a deletion (shown as a shaded circle), which is initially catenated with the chromosome but which is released by the activity of topo IV.

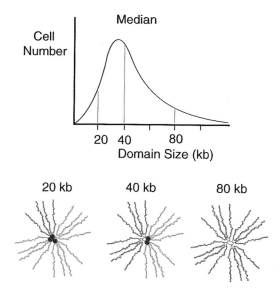

Median

Cell Number

20 40   80
Domain Size (kb)

20 kb   40 kb   80 kb

● Domain Barrier

FIGURE 3 The domain size distribution of exponentially growing cultures of *S. typhimurium* in the 43- to 45-min (*his-cob*) segment of the chromosome. The curve is a right skewed distribution, with a median size of 40 kb; less than 10% of the cells have domains under 2 kb, and about 5% of the cells have 80-kb domains. Below the graph is an illustration of how the domains might be cordoned off by barriers. For a 20-kb domain with 60 to 75 interwound supercoils, all sites in the red zone communicate and sequences in the blue zone do not. A 40-kb domain has at least one less barrier, and the 80-kb domain has no barriers, so all sites can communicate through slithering and supercoil branch migration.

proteins were hypothesized to attach the chromosome to intracellular partition machinery (49). Recently, this class of proteins has been discovered by using high-resolution fluorescent microscopy. In *E. coli,* the SeqA protein binds to nascent hemimethylated DNA made by the replication fork; the SeqA protein is aggregated, and it occupies a unique position near the cell midpoint throughout the cell cycle (46). Another interesting protein attachment was found in *Bacillus subtilis,* where the SpoOJ protein was discovered to bind six to eight specific chromosomal sites in vivo. SpoOJ is aggregated in living cells, and through its DNA attachments, this complex causes the origin-proximal one-fourth of the chromosome to move to specific cellular locations during cell division (66). Although no protein with SpoOJ homology has been found in the *E. coli* genome, similar observations of the behavior of the *E. coli* *oriC*-proximal one-fourth of the chromosome were made with fluorescent *lacI* repressors bound to *lac* operators positioned near *oriC* (132).

Second, attachment of DNA to the membrane causes supercoil barriers. Although supercoil barriers were originally proposed as a mechanism for chromosome sorting (49), later work showed that about 40 membrane-bound sites could be detected when cells were disrupted in special ways (63). More recent studies showed that cotranscribed and translated integral membrane proteins and outer membrane proteins temporarily restricted DNA rotation and stimulated hypernegative supercoiling by tethering transcription machinery to the membrane (74).

A third hypothetical type of barrier involves a topoisomerase bound to a high-affinity site. Two gyrase sites with measurable physiological phenotypes are known. First, a site near the origin of replication of plasmid pSC101 (128) has an important role both for initiation of replication and for efficient plasmid partitioning (8). Second, a site at the center of bacteriophage Mu is important for viral transposition and replication (93). In this case, the central location of gyrase is critical and the gyrase site acts as a symmetry element to stimulate synapse of the ends for transposition (92, 94, 95).

One important question is whether critical gyrase sites exist in the bacterial chromosome. Cells treated with oxolinic acid and then with sodium dodecyl sulfate were broken into large pieces (13, 28). Some breaks appeared to be specific and changed with the transcriptional activity of the genome (13). However, unlike the sites in pSC101 and in Mu, these sites were not efficiently cleaved when cloned on plasmids. Reports that gyrase binds to short repeat elements called *rep* sequences or bacterial interspersed mosaic elements (BIMEs) have also appeared (26, 137). However, the significance of gyrase interaction at *rep* or BIME sites remains untested. In vivo, gyrase cleaves more than 30 sites in plasmid pBR322 (69).

Most models for topoisomerase modulation of domain structure propose binding of the enzymes either to the base or to the tips of supercoil loops (33, 92, 102). However, recent genetic studies with gyrase and topo IV mutants show that topoisomerases may control long-range interactions in DNA by a different mechanism. Some temperature-sensitive DNA gyrase and topo IV mutants were found to have domains much smaller than that of the WT. One hypothesis to explain this result is that the chromosomes of strains bearing these mutations were more tangled than normal. Tangles could include intramolecular snarls (knots) and intermolecular links between sister chromosomes and plasmids (catenanes). This hypothesis invokes domain-modulating roles for gyrase and topo IV that are consistent with their known biochemical activities. Both gyrase and topo IV must work together to keep DNA unsnarled. Topo IV is efficient at decatenating and unknotting DNA in vitro (39, 41) and in vivo (54, 72, 126, 127, 139, 140). However, to unknot and decatenate molecules completely, topo IV requires a negatively supercoiled substrate (106, 107, 140). The tangle hypothesis explains why both gyrase and topo IV mutants have a chromosome segregation phenotype (88, 89, 118) and why gyrase mutants have an increased knot frequency in plasmids (112).

PROSPECTIVE

After many years in which the questions of bacterial genome structure seemed intractable, several productive approaches have evolved in the last 2 years. Experiments with the green fluorescent protein (GFP) fused to various site-specific DNA binding proteins now allow direct observation of chromosomal movement in living cells, and such studies have provided compelling evidence in *B. subtilis*, *E. coli*, and *Caulobacter* that chromosomes attach to specific structures as cells progress through the cell cycle (111, 132, 133). Two-color fluorescent in situ hybridization can be done on fixed preparations of bacteria to distinguish plasmid and chromosome, or even origin- and terminus-specific regions of a chromosome (46, 87). Recombination assays now provide information about plectonemic structure and the locations and distribution of topological barriers (14, 15, 45, 60, 115, 140).

Understanding supercoil movement in the bacterial genome is important for many reasons. Our models of structure shape how we think about fundamental genetic processes. Homologous and site-specific genetic recombination, adaptive mutation, "supercoil-regulated" gene transcription, gene order on chromosomes, and plasmid-chromosome replication segregation are all phenomena that are likely to be influenced by DNA dynamics. The lessons of bacterial chromosomes, like the lessons of bacterial DNA replication, transposition, and recombination, will apply to eukaryotic chromosomes. Unknotting and untangling of DNA strands and replication fork convergence are everybody's problems. Whereas the physical consequences of the plectonemic double helix will continue to challenge the imagination of biochemists and geneticists alike, Watson and Crick were fundamentally right. The problem isn't insuperable.

ACKNOWLEDGMENTS

Work in the author's laboratory is supported by grants GM-33143 from the National Institutes of Health and MCB-9604875 from the National Science Foundation.

REFERENCES

1. **Adachi, T., K. Mizuuchi, R. Menzel, and M. Gellert.** 1984. DNA sequence and transcription of the region upstream of the *E. coli gyrB* gene. *Nucleic Acids Res.* **12**:6389–6395.

2. **Adams, D. E., E. M. Shekhtman, E. L. Zechiedrich, M. B. Schmid, and N. R. Cozzarelli.** 1992. The role of topoisomerase IV in partitioning bacterial replicons and the structure of catenated intermediates in DNA replication. *Cell* **71**:277–288.

3. **Albert, A.-C., A.-M. Roman, G. Bouche, M. Leng, and A. R. Rahmouni.** 1994. Gradual and oriented B-Z transition in 5'-untranscribed region of mouse ribosomal DNA. *J. Biol. Chem.* **269**:19238–19244.

4. **Alfano, C., and R. McMacken.** 1988. The role of template superhelicity in the initiation of bacteriophage l DNA replication. *Nucleic Acids Res.* **16**:9611–9630.

5. **Baker, T., K. Sekimizu, B. E. Funnell, and A. Kornberg.** 1986. Extensive unwinding of the plasmid template during staged enzymatic initiation of DNA replication from the origin of the Escherichia coli chromosome. *Cell* **45**:53–64.

6. **Bauer, W. R.** 1978. Structure and reactions of closed duplex DNA. *Annu. Rev. Biophys. Bioeng.* **7**:287–313.

7. **Benjamin, H. W., M. M. Matzuk, M. A. Krasnow, and N. R. Cozzarelli.** 1985. Recombination site selection by Tn3 resolvase: topological tests of a tracking mechanism. *Cell* **40**:147–158.

8. **Biek, D. P., and J. Strings.** 1995. Partition functions of mini-F affect plasmid DNA topology in *Escherichia coli. J. Mol. Biol.* **246**:388–400.

9. **Bliska, J. B., and N. R. Cozzarelli.** 1987. Use of site-specific recombination as a probe of DNA structure and metabolism in vivo. *J. Mol. Biol.* **194**:205–218.

10. **Chalker, A. F., D. R. Leach, and R. G. Lloyd.** 1988. *Escherichia coli sbcC* mutants permit stable propagation of DNA replicons containing a long palindrome. *Gene* **71**:201–205.

11. **Chen, D., S. Bachellier, and D. M. Lilley.** 1998. Activation of the *leu-500* promoter by a reversed polarity *tetA* gene. Response to global plasmid supercoiling. *J. Biol. Chem.* **273**:653–659.

12. **Chen, D., R. Bowater, and D. M. J. Lilley.** 1993. Activation of the *leu-500* promoter; a topological domain generated by divergent transcription in a plasmid. *Biochemistry* **32**:13162–13170.

13. **Condemine, G., and C. L. Smith.** 1990. Transcription regulates oxolinic acid-induced DNA gyrase cleavage at specific sites on the *E. coli* chromosome. *Nucleic Acids Res.* **18**:7389–7396.

14. **Cornet, F., J. Louarn, J. Patte, and J.-M. Louarn.** 1996. Restriction of the activity of the recombination site *dif* to a small zone of the *Escherichia coli* chromosome. *Genes Dev.* **10**:1152–1161.

15. **Corre, J., F. Cornet, J. Patte, and J.-M. Louarn.** 1997. Unraveling a region-specific hyperrecombination phenomenon: genetic control and modalities of terminal recombination in *Escherichia coli. Genetics* **147**:979–989.

16. **Cozzarelli, N. R.** 1980. DNA gyrase and the supercoiling of DNA. *Science* **207:**953–960.

17. **Craigie, R., and K. Mizuuchi.** 1986. Role of DNA topology in Mu transposition: mechanism of sensing the relative orientation of two DNA segments. *Cell* **45:**793–800.

18. **Critchlow, S. E., and A. Maxwell.** 1996. DNA cleavage is not required for the binding of quinolone drugs to the DNA gyrase-DNA complex. *Biochemistry* **35:**7387–7393.

19. **DiGate, R. J., and K. J. Marians.** 1988. Identification of a potent decatenating enzyme from Escherichia coli. *J. Biol. Chem.* **263:**13366–13373.

20. **DiGate, R. J., and K. J. Marians.** 1989. Molecular cloning and DNA sequence analysis of Escherichia coli topB, the gene encoding topoisomerase III. *J. Biol. Chem.* **264:**17924–17930.

21. **DiNardo, S., K. A. Voelkel, R. Sternglanz, A. E. Reynolds, and A. Wright.** 1982. *Escherichia coli* DNA topoisomerase I mutants have compensatory mutations in DNA gyrase genes. *Cell* **31:**43–51.

22. **Dorman, C. J., G. C. Barr, N. N. Bhriain, and C. F. Higgins.** 1988. DNA supercoiling and the anaerobic growth phase regulation of *tonB* gene expression. *J. Bacteriol.* **170:**2816–2826.

23. **Drlica, K.** 1992. Control of bacterial supercoiling. *Mol. Microbiol.* **6:**425–433.

24. **Drlica, K., and X. Zhao.** 1997. DNA gyrase, topoisomerase IV, and the 4-quinolones. *Microbiol. Mol. Biol. Rev.* **61:**377–392.

25. **Drolet, M., P. Phoenix, R. Menzel, E. Masse, L. F. Liu, and R. J. Crouch.** 1995. Overexpression of RNase H partially complements the growth defect of an *Escherichia coli* delta *topA* mutant: R-loop formation is a major problem in the absence of DNA topoisomerase I. *Proc. Natl. Acad. Sci. USA* **92:**3526–3530.

26. **Espeli, O., and F. Boccard.** 1997. In vivo cleavage of *Escherichia coli* BIME-2 repeats by DNA gyrase: genetic characterization of the target and identification of the cut site. *Mol. Microbiol.* **26:**767–777.

27. **Figueroa, N., and L. Bossi.** 1988. Transcription induces gyration of the DNA template in *Escherichia coli. Proc. Natl. Acad. Sci. USA* **85:**9416–9420.

28. **Franco, R. J., and K. Drlica.** 1988. DNA gyrase on the bacterial chromosome. Oxolinic acid-induced DNA cleavage in the *dnaA-gyrB* region. *J. Mol. Biol.* **201:**229–233.

29. **Frank-Kamenetskii, M. D., and S. M. Mirkin.** 1995. Triplex DNA structures. *Annu. Rev. Biochem.* **64:**65–96.

30. **Froelich-Ammon, S. J., and N. Osheroff.** 1995. Topoisomerase poisons: harnessing the dark side of enzyme mechanism. *J. Biol. Chem.* **270:**21429–21432.

31. **Gampar, H. B., and J. E. Hearst.** 1982. A topological model for transcription based on unwinding angle analysis of *E. coli* RNA polymerase binary, initiation and ternary complexes. *Cell* **29:**81–90.

32. **Gangloff, S., J. P. McDonald, C. Bendixen, L. Arthur, and R. Rothstein.** 1994. The yeast type I topoisomerase Top3 interacts with Sgs1, a DNA helicase homolog: a potential eukaryotic reverse gyrase. *Mol. Cell. Biol.* **14:**8391–8398.

33. **Gasser, S. M., and U. K. Laemmli.** 1987. A glimpse at chromosomal order. *Trends Genet.* **3:**16–21.

34. **Gellert, M., K. Mizuuchi, M. H. O'Dea, and H. A. Nash.** 1976. DNA gyrase: an enzyme that introduces superhelical turns into DNA. *Proc. Natl. Acad. Sci. USA* **73:**3872–3876.

35. **Gellert, M., and H. Nash.** 1987. Communication between segments of DNA during site-specific recombination. *Nature* **325:**401–404.

36. **Goldstein, E., and K. Drlica.** 1984. Regulation of bacterial DNA supercoiling: plasmid linking numbers vary with growth temperature. *Proc. Natl. Acad. Sci. USA* **81:**4046–4050.

37. **Heichman, K. A., and R. C. Johnson.** 1990. The Hin invertasome: protein-mediated joining of distant recombination sites at the enhancer. *Science* **249:**511–517.

38. **Herbert, A., and A. Rich.** 1996. The biology of left-handed Z-DNA. *J. Biol. Chem.* **271:**11595–11598.

39. **Hiasa, H., R. J. DiGate, and K. J. Marians.** 1994. Decatenating activity of *Escherichia coli* DNA gyrase and topoisomerases I and III during oriC and pBR322 DNA replication in vitro. *J. Biol. Chem.* **269:**2093–2099.

40. **Hiasa, H., and K. J. Marians.** 1994. Topoisomerase III, but not topoisomerase I, can support nascent chain elongation during theta-type DNA replication. *J. Biol. Chem.* **269:**32655–32659.

41. **Hiasa, H., and K. J. Marians.** 1994. Topoisomerase IV can support *oriC* DNA replication in vitro. *J. Biol. Chem.* **269:**16371–16375.

42. **Hiasa, H., and K. J. Marians.** 1996. Two distinct modes of strand unlinking during q-type DNA replication. *J. Biol. Chem.* **271:**21529–21535.

43. **Higgins, C. F.** 1994. The bacterial chromosome: DNA topology, chromatin structure and gene expression, p. 11–23. *In* J. S. Heslop-Harrison and R. Flavel (ed.), *The Chromosome.* Bios, Oxford, England.

44. **Higgins, C. F., C. J. Dorman, D. A. Stirling, L. Waddell, I. R. Booth, G. May, and E. Bremer.** 1988. A physiological role for DNA supercoiling in the osmotic regulation of gene expression in S. typhimurium and E. coli. *Cell* **52:** 569–584.

45. **Higgins, N. P., X. Yang, Q. Fu, and J. R. Roth.** 1996. Surveying a supercoil domain by using the γδ resolution system in *Salmonella typhimurium*. *J. Bacteriol.* **178:**2825–2835.

46. **Hiraga, S., C. Ichinose, N. Hironori, and M. Yamazoe.** 1998. Cell cycle-dependent duplication and bidirectional migration of SeqA-associated DNA-protein complexes in *E. coli*. *Mol. Cell.* **1:**381–387.

47. **Holmes, F. L.** 1998. The DNA replication problem, 1953–1958. *Trends Biochem. Sci.* **23:** 117–120.

48. **Htun, H., and J. E. Dahlberg.** 1989. Topology and formation of triple-stranded H-DNA. *Science* **243:**1571–1576.

49. **Jacob, F., S. Brenner, and F. Cuzin.** 1963. On the regulation of DNA replication in bacteria. *Cold Spring Harbor Symp. Quant. Biol.* **28:**329–348.

50. **Jaworski, A., N. P. Higgins, R. D. Wells, and W. Zacharias.** 1991. Topoisomerase mutants and physiological conditions control supercoiling and Z-DNA formation in vivo. *J. Biol. Chem.* **266:**2576–2581.

51. **Jaworski, A., W.-T. Hsieh, J. A. Blaho, J. E. Larson, and R. D. Wells.** 1987. Left handed DNA in vivo. *Science* **238:**773–777.

52. **Johnson, R. C., A. C. Glasgow, and M. I. Simon.** 1987. Spatial relationship of the Fis binding sites for Hin recombinational enhancer activity. *Nature* **329:**462–465.

53. **Kanaar, R., A. Klippel, E. Shekhtman, J. M. Dungan, R. Kahmann, and N. R. Cozzarelli.** 1990. Processive recombination by the phage Mu Gin system: implications for the mechanisms of DNA strand exchange, DNA site alignment, and enhancer action. *Cell* **62:**353–366.

54. **Kato, J., Y. Nishimura, R. Imamura, H. Niki, S. Hiraga, and H. Suzuki.** 1990. New topoisomerase essential for chromosome segregation in *E. coli*. *Cell* **63:**393–404.

55. **Kavenoff, R., and B. Bowen.** 1976. Electron microscopy of membrane-free folded chromosomes from *Escherichia coli*. *Chromosoma* **59:**89–101.

56. **Kim, R. A., P. R. Caron, and J. C. Wang.** 1995. Effects of yeast DNA topoisomerase III on telomere structure. *Proc. Natl. Acad. Sci. USA* **92:** 2667–2671.

57. **Kornberg, A., and T. Baker.** 1991. *DNA Replication*. W. H. Freeman & Co., New York, N.Y.

58. **Krause, H. M., and N. P. Higgins.** 1986. Positive and negative regulation of the Mu operator by Mu repressor and *Escherichia coli* integration host factor. *J. Biol. Chem.* **261:**3744–3752.

59. **Kuempel, P., J. Henson, L. Dircks, M. Tecklenburg, and D. Lim.** 1991. dif, a recA-independent recombination site in the terminus region of the chromosome of *Escherichia coli*. *New Biol.* **3:**799–811.

60. **Kuempel, P., A. Hogaard, M. Nielsen, O. Nagappan, and M. Tecklenburg.** 1996. Use of a transposon (Tndif) to obtain suppressing and nonsuppressing insertions of the dif resolvase site of *Escherichia coli*. *Genes Dev.* **10:**1162–1171.

61. **Leach, D., J. Lindsey, and E. Okely.** 1987. Genome interactions which influence DNA palindrome mediated instability and inviability in *Escherichia coli*. *J. Cell Sci.* **7:**33–40.

62. **Leach, D. R., E. A. Okely, and D. J. Pinder.** 1997. Repair by recombination of DNA containing a palindromic sequence. *Mol. Microbiol.* **26:**597–606.

63. **Leibowitz, P. J., and M. Schaechter.** 1975. The attachment of the bacterial chromosome to the cell membrane. *Int. Rev. Cytol.* **41:**1–28.

64. **Li, T., Y. A. Panchenko, M. Drolet, and L. F. Liu.** 1997. Incompatibility of the *Escherichia coli rho* mutants with plasmids is mediated by plasmid-specific transcription. *J. Bacteriol.* **179:** 5789–5794.

65. **Lilley, D. M. J., R. M. Clegg, S. Diekmann, N. C. Seeman, E. von Kitzing, and P. J. Hagerman.** 1995. A nomenclature of junctions and branchpoints in nucleic acids. *Nucleic Acids Res.* **23:**3363–3364.

66. **Lin, D. C.-H., P. A. Levin, and A. D. Grossman.** 1998. Bipolar localization of a chromosome partition protein in *Bacillus subtilis*. *Proc. Natl. Acad. Sci. USA* **94:**4721–4726.

67. **Liu, L. F., and J. C. Wang.** 1987. Supercoiling of the DNA template during transcription. *Proc. Natl. Acad. Sci. USA* **84:**7024–7027.

68. **Lockshon, D., and D. R. Morris.** 1983. Positively supercoiled plasmid DNA is produced by treatment of Escherichia coli with DNA gyrase inhibitors. *Nucleic Acids Res.* **11:**2999–3017.

69. **Lockshon, D., and D. R. Morris.** 1985. Sites of reaction of Escherichia coli DNA gyrase on pBR322 in vivo as revealed by oxolinic acid-induced plasmid linearization. *J. Mol. Biol.* **181:** 63–74.

70. **Lodge, J. K., T. Kazic, and D. E. Berg.** 1989. Formation of supercoiling domains in plasmid pBR322. *J. Bacteriol.* **171:**2181–2187.

71. **Louarn, J.-M., J. Louarn, V. Francois, and J. Patte.** 1991. Analysis and possible role of hyperrecombination in the termination region of

the *Escherichia coli* chromosome. *J. Bacteriol.* **173:** 5097–5104.

72. **Luttinger, A. L., A. L. Springer, and M. B. Schmid.** 1991. A cluster of genes that affects nucleoid segregation in Salmonella typhimurium. *New Biol.* **3:**687–697.

73. **Lydersen, B. K., and D. E. Pettijohn.** 1977. Interactions stabilizing DNA tertiary structure in *Escherichia coli* chromosome investigated with ionizing radiation. *Chromosoma* **62:**199–215.

74. **Lynch, A. S., and J. C. Wang.** 1993. Anchoring of DNA to the bacterial cytoplasmic membrane through cotranscriptional synthesis of polypeptides encoding membrane proteins or proteins for export: a mechanism of plasmid hypernegative supercoiling in mutants deficient in DNA topoisomerase I. *J. Bacteriol.* **175:**1645–1655.

75. **Marians, K. J.** 1992. Prokaryotic DNA replication, p. 673–719. *In* C. Richardson, J. Abelson, A. Meister, and C. Walsh (ed.), *Annual Review of Biochemistry.* Annual Reviews Inc., Palo Alto, Calif.

76. **Marians, K. J., and H. Hiasa.** 1997. Mechanism of quinolone action. A drug-induced structural perturbation of the DNA precedes strand cleavage by topoisomerase IV. *J. Biol. Chem.* **272:** 9401–9409.

77. **Maxwell, A., and M. Gellert.** 1984. The DNA dependence of the ATPase activity of DNA gyrase. *J. Biol. Chem.* **259:**14472–14480.

78. **McClellan, J. A., P. Boublikova, E. Palecek, and D. M. J. Lilley.** 1990. Superhelical torsion in cellular DNA responds directly to environmental and genetic factors. *Proc. Natl. Acad. Sci. USA* **87:**8373–8377.

79. **McClure, W.** 1985. Mechanism and control of transcription initiation in prokaryotes. *Annu. Rev. Biochem.* **54:**171–204.

80. **McNairn, E., N. Ni Bhriain, and C. J. Dorman.** 1995. Overexpression of the *Shigella flexneri* genes coding for DNA topoisomerase IV compensates for loss of DNA topoisomerase I: effect on virulence gene expression. *Mol. Microbiol.* **15:**507–517.

81. **Menzel, R., and M. Gellert.** 1983. Regulation of the genes for E. coli DNA gyrase: homeostatic control of DNA supercoiling. *Cell* **34:**105–113.

82. **Menzel, R., and M. Gellert.** 1987. Fusions of the *Escherichia coli gyrA* and *gyrB* control regions to the galactokinase gene are inducible by coumermycin treatment. *J. Bacteriol.* **169:**1272–1278.

83. **Miller, H. I., A. Kikuchi, H. A. Nash, R. A. Weisberg, and D. I. Friedman.** 1981. Site-specific recombination of bacteriophage l: the role of host gene products. *Cold Spring Harbor Symp. Quant. Biol.* **45:**1121–1126.

84. **Miller, W. G., and R. W. Simons.** 1993. Chromosomal supercoiling in *Escherichia coli. Mol. Microbiol.* **10:**675–684.

85. **Mirkin, S. M., and M. D. Frank-Kamenetskii.** 1994. H-DNA and related structures. *Annu. Rev. Biophys. Biomol. Struct.* **23:** 541–576.

86. **Mojica, F. J. M., and C. F. Higgins.** 1997. In vivo supercoiling of plasmid and chromosomal DNA in an *Escherichia coli hns* mutant. *J. Bacteriol.* **179:**3528–3533.

87. **Niki, H., and S. Hiraga.** 1997. Subcellular distribution of actively partitioning F plasmid during the cell division cycle in *E. coli. Cell* **90:**951–957.

88. **Ogura, T., H. Niki, H. Mori, M. Morita, M. Hasegawa, C. Ichinose, and S. Hiraga.** 1990. Identification and characterization of *gyrB* mutants of *Escherichia coli* that are defective in partitioning of mini-F plasmids. *J. Bacteriol.* **172:** 1562–1568.

89. **Orr, E., N. F. Fairweather, I. B. Holland, and R. H. Pritchard.** 1979. Isolation and characterization of a strain carrying a conditional lethal mutation in the *cou* gene of *Escherichia coli* K12. *Mol. Gen. Genet.* **177:**103–112.

90. **Palecek, E.** 1992. Probing of DNA structure in cells with osmium tetroxide—2,2′-bipyridine, p. 305–318. *In* D. M. J. Lilley and J. E. Dahlberg (ed.), *Methods in Enzymology.* Academic Press, Inc., San Diego, Calif.

91. **Parker, C. N., and S. E. Halford.** 1991. Dynamics of long-range interactions on DNA: the speed of synapsis during site-specific recombination by resolvase. *Cell* **66:**781–791.

92. **Pato, M., and M. Banerjee.** 1996. The Mu strong gyrase-binding site promotes efficient synapsis of the prophage termini. *Mol. Microbiol.* **22:**283–292.

93. **Pato, M., M. M. Howe, and N. P. Higgins.** 1990. A DNA gyrase binding site at the center of the bacteriophage Mu genome required for efficient replicative transposition. *Proc. Natl. Acad. Sci. USA* **87:**8716–8720.

94. **Pato, M. L., and M. Karlock.** 1994. Central location of the Mu strong gyrase binding site is obligatory for optimal rates of replicative transposition. *Proc. Natl. Acad. Sci. USA* **91:**7056–7060.

95. **Pato, M. L., M. Karlock, C. Wall, and N. P. Higgins.** 1995. Characterization of Mu prophage lacking the central strong gyrase binding site: location of the block in replication. *J. Bacteriol.* **177:**5937–5942.

96. **Pavitt, G. D., and C. F. Higgins.** 1993. Chromosomal domains of supercoiling in *Salmonella typhimurium. Mol. Microbiol.* **10:**685–696.

97. **Peck, L. J., and J. C. Wang.** 1983. Energetics of B-to-Z transition in DNA. *Proc. Natl. Acad. Sci. USA* **80:**6206–6210.

98. **Pettijohn, D. E., and O. Pfenninger.** 1980. Supercoils in prokaryotic DNA restrained in vivo. *Proc. Natl. Acad. Sci. USA* **77:**1331–1335.

99. **Pruss, G. J., S. H. Manes, and K. Drlica.** 1982. Escherichia coli DNA topoisomerase I mutants: increased supercoiling is corrected by mutations near gyrase genes. *Cell* **31:**35–42.

100. **Rahmouni, A. R., and R. D. Wells.** 1992. Direct evidence for the effect of transcription on local DNA supercoiling *in vivo. J. Mol. Biol.* **223:**131–144.

101. **Raji, A., D. J. Zabel, C. S. Laufer, and R. E. Depew.** 1985. Genetic analysis of mutations that compensate for loss of *Escherichia coli* DNA topoisomerase I. *J. Bacteriol.* **162:**1173–1179.

102. **Razin, S. V., P. Petrov, and R. Hancock.** 1991. Precise localization of the globin gene cluster within one of the 20 to 300-kbp DNA fragments released by cleavage of chicken chromosomal DNA at topoisomerase II sites in vivo: evidence that the fragments are DNA loops or domains. *Proc. Natl. Acad. Sci. USA* **88:**8515–8519.

103. **Richardson, S. M. H., C. F. Higgins, and D. M. J. Lilley.** 1984. The genetic control of DNA supercoiling in *Salmonella typhimurium. EMBO J.* **3:**1745–1752.

104. **Richardson, S. M. H., C. F. Higgins, and D. M. J. Lilley.** 1988. DNA supercoiling and the *leu-500* promoter mutation of *Salmonella typhimurium. EMBO J.* **7:**1863–1869.

105. **Rudd, K. E., and R. Menzel.** 1987. *his* operons of *Escherichia coli* and *Salmonella typhimurium* are regulated by DNA supercoiling. *Proc. Natl. Acad. Sci. USA* **84:**517–521.

106. **Rybenkov, V., C. Ullsperger, A. Vologodskii, and N. R. Cozzarelli.** 1997. Simplification of DNA topology below equilibrium values by type II topoisomerases. *Science* **277:**690–693.

107. **Rybenkov, V. V., A. V. Vologodskii, and N. R. Cozzarelli.** 1997. The effect of ionic conditions on the conformations of supercoiled DNA. II. Equilibrium catenation. *J. Mol. Biol.* **267:**312–323.

108. **Scheirer, K. E., and N. P. Higgins.** 1997. The DNA cleavage reaction of DNA gyrase. Comparison of stable ternary complexes formed with enoxacin and CcdB protein. *J. Biol. Chem.* **272:**27202–27209.

109. **Schmid, M. B.** 1990. A locus affecting nucleoid segregation in *Salmonella typhimurium. J. Bacteriol.* **172:**5416–5424.

110. **Schofield, M., R. Agbunag, and J. Miller.** 1992. DNA inversions between short inverted repeats in *Escherichia coli. Genetics* **132:**295–302.

111. **Shapiro, L., and R. Losick.** 1997. Protein localization and cell fate in bacteria. *Science* **276:**712–718.

112. **Shishido, K., N. Komiyama, and S. Ikawa.** 1987. Increased production of a knotted form of plasmid pBR322 DNA in *Escherichia coli* DNA topoisomerase mutants. *J. Mol. Biol.* **195:**215–218.

113. **Sinden, R. R., and D. E. Pettijohn.** 1981. Chromosomes in living *Escherichia coli* cells are segregated into domains of supercoiling. *Proc. Natl. Acad. Sci. USA* **78:**224–228.

114. **Spirito, F., and L. Bossi.** 1996. Long-distance effect of downstream transcription on the activity of supercoiling-sensitive *leu-500* promoter in a *topA* mutant of *Salmonella typhimurium. J. Bacteriol.* **178:**7129–7137.

115. **Staczek, P., and N. P. Higgins.** 1998. DNA gyrase and Topoisomerase IV modulate chromosome domain size in vivo. *Mol. Microbiol.* **29:**1435–1448.

116. **Stark, M. W., D. J. Sherratt, and M. R. Boocock.** 1989. Site specific recombination by Tn*3* resolvase: topological changes in the forward and reverse reactions. *Cell* **58:**779–790.

117. **Stark, W. M., C. N. Parker, S. E. Halford, and M. R. Boocock.** 1994. Stereoselectivity of DNA catenane fusion by resolvase. *Nature* **368:**76–78.

118. **Steck, T. R., and K. Drlica.** 1985. Involvement of DNA gyrase in bacteriophage T7 growth. *J. Virol.* **53:**296–298.

119. **Steck, T. R., G. J. Pruss, S. H. Manes, L. Burg, and K. Drlica.** 1984. DNA supercoiling in gyrase mutants. *J. Bacteriol.* **158:**397–403.

120. **Stewart, L., M. R. Redinbo, X. Qiu, W. G. J. Hol, and J. J. Champoux.** 1998. A model for the mechanism of human topoisomerase I. *Science* **279:**1534–1541.

121. **Sugino, A., and N. R. Cozzarelli.** 1980. The intrinsic ATPase of DNA gyrase. *J. Biol. Chem.* **255:**6299–6306.

122. **Sugino, A., N. P. Higgins, P. O. Brown, C. L. Peebles, and N. R. Cozzarelli.** 1978. Energy coupling in DNA gyrase and the mechanism of action of novobiocin. *Proc. Natl. Acad. Sci. USA* **74:**4838–4842.

123. **Tan, J., L. Shu, and H.-Y. Wu.** 1994. Activation of the *leu-500* promoter by adjacent transcription. *J. Bacteriol.* **176:**1077–1086.

124. **Trucksis, M., E. I. Golub, D. J. Zabel, and R. E. Depew.** 1981. *Escherichia coli* and *Salmonella typhimurium supX* genes specify DNA topoisomerase I. *J. Bacteriol.* **147:**679–681.

125. **Tse-Dinh, Y.-C., and R. K. Beran.** 1988. Multiple promoters for transcription of the Escherichia coli DNA topoisomerase I gene and their regulation by DNA supercoiling. *J. Mol. Biol.* **202:**735–742.

126. **Ullsperger, C., and N. R. Cozzarelli.** 1996. Contrasting enzymatic activities of topoisomerase IV and DNA gyrase from *Escherichia coli. J. Biol. Chem.* **271:**31549–31555.

127. **Ullsperger, C. J., A. V. Vologodskii, and N. R. Cozzarelli.** 1995. Unlinking of DNA by topoisomerases during DNA replication, p. 115–142. *In* F. Eckstein and D. M. J. Lilley (ed.), *Nucleic Acids and Molecular Biology.* Springer-Verlag, Berlin, Germany.

128. **Wahle, E., and A. Kornberg.** 1988. The partition locus of plasmid pSC101 is a specific binding site for DNA gyrase. *EMBO J.* **7:**1889–1895.

129. **Wang, J. C.** 1971. Interaction between DNA and an *Escherichia coli* protein omega. *J. Mol. Biol.* **55:**523–533.

130. **Wang, J. C.** 1996. DNA topoisomerases. *Annu. Rev. Biochem.* **65:**635–692.

131. **Watson, J. D., and F. H. C. Crick.** 1953. Genetic implications of the structure of deoxyribonucleic acid. *Nature* **171:**964–967.

132. **Webb, C. D., A. Teleman, S. Gordon, A. Straight, A. Belmont, D. C.-H. Lin, A. D. Grossman, A. Wright, and R. Losick.** 1997. Bipolar localization of the replication origin regions of chromosomes in vegetative and sporulating cells of *B. subtilis. Cell* **88:**667–674.

133. **Wheeler, R. T., and L. Shapiro.** 1997. Bacterial chromosome segregation: is there a mitotic apparatus? *Cell* **88:**577–579.

134. **Worcel, A., and E. Burgi.** 1972. On the structure of the folded chromosome of *Escherichia coli. J. Mol. Biol.* **71:**127–147.

135. **Wu, H.-Y., S. Shyy, J. C. Wang, and L. F. Liu.** 1988. Transcription generates positively and negatively supercoiled domains in the template. *Cell* **53:**433–440.

136. **Wu, H.-Y., J. Tan, and M. Fang.** 1995. Long-range interaction between two promoters: activation of the *leu-500* promoter by a distant upstream promoter. *Cell* **82:**445–451.

137. **Yang, Y., and G. F.-L. Ames.** 1988. DNA gyrase binds to the family of prokaryotic repetitive extragenic palindromic sequences. *Proc. Natl. Acad. Sci. USA* **85:**8850–8854.

138. **Zacharias, W., A. Jaworski, J. E. Larson, and R. D. Wells.** 1988. The B- to Z-DNA equilibrium in vivo is perturbed by biological processes. *Proc. Natl. Acad. Sci. USA* **85:**7069–7073.

139. **Zechiedrich, E. L., and N. R. Cozzarelli.** 1995. Roles of topoisomerase IV and DNA gyrase in DNA unlinking during replication in *Escherichia coli. Genes Dev.* **9:**2859–2869.

140. **Zechiedrich, E. L., A. B. Khodursky, and N. R. Cozzarelli.** 1997. Topoisomerase IV, not gyrase, decatenates products of site-specific recombination in *Escherichia coli. Genes Dev.* **11:**2580–2592.

LOCAL GENETIC CONTEXT, SUPERCOILING, AND GENE EXPRESSION

Andrew St. Jean

12

A great part of understanding how a cell works is wrapped up in understanding the mechanisms controlling the expression of its genetic content. As a result, a great deal of effort has been and is devoted to investigating gene expression at both the transcriptional and translational levels. Most of this work considers the gene (or operon, as the case may be) in isolation from the rest of the genome and focuses on proteins that may act as activators or repressors of expression, the specific DNA sequences or regions of DNA that these proteins bind to, or how RNA polymerase itself interacts with DNA through promoter sequences.

While these sorts of studies are vitally important to our understanding of how cells work, they are incomplete. For some time investigators have been wrestling with the fact that genes do not exist in isolation but are part of much larger units called genomes. As such, the forces influencing a gene's expression will not consist solely of the need for that gene's product in a specific amount at a specific time but also for every other gene's product, as well

as the organizational and structural constraints imposed by the sheer number of genes present.

In this chapter, I will discuss the influence a gene's neighbors can have on that gene's expression and look at the mechanisms by which this can occur. In keeping with the spirit of this book, the possible consequences of gene expression on genomic structure and dynamics will be discussed.

GENES AS COMPONENTS OF THE NUCLEOID

The mechanisms by which a gene is regulated and expressed must take into account one salient feature in all cases: that there are hundreds to thousands of other genes in the genome that must also be regulated in particular ways. This fact will have a profound impact on how any one gene will operate, and because of this, it is instructive to review the arrangement in space of prokaryotic genes as they undergo transcription.

DNA in a prokaryotic cell must be arranged in a way that allows its efficient packaging while leaving the genes accessible for expression. This is accomplished by the nucleoid. Studies of the nucleoid indicate that it is divided into two structurally and functionally different parts; a ribosome-free

Andrew St. Jean, Department of Biology, University of Ottawa, Ottawa, Ontario K1N 6N5, Canada.

Organization of the Prokaryotic Genome, Edited by Robert L. Charlebois,
© 1999 American Society for Microbiology, Washington, D.C.

core, which for the most part is metabolically quiescent, and the surface, which is in contact with other cellular components (for a review, see reference 40). The localization of nascent mRNA transcripts (42) and RNA polymerase (13) to the surface of the nucleoid suggests that DNA must move out of the core in order to be processed.

Both the packaging of DNA and its metabolic functioning are greatly facilitated by supercoiling. Supercoiling imparts torsional stress to DNA, which influences its interaction with RNA polymerase and other DNA-binding proteins as well as contributing to its compaction. In all prokaryotes, with the exception of certain hyperthermophilic archaea, the DNA is maintained in a state of negative superhelical tension. A specific linking difference of about −0.06 has been measured in bacterial DNA (2). The specific linking difference (σ) represents the change in linking number (ΔLk) of a DNA molecule from that of its most relaxed topoisomer, normalized to allow comparison of different-sized molecules (2). It is known that in bacteria, roughly half of the genomic supercoils in a cell are constrained and not free to diffuse along the DNA strand (5, 36). The above figure refers to the total σ of the genome; a value for the unconstrained portion of genomic supercoils—referred to as the effective σ—is approximately −0.025 (5). The distinction between total and effective σ is important, as will be seen later in the chapter.

Supercoiling can induce different higher-order structures in DNA; the two most relevant are illustrated in Fig. 1. Of the two, solenoidal supercoils are much better at compacting DNA than plectonemic supercoils. Solenoids—in the form of nucleosomes—are used by eukaryotes to store their DNA, and there is some evidence that euryarchaeota also use solenoidal supercoils to some extent (48). No evidence exists for nucleosomes in bacteria. This does not preclude the existence of solenoidal supercoils within the bacterial nucleoid (24), although firm evidence for or against this possibility is still lacking.

On the other hand, evidence does exist for the plectonemic arrangement of DNA within bacteria. It has been found that small, high-copy-number plasmids in *Escherichia coli* are maintained as tightly packed liquid crystals of plectonemically supercoiled molecules, with the implication that they are held in this topology throughout the cell cycle (38). Extensive work on the mechanism of transposition in the $\gamma\delta$ transposon has led to the elucidation of the three-dimensional structure needed to effect recombination (4). Direct copies of the *res* locus located on a single molecule must be brought into close opposition along a plectonemic supercoil. This recombination system has been used to show that a 90-kbp stretch of the *Salmonella enterica* serovar Typhimurium chromosome is at least transiently held in a plectonemically supercoiled form (20).

TRANSCRIPTION, SUPERCOILING, AND GENE EXPRESSION

Supercoiling, whatever its particular effect on a gene, influences expression by imparting free energy to the DNA, which can be used to promote its melting by RNA polymerase, and by inducing a three-dimensional conformation to the DNA that will enhance or inhibit the binding of accessory proteins. The importance of supercoiling to cellular functioning is illustrated by the tight control prokaryotes maintain over this property of their genomes. In bacteria, this control is implemented through the opposing actions of topoisomerase I, which relaxes negatively supercoiled DNA, and topoisomerase II (gyrase), which introduces negative supercoils into DNA. While the maintenance of genomic supercoiling at levels conducive to cellular functions seems straightforward, it is complicated by the fact that at least one important cellular function—transcription—can alter DNA supercoiling levels. The twin-supercoil domain model elaborated by Liu and Wang states that a migrating transcriptional bubble will induce positive supercoils downstream and negative

A

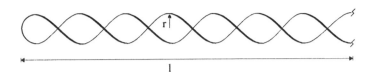

B

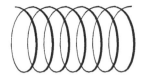

FIGURE 1 Two alternate conformations that may be adopted by supercoiled DNA. (A) DNA wrapped around itself in plectonemic supercoils with a terminal loop to the left. The radius of the plectoneme—indicated by the letter *r*—varies greatly with even modest changes in the linking number of the DNA, decreasing as the number of supercoils increases (6). The length of the plectoneme—indicated by the letter *l*—remains relatively constant over the range of linking numbers normally experienced in vivo and is 41% of the length of the linear DNA strand (6). Bacterial plasmids and at least portions of chromosomal DNA regularly adopt this conformation. (B) DNA wound in solenoidal supercoils, forming a toroidal shape. The best known example of this conformation is the eukaryotic nucleosome, with its DNA wrapped around the core histone proteins.

supercoils upstream of its direction of travel (26). This effect has since been confirmed by different groups (30, 37, 54), and its significance will become apparent when we consider the effects gene expression can have on neighboring genes. A second observation made by Liu and Wang is that to prevent the nascent RNA transcript from becoming entangled with the DNA template, RNA polymerase must be prevented from rotating around the DNA axis (26). It was suggested that the viscous drag induced by the RNA strand and bound ribosomes being pulled along behind the transcribing RNA polymerase would prevent this rotation (53). Anchoring of nascent proteins from these ribosomes in the cytoplasmic membrane in the case of membrane or periplasmic proteins would further enhance this effect (27). As has been pointed out, however, the newly created RNA strand increases in length at the same rate that the polymerase

travels down the DNA template (50). Since the RNA strand is not, in fact, pulled along behind the polymerase, it and any bound ribosomes could not generate any viscous drag. Instead, it was found that RNA polymerase positions itself at the terminal loop of a plectonemically supercoiled DNA molecule and that it maintains this position during the entire course of transcribing a gene (50). In this scenario, it is the specific three-dimensional conformation of the DNA template-RNA polymerase complex that maintains the polymerase at the end of the plectoneme and causes the DNA to migrate through it. This means not only that the DNA will be rotating around its long axis during the course of transcription, but that the interwound strands of the plectoneme will be slithering past one another, as shown in Fig. 2. Most of the data presented by ten Heggeler-Bordier et al. (50) show only a single RNA polymerase occupying a ter-

FIGURE 2 Transcription on a plectonemically supercoiled DNA molecule. The RNA polymerase (black oval) positions itself at the terminal loop of the plectoneme and maintains this position throughout the transcription process. The polymerase enzyme remains on one side of the loop to prevent the nascent RNA strand from becoming entangled in the DNA molecule. This necessitates the rotation of the DNA around its long axis. The maintenance of the polymerase enzyme at the terminal loop also means that the interwound strands of the plectoneme slither past one another, allowing distantly positioned sequences to come into close apposition. Both movements of the DNA molecule are indicated by arrows.

minal loop, although two transcribing polymerases were also found within the same loop. It is unknown what conformation the DNA takes when many RNA polymerase molecules are transcribing the same gene simultaneously.

How does the RNA polymerase maintain its position at the end of a plectoneme? Intrinsically curved DNA spontaneously migrates to such ends, as this minimizes the free energy of the DNA molecule (25). It is known that bent DNA is present in the upstream regions of many bacterial genes, and it is thought that RNA polymerase itself bends DNA as part of the initiation sequence (for a review, see reference 35). A process where the bending action of RNA polymerase itself promotes its migration to the terminal loop of a plectoneme and preserves its position there is favored by ten Heggeler-Bordier et al. (50). However, given the bending properties of the major DNA-binding proteins of bacteria (HU, H-NS, and IHF), it seems likely that these proteins are involved as well, especially when one considers that H-NS and IHF seem to function as general transcription factors (3, 17). One can envision any of these proteins binding to the promoter region of a gene, causing that DNA to bend and thus initiating its migration to a terminal loop, where RNA polymerase can begin transcription. Depending on the orientation of protein binding sites and the gene in question, DNA bending may exclude the promoter region from the termi-

nal loop, inhibiting rather than enhancing transcription.

Although most of the work discussed above deals with the conformation of transcribing DNA on plasmids, evidence from the γδ recombination system suggests that a similar process occurs on the chromosomes of bacteria. The picture of transcription we have, then, is of RNA polymerase maintaining a position at the terminal loop of a plectonemically supercoiled DNA strand, causing the DNA to rotate around its long axis and at the same time slithering the two intertwined strands past one another. The rotation of the DNA causes local changes in supercoiling levels which need to be ameliorated to some degree either by diffusion along the DNA or by the action of topoisomerases. The greatest difference between small plasmids and larger molecules may be that plasmids can rotate the entire molecule around its long axis during transcription, relieving local changes in supercoiling levels. Obviously, this would not be an option for a chromosome, which must use other mechanisms to dissipate the supercoiling effects of transcription, perhaps relying more on topoisomerases. For either plasmid or chromosome, however, the simple scenario of a single gene being transcribed in isolation will rarely hold, and there is an increasing amount of data that suggests neighboring genes have the potential to influence one another's expression.

LOCAL CONTEXT AND GENE EXPRESSION

The twin-supercoil domain model of Liu and Wang (26) predicts that genes sensitive to changes in DNA supercoiling will have their expression influenced by the transcription of nearby genes. This assumes that the changes in supercoiling induced by transcription can diffuse far enough along the DNA strand to reach neighboring genes and that once they do, these changes are strong enough to affect these genes. A great deal of work has been devoted to one locus in particular, the *leu-500* promoter of *S. enterica*, in order to examine these assumptions.

The *leu-500* promoter contains a single A → G transition which abolishes transcription and results in leucine auxotrophy (32). The activity of this promoter can be restored, however, by introducing mutations into the *topA* gene (coding for topoisomerase I) of *S. enterica* (12). This reactivation of the promoter does not seem to depend on the global level of supercoiling of genomic DNA but only on the absence of functional TopA (39). Reactivation was found to occur when *leu-500* was located in its native position on the chromosome but not when it was cloned into a plasmid (39). These observations suggest that it is local changes in supercoiling levels that influence expression from this promoter. Indeed, it was later found that positioning a second gene upstream of a plasmid-borne *leu-500* and oriented divergently could activate *leu-500* in a *topA* mutant strain (8). One explanation for this is that negative supercoils propagated upstream from the second gene diffuse into the *leu-500* promoter region, lowering the energy required to initiate transcription and allowing expression from the mutant promoter. Further investigation showed that activation of the *leu-500* promoter could be enhanced by transcription from a promoter positioned downstream from *leu-500* and oriented in the same direction (9). The effects of both upstream and downstream transcription on the activity of *leu-500* have since been confirmed (49). Unlike that of Chen et al. (9), this study did not find any difference in *leu-500* expression when the upstream gene was membrane bound or cytosolic, arguing against a role for membrane anchoring in contributing to the accumulation of excess supercoiling (49). One of the properties noted about the interactions between *leu-500* and adjacent promoters was the short range on which they operated: less than 250 bp (49). This limit in range was tested only for a promoter positioned upstream of *leu-500*, however. A much longer range effect was found for downstream-positioned promoters (45). Here, the *leu-500* promoter was fused to a *lacZ* gene and a second promoter—either the IPTG (isopropyl-β-D-thiogelactopyranoside)-inducible Tac or constitutive *hisR*[hpa] promoters—positioned at the 3′ end of this gene. Transcription from *leu-500* was enhanced by transcription from both of these downstream promoters (45).

It was also found that the *leu-500* promoter could apparently be influenced by upstream promoters 1.9 kbp away (54). It was confirmed that expression from *leu-500* was dependent on expression from the upstream promoter and that at least two different promoters (*ilvIH* and *lac*) could be used with equal efficacy (54). The intervening sequence—which, together with the *ilvIH* promoter, is the native upstream sequence of the *leu-500* promoter in *S. enterica*—was found to be important, since deleting 500 bp of this sequence abolished activation of *leu-500* (54).

More recent work has revealed complex interactions between *ilvIH* and *leu-500* mediated by this intervening sequence (14). An open reading frame (ORF), *leuO*, is located between the *leu-500* and *ilvIH* promoters, positioned 410 bp upstream of *leu-500* and oriented divergently from it and in the same direction as *ilvIH*. Transcription of *leuO* is dependent on transcription from *ilvIH*, as mutations in this promoter or its complete removal eliminate transcription of *leuO* (14). In turn, the product of *leuO* is required for activation of *leu-500*. It is postulated, though not confirmed, that LeuO is a DNA-binding protein, due to its possession of a helix-turn-helix motif, and that it binds to the *leu-500* promoter region to effect activation. The action of LeuO is not solely responsible for *leu-500* activation, however, as supplying LeuO from a gene on a second plasmid did not lead to transcription from *leu-500*. Both transcription from a divergent promoter positioned 410 bp upstream of *leu-500* and functional LeuO protein are necessary for *leu-500* activation. The authors describe their results in terms of a relay system in which transcription from *ilvIH* activates transcription of *leuO*—presumably by the diffusion of supercoils into the *leuO* promoter—which then activates *leu-500* via possible protein-DNA interactions in addition to diffusion of supercoils (14).

The discovery of this relay system shortens the maximum distance demonstrated for direct interaction between divergently oriented promoters from 1.9 kbp to 410 bp. The *ilvIH* promoter is located approximately 1.5 kbp downstream of the *leuO* promoter and 540 bp downstream from its coding region (14). Together with the results of Spirito and Bossi (45), this suggests that interactions at ranges significantly longer than 250 bp are possible.

The studies discussed above use transcription from inducible and constitutive promoters to measure effects on expression from the *leu-500* promoter. This information, together with measurements of the total specific linking difference of the plasmids determined by gel electrophoresis, is used to infer local domains of altered supercoiling in various regions upstream and downstream of the transcribing promoters. Another group has applied the use of the intercalating agent 4,5′,8-trimethylpsoralen (TMP) to directly measure supercoiling levels in the different regions of their plasmid constructs (30).

Psoralen and its derivatives are chemicals which can penetrate cells and intercalate into a DNA strand in a supercoil-dependent manner. As negative supercoils increase, so too does psoralen intercalation (23). This reaction is reversible and does not appear to alter the DNA's geometry appreciably. When treated cells are exposed to measured doses of UV light, psoralen forms both intra- and interstrand covalent bonds at a rate linearly proportional to the level of negative supercoiling of the DNA (43). DNA so treated is digested, denatured briefly, and immediately placed at low temperature (4°C), allowing only cross-linked fragments of DNA to reanneal. By separating the resulting single- and double-stranded DNA on an electrophoretic gel, preparing a Southern blot, and using appropriate probes, a method is provided for determining the supercoiling levels of the specific regions of DNA under study (10).

The TMP experiments applied to the *leu-500* promoter for the most part confirmed results from previous studies (30). The real value of this work is that by using this methodology, the twin-supercoil domain model can be tested by looking at the supercoil levels of the relevant DNA regions directly without the need to infer behavior from changes in plasmid-specific linking differences. One additional finding that further supports the twin-supercoil domain model is that the length of the transcript located upstream from the *leu-500* promoter affects the level of supercoiling changes. As the transcript length increases, so too does the negative supercoiling in the *leu-500* region (30). Thus, longer transcripts may have stronger effects on neighboring genes—a significant finding when one considers the prevalence of operons within prokaryotic genomes. An enhanced effect caused by longer transcripts was also noted by Fang and Wu (14).

Although the *leu-500* promoter has been shown to be influenced by transcription both upstream and downstream of its position, the most pronounced findings have always been in strains of *S. enterica* that contain a defective *topA* gene. Indeed, Spirito and Bossi (45) have questioned whether supercoiling interactions between transcriptional units have an appreciable effect in wild-type cells. However, the effects they observed in a *topA* mutant strain were very strong, even exaggerated. Spirito and Bossi (45) found *leu-500* expression to increase from 24 to 38 times upon activating transcription both upstream and downstream of the promoter in a *topA* mutant. This is contrasted with only a doubling in expression from *leu-500* in a wild-type strain. The large difference in transcription rates between mutant and wild-type strains may here be obscuring the fact that doubling the output of a gene can have significant consequences for a cell, either good or bad, depending on the situation. Simply because the results from wild-type cells are not as great as those from mutant cells does not mean they can be dismissed as being inconsequential. Presumably, it is the results from the mutant strain that are unusual as far as the cell is concerned, and the use of this strain simply makes an effect that does operate in wild-type cells more noticeable to the researcher. Other studies mentioned above also found reduced but detectable changes in *leu-500* activity in wild-type strains (15, 30, 54), confirming that topological coupling of transcriptional units in such cells is real and not due to experimental error.

EVIDENCE FOR THE CHROMOSOME

All these studies, while providing support for the idea that the expression of one gene can indeed affect that of neighboring genes, look at such interactions on plasmids that have been specially constructed for the purpose. What evidence exists that such interactions also occur on the chromosomes of bacteria?

The answer is, surprisingly little. This is largely because plasmids are much easier to work with when constructing arrangements of promoters and reporter genes and it is more convenient to maintain these constructs as plasmids when experiments with different host strains are to be performed. Also, the number and sizes of transcripts are usually known for an entire plasmid and changes in linking number can be determined by gel electrophoresis. For these reasons, the chromosomes of bacteria have been largely avoided for the types of investigations into the local effects of supercoiling discussed above. However, studies of the global regulation of supercoiling levels

in the chromosome may shed some light on this question.

It is generally agreed that the genome of *E. coli* is divided into 40 to 50 domains of approximately 100 kbp each which are topologically independent of one another; relaxing one domain by nicking the DNA does not relax the others (44, 52). It has been suggested that topologically independent domains are a good way to organize genes whose expression is sensitive to supercoiling. Large numbers of genes could be coordinately regulated if they responded to supercoiling in a similar manner and were located on the same domain. However, it was unclear whether the different domains were, in fact, maintained at different supercoiling levels, as all estimates of chromosomal linking number include the entire chromosome as a single unit. To answer this question, two groups independently proceeded to insert expression cassettes consisting of supercoil-sensitive promoters fused to reporter genes into different locations on the *E. coli* (29) and *S. enterica* (34) chromosomes. The rationale behind these experiments was that if the different chromosomal domains are maintained at different supercoiling levels, this should show up in the level of expression of the reporter genes. The main finding of both groups was that no stable differences in supercoiling level could be found in the chromosome of either organism (29, 34). The differences in expression that were observed were attributed to varying copy numbers of the cassettes due to their positions relative to *oriC*. However, both groups also acknowledge additional variations in their data that cannot be explained by copy number or stably maintained domains of supercoiling. Miller and Simons (29) dismiss these variations in the ratio of gene expression between the two promoters used in this study (which were as high as 45%) as representing insignificant differences in supercoiling. Pavitt and Higgins (34), on the other hand, actually state that the variations in gene expression seen in their experiments were likely due to local variations in supercoiling level. In both cases, these variations were only touched upon and quickly dismissed, as the emphasis of both studies was on large, stable domains of supercoiling. The coupling of gene expression due to transcription is necessarily local. Therefore, the experimental noise of both studies may represent just the sorts of interactions one might expect if coupling occurs on the chromosome in bacteria.

CAVEATS AND CONSIDERATIONS

Because supercoiling is such a dynamic property of DNA, varying from region to region within a molecule, it is very difficult to study directly. Therefore, a number of methods have been developed to take advantage of more easily measured properties which can then be used to infer the supercoiling status of DNA. Because our understanding of the role and effects of supercoiling in the prokaryotic genome is still incomplete, however, these methods can introduce problems with the interpretation of results that can obscure more than they reveal.

As was mentioned above, much of the work involving DNA supercoiling in prokaryotes has been done with plasmids, particularly the use of reporter plasmids to measure σ, which is then inferred to accurately represent σ values for the chromosome as well. Plasmids in these situations are often considered to be little versions of the chromosome, but this assumption has never been systematically tested. Another problem with using plasmids to measure σ is that it can be determined only for the entire molecule and not specific regions, an important consideration when local domains of supercoiling are to be investigated. Also, the necessity of using gel electrophoresis to view the results precludes its use when studying the chromosome.

Perhaps a more important consequence of using σ is that the influence of DNA-binding proteins is completely masked. It was mentioned earlier that approximately half of the genomic supercoils of bacteria are constrained in vivo (5, 36). It is known that both HU and H-NS can constrain DNA supercoils, and both these proteins are present in bacteria in great abundance (11, 51). Altering the amount of such constraining protein on a DNA molecule induces changes in effective σ without the necessity of altering total σ. Alternatively, effective σ could be maintained at relatively constant levels by topoisomerases while total σ could change due to changes in the amount of bound protein. This property of supercoil-constraining proteins has been posited as the cause of a number of anomalies in the results of various studies (18). These include the supercoil sensitivity—or lack thereof—of the proU operon, the inconsistency of ΔLk in different plasmids exposed to the same hns mutant, and the opposite effect on ΔLk that the same change in environmental conditions can have on E. coli and S. enterica versus Shigella flexneri.

Even the use of reporter genes to measure the expression from supercoil-sensitive promoters is not free of problems. The luxAB reporter gene can itself influence expression from certain promoters, including leu-500 and proU (16). There exists a region of naturally curved DNA within the lux coding region (33), and this appears to alter the expression from certain promoters. However, this effect does not simply apply to all supercoil-sensitive promoters, as the gyrB promoter does not seem to be similarly affected (16).

The use of TMP avoids most of the pitfalls described above by measuring supercoiling levels directly. This system works equally well for chromosomal (31) and plasmid (10, 30) DNA, making reliance on reporter plasmids unnecessary. TMP is a sensitive indicator of supercoiling, and it is claimed to be able to detect changes in supercoil levels of 15% (10) and 12% (30). Changes of this magnitude are regularly experienced by the genome of E. coli (1, 21). Because TMP intercalation varies with the level of unconstrained supercoils and does not measure total σ, effects caused by changes in protein binding to DNA are not masked as they can be with reporter plasmids. Best of all, TMP allows the measurement of supercoil levels in specific regions of DNA without relying on the insertion of known supercoil-sensitive promoters fused to reporter genes. Coupled with Northern analysis, TMP could facilitate the investigation of transcription on local supercoil levels for large numbers of genes.

SUPERCOILING AND THE GENOME

So what is the role of supercoiling in the genome of prokaryotes? Is it used simply as a way to package and structure the genome, or is it also involved in the expression of genes? Supercoiling is certainly important to prokaryotes, given its ubiquity and the tight control normally exercised by topoisomerases. Changing supercoil levels by using topoisomerase mutants—even by as little as 20%—alters the expression of many genes in *E. coli* (47). In this light, supercoiling does indeed influence gene expression, but that statement does not answer a more interesting and far-reaching question. Is this influence an undesirable consequence of abnormal fluctuations in a property the cell is going to great lengths to control, or are changes in supercoiling level actively used by cells to coordinate gene activity? Even now, this question is still very difficult to answer.

A great deal of work has been stimulated by the observation that the *E. coli* genome is divided into topologically independent domains of approximately 100 kbp each. This structural base of the genome was linked to the discovery, made by many different groups, that bacteria alter the level of genomic supercoiling in response to various transitions in growth conditions, such as aerobic-to-anaerobic growth (21), osmotic upshift (22), temperature changes (41), and nutrient status of the growth medium (1). It was posited that the cells were responding to these changes by altering the level of supercoiling in order to change the expression of large numbers of genes; turning off those genes no longer needed and turning on other genes useful for the new conditions (19). Given the domain structure of the genome, it seemed only logical to arrange genes so that those responding

to supercoiling in similar ways would be located on the same domain, making coordinate expression easier for the cell. While a large amount of data was accumulated which seemed to support this interpretation, these studies were invariably done with reporter plasmids and gel electrophoresis to measure ΔLk, the pitfalls of which have already been noted. As was mentioned, studies with expression cassettes on chromosomal DNA indicate that domains are not stably maintained at different supercoiling levels (29, 34) and work with the $\gamma\delta$ recombination system suggests that the domain boundaries themselves are not constant for all cells in a population (20, 46). Also, investigation of the photosynthesis operons of *Rhodobacter capsulatus*—put forth as an example of regulation by a stable change in supercoiling—found no evidence for such a change when TMP was used to directly measure supercoiling levels in the region of the two operons (10). Given that changes in the extent of protein binding to DNA can introduce changes in total σ without conferring similar changes in effective σ, the evidence provided by these more direct methods seems the more reliable. Because supercoil-constraining DNA-binding proteins function in the regulation of large numbers of genes, it may be that the ΔLk studies were detecting changes in the pattern of protein binding which would not necessarily lead to changes in unconstrained supercoiling levels. Altered patterns of protein binding, in this case, would be the cause of changing gene expression, and changes in ΔLk would be only a consequence.

If stably maintained global changes in supercoiling levels are not used by cells to change gene expression, this leaves local changes as an avenue for supercoiling to exert its influence. To date, the accumulated evidence suggests that coordinated expression of genes through transcription-induced changes in supercoiling is possible but does not confirm that it actually occurs in cells outside an experimental setting. Most of the work done makes use of the *leu-500* promoter and specially constructed plasmids. A much wider range of promoters need to be investigated to determine whether the effect seen with this promoter can be generalized. It is also important to investigate local effects of supercoiling on gene arrangements as they naturally occur, preferably on chromosomal DNA. This will be greatly facilitated by using TMP or another method with the same characteristic of measuring supercoiling changes directly. Most evidence so far deals with interactions between transcriptional units over very short ranges (<250 bp), which could be seen to limit the potential of such interactions in genomes. However, data from 12 completely sequenced prokaryotic chromosomes suggests this is not the case, as shown in Table 1. In all chromosomes presented, a significant proportion of adjacent ORFs occur within 250 bp of each other—often well over half of the ORFs analyzed. In this analysis, only adjacent ORFs occurring on opposite strands were used to ensure that none would be cotranscribed. Even genes separated by longer distances may not be free from the influence of supercoiling. Given the prevalence of DNA loops and curved DNA in gene regulation—both of

which may involve sequences hundreds of base pairs upstream of the affected gene or operon (28, 35)—changes in supercoiling level may not need to reach an adjacent gene itself to affect its expression.

If local changes in supercoiling can influence neighboring genes, what does this mean for the cell? The local-context model has been elaborated as an answer to this question (7). This model takes as assumptions that the regulation of gene expression is crucial to cell survival, that expression of a gene evolves to maximize its chances of survival and propagation, and that most changes in the expression of genes so adapted will be deleterious. Given these conditions, if a gene's expression is influenced by the expression of its neighbors, that gene will tend to become dependent on its neighbors to condition the local environment—in this case the local supercoiling level—in order to maintain its own normal pattern of expression. This has the consequence that if genes are rearranged by any means, their pattern of expression will likely become suboptimal and the cell will suffer a competitive disadvantage compared to the nonrearranged strain. Therefore, genes will

TABLE 1 Counts of distances between adjacent ORFs separated by less than 250 bp of DNA occurring on opposite strands[a]

Organism	ORFs transcribed convergently		ORFs transcribed divergently	
	No.	% of all convergent ORFs	No.	% of all divergent ORFs
Aquifex aeolicus	229	92.71	171	69.51
Archaeoglobus fulgidus	342	93.44	281	76.78
Bacillus subtilis	470	86.24	349	64.15
Borrelia burgdorferi	108	93.91	99	86.09
Chlamydia trachomatis	118	94.40	63	50.00
Escherichia coli	562	88.64	360	56.78
Helicobacter pylori J99	120	72.73	126	76.36
Methanococcus jannaschii	217	85.43	180	70.87
Mycobacterium tuberculosis	591	89.95	454	69.21
Pyrococcus horikoshii	477	95.40	279	55.91
Rickettsia prowazekii	41	40.20	58	57.43
Synechocystis sp. strain PCC6803	476	87.82	323	59.59

[a] From the genomes of selected prokaryotes. The data on ORF position and orientation were derived from the annotated GenBank files for each genome. All figures are for chromosomal DNA.

tend to stay put unless a rearrangement confers a definite advantage (such as the insertion of a useful metabolic function into a chromosome).

The growing stockpile of prokaryotic genomic sequence provides an unprecedented resource with which to test the local–context model as well as the basic concept of supercoil-mediated coupling of gene expression. It represents a huge number of genes and operons of known length, orientation, and distance from one another partitioned into an ever-increasing number of genomes. Given the extensive characterization of *E. coli* genes, this genome seems like a reasonable place to start. A gene with an inducible promoter would be selected; its expression and that of neighboring genes would be determined and correlated with changes in local supercoiling levels. This would be repeated in the presence and absence of inducer and with different inducible genes with neighbors in different orientations. It would be important to include controls; for example, to ensure that the inducer itself was not altering expression of the neighboring genes directly. Such experiments would also preferably be done in a strain free of mutations in its topoisomerases or supercoil-constraining DNA-binding proteins to ensure that whatever results were observed could be related to cells growing outside the laboratory.

In contrast to the relatively clear-cut operation of regulatory proteins and their binding sites in gene expression, the role of supercoiling is characterized by shades of gray. It has still not been conclusively shown that supercoiling is of general utility in regulating gene expression in cells, only that the potential for this function is present. Detailed work, with special emphasis on the strengths and shortcomings of the methodologies used, will be needed to determine the importance of transcription-induced supercoiling on the coupling of gene expression.

REFERENCES

1. **Balke, V., and J. Gralla.** 1987. Changes in the linking number of supercoiled DNA accompany growth transitions in *Escherichia coli. J. Bacteriol.* **169:**4499–4506.

2. **Bates, A. D., and A. Maxwell.** 1993. DNA supercoiling, p. 17–45. *In* D. Rickwood, (ed.), *DNA Topology* (*In Focus* series). IRL Press, Oxford, England.

3. **Bertin, P., P. Lejeune, C. Laurent-Winter, and A. Danchin.** 1990. Mutations in *bglY*, the structural gene for the DNA-binding protein H1, affect expression of several *Escherichia coli* genes. *Biochimie* **72:**889–891.

4. **Bliska, J. B., H. W. Benjamin, and N. R. Cozzarelli.** 1991. Mechanism of Tn3 resolvase recombination in vivo. *J. Biol. Chem.* **266:**2041–2047.

5. **Bliska, J. B., and N. R. Cozzarelli.** 1987. Use of site-specific recombination as a probe of DNA structure and metabolism in vivo. *J. Mol. Biol.* **194:**205–218.

6. **Boles, T. C., J. H. White, and N. R. Cozzarelli.** 1990. Structure of plectonemically supercoiled DNA. *J. Mol. Biol.* **213:**931–951.

7. **Charlebois, R. L., and A. St. Jean.** 1995. Supercoiling and map stability in the bacterial chromosome. *J. Mol. Evol.* **41:**15–23.

8. **Chen, D., R. P. Bowater, C. J. Dorman, and D. M. J. Lilley.** 1992. Activity of a plasmid-borne *leu-500* promoter depends on the transcription and translation of an adjacent gene. *Proc. Natl. Acad. Sci. USA* **89:**8784–8788.

9. **Chen, D., R. P. Bowater, and D. M. J. Lilley.** 1993. Activation of the *leu-500* promoter: a topological domain generated by divergent transcription in a plasmid. *Biochemistry* **32:**13162–13170.

10. **Cook, D. N., G. A. Armstrong, and J. E. Hearst.** 1989. Induction of anaerobic gene expression in *Rhodobacter capsulatus* is not accompanied by a local change in chromosomal supercoiling as measured by a novel assay. *J. Bacteriol.* **171:**4836–4843.

11. **Drlica, K.** 1992. Control of bacterial DNA supercoiling. *Mol. Microbiol.* **6:**425–433.

12. **Dubnau, E., and P. Margolin.** 1972. Suppression of promoter mutations by the pleiotropic *supX* mutations. *Mol. Gen. Genet.* **117:**91–112.

13. **Dürrenberger, M., M.-A. Bjornsti, T. Uetz, J. A. Hobot, and E. Kellenberger.** 1988. Intracellular location of the histonelike protein HU in *Escherichia coli. J. Bacteriol.* **170:**4757–4768.

14. **Fang, M., and H.-Y. Wu.** 1998. A promoter relay mechanism for sequential gene activation. *J. Bacteriol.* **180:**626–633.

15. **Fang, M., and H.-Y. Wu.** 1998. Suppression of *leu-500* mutation in topA+ *Salmonella typhimurium* strains. *J. Biol. Chem.* **273:**29929–29934.

16. **Forsberg, Å. J., G. D. Pavitt, and C. F. Higgins.** 1994. Use of transcriptional fusions to

monitor gene expression: a cautionary tale. *J. Bacteriol.* **176:**2128–2132.

17. **Friedman, D. I.** 1988. Integration host factor: a protein for all reasons. *Cell* **55:**545–554.

18. **Gowrishankar, J., and D. Manna.** 1996. How is osmotic regulation of transcription of the *Escherichia coli proU* operon achieved? *Genetica* **97:**363–378.

19. **Higgins, C. F., C. J. Dorman, and N. Ní Bhriain.** 1990. Environmental influences on DNA supercoiling: a novel mechanism for the regulation of gene expression, p. 421–432. *In* K. Drlica, and M. Riley (ed.), *The Bacterial Chromosome.* American Society for Microbiology, Washington, D.C.

20. **Higgins, N. P., X. Yang, Q. Fu, and J. R. Roth.** 1996. Surveying a supercoil domain by using the γδ resolution system in *Salmonella typhimurium. J. Bacteriol.* **178:**2825–2835.

21. **Hsieh, L.-S., R. M. Burger, and K. Drlica.** 1991. Bacterial DNA supercoiling and [ATP]/[ADP]: changes associated with a transition to anaerobic growth. *J. Mol. Biol.* **219:**443–450.

22. **Hsieh, L.-S., J. Rouviere-Yaniv, and K. Drlica.** 1991. Bacterial DNA supercoiling and [ATP]/[ADP] ratio: changes associated with salt shock. *J. Bacteriol.* **173:**3914–3917.

23. **Hyde, J. E., and J. E. Hearst.** 1978. Binding of psoralen derivatives to DNA and chromatin: influence of the ionic environment on dark binding and photoreactivity. *Biochemistry* **17:**1251–1257.

24. **Kellenberger, E., and B. Arnold-Schulz-Gahmen.** 1992. Chromatins of low-protein content: special features of their compaction and condensation. *FEMS Microbiol. Lett.* **100:**361–370.

25. **Laundon, C. H., and J. D. Griffith.** 1988. Curved helix segments can uniquely orient the topology of supertwisted DNA. *Cell* **52:**545–549.

26. **Liu, L. F., and J. C. Wang.** 1987. Supercoiling of the DNA template during transcription. *Proc. Natl. Acad. Sci. USA* **84:**7024–7027.

27. **Lodge, J. K., T. Kazic, and D. E. Berg.** 1989. Formation of supercoiling domains in plasmid pBR322. *J. Bacteriol.* **171:**2181–2187.

28. **Matthews, K. S.** 1992. DNA looping. *Microbiol. Rev.* **56:**123–136.

29. **Miller, W. G., and R. W. Simons.** 1993. Chromosomal supercoiling in *Escherichia coli. Mol. Microbiol.* **10:**675–684.

30. **Mojica, F. J. M., and C. F. Higgins.** 1996. Localized domains of DNA supercoiling: topological coupling between promoters. *Mol. Microbiol.* **22:**919–928.

31. **Mojica, F. J. M., and C. F. Higgins.** 1997. In vivo supercoiling of plasmid and chromosomal DNA in an *Escherichia coli hns* mutant. *J. Bacteriol.* **179:**3528–3533.

32. **Mukai, F. H., and P. Margolin.** 1963. Analysis of unlinked suppressors of an O° mutation in *Salmonella. Proc. Natl. Acad. Sci. USA* **50:**140–148.

33. **Owen-Hughes, T., G. D. Pavitt, D. S. Santos, J. Sidebotham, C. S. J. Hulton, J. C. D. Hinton, and C. F. Higgins.** 1992. Interaction of H-NS with curved DNA influences DNA topology and gene expression. *Cell* **71:**255–265.

34. **Pavitt, G. D., and C. F. Higgins.** 1993. Chromosomal domains of supercoiling in *Salmonella typhimurium. Mol. Microbiol.* **10:**685–696.

35. **Pérez-Martín, J., F. Rojo, and V. Lorenzo.** 1994. Promoters responsive to DNA bending: a common theme in prokaryotic gene expression. *Microbiol. Rev.* **58:**268–290.

36. **Pettijohn, E., and O. Pfenninger.** 1980. Supercoils in prokaryotic DNA restrained in vivo. *Proc. Natl. Acad. Sci. USA* **77:**1331–1335.

37. **Rahmouni, A. R., and R. D. Wells.** 1992. Direct evidence for the effect of transcription on local DNA supercoiling in vivo. *J. Mol. Biol.* **223:**131–144.

38. **Reich, Z., S. Levin-Zaidman, S. B. Gutman, T. Arad, and A. Minsky.** 1994. Supercoiling-regulated liquid-crystalline packaging of topologically-constrained, nucleosome-free DNA molecules. *Biochemistry* **33:**14177–14184.

39. **Richardson, S. M. H., C. F. Higgins, and D. M. J. Lilley.** 1988. DNA supercoiling and the *leu-500* promoter mutation of *Salmonella typhimurium. EMBO J.* **7:**1863–1869.

40. **Robinow, C., and E. Kellenberger.** 1994. The bacterial nucleoid revisited. *Microbiol. Rev.* **58:**211–232.

41. **Rohde, J. R., J. M. Fox, and S. A. Minnich.** 1994. Thermoregulation in *Yersinia enterocolitica* is coincident with changes in DNA supercoiling. *Mol. Microbiol.* **12:**187–199.

42. **Ryter, A., and A. Chang.** 1975. Localization of transcribing genes in the bacterial cell by means of high resolution autoradiography. *J. Mol. Biol.* **98:**797–810.

43. **Sinden, R. R., J. O. Carlson, and D. E. Pettijohn.** 1980. Torsional tension in the DNA double helix measured with trimethylpsoralen in living *E. coli* cells: analogous measurements in insect and human cells. *Cell* **21:**773–783.

44. **Sinden, R. R., and D. E. Pettijohn.** 1981. Chromosomes in living *E. coli* cells are segregated into domains of supercoiling. *Proc. Natl. Acad. Sci. USA* **78:**224–228.

45. **Spirito, F., and L. Bossi.** 1996. Long-distance effect of downstream transcription on activity of the supercoiling-sensitive *leu-500* promoter in a

topA mutant of *Salmonella typhimurium. J. Bacteriol.* **178:**7129–7137.

46. **Staczek, P., and P. Higgins.** 1998. Gyrase and topo IV modulate chromosome domain size *in vivo. Mol. Microbiol.* **29:**1435–1448.

47. **Steck, T. R., R. J. Franco, J.-Y. Wang, and K. Drlica.** 1993. Topoisomerase mutations affect the relative abundance of many *Escherichia coli* proteins. *Mol. Microbiol.* **10:**473–481.

48. **Takayanagi, S., S. Morimura, H. Kusaoke, Y. Yokoyama, K. Kano, and M. Shioda.** 1992. Chromosomal structure of the halophilic archaebacterium *Halobacterium salinarium. J. Bacteriol.* **174:**7207–7216.

49. **Tan, J., L. Shu, and H.-Y. Wu.** 1994. Activation of the *leu-500* promoter by adjacent transcription. *J. Bacteriol.* **176:**1077–1086.

50. **ten Heggeler-Bordier, B., W. Wahli, M. Adrian, A. Stasiak, and J. Dubochet.** 1992. The apical localization of transcribing RNA polymerases on supercoiled DNA prevents their rotation around the template. *EMBO J.* **11:**667–672.

51. **Tupper, A. E., T. A. Owen-Hughes, D. W. Ussery, D. S. Santos, D. J. P. Ferguson, J. M. Sidebotham, J. C. D. Hinton, and C. F. Higgins.** 1994. The chromatin-associated protein H-NS alters DNA topology *in vitro. EMBO J.* **13:**258–268.

52. **Worcel, A., and E. Burgi.** 1972. On the structure of the folded chromosome of *E. coli. J. Mol. Biol.* **71:**127–147.

53. **Wu, H.-Y., S. Shyy, J. C. Wang, and L. F. Liu.** 1988. Transcription generates positively and negatively supercoiled domains in the template. *Cell* **53:**433–440.

54. **Wu, H.-Y., J. Tan, and M. Fang.** 1995. Long-range interaction between two promoters: activation of the *leu-500* promoter by a distant upstream promoter. *Cell* **82:**445–451.

"STABLE" GENOMES

Kenneth E. Sanderson, Michael McClelland, and Shu-Lin Liu

13

The order of genes in the chromosomes of enteric bacteria, such as *Escherichia coli* K-12 and *Salmonella typhimurium* LT2, has long been recognized to be strongly conserved (34, 69). This conclusion was based on the alignments of genetic maps which were originally determined from classical genetic data, primarily F factor-mediated conjugation and phage-mediated transduction (3, 70). It is postulated that these genera diverged 120 to 160 million years ago (60), yet they retained genomes of very similar size and the same order of genes. Over the same period extensive evolution occurred at the base pair level, for the homologous genes which retained identical gene locations had diverged by an average of 15% (73); this presented a paradox as to how the overall chromosomes could have remained so conserved while so many base pair mutations were occurring. This extreme conservation in enteric bacteria is especially surprising because during growth in culture, and presumably in nature, chromosomes are frequently rearranged; duplications of segments of the chromosome occur at high frequencies (10^{-3} to 10^{-5}) (1, 29, 68), and some inversions and translocations, especially those with endpoints in the *rrn* operons, are common (1, 28). The conservation of gene order reported earlier has been confirmed by extensive recent work, but in addition it has been revealed that several types of genetic changes are superimposed on this conservation which were either undetectable by the methods used earlier or which were simply overlooked. In contrast to the relative stability of the enteric bacteria, the chromosomes of some other genera, such as *Bacillus* and *Streptomyces* (21, 33), and those of *Neisseria gonorrhoeae* and *Neisseria meningitidis* (19, 23) show a striking degree of rearrangement.

Strains in the genus *Salmonella* are distinguished by their surface antigens, the O (somatic) antigen and H (flagellar) antigen, and each different combination, or serovar, was given a different species name in the Kauffmann-White scheme (31). Members of this genus have been shown to belong to the same DNA-DNA reassociation group (18), and therefore many workers now designate almost all members of the genus as a single species, *Salmonella enterica*, with the serovar listed after

Kenneth E. Sanderson, Salmonella Genetic Stock Centre, Department of Biological Sciences, University of Calgary, Calgary, Alberta, Canada T2N 1N4.   *Michael McClelland*, Sydney Kimmel Cancer Center, Room 300, 3099 Science Park Road, San Diego, CA 92121.   *Shu-Lin Liu*, Department of Medical Biochemistry, University of Calgary, Calgary, Alberta, Canada T2N 1N4.

Organization of the Prokaryotic Genome, Edited by Robert L. Charlebois,
© 1999 American Society for Microbiology, Washington, D.C.

the species name, e.g., *S. enterica* serovar Typhimurium (37, 63). We have retained the use of the historical species names in *Salmonella* in this review, but it must be recognized that there is about as much genetic variation in the species *E. coli* of the genus *Escherichia* as there is in the genus *Salmonella* (57, 72).

The observed stability of the chromosome during evolution, in spite of the high frequency of rearrangements in culture, indicates that major forces must act to eliminate cells in which the genome has been rearranged. The purposes of this review are to evaluate our present knowledge about the degree of stability of the genome of enteric bacteria, to discuss the forces which have contributed to maintaining stability, and to consider the types of rearrangements which can and do occur even in those genomes generally considered to be stable.

THE ORDER OF GENES OF MOST ENTERIC BACTERIA IS STRONGLY CONSERVED

The conservation of the chromosomes of *E. coli* K-12 and *S. typhimurium* LT2 has been known for a long time, but because genetic analysis can be applied to only a few strains, it was not certain if other enteric bacteria were similarly conserved. With the use of methods of physical analysis of DNA, and especially the introduction of the use of pulsed-field gel electrophoresis (PFGE), used first in *E. coli* by Smith and colleagues (77), the genomes of many strains were determined and conservation of gene order was shown to be the rule.

The following two modifications of PFGE methods, which allow the determination of genome structure in many strains, have been used in representative enteric bacteria.

I-*Ceu*I Analysis

Digestion with the endonuclease I-*Ceu*I followed by separation of the DNA fragments by PFGE is especially suitable for construction of a genomic cleavage map showing the number and locations of the *rrn* genes for rRNA (the "*rrn* skeleton") (30, 41, 46) (Fig. 1 and 2). I-

*Ceu*I is encoded by a class I mobile intron which is inserted into the *rrl* gene coding for the large-subunit rRNA (23S rRNA) in the chloroplast DNA of *Chlamydomonas eugametos* (52). I-*Ceu*I is specific for, and cuts in, a 19-bp sequence in the *rrl* gene (51). Because ribosomal DNA sequences are strongly conserved, the I-*Ceu*I site is present in all seven *rrl* genes of enteric bacteria but at no other site so far detected (41). The order of I-*Ceu*I fragments is shown in Fig. 1, and their approximate sizes for *S. typhimurium* LT2 are shown in Fig. 2; the order is the same in *E. coli* K-12, and the sizes are similar (3, 5).

In a study of 17 independent strains of *S. typhimurium*, all of which had seven I-*Ceu*I fragments, all seven of the fragments in 15 of the strains were of indistinguishable sizes; a few of the strains each had one fragment which was slightly larger (Fig. 2). In addition, the analysis of partial digestion patterns revealed that at least the fragments CDEF are always retained in that order. These data show very clearly that the chromosomes of independent strains of *S. typhimurium* are strongly conserved. The fact that each I-*Ceu*I fragment (usually) retains the same length shows that this fragment has not undergone genomic rearrangements, for insertions, deletions, or translocations would change fragment lengths, as would inversions which include an *rrn* operon. I-*Ceu*I-A, the large fragment including the *Ter* region, is about 2,500 kb and was not accurately determined in our study, so rearrangements in this part of the chromosome cannot be efficiently detected. With the equipment and methods used in our laboratories, the sizes of molecules over 2 Mb could not be determined.

In a further analysis by the same methods, most of the 72 strains of the Salmonella reference B (SARB) set, established by Selander and colleagues by multilocus enzyme electrophoresis (12) and representing many of the species of subgenus I of *Salmonella*, were shown to be highly conserved (48a). Each strain had seven I-*Ceu*I fragments, indicating the possession of seven *rrn* operons, and most

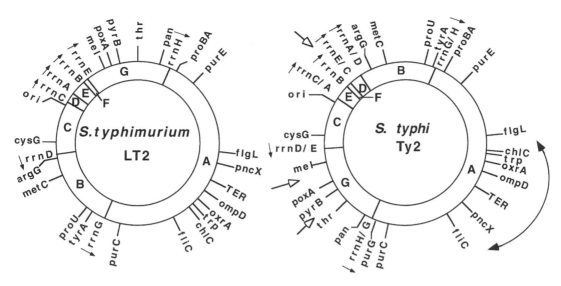

FIGURE 1 Comparison of the genomic maps for the endonuclease I-*Ceu*I and positions of selected genes determined by analysis of the locations of insertions of Tn*10* in *S. typhimurium* LT2 (40) and *S. typhi* Ty2 (47). The arrows beside the *rrn* operons indicate the direction of transcription. There are three types of rearrangements in *S. typhi* with respect to *S. typhimurium*. (i) The arc with arrowheads at both ends indicates a segment of the *S. typhi* Ty2 genome within the I-*Ceu*I-A fragment which is inverted relative to LT2. (ii) The open arrows indicate three regions in which the intervals between homologous genes are much longer in *S. typhi* than in *S. typhimurium*; these are postulated to be insertions of DNA (pathogenicity islands). (iii) The I-*Ceu*I fragments, which are in the order ABCDEFG in *S. typhimurium* (and most others enteric bacteria studied), are in the order AGCEFDB in *S. typhi*, presumably due to homologous recombination among the *rrn* operons.

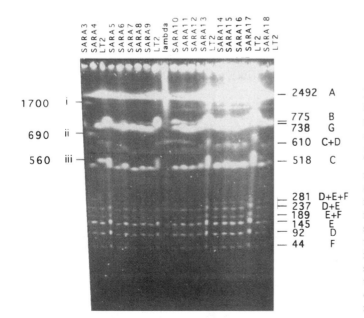

FIGURE 2 Separation by PFGE of fragments from I-*Ceu*I digestion of DNA from independent wild-type strains of *S. typhimurium* from the SARA set of strains (4), taken from Fig. 1 of reference 46. The lanes marked LT2 represent DNA from strain LT2 (which is SARA2); other lanes show the number of the strain from the SARA set. The normal I-*Ceu*I fragments and their sizes in kilobases are shown as single letters on the right. Unusual bands for a few of the strains are indicated on the left, along with their sizes. Some of the partial digestion products, such as D+E and E+F, are shown on the right; these show that the fragment order is DEF.

of the fragments could be assigned to a specific chromosome region based only on size. For example, the size of a specific fragment, e.g., fragment I-*Ceu*I-C, which includes the *oriC* gene, varies from 500 kb in some species up to 550 kb in other species, but in different strains of a species the size of the fragment is usually the same; this presumably indicates that there have been insertions and deletions of DNA during the evolution of different species. In addition, the I-*Ceu*I fragments normally retained the same order on the chromosome (I-*Ceu*I-ABCDEFG), with a few exceptions which are detailed below.

Tn*10* Insertions

The transposon Tn*10* has one site for the endonuclease *Xba*I and two sites for *Bln*I (equivalent to *Avr*II); thus, insertion of Tn*10* into specific genes enables the mapping of these genes on the physical chromosome. *Xba*I and *Bln*I digest with hexanucleotide specificity, but since the sequence CTAG is rare in bacterial genomes, they cut enteric bacterial genomes infrequently (6, 53). Many genes which contained Tn*10* insertions were mapped on the chromosome of *S. typhimurium* LT2 (44, 80); these transposons were transduced by bacteriophage P22 into related species, such as *S. enteritidis* (42), *Salmonella paratyphi* A (45), and *S. paratyphi* B (39), thus permitting the mapping of the locations of specific genes by physical analysis of the locations of the sites for *Xba*I and *Bln*I. Transductants were detected by selection for tetracycline resistance on Tn*10*. All of these strains showed a high degree of conservation of the order of genes on the chromosome, though there were differences in the lengths of intervals between genes (suggesting insertions and deletions in different species) and also inversions over the *Ter* region (see below).

Thus, it is clear that conservation of the order of genes is the norm within those enteric bacteria, such as *Escherichia* and *Salmonella*, which have been studied in sufficient detail. Using PFGE, it was previously shown that the chromosome map of *Shigella flexneri*

2a closely resembles that of *E. coli* K-12 (with the exceptions of a 30-kb deletion from *S. flexneri* which removes the *lac* genes and an inversion of about 450 kb over the *Ter* region) (61). The genomes of *Proteus, Morganella*, and other enteric bacteria are as yet unreported.

WHAT IS THE BASIS FOR CONSERVATION OF GENE ORDER?

In spite of a whole series of genetic changes which would be expected to rearrange the genome, the order of genes on the chromosomes of independently isolated wild-type strains remains surprisingly similar. The following are forces which would be expected to act in a conservative way to maintain gene order.

Genome Balance

Bidirectional replication of the chromosome from *oriC* terminates at *Ter*; cells with rearrangements in which *oriC* and *Ter* are not 180° apart on the circular chromosome may be defective in termination (29). The role of *Ter* has been difficult to precisely define; with the exception of the *dif* locus, the entire *E. coli Ter* region can be deleted without significantly affecting cell growth (29a). The *dif* region, however, is very position dependent; if the *dif* locus is moved from its usual site close to *Ter*C, the cell gains the filamentous phenotype (78a); thus, the *dif* locus might have an important role in genomic balance. An inversion of *E. coli* K-12 which displaces the origin of replication (*oriC*) by 6% with respect to *Ter* gave a small reduction in growth rate, while a large inversion (19%) resulted in a major growth rate reduction; reversion to a genomically balanced strain restored the normal growth rate (28). Strains which are not genomically balanced may not survive in nature because of reduced growth rates. Inversions and other events which may disrupt chromosome balance occur frequently and yet survive in culture (1, 28), though such inversions are seldom found in nature in most species.

Gene Dosage

During replication, genes which are close to *oriC* are present in extra copies, causing increased gene expression (71); this may result in adaptation of genes to the positions they occupy and thus selection against cells with rearrangements which result in inappropriate levels of gene expression. It is not easy to distinguish between the forces of genome balance and gene dosage; many of the changes in the genome would affect both, as pointed out by Hill and Gray (28). Data on genomic rearrangements in *Salmonella typhi*, however, particularly support the role of gene dosage, for in spite of many rearrangements which occur in wild-type strains of *S. typhi* (genomic rearrangements are unusually common in this species; see below), involving recombination among the *rrn* operons, there are few which dramatically alter the distances of genes from *oriC*; we assume that translocations of this type occur but that shifting with respect to *oriC* results in altered gene expression and therefore failure to survive in evolution (48). Some of these translocations which are not detected would not affect the genomic balance but would affect gene dosage; this makes the gene dosage theory more likely in this situation.

Operon Structure

Genes within the operon, usually for related functions, must remain together to retain operon regulation. This could explain the absence of rearrangements within operons; indeed, the similarity of operon structure in *E. coli* K-12 and *S. typhimurium* LT2 is striking. However, genes for related functions which are part of regulons (such as the SOS system) can still be regulated even though they are separated, and therefore rearrangements should be possible. In addition, operon structure and regulation do not explain conserved chromosome order for genes for unrelated functions.

Transcription Orientation

Genes, such as *rrn* operons, with high transcription rates are usually oriented in the same direction for both replication and transcription (13); this will be a conservative force, but it works only for the few genes which are highly

transcribed. In addition, it cannot explain the failure to find wild-type strains which have translocations of I-*Ceu*I fragments resulting from homologous recombination between *rrn* operons (48); these translocations normally retain the same orientation for all genes and occur commonly in culture, yet they are seldom found in wild-type strains.

Chi Sequences

The chi sequences which activate the RecBCD recombinase are present at a 60-fold excess of the numbers expected according to base composition, and 80% of these sequences are oriented so they will activate a RecBCD complex moving toward the origin of replication from either side (14); this should be a conservative force, because rearrangements may reduce the degree of overabundance of the specific orientation (68).

Superhelical Context

There are about 50 domains of localized supercoiling in the *E. coli* chromosome (75, 81). The degree of supercoiling is an important force in controlling gene expression (27, 64). If the domains have relatively fixed ends, and if each domain has a characteristic level of supercoiling which influences gene expression, then this differential gene expression could be a conservative force which acts to retain blocks of genes with otherwise-unrelated functions within the same domain in the chromosome, because the level of gene expression determined by the supercoiling is adaptive; rearrangements which move the domain would be at a selective disadvantage (16). Experimental studies indicate that the level of supercoiling does not differ significantly among chromosomal domains, as measured by gene expression of supercoiling-sensitive promoters, in either *S. typhimurium* (62) or *E. coli* (55), indicating that the domains do not have fixed ends. It can be argued that this makes it less likely that the superhelical structure of domains is a significant force in conserving gene order. However, levels of supercoiling within the domain can be greatly influenced by local gene expression, as RNA polymerase movement causes overwinding and underwinding of the DNA as it moves along the chromosome, so superhelical context could have a substantial role even in the absence of fixed ends for the domains (16).

Sequence Features Required for Chromosome Mechanics

Since specific sequences, in specific places, may be needed for orderly folding, replication timing, and partitioning of the chromosome, Roth et al. (68) postulated that large rearrangements might alter the positions of these sequence features and thus reduce the competitive advantage of a cell. This idea was supported by studies of *S. typhimurium*, which indicate that inversions occur preferentially in regions which are approximately 120 kb apart, corresponding to the size of nucleoid domains; these data predict that interactions occur between distant chromosome regions which have spacial proximity due to nucleoid structure (35).

Maintaining Colinearity as a Site for Recombination To Insert Loops of Foreign DNA

It is possible that transspecies recombination due to conjugation or transduction followed by homologous recombination, though very rare, is so important that colinearity is an important advantage. Ironically, rather than gene-converting shared loci, which is the main result intraspecifically, conjugation and transduction could be critical in providing patches of homology flanking loops so that the loops can be inserted into new sites. Without extensive colinearity of flanking regions, loops require mechanisms of insertion which may occur less frequently; thus, colinear genomes would be more fit in the long run. In this way, the genomes might form a more efficient coevolving cluster of exchangeable and partly overlapping blocks of genes.

Each of these forces could reduce the competitiveness of strains with rearranged genomes, but even with all these forces mobilized, it still seems remarkable that the genomes of strains in the genera *Escherichia* and *Salmonella*, which diverged more than 100 million years ago and are separated by many billions of cell divisions, have retained such similarity of gene order. For example, the operons controlling *ilv* synthesis and the *metE* gene are close to each other in the same order on the chromosome of both genera. In addition, the I-*Ceu*I-F fragment, which includes the genes between the *rrn*B and *rrn*E operons, has retained a size of about 44 kb that is indistinguishable in hundreds of different strains of *Salmonella* and *E. coli* which we have analyzed, indicating that there have been no detectable insertions or deletions in any of these strains (48a). The forces described above do not seem powerful enough to explain this degree of conservation; there may be other forces not yet discovered.

Genome maps of the cat (*Felis catus*) and of humans reveal a high degree of conservation of linked homologous genes on the same chromosome in both species (called "syntenic conservation") (58). The genes present on two chromosomes seem to be totally conserved; the actual order of genes on the chromosome is not yet known in cats (58). Human-mouse conservation is also striking, though it is not as extensive. After determining the number of homology segments in several animal species, and considering that carnivores and primates shared a common ancestor 65 to 80 million years ago, it was calculated that a single chromosome break was established in the genome every 10 to 12 million years; this is an extremely slow rate of chromosome evolution (58). Thus, the conservation of gene order found in enteric bacteria during a divergence of up to 160 million years is also present in carnivores and primates during a divergence of 65 to 80 million years; it is not yet possible to determine if the forces maintaining gene order in these two very unrelated groups are the same.

THERE ARE IMPORTANT GENETIC REARRANGEMENTS EVEN IN THE "STABLE" ENTERIC BACTERIA

The order of homologous genes in the linkage maps of *S. typhimurium* LT2 (70) and *E. coli*

K–12 (3) has long been recognized to be almost identical (69). The similarities were emphasized because initial studies, done in the tradition of the early studies of Beadle and Tatum (20), emphasized housekeeping genes for biosynthesis of amino acids, purines and pyrmidines, and vitamins; these genes were usually present in both genera, and they were found to be in the same order on the chromosome. It is now clear that this identity of gene order concealed many other important differences, especially in nonhousekeeping genes. Rearrangements may produce new forms at such high frequency that they are detectable within a culture; more commonly it results in changes which are detectable only on a longer, evolutionary timescale. The following are some of the types of rearrangements which have been found in the enteric bacteria.

Frequently Rearranged Cassettes of Genes or Parts of Genes Which Control Variable Surface Properties

The genus *Salmonella* is separated into many species in the Kauffmann-White scheme, which is based on variable surface properties; these properties are determined by the O antigen, or lipopolysaccharide (LPS), and by the H antigen, or flagella (31). *E. coli* and *Salmonella* populations were shown to be basically clonal by the following data: the presence of strong linkage disequilibrium among alleles for genes which control housekeeping enzymes (established by work with multilocus enzyme electrophoresis [38, 57, 72]), the association of specific O and H serotypes with only one or a small number of multilocus enzyme genotypes, and the global distribution of specific genotypes. The rate of recombination which would give rise to allelic exchange is sufficiently low that most of the species are single electrophoretic types or families of closely related electrophoretic types. Thus, the existence of these clones shows why serotyping is a valuable system in defining groups of strains with related properties of host range and pathogenicities. In addition, the sequences of the allelic forms of the genes, such as *putP, aceK,*

gapA, mdh, and *icd,* which control housekeeping functions (normally by means of cytoplasmic proteins) are highly related within *Salmonella* or *E. coli* populations, or even between these populations (57, 72).

However, though the alleles of most genes retain similar sequences in different strains, the two groups of genes which control surface properties for LPS and for flagella show significant differences within each genus. Cassettes of genes which control the synthesis of specific nucleotide sugars and their transfer onto the LPS core result in the synthesis of many unique LPS molecules, resulting in many different O (somatic) antigens recognized through unique sugars on the cell surfaces. These cassettes are inserted at the *rfb* gene cluster at a specific region on the chromosome; species with different LPSs have quite different cassettes, with the result that even closely related species of *Salmonella* may have completely nonhomologous genes in this specific part of the chromosome (65, 82).

Variation in the flagellar antigens in *Salmonella* is controlled in a different but related way. *fliC*, which is the structural gene for the phase-1 flagellin, has a region of high variability in the center of the gene, surrounded by less variable regions; this variability is the basis for the antigenic variability of the flagella which are found in different species of *Salmonella* (78). In some of the species, this antigenic variation has been shown to be due to mutation in *fliC* (22, 72); in other cases it apparently results from lateral transfer of the *fliC* gene to different species within the genus (78). The central region of the gene is much more variable than those of other housekeeping genes in both length and sequence (22, 72).

Thus, the chromosomes of the many strains within a genus such as *Salmonella* are a mosaic in which housekeeping genes are functionally indistinguishable and usually show low variability in the corresponding amino acid sequences, but other blocks of genes, especially those controlling surface properties such as LPS or flagella, are related but often quite different. Thus, every strain of *Salmonella* examined so far has a form of the *rfb* gene (which controls the structure of the LPS O antigen) at the same location on the chromosome; these could be called alleles, in terms of their locations and overall functions, but in some cases they are almost entirely nonhomologous and so are not allelic in the normal sense. In the same fashion, very different "alleles" of the same genes are present at the same locus on the *E. coli* chromosome.

Site-Specific DNA Inversion Systems

Site-specific DNA inversion systems are widespread, occurring in genera in which the overall gene order is stable, and they result in phenotypic switches at rates as high as 10^{-3} to 10^{-4}. The Hin recombinase system for flagellar phase variation in *S. typhimurium* (and in many other species of *Salmonella*) (74), systems for phase variation in fimbriae of *E. coli* (10), the "shufflon" (Rci) system in IncI plasmids of enteric bacteria (32), and the Gin and Cin systems for phase variation in phages of *E. coli* (summarized in reference 20) all show that systems resulting in localized switches in the genome, producing phenotypic changes, are common in enteric bacteria.

Rearrangments in the *Ter* Region

The *Ter* region is the only region in which inversions in wild-type strains of *Salmonella* are common (other than inversions involving *rrn* genes, observed in a few species and discussed below). Inversions overlap the *Ter* region in the following species, compared with *S. typhimurium* (the size of the inversion in kilobases is shown): *E. coli*, 480 kb (3, 5); *S. enteritidis*, 750 kb (42); *S. typhi*, 500 kb; and *S. paratyphi* C, 700 kb (26) (Fig. 3). No inversion was detected in *S. paratyphi* B (39). All these inversions differ by at least one endpoint. Inversions at many points in the genome are detected in experimental studies of *E. coli* K-12 and *S. typhimurium* LT2, though some endpoints are much less common than others (50, 68). It has been shown that the *E. coli* chromosome contains recombination sites that are targets for the recombinases XerC and XerD of the lambda integrase family (7); these sites

for recombination are at *dif* in the terminus region of the genetic map between *TerA* and *TerC* (17, 36), but the endpoints for the inversions reported above do not seem to be in *dif*. There is also RecA-dependent homologous recombination in the *Ter* region, resulting in deletions (49). It is likely that these types of recombination are responsible for the inversions detected in the *Ter* region in wild-type strains of *Salmonella* species.

Gene Amplification

Chromosomal duplications are common in groups such as *S. typhimurium* (1, 68) and *E. coli* (29), which are normally considered to be stable, as well as in others, such as *Bacillus subtilis* and *Haemophilus influenzae* (67); they are readily selected in the laboratory and are presumably present in nature. Repeated sequences, such as *rrn* operons (1, 29), insertion sequences like IS*200* (25), repetitive extragenic palindromic (REP) elements, and *rhs* genes, provide the substrates for *recA*-dependent recombination, producing tandem duplications (reviewed in reference 67). These tandem duplications can be found in almost all parts of the chromosome; in *S. typhimurium* they may be as frequent as 3% of the cells, and the duplications can be as large as 50% of the chromosome (2). The data summarized by Roth et al. (68) indicate that duplications, especially in regions bounded by repeated sequences such as *rrn* operons, are rather common in natural populations of some species, but they are not thought to be stable and usually revert to the haploid form.

Rearrangement in the *rrn* Operon

Based on genomic maps for the enzymes *Xba*I, *Bln*I, *Spe*I, and I-*Ceu*I, rearrangements due to inversions and translocations resulting from homologous recombination between the *rrn* operons are very rare in most species; the I-*Ceu*I fragments are in the order ABCDEFG in each of the species *S. typhimurium* (40) (Fig. 1), *S. paratyphi* B (39), and *S. enteritidis* (42), the same order as in *E. coli* K-12 (3). Because I-*Ceu*I cleaves only *rrn* operons and because

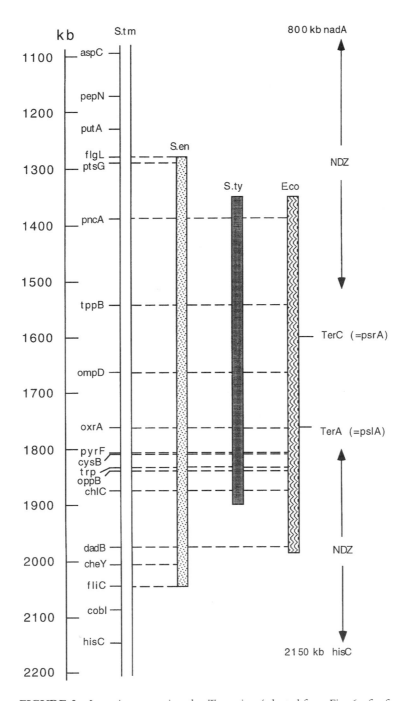

FIGURE 3 Inversions covering the *Ter* region (adapted from Fig. 6 of reference 47). The open vertical bar and genes on the left (S.tm) show the order of genes in *S. typhimurium* LT2, with the positions of these genes shown in kilobases (40). The shorter bars to the right indicate segments of the chromosomes of *S. enteritidis* (S.en) (42), *S. typhi* (S.ty) (47), and *E. coli* K-12 (E.co) (5), which are inverted. The hatched horizonal lines join homologous genes. *TerA* (equivalent to *pslA*) and *TerC* (equivalent to *plrA*) indicate the locations of replication termination in *E. coli* K-12 (7, 17, 36, 49).

the *rrn* skeleton is highly conserved in enteric bacteria, related wild-type strains usually yield identical fingerprints (Fig. 2) (46). However, in some of the species of *Salmonella*, especially those which are host specialized, the order of I-*Ceu*I fragments is rearranged, although the order of genes on individual fragments is similar (47). These rearrangements are attributed to homologous recombination between pairs of the seven *rrn* operons, which are 6-kb blocks of repetitive DNA in a chromosome in which most of the DNA is not similar enough to recombine efficiently. For example, the order of I-*Ceu*I fragments in *S. typhi* Ty2 (limited to growth in humans) is changed to AGCEFDB. The *rrn* skeletons of 127 wild-type strains of *S. typhi* were analyzed by partial I-*Ceu*I digestion (Fig. 4) and showed 21 different orders, which we called "genome types," postulated to be due to inversions and translocations with *rrn* endpoints (Fig. 5) (48). The genomes of strains of *S. paratyphi* A (45) and C (26) (both also limited to growth in humans) are similarly rearranged, as are the genomes of *Salmonella gallinarum* and *Salmonella pullorum* (limited to growth in fowl). The order of I-*Ceu*I fragments in the chromosomes of other species of *Salmonella*, most of which are generalists able to grow in many hosts, is strongly conserved (48a). The basis for this apparent correlation between genomic rearrangements and host specialization in *Salmonella* spp. is not clear.

Other enteric bacteria also show some rearrangements. The order of genes on the chromosome of *Klebsiella oxytoca* M51a resembles that of *S. typhimurium*, but the chromosome differs as follows: there are eight *rrn* operons instead of seven, one of the I-*Ceu*I fragments is translocated to a new location, and the total size of the chromosome is 5,200 rather than 4,800 kb as in *S. typhimurium*. Forty other wild-type strains of *Klebsiella* spp. all have eight *rrn* operons, and in a few cases there are further translocations of the I-*Ceu*I fragments (48a).

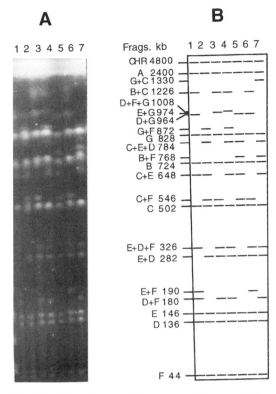

FIGURE 4 (A) Partial digestion of DNA of independent strains of *S. typhi* with the endonuclease I-*Ceu*I, separation by PFGE, and staining with ethidium bromide (from Fig. 2 of reference 48). Lanes: 1, strain 26T4; 2, strain 26T9; 3, strain 26T12; 4, strain 26T19; 5, strain 26T38; 6, strain 26T48, 7, strain 26T49. (B) The fragments shown in panel A are indicated by bars, and their sizes and the fragments they are inferred to include are labelled on the left.

Unique Blocks of DNA from Lateral Transfer (Loops or Pathogenicity Islands)

Though gene order is conserved, the linkage maps of *E. coli* and *S. typhimurium* reveal differences in the lengths of intervals between genes, indicating excess nonhomologous DNA in one of the genera; these are called "loops," reminiscent of the terminology for heteroduplex D loops (34, 66). These differences could have arisen by a gain of genes in one genome or a loss from the other. For example, *E. coli* is Lac[+] and *S. typhimurium* is

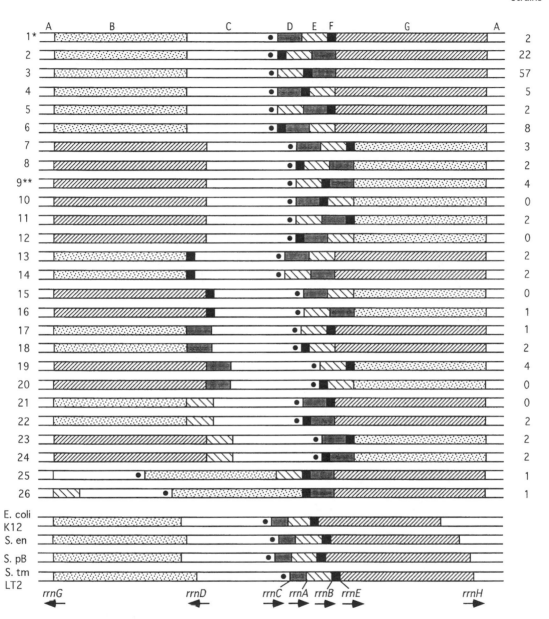

FIGURE 5 The order of I-*Ceu*I fragments in 127 strains of *S. typhi* and in other enteric bacteria, showing 26 different genome types (adapted from Fig. 2 of reference 48). The order of I-*Ceu*I fragments B to G was determined from data of the type shown in Fig. 2. The order of I-*Ceu*I fragments from strain Ty2 (genomic type 9) had been previously determined by analysis with Tn*10* insertions (47). The same order was confirmed by partial I-*Ceu*I digestion (as shown in Fig. 3). The I-*Ceu*I-A fragment joins the left ends of the fragments shown to the right end, forming circles, but their orientations are not known. The orientations of I-*Ceu*I-fragments B, D, E, F, and G can be inferred from the polarities of the *rrn* operons. The solid dots in the I-*Ceu*I fragments indicate the location of *oriC*. The number of strains of each genome type is shown; some of the theoretical genome types were not detected. The order and sizes of the fragments between *rrn* operons were previously determined for *E. coli* K-12 (3), *S. enteritidis* (42), *S. paratyphi* B (39), and *S. typhimurium* LT2 (40). These are illustrated at the bottom, drawn to scale, and all have the same order of fragments as genome type 1 of *S. typhi*.

Lac⁻; the *lac* operon at 8 min in *E. coli* lies within a loop which contains DNA that has no homologue in *S. typhimurium* (15). These loops have been called pathogenicity islands because in many cases they include a block of genes contributing to a specific virulence phenotype (though not all such loops function in pathogenicity). Two *Salmonella* pathogenicity islands (SPI), both about 40 kb, have been identified; SPI-1 governs the ability of *Salmonella* to invade epithelial cells (56), and SPI-2 mediates survival within macrophages (24, 59). SPI-1 is present in all species of *Salmonella*, while SPI-2 is present in all except *Salmonella bongori* (24). Pathogenicity islands have also been identified in *E. coli* strains of the enteropathogenic *E. coli* group and the uropathogenic *E. coli* group, as well as in *Yersinia pestis* and *Vibrio cholerae* (reviewed in reference 43). In some cases the islands are targeted to regions which contain a tRNA gene, such as the *selC* gene for selenocysteine-specific tRNA for PAI-1 in uropathogenic *E. coli* (11), locus for enterocyte effacement in enteropathogenic *E. coli* (54) strains, or the *leuX* gene for leucine tRNA in uropathogenic *E. coli* (76). In other cases there is no repeat sequence or gene which could explain the basis for the location (56).

Chromosomes of individual species can differ by loops of this type, and PFGE analysis has identified these loops. Firstly, genetic maps of *Salmonella* show that intervals between the same pair of genes in different species may be very different. For example, in *S. typhi* the *mel-poxA* interval is 120 kb longer than those in *S. typhimurium* and other species; this loop was found to carry the *viaB* gene for the Vi antigen, missing from most other *Salmonella* species (47, 48). This suggests the existence of a pathogenicity island; the existence of unique DNA in this region was confirmed, and it was shown to carry *rci* genes for site-specific recombination (83). Secondly, changes in the lengths of the I-*Ceu*I fragments indicate insertions or deletions; e.g., the I-*Ceu*I-D fragment is 136 kb in *S. typhi* but only 92 kb in all other *Salmonella* species examined so far, indicating

an insertion of about 44 kb in this region (47). Bloch and Rode (9) identified pathogenicity islands in *E. coli* K1 by crossing it with *E. coli* K-12; they showed that a segment of the chromosome of K1 controlled invasiveness. Thus, lateral transfer resulting in loops of species-specific or genus-specific DNA is a major genetic mechanism resulting in species adaptation in bacteria. It involves entry of nonhomologous genes, presumably by classical genetic transfer methods, such as conjugation and transduction, and integration of these genes into the chromosome.

HOW WIDESPREAD IS GENOME STABILITY?

It is still difficult to reconcile all the conflicting data on genome stability versus plasticity. Within the enteric bacteria, many investigators, including ourselves, have emphasized the overall stability of the genomic maps in different species of *Salmonella* and in different genera, such as *Salmonella* and *Escherichia*. This stability was first seen in genetic maps and later confirmed by PFGE. But when a wider range of genera is considered, the picture changes. Watanabe et al. (79) identified and compared orthologous gene sets among the sequenced DNA of *H. influenzae*, *Mycoplasma genitalium*, *E. coli*, and *B. subtilis*. These comparisons of the gene order indicate that rearrangements have occurred so frequently during evolution that the overall genomic structures of the different genera look unrelated. In fact, even the operons have usually become reorganized, even between the most closely related pair, *H. influenzae* and *E. coli*. There are a few exceptional regions showing almost complete conservation in all genera, especially the S10 region, which codes for 26 ribosomal proteins; the RNA polymerase alpha subunit; and the *secY* subunit (79).

The apparent stability of the genome in *E. coli* and *S. typhimurium* detected from genetic data has been confirmed by physical analysis by PFGE (as described above), but it has not yet been rigorously tested by comparisons of all the nucleotide sequences in the two genera.

The genome of *E. coli* was completely sequenced in 1997 (8), though much of the sequence was available for several years before that. StySeq1 is a nonredundant DNA sequence database of *S. typhimurium* LT2 assembled by Kenn Rudd in 1995 and reported in reference 70; it represented 548 kb of nonredundant DNA (about 11% of the entire genome) and comprised 197 contigs. Orthologous genes in the aligned DNA sequences were compared in the two genera, and most comparisons indicated a close correlation between the orientations and map positions of genes in *E. coli* and *S. typhimurium* (70). However, a comparison of the complete nucleotide sequences of both genera is urgently needed to reveal the degree of conservation of the two genera. At this time, a complete genomic sequence is not available for any of the species of *Salmonella*, though the portion of the genome of *S. typimurium* summarized as StySeq1 (70) has been deposited in GenBank, and a significant amount of random sequence of *S. typhimurium* and *S. typhi* has been collected and deposited by the Genome Sequencing Center at Washington University, St. Louis, Mo., at http://genome.wustl.edu/gsc/bacterial/salmonella.html.

SUMMARY

We conclude that the overall conservation of gene order within the enteric bacteria which was reported many years ago in comparisons of *E. coli* and *S. typhimurium* has been confirmed by the analysis of the physical maps of many strains of *Salmonella* and other enteric bacteria determined by PFGE and by the comparison of nucleotide sequences. However, data from many laboratories show that this conservation does not extend to other eubacteria; the chromosomes of even the relatively closely related species *E. coli* and *H. influenzae* are substantially rearranged.

Genetic rearrangements occur at suprisingly high rates (up to 10^{-3} to 10^{-4}) in cells of the enteric bacteria cultured in laboratory medium, and presumably in nature; therefore, infrequent detection of rearrangements in wild-type strains must be due to failure of the rearranged types to compete successfully in nature. Several forces have been proposed which could be responsible for the conservation of gene order in the enteric bacteria. These include the following: the need for genomic balance in chromosome replication, an influence of gene dosage on gene expression, the effect of operon structure on gene expression, the influence of transcription orientation, the effect of superhelicity within chromosome domains on gene expression, the influence on chromosome mechanics of sequence features required for folding, and a need for colinearity to provide a site for recombination to insert loops of foreign DNA. Each of these forces could have an influence, but the degree of conservation of gene order of apparently unrelated genes may not be fully explained, even by all these forces.

We should not ignore the many types of genomic instability that operate in the enteric bacteria. Site-specific DNA inversion systems result in switching from phase 1 to phase 2 flagella at a rate up to 10^{-3}, so that all cell cultures are genetically and phenotypically mixed for genes controlled by these systems, and there are several such systems. The different species (or serovars) of *Salmonella*, which contain mostly very homologous genes, frequently have nonhomologous genes for control of the LPS antigens. Gene amplification can result in multiple copies of parts of the genome. Rearrangements are common in the *Ter* region for termination of replication. Inversions and translocations due to homologous recombination between *rrn* operons are commonly found in wild-type strains of a few species of *Salmonella*, such as *S. typhi*. And most importantly, unique blocks of DNA (called loops or pathogenicity islands) are inserted into the chromosome, presumably as a result of lateral transfer from other genera. All of these instabilities are undoubtedly important as genetic mechanisms which result in species and genera which are better adapted to new or changing environments.

ACKNOWLEDGMENTS

This work was supported by a grant from the Natural Sciences and Engineering Research Council of Canada and by grants RO1AI34829 from the National Institute of Allergy and Infectious Diseases of the National Institutes of Health (Bethesda, Md.) (to M.M. and K.E.S.) and RO1AI43283 (from the same agency).

REFERENCES

1. **Anderson, R. P., and J. R. Roth.** 1978. Gene duplication in bacteria: alteration of gene dosage by sister chromosome exchange. *Cold Spring Harbor Symp. Quant. Biol.* **43:**1083–1087.
2. **Anderson, R. P., and J. R. Roth.** 1981. Spontaneous tandem genetic duplications in *Salmonella typhimurium* arise by unequal recombination between ribosomal RNA (*rrn*) cistrons. *Proc. Natl. Acad. Sci. USA* **78:**3113–3117.
3. **Bachmann, B. J.** 1990. Linkage map of *Escherichia coli* K-12, edition 8. *Microbiol. Rev.* **54:** 130–197.
4. **Beltran, P., S. A. Plock, N. H. Smith, T. S. Whittam, D. C. Old, and R. K. Selander.** 1991. Reference collection of strains of the *Salmonella typhimurium* complex from natural populations. *J. Gen. Microbiol.* **137:**601–606.
5. **Berlyn, M. B., K. B. Low, and K. E. Rudd.** 1996. Integrated linkage map of *Escherichia coli* K-12, edition 9, p. 1715–1902. In F. C. Neidhardt, R. Curtiss III, J. L. Ingraham, E. C. C. Lin, K. B. Low, B. Magasanik, W. Reznikoff, M. Riley, M. Schaechter, and H. E. Umbarger (ed.), *Escherichia coli* and Salmonella: *Cellular and Molecular Biology.* 2nd ed. American Society for Microbiology, Washington, D.C.
6. **Bhagwat, A. S., and M. McClelland.** 1992. DNA mismatch correction by very short patch repair may have altered the abundance of oligonucleotides in the *E. coli* genome. *Nucleic Acids Res.* **20:**1663–1668.
7. **Blakely, G., G. May, R. McCulloch, and L. K. Arciszewaska.** 1993. Two related recombinases are required for chromosomal segregational cell division. *New Biol.* **8:**789–798.
8. **Blattner, F. R., G. Plunkett III, C. A. Bloch, N. T. Perna, V. Burland, M. Riley, J. Collado-Vides, J. D. Glasner, C. K. Rode, G. F. Mayhew, J. Gregor, N. W. Davis, H. A. Kirkpatrick, M. A. Goeden, D. J. Rose, B. Mau, and Y. Shao.** 1997. The complete genome sequence of *Escherichia coli* K-12. *Science* **277:**1453–1462.
9. **Bloch, C. A., and C. K. Rode.** 1996. Pathogenicity island evaluation in *Escherichia coli* K1 by crossing with laboratory strain K-12. *Infect. Immun.* **64:**3218–3223.
10. **Blomfield, I. C., M. S. McClain, J. A. Princ, P. J. Calie, and B. A. Eisenstein.** 1991. Type 1 fimbriation and *fimE* mutants of *Escherichia coli* K-12. *J. Bacteriol.* **173:**5298–5307.
11. **Blum, G., M. Ott, A. Lischewski, A. Ritter, H. Imrich, H. Tschape, and J. Hacker.** 1994. Excision of large DNA regions termed pathogenicity islands from tRNA-specific loci in the chromosome of an *Escherichia coli* wild-type pathogen. *Infect. Immun.* **62:**606–614.
12. **Boyd, E. F., F. S. Wand, P. Beltran, S. A. Plock, K. Nelson, and R. K. Selander.** 1993. *Salmonella* reference collection B (SARB): strains of 37 serovars of subspecies I. *J. Gen. Microbiol.* **139:**1125–1132.
13. **Brewer, B. J.** 1988. When polymerases collide: replication and the transcriptional organization of the *E. coli* chromosome. *Cell* **53:**679–686.
14. **Burland, V., F. Plunkett III, D. Daniels, and F. R. Blattner.** 1993. DNA sequence and analysis of 136 kilobases of *Escherichia coli* genome: organizational symmetry around the origin of replication. *Genomics* **16:**551–561.
15. **Buvinger, W. E., K. A. Lampel, R. J. Bojanowski, and M. Riley.** 1984. Location and analysis of nucleotide sequences at one end of a putative *lac* transposon in the *Escherichia coli* chromosome. *J. Bacteriol.* **159:**618–623.
16. **Charlebois, R. L., and A. St. Jean.** 1995. Supercoiling and map stability in the bacterial chromosome. *J. Mol. Evol.* **41:**5–23.
17. **Cornet, F., J. Louarn, J. Patte, and J.-M. Louarn.** 1996. Restriction of the activity of the recombination site *dif* to a small zone of the *Escherichia coli* chromosome. *Genes Dev.* **10:**1152–1161.
18. **Crosa, J. H., D. J. Brenner, W. H. Ewing, and S. Falkow.** 1973. Molecular relationships among the *Salmonellae. J. Bacteriol.* **115:**307–315.
19. **Dempsey, J. A., A. B. Wallace, and J. G. Cannon.** 1995. The physical map of the chromosome of a serogroup A strain of *Neisseria meningitidis* shows complex rearrangements relative to the chromosomes of the two mapped strains of the closely related species *N. gonorrhoeae. J. Bacteriol.* **177:**6390–6400.
20. **Dybvig, K.** 1993. DNA rearrangements and phenotypic switching in prokaryotes. *Mol. Microbiol.* **10:**465–471.
21. **Fonstein, M., and R. Haselkorn.** 1995. Physical mapping of bacterial genomes. *J. Bacteriol.* **177:**3361–3369.
22. **Frankel, G., S. M. C. Newton, G. K. Schoolnik, and B. A. D. Stocker.** 1989. Intragenic recombination in a flagellin gene: char-

acterization of the H1-j gene of *Salmonella typhi*. *EMBO J.* **8**:3149–3152.

23. **Gibbs, C. P. and T. F. Meyers.** 1996. Genome plasticity in *Neisseria gonorrhoeae*. *FEMS Microbiol. Lett.* **145**:173–179.

24. **Groisman, E. A., and H. Ochman.** 1996. Pathogenicity islands: bacterial evolution in quantum leaps. *Cell* **87**:791–794.

25. **Haack, K., and J. Roth.** 1995. Recombination between chromosomal IS*200* elements supports frequent duplication formation in *Salmonella typhimurium*. *Genetics* **141**:1231–1243.

26. **Hessel, A., S.-L. Liu, and K. E. Sanderson.** 1995. The chromosome of *Salmonella paratyphi* C contains an inversion and is rearranged relative to *S. typhimurium* LT2, p. 503. In *Abstracts of the 95th General Meeting of the American Society for Microbiology. 1995.* American Society for Microbiology, Washington, D.C.

27. **Higgins, C. F., C. J. Dorman, D. A. Stirling, L. Waddell, I. R. Booth, G. May, and E. Bremer.** 1988. A physiological role for DNA supercoiling in the osmotic regulation of gene expression in *S. typhimurium* and *E. coli. Cell* **52**:569–584.

28. **Hill, C. W., and J. A. Gray.** 1988. Effects of chromosomal inversion on cell fitness in *Escherichia coli* K-12. *Genetics* **119**:771–778.

29. **Hill, C. W., and B. W. Harnish.** 1981. Inversions between ribosomal RNA genes of *Escherichia coli. Proc. Natl. Acad. Sci. USA* **78**:7069–7072.

29a.**Hill, T. M.** 1996. Features of the chromosomal terminus region, p. 1602–1614. *In* F. C. Neidhardt, R. Curtiss III, J. L. Ingraham, E. C. C. Lin, K. B. Low, B. Magasanik, W. S. Reznikoff, M. Riley, M. Schaechter, and H. E. Umbarger (ed.), Escherichia coli *and* Salmonella: *Cellular and Molecular Biology*, 2nd ed. American Society for Microbiology, Washington, D.C.

30. **Honeycutt, B. W., M. McClelland, and B. W. Sobral.** 1993. Physical map of the genome of *Rhizobium meliloti* 1021. *J. Bacteriol.* **175**:6945–6952.

31. **Kauffman, F.** 1966. The bacteriology of Enterobacteriaceae. Williams and Wilkens, Baltimore, Md.

32. **Kim, S.-R., and T. Komano.** 1992. Nucleotide sequence of the R721 shufflon. *J. Bacteriol.* **174**:7053–7058.

33. **Kolsto, A.-B.** 1997. Dynamic bacterial genome organization. *Mol. Microbiol.* **24**:241–248.

34. **Krawiec, S., and M. Riley.** 1990. Organization of the bacterial genome. *Microbiol. Rev.* **54**:502–539.

35. **Krug, P. J., A. Z. Gileski, R. J. Code, A. Torjussen, and M. B. Schmid.** 1994. End-

point bias in large Tn*10*-catalyzed inversions in *Salmonella typhimurium. Genetics* **136**:747–756.

36. **Kuempel, P. L., J. M. Hensen, L. Dircks, M. Tecklenburg, and D. F. Lim.** 1991. *dif*, a *recA*-independent recombination site in the terminus of the chromosome of *Escherichia coli. New Biol.* **3**:799–811.

37. **Le Minor, L.** 1988. Typing of *Salmonella* species. *Eur. J. Clin. Microbiol. Infect. Dis.* **7**:214–218.

38. **Li, J., K. Nelson, A. C. McWhorter, T. S. Whittam, and R. K. Selander.** 1994. Recombinational basis of serovar diversity in Salmonella enterica. *Proc. Natl. Acad. Sci. USA* **91**:2552–2556.

39. **Liu, S.-L., A. Hessel, H.-Y. M. Cheng, and K. E. Sanderson.** 1994. The *XbaI-BlnI-CeuI* genomic cleavage map of *Salmonella paratyphi* B. *J. Bacteriol.* **176**:1014–1024.

40. **Liu, S.-L., A. Hessel, and K. E. Sanderson.** 1993. The *XbaI-BlnI-CeuI* genomic cleavage map of *Salmonella typhimurium* LT2 determined by double digestion, end-labelling, and pulsed-field gel electrophoresis. *J. Bacteriol.* **175**:4104–4120.

41. **Liu, S.-L., A. Hessel, and K. E. Sanderson.** 1993. Genomic mapping with I-*CeuI*, an intron-encoded endonuclease, specific for genes for ribosomal RNA, in *Salmonella* spp., *Escherichia coli*, and other bacteria. *Proc. Natl. Acad. Sci. USA* **90**:6874–6878.

42. **Liu, S.-L., A. Hessel, and K. E. Sanderson.** 1993. The *XbaI-BlnI-CeuI* genomic cleavage map of *Salmonella enteritidis* shows an inversion relative to *Salmonella typhimurium* LT2. *Mol. Microbiol.* **10**:655–664.

43. **Liu, S.-L., C. P.-F. Qi, V. Stewart, and K. E. Sanderson.** 1997. A genome map of *Klebsiella oxytoca* M51a, p. 319. *In Abstracts of the 97th General Meeting of the American Society for Microbiology 1997.* American Society for Microbiology, Washington, D.C.

44. **Liu, S.-L., and K. E. Sanderson.** 1992. A physical map of the *Salmonella typhimurium* LT2 genome made by using *XbaI* analysis. *J. Bacteriol.* **174**:1662–1672.

45. **Liu, S.-L., and K. E. Sanderson.** 1995. The chromosome of *Salmonella paratyphi* A is inverted by recombination between *rrnH* and *rrnG. J. Bacteriol.* **177**:6585–6592.

46. **Liu, S.-L., and K. E. Sanderson.** 1995. I-*CeuI* reveals conservation of the genome of independent strains of *Salmonella typhimurium. J. Bacteriol.* **177**:3355–3357.

47. **Liu, S.-L., and K. E. Sanderson.** 1995. The genomic cleavage map of *Salmonella typhi* Ty2. *J. Bacteriol.* **177**:5099–5107.

48. **Liu, S.-L., and K. E. Sanderson.** 1996. Highly plastic chromosomal organization in *Salmonella typhi. Proc. Natl. Acad. Sci. USA* **93:**10303–10308.

48a.**Liu, S.-L., and K. E. Sanderson.** Unpublished data.

49. **Louarn, J., F. Cornet, V. Francois, J. Patte, and J.-M. Louarn.** 1994. Hyperrecombination in the terminus region of the *Escherichia coli* chromosome: possible relation to nucleoid organization. *J. Bacteriol.* **176:**7524–7531.

50. **Mahan, M. J., A. M. Segall, and J. R. Roth.** 1990. Recombination events that rearrange the chromosome: barriers to inversion, p. 341–349. In K. Drlica and M. Riley (ed.), *The Bacterial Chromosome.* American Society for Microbiology, Washington, D.C.

51. **Marshall, P., T. B. Davis, and C. Lemieux.** 1994. The I-*Ceu*I endonuclease: purification and potential role in the evolution of *Chlamydomonas* group I introns. *Eur. J. Bacteriol.* **220:**855–859.

52. **Marshall, P., and C. Lemieux.** 1991. Cleavage pattern of the homing endonuclease encoded by the fifth intron in the chloroplast subunit rRNA-encoding gene of *Chlamydomonas eugametos. Gene* **104:**1241–1245.

53. **McClelland, M., R. Jones, Y. Patel, and M. Nelson.** 1987. Restriction endonucleases for pulsed field mapping of bacterial genomes. *Nucleic Acids Res.* **15:**5085–6005.

54. **McDaniel, T. K., K. G. Jarvis, M. S. Donnenberg, and J. B. Kaper.** 1995. A genetic locus of enterocyte effacement conserved among diverse enterobacterial pathogens. *Proc. Natl. Acad. Sci. USA* **92:**1664–1668.

55. **Miller, W. G., and R. W. Simons.** 1993. Chromosomal supercoiling in *Escherichia coli. Mol. Microbiol.* **10:**675–684.

56. **Mills, D. M., V. Balaj, and C. A. Lee.** 1995. A 40 kilobase chromosomal fragment encoding *Salmonella typhimurium* invasion genes is absent from the corresponding region of the *Escherichia coli* chromosome. *Mol. Microbiol.* **15:**749–759.

57. **Nelson, K., F.-S. Wang, E. F. Boyd, and R. K. Selander.** 1997. Size and sequence polymorphism in the isocitrate dehydrogenase kinase/phosphatase gene (*aceK*) and flanking regions in *Salmonella enterica* and *Escherichia coli. Genetics* **147:**1509–1520.

58. **O'Brien, S. J., J. Wienberg, and L. A. Lyons.** 1997. Comparative genomics: lessons from cats. *Trends Genet.* **13:**393–398.

59. **Ochman, H., F. C. Soncini, F. Solomon, and E. A. Groisman.** 1996. Identification of a pathogenicity island required for *Salmonella* survival in host cells. *Proc. Natl. Acad. Sci. USA* **93:**7800–7804.

60. **Ochman, H., and A. C. Wilson.** 1987. Evolutionary history of enteric bacteria, p. 1649–1654. In F. C. Neidhardt, J. L. Ingraham, K. B. Low, B. Magasanik, M. Schaechter, and H. E. Umbarger (ed.), Escherichia coli and Salmonella typhimurium: *Cellular and Molecular Biology.* American Society for Microbiology, Washington, D.C.

61. **Okada, N., C. Sasakawa, T. Tobe, K. A. Talukder, K. Komatsu, and M. Yoshikawa.** 1991. Construction of a physical map of the chromosome of *Shigella flexneri* 2a and the direct assignment of nine virulence-associated loci identified by Tn*5* insertions. *Mol. Microbiol.* **5:**2171–2180.

62. **Pavitt, G. D., and C. F. Higgins.** 1993. Chromosomal domains of supercoiling in *Salmonella typhimurium. Mol. Microbiol.* **10:**685–696.

63. **Popoff, M. Y., J. Bockemuel, and A. McWhorter-Murlin.** 1993. Supplement 1992 (no. 36) to the Kauffmann-White Scheme. *Res. Microbiol.* **144:**495–498.

64. **Pruss, G. J., and K. Drlica.** 1989. DNA supercoiling and prokaryotic transcription. *Cell* **56:**521–523.

65. **Reeves, P. R.** 1993. Evolution of O antigen variation by interspecific gene transfer on a large scale. *Trends Genet.* **9:**17–22.

66. **Riley, M., and K. E. Sanderson.** 1990. Comparative genetics of *Escherichia coli* and *Salmonella typhimurium*, p. 85–95. In K. Drlica and M. Riley (ed.), *The Bacterial Chromosome.* American Society for Microbiology, Washington, D.C.

67. **Romero, D., and R. Palacios.** 1997. Gene amplification and genomic plasticity in prokaryotes. *Annu. Rev. Genet.* **31:**91–111.

68. **Roth, J. R., N. Benson, T. Galitski, K. Haack, J. G. Lawrence, and L. Miesel.** 1996. Rearrangements of the bacterial chromosome: formation and applications, p. 2256–2276. In F. C. Neidhardt, R. Curtiss III, J. L. Ingraham, E. C. C. Lin, K. B. Low, B. Magasanik, W. S. Reznikoff, M. Riley, M. Schaechter, and H. E. Umbarger (ed.), Escherichia coli and Salmonella typhimurium: *Cellular and Molecular Biology*, 2nd ed. American Society for Microbiology, Washington, D.C.

69. **Sanderson, K. E.** 1976. Genetic relatedness in the family *Enterobacteriaceae. Annu. Rev. Microbiol.* **30:**327–349.

70. **Sanderson, K. E., A. Hessel, and K. E. Rudd.** 1995. The genetic map of *Salmonella typhimurium* LT2, edition VIII. *Microbiol. Rev.* **59:** 241–303.

71. **Schmid, M., and J. R. Roth.** 1987. Gene location affects expression level in *Salmonella typhimurium. J. Bacteriol.* **169:**2872–2875.

72. **Selander, R. K., J. Li, and K. Nelson.** 1996. Evolutionary genetics of *Salmonella enterica*, p. 2691–2707. *In* F. C. Neidhardt, R. Curtiss III, J. L. Ingraham, E. C. C Lin, K. B. Low, B. Magasanik, W. S. Reznikoff, M. Riley, M. Schaechter, and H. E. Umbarger (ed.), Escherichia coli *and* Salmonella*: Cellular and Molecular Biology*, 2nd ed. American Society for Microbiology, Washington, D.C.

73. **Sharp, P.** 1991. Determinants of DNA sequence divergence between *Escherichia coli* and *Salmonella typhimurium*: codon usage, map position, and concerted evolution. *J. Mol. Evol.* **33:**23–33.

74. **Silverman, M., and M. Simon.** 1980. Phase variation: genetic analysis of switching mutants. *Science* **19:**845–854.

75. **Sinden, R. R., and D. E. Pettijohn.** 1981. Chromosomes in living *Escherichia coli* cells are segregated into domains of supercoiling. *Proc. Natl. Acad. Sci. USA* **78:**224–228.

76. **Sirisena, D. M., P. R. MacLachlan, S. L. Liu, A. Hessel, and K. E. Sanderson.** 1994. Molecular analysis of the *rfaD* gene, for heptose synthesis, and the *rfaF* gene, for heptose transfer, in lipopolysaccharide synthesis in *Salmonella typhimurium*. *J. Bacteriol.* **176:**2379–2385.

77. **Smith, C. L., J. Econome, A. Schutt, S. Klco, and C. R. Cantor.** 1987. A physical map of the *Escherichia coli* K-12 genome. *Science* **236:**1446–1453.

78. **Smith, N. H., P. Beltran, and R. K. Selander.** 1992. Recombination of *Salmonella* phase-1 flagellin genes generates new serovars. *J. Bacteriol.* **172:**2209–2216.

78a.**Techlenburg, M., A. Maummer, O. Nagappan, and P. L. Kuempel.** 1995. The *dif*-resolvase can be replaced by a 33 basepair sequence, but function depends on location. *Proc. Natl. Acad. Sci. USA* **92:**1352–1356.

79. **Watanabe, H., H. Mori, T. Itoh, and T. Gojobori.** 1997. Genome plasticity as a paradigm of eubacterial evolution. *J. Mol. Evol.* **44(suppl.1):**S57–S64.

80. **Wong, K. K., and M. McClelland.** 1992. A *Bln*I restriction map of the *Salmonella typhimurium* LT2 genome. *J. Bacteriol.* **174:**1656–1661.

81. **Worcel, A., and E. Burgi.** 1972. On the structure of the folded chromosome of *Escherichia coli*. *J. Mol. Biol.* **71:**127–147.

82. **Xiang, S.-H., A. M. Haase, and P. R. Reeves.** 1993. Variation in the *rfb* gene clusters in *Salmonella enterica*. *J. Bacteriol.* **175:**4877–4884.

83. **Zhang, X.-L., C. Morris, and J. Hackett.** 1997. Molecular cloning, nucleotide sequence, and function of a site-specific recombinase encoded in the major 'pathogenicity island' of *Salmonella typhi*. *Gene* **202:**139–146.

UNSTABLE LINEAR CHROMOSOMES: THE CASE OF *STREPTOMYCES*

Pierre Leblond and Bernard Decaris

14

The classical distinction between bacteria possessing stable or unstable genomes is based on the comparison of the genetic organizations of strains belonging to the same species. During evolution, the genome has been relatively stable for some species; however, DNA rearrangements are observed at high frequencies in laboratory culture for all species. The stability of the genome is dependent on the strength of the selective pressures that eliminate chromosome rearrangements. This pressure may also vary along the length of the chromosome and give rise to regions that are more stable than others.

In this chapter, large-scale DNA rearrangements, including deletions, amplifications, and other DNA alterations such as interchromosomal interactions, will be dealt with and special attention will be given to the instability of *Streptomyces* species. Specific interest in *Streptomyces* species is merited by the extent of intraclonal instability seen and also the strong intra- and interspecific polymorphism at the ends of the linear chromosomal DNA. Fur-

thermore, genetic instability is shared by all the *Streptomyces* species so far studied and seems to be a characteristic of the genus. By contrast, no such instability or favorable experimental context is available for other bacteria harboring linear chromosomes (*Borrelia* and *Agrobacterium*).

Site-specific recombination, such as phase inversion, and integrative elements (plasmids or bacteriophages) will not be considered, nor will generalized increased mutagenesis (Mut phenotypes).

Questions about the plasticity found at the ends of the *Streptomyces* chromosomal DNA will be addressed. Does the selective pressure for the maintenance of chromosome stability decrease from the origin of replication toward the replication termini, thereby creating a gradient of selective pressure? Are recombination frequencies increased at the ends of the chromosomal DNA compared to the rest of the genome? What could cause this increase in recombination frequency? Is a specific phenomenon taking place at the termini of the linear replicon (e.g., termination of replication or partitioning) important for maintaining the stability of the replicon, or is a specific structure keeping the ends associated?

We will discuss these phenomena by considering the important mechanisms responsible

Pierre Leblond and Bernard Decaris, Laboratoire de Génétique et Microbiologie UA INRA 952, Université Henri Poincaré, Nancy 1, Faculté des Sciences BP 239, 54506 Vandoeuvre-lès-Nancy, France.

Organization of the Prokaryotic Genome, Edited by Robert L. Charlebois,
© 1999 American Society for Microbiology, Washington, D.C.

for chromosomal integrity, notably replication and recombination, and draw some conclusions about their role in the DNA plasticity of the bacterial chromosome.

These large-scale rearrangements may also have evolutionary consequences for the chromosome architecture, and in *Streptomyces*, the high level of localized DNA plasticity could be responsible for the diversity seen at the ends of the linear chromosomal DNA.

STREPTOMYCES AS A MODEL FOR THE STUDY OF GENETIC INSTABILITY AND GENOME REARRANGEMENTS IN BACTERIA

Genetic instability as indicated by the rate of spontaneous mutations of specific traits is much higher in *Streptomyces* (10^{-3} to 10^{-2}) than the classical rates found in other organisms. This phenomenon is ubiquitous within *Streptomyces* species, and the resultant phenotypic variability was described as early as the beginning of the century by Beijerinck (9).

Streptomyces species are distinguished from other bacterial species by their complex biological cycle, the impressive and economically important products of their secondary metabolism, and their genomic characteristics.

Streptomyces species belong to the order *Actinomycetales* and are filamentous gram-positive bacteria living in the soil. They possess a complex life cycle that begins on solid medium by the germination of spores to form the vegetative mycelium. This ramping mycelium produces aerial hyphae as nutrients become limiting. Spores are produced by septation from the apex of the aerial mycelium. While mycelia contain multiple nucleoids, spores harbor a single chromosome (21).

Streptomyces species represent the most impressive reservoir of antibiotic resistance and biosynthesis genes. The genes available for biotechnological use are of considerable economic interest. In addition, the activities of the metabolites produced by the streptomycetes are extremely varied: antibiotics, immunomodifiers, pigments, herbicides, antiviral compounds, and enzyme inhibitors are pro-

duced (87, 113). The availability of numerous resistance genes is also of considerable importance, as they could participate in the spreading of antibiotic resistance in bacteria.

Studying phenotypic instability in *Streptomyces* is particularly interesting due to the organism's genomic characteristics. *Streptomyces* species possess one of the largest genomes among eubacteria, approximately 8 Mb. The chromosomal DNA is linear and shows the typical invertron structure (i.e., the presence of large terminal inverted repeats [TIRs] associated with terminal proteins) (72, 75, 77, 90). The chromosomal DNA molecule is often accompanied by circular plasmids or linear invertronic plasmids. The G+C content is high: 70 to 74% (22), which is at the upper limit of the range seen in bacteria (25 to 75% [67]). This trait leads to a strong bias for the utilization of codons, which allows the rapid identification of genes from a nucleotide sequence of potential open reading frames (11). The genome size of the *Streptomyces* chromosome has been estimated to be approximately 8 Mb for all species studied [*S. coelicolor* A3(2) (62), *S. lividans* 66 (74), *S. griseus* (75), *S. ambofaciens* (72), and *S. rimosus* (90)]. This makes the *Streptomyces* chromosome one of the largest among the bacteria. *Borrela burgdorferi* harbors a 950-kb linear chromosome (18), while the *Agrobacterium tumefaciens* linear replicon is estimated to be about 3 Mb in length (3). Questions can be asked about the biological significance of the size of the genome. It was proposed that genome size is related to the biological complexity of the life cycle. *Streptomyces* can undergo complex cycles of morphological differentiation, and the large genome could help to adapt to fluctuating environments (53).

Specific mutable genes are reproducibly affected by genetic instability (13, 68). The characteristics affected include aerial mycelium formation, spore production, antibiotic resistance and/or production, and formation of extracellular enzymes (i.e., tyrosinase, known to be involved in the pigmentation process in some species). Therefore, instability preferentially affects traits belonging to the differentiation process or to secondary metabolism. Genetic instability is particularly easy to observe under laboratory conditions because of the numerous characters that are not essential and so can be lost without affected viability.

Streptomyces has a further advantage in that spontaneous instability can be increased to 100% in the population by treatments such as UV light exposure, inhibitors of gyrase, cold storage, and protoplast regeneration.

The phenotypic instability is closely associated with genomic rearrangements, such as large deletions and intense tandem DNA amplifications. These two phenomena are seen only at the ends of the linear chromosomal DNA. These regions can make up approximately 25% of the genomic DNA in *S. ambofaciens* under laboratory conditions. In contrast, an increasing number of genes have been identified in these regions [see reference 97 for a review of the genetically best-characterized species, *S. coelicolor* A3(2)]. Laboratory conditions, i.e., rich medium and constant temperature, may reduce the selective pressure for a large genome. This is in contrast to all *Streptomyces* strains isolated from soil, which contain 8-Mb linear chromosomal DNA. These data suggest that the loss of large regions of the chromosome is selected against under natural conditions.

In addition, the structural instability of the chromosome leads to conformation changes: the chromosome becomes circular in some mutant strains. These alterations are interpreted as intrachromosomal recombination events. However, interchromosomal events have been characterized that trigger translocation of the terminal parts of the linear DNA. Until recently, only illegitimate-recombination mechanisms were used to explain this instability; now several new lines of evidence suggest that some of the rearrangements could be due to homologous recombination. Chromosomal alterations are thought to result from a general increase of structural instability, leading to a cascade of rearrangement events.

The forces that favor these DNA rearrangements are discussed here in relation to the characteristics of the *Streptomyces* genome. The linear structure of the chromosomal DNA raises questions about the replication mechanisms, the unstable region corresponding to natural termini of chromosomal replication. Secondly, interactions between chromosomes or between linear plasmids and chromosomes within polyploid hyphae or during the replication process might be involved in these phenomena. Whatever the force triggering this structural instability, genomic instability reshapes the chromosome and might directly participate in the construction of the actual chromosome structure.

DYNAMIC FORCES WITHIN THE GENOME: INTERACTIONS BETWEEN PLASMID AND CHROMOSOME DNAs

All the wild-type (WT) *Streptomyces* species studied so far have linear chromosomal DNA (Fig. 1) (77). Several of these species also possess linear plasmids (the so-called giant linear plasmids) that also show an invertron structure (101). This similarity suggests that exchanges have occurred between these molecules. Comparison of the structures and sequences of the terminal repeats of the chromosomal DNAs of different species reveals a paradox. All *Streptomyces* WT strains (isolated from the soil) have TIRs but appear to be highly polymorphic at the inter- and intraspecific levels.

Streptomyces Chromosomes and Plasmids Share the Invertron Structure

The historical dogma that all prokaryotes possess single circular chromosomes has been refuted during the last decade. The first demonstration of the linearity of a bacterial chromosome was in *B. burgdorferi*, the Lyme disease agent (34). The presence of linear chromosomes was then reported in *A. tumefaciens* (3) and *Streptomyces* species (77) and suggested in *Rhodoccocus fascians* (27). In addition, bacterial genomics has revealed complex genome organizations in several of the eubac-

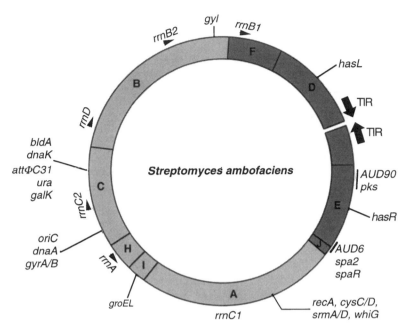

FIGURE 1 Invertron structure of the *Streptomyces* linear chromosomal DNA. The physical (*Ase*I) and genetic map of the *S. ambofaciens* DSM40697 chromosome typifies the invertron structure of the *Streptomyces* chromosomes. The 8-Mb chromosomal DNA is linear and possesses TIRs associated with proteins. The *oriC* locus is approximately opposite the ends of the DNA. Most gene localizations are from Leblond et al. (72) or unpublished data; the transcriptional orientation of the *rrn* loci are from Berger et al. (10), and the localization of *hasL/R* are from Fischer et al. (39). The unstable and deletable region, including *Ase*I fragments F, D, G, E, and J, is localized at the ends of the DNA and is indicated by dark shading.

terial phyla. Further evidence has come from reports of the presence of more than one chromosome in several unrelated bacterial genera (*Rhodobacter* [112], *Brucella* [60, 82], *Agrobacterium* [3], and *Leptospira* [123]).

In *S. lividans* 66, Lin et al. (77) found two copies of sequences homologous to the ends of the linear plasmid SLP2 present in the chromosomal DNA. The presence of proteins covalently associated with the ends was shown by the comparison of electrophoretic migration of DNA samples treated or untreated with a proteinase. The untreated sample was incubated with a detergent in order to keep intact only those proteins covalently bound to the DNA. C. W. Chen's and D. A. Hopwood's groups have described a linear chromosomal DNA for seven other species, including *S. coelicolor* A3(2) and *S. parvulus* (77).

A discontinuity in the physical maps of *S. ambofaciens* DSM40697, ATCC 15154, and ETH9427 was investigated by the methodology developed by Lin et al. (77) (Fig. 1). This discontinuity corresponds to a missing *Ase*I linking clone (72). The two large *Ase*I restriction fragments flanking the break showed a retarded pulsed-field gel electrophoresis (PFGE) mobility, indicating the presence of covalently bound proteins. In addition, the undigested chromosomal DNA migrated as an 8-Mb linear molecule in PFGE. The terminal structure was found to contain long terminal repeats, for example, 210 kb in strain DSM40697. The presence of repeats was reported in all linear *Streptomyces* replicons and seems to be a common feature. These repeats show a large degree of polymorphism in their sizes and sequences. For example, the TIR sizes range from 24 kb in *S. griseus* (75) to 31 kb in *S. lividans* 66 (76) and 550 kb in *S. rimosus* (90). Sequences corresponding to the ends of the TIR of *S. ambofaciens* DSM40697 (about the last 30 kb) were found not to be homologous to the total DNA of strain ATCC 23877 (ATCC 15154 ancestor). There was also no DNA homology between the TIRs of the two related species *S. lividans* 66 and *S. ambofaciens* DSM40697 or ATCC 23877 (our unpublished data). Contrasting with this polymorphism, the phylogenetic tree constructed from 16S rDNA sequence analysis showed a strong relationship among the tree species *S. lividans* 66, *S. coelicolor* A3(2), and *S. ambofaciens* (110).

These data suggest that the terminal region of the chromosomal DNA, and in particular the TIR, did not evolve by the same mechanisms as the rest of the DNA (i.e., mainly sequence divergence). Is gene replacement responsible for this polymorphism rather than loss of DNA sequences (see "Chromosome arm replacement" below)? In addition, these data favor the existence of a specific structure at the ends of the invertrons, whatever the length and sequence of the TIR. This structure could include a synaptic association of the terminal repeats, as in the "racket frame" model proposed by Sakaguchi (101) for the linear plasmid SCP1.

Linear Plasmids

Most of the plasmids isolated from *Streptomyces* are circular (53). However, linear plasmids have been reported, varying in size from 9 to more than 600 kb (49). The first linear plasmid to be reported was pSLA2 (17 kb) in *S. rochei* (46, 50). Later, large linear replicons were characterized in *S. coelicolor* A3(2) (63) by the orthogonal-field-alternation gel electrophoresis technique. This approach led to the identification of plasmid SCP1 (350 kb), which contains the methylenomycin biosyn-

thesis and resistance genes. Giant linear plasmids were then detected in several other antibiotic-producing species, although no direct correlation was found between the presence of a linear plasmid and production of an antibiotic.

The structures of the linear replicons revealed that they were members of the invertron group (101). The range of sizes of the TIRs in plasmids is as wide as for chromosomal DNA, with 44 bp at the lower end of the range in SLP2 in *S. lividans* 66 (24) and, at the upper limit, the 80-kb SCP1 (64). Recently, 180-kb TIRs were characterized in pPZG101, found in the bacterium *S. rimosus*, forming a total length of 387 kb (43). The 5′ end of the DNA is covalently associated with a protein, as seen in pSLA2 of *S. rochei*. This is demonstrated by resistance to 5′ exonuclease activity but sensitivity to exonuclease III (51). In *S. ambofaciens* DSM40697, the plasmid pSAM1 (80 kb [88]) was found in both linear (concatemers) and circular forms (73).

The presence of linear extrachromosomal molecules showing the same characteristics as the chromosomal DNA raises questions about their origins and their interactions. Two hypotheses are proposed: firstly, that the linear chromosomal DNA gave rise to smaller linear derivatives, and secondly, that integration of linear plasmids into a circular chromosome generated linearity in *Streptomyces*.

Irrespective of the origin of linear plasmids and chromosomes, several genera have been described with both a linear chromosome and linear plasmids. These were found to share the same structural characteristics. For example, in *B. burgdorferi* the chromosome exhibits covalently closed extremities (hairpins), as do the *B. burgdorferi* plasmids (19). It can therefore be hypothesized that a specific mechanism (exchange) or pressure led to the coevolution of both structures.

The frequent exchanges of DNA sequences between chromosome and plasmid is supported by several lines of evidence in different species and genera. For example, in *S. coelicolor* A3(2), sequences belonging to the plasmid

SCP1 (350 kb) were found to be integrated into the chromosomal DNA. The integration occurred close to the origin of replication (97). The integration was mediated by at least two recombination events (to maintain the integrity of the chromosome) and was presumably followed by an additional deletion event and homologous recombination between two copies of IS466, one internal to SCP1 and one on the chromosome (44). The second deletion event removed the agarase gene. The evolutionary consequences of such interactions were considered in the framework of genetic exchange of biosynthesis and resistance genes among *Streptomyces* species. Hence, SCP1 integrated forms can pick chromosomal sequences to give SCP1 prime plasmids (54, 114).

Other examples of putative interactions between plasmids and chromosomes have been reported. In *S. bambergiensis* S712, cross hybridizations of DNA sequences were found among the linear plasmid pBL1 (43 kb), the large linear plasmid PBS1 (640 kb), and the chromosomal DNA. The authors suggested that the linear pBL1 plasmid was derived from one end of PBS1 and that PBS1 was integrated into the chromosome or shared DNA sequences with it and pBL1 (122).

Two other examples of intermolecular interactions were investigated by C. W. Chen's and D. A. Hopwood's groups. The study of the structure of the *S. lividans* 66 linear plasmid SLP2 (50 kb) led to the discovery of a linear chromosome by the detection of two DNA extremities homologous to a part of the linear plasmid SLP2. It was suggested that SLP2 had integrated into the chromosome to give rise to a linear molecule (77, 96).

More recently, in *S. rimosus*, exchanges between the linear plasmid pPZG101 and the chromosome were characterized by J. Cullum's group. Plasmid pPZG101 is an invertron of 387 kb (43). A recombination event at a 4-bp common sequence exchanged one end of the plasmid with that of the chromosome. This event led to the formation of a hybrid chromosome with one original chromosomal end and one plasmid end. Another

strain contained a pPZG101 prime plasmid of approximately 1 Mb, including the oxytetracyclin biosynthesis cluster (91). In this case, it was demonstrated that a single recombination event was sufficient to integrate part of the linear plasmid into the chromosomal DNA. As pointed out by the authors, the stability of these hybrid molecules is questionable unless the homologous sequences found at the very ends of the chromosomal DNA in *S. lividans* 66 (there is a high degree of conservation in the last 166 to 168 bp of the DNA [77]) are sufficient to stabilize the structure of the chromosome.

It should be noted that whatever screening strategy was used to isolate *Streptomyces* species from the soil, TIRs were always found at the ends of all linear replicons. This suggests that a strong selective pressure is applied for the presence of such a structure. Thus, these hybrid molecules are selected against in the natural environment and/or there is a mechanism responsible for the homogeneity of the two terminal parts of a single chromosome.

Similar interactions have recently been reported in *Borrelia* (19). Analysis of the polymorphism in the right arm of the chromosome revealed that sequences belonging to the telomere in some species are derived from plasmid sequences. The closer the sequences are to the end of the chromosome the more likely they are to be located on a linear plasmid. Casjens et al. (19) have suggested that historically exchanges have taken place via a single crossover between the chromosome and linear plasmids. The origin of the linear chromosome in *Borrelia* is likely to have been a circular chromosome, since all other members of the spirochete family possess one (19).

REPLICATION OF THE INVERTRONS: IS RECOMBINATION IN THE TERMINAL REGIONS OF THE CHROMOSOMAL DNA ASSOCIATED WITH REPLICATION TERMINATION?

At present, two different replication mechanisms have been demonstrated for invertrons. In *Bacillus subtilis*, replication of bacteriophage Φ29 is primed at both ends by the terminal proteins (bacteriophage-encoded DNA polymerase). It proceeds over the whole length of the phage DNA, resulting in semi-conservative replication without formation of Okazaki fragments (reviewed in reference 102). In contrast, the linear plasmid pSLA2 (17 kb) of *S. rochei* has an origin of replication that is centrally located and bidirectional (20). The majority of pSLA2 is replicated from this central point. The replication intermediates are linear DNA molecules that possess recessed 5′ ends of approximately 280 nucleotides (nt). The synthesis of the 5′-terminal DNA strand (280 nt) is presumably primed on the template, consisting of the 3′ overhang of the leading strand, by a protein attached to the 5′ termini of the complete pSLA2 (20).

Similarly, the *Streptomyces* chromosome has an origin of replication (*oriC* locus) that maps close to the middle of the linear DNA (10, 17, 62, 74, 84, 121). However, in *S. rimosus oriC* is not centrally placed (it is distant from one end by 34 to 44% of the total chromosome size [90]).

The activity of this centrally located replication origin was first demonstrated by gene dosage experiments. Genes close to the origin of replication have been found to be over-represented in comparison to those close to the terminus. Hybridization experiments were therefore carried out with gene probes corresponding to scattered loci on the genome. It could be shown that the degree of over-representation decreased from the centrally located origin to a minimum close to the ends of the linear chromosome (84).

In *S. ambofaciens*, PFGE time course experiments showed that when mycelia were harvested from rapidly growing cultures, the regions flanking the centrally located *oriC* locus were overrepresented (37). These data strongly suggested that reinitiation of replication takes place at this *oriC* during exponential growth and that the *oriC* is bidirectionally active. In addition, the trancriptional orientation of the *rrn* loci was found to be divergent from the *oriC* locus (10). Considering that highly expressed genes are usually transcribed in the direction of the migration of the replication

forks (16), these results indicate that the linear chromosome of *S. ambofaciens* is bidirectionally replicated from the *oriC* locus toward the chromosome ends.

Together these data show that the replication termini lie in the terminal parts of the chromosomal DNA in *Streptomyces*. In addition, the unstable region in the streptomycetes has been found to share characteristics common to the replication termini of *Escherichia coli* and *B. subtilis* (68). One common characteristic is that these chromosomal regions can be removed under laboratory conditions (47, 58) with no adverse effects. This shows that no essential genes are present in either of these regions. This is of great significance when it is considered that the high frequencies of rearrangement are observed in these regions. This raises questions about the role of the termination process in the structural instability of the genome (see "Hypotheses: chromosome rearrangements result from a cascade of structural instability events" below). Recent in vivo data support the existence of a specific organization for the terminal regions, for example, a synaptic structure keeping the TIRs associated (derived from the racket frame model [101]). At the onset of stationary phase, a slight overrepresentation of the terminal fragments (flanking the ends of the chromosome) was noticed (37). This overrepresentation could be interpreted as a reinitiation of replication in the terminal regions. The loci implicated could be located either at the DNA extremities, with the terminal proteins acting as primers, or in the preterminal regions, initiating replication bidirectionally. On the other hand, overrepresentation of a DNA fragment might indicate its relative resistance to degradation.

GENETIC INSTABILITY IS UBIQUITOUS IN *STREPTOMYCES* SPECIES

Spontaneous Phenotypic Instability—Hypervariability

Genetic instability has been described in all *Streptomyces* species studied so far and is a characteristic of this genus. The high spontaneous-mutation rate mostly affects genes involved in differentiation processes, both the morphological and physiological aspects. The unstable mutants appear by apparently distinct pathways that depend on the relative localization of the genes in the unstable region. The screening method used to isolate the mutant strains can reveal these different pathways. For example, in *S. violaceus-ruber* and *S. cattleya*, pleiotropic mutants appear directly in WT platings (25, 98), while in *S. lividans* 66, two steps are needed to give rise to pleiotropic mutants (4, 33). This two-step mechanism corresponds to two levels of instability: the first produced about 1% mutants; the second showed a dramatic increase in the frequencies (from 25% in *S. lividans* 66 [4] to 87% in *S. ambofaciens* [69]). In *S. ambofaciens*, the first level of instability, called basic genetic instability, produced pigment-defective colonies at a frequency of approximately 1% (30). When these pigment-defective colonies were subcloned, they generated progeny mostly of the parental phenotype. The second level of instability was called hypervariability and was observed in progeny of most of these pigment-defective mutants (69). This second level of instability is typified by the production of progeny in which no preponderant phenotype could be distinguished (Fig. 2).

Within a given species the genetic instability appears to affect certain characteristics more than others. Examples of this are the argininosuccinate synthetase gene in the Arg⁻ mutants derived from *S. lividans* 66 (4) and the tyrosinase gene (encoding an extracellular enzyme) in *S. glaucescens* and *S. reticuli* (15, 106). In *S. ambofaciens*, several unstable characteristics were described: pigmentation, colony morphology, sporulation, aerial mycelium production, antibiotic production (spiramycin), auxotrophy to arginine, sensitivity to UV light or mitomycin C, and resistance to novobiocin or nosiheptide (69, 116). These hot spots of instability reflect the localization of the genes encoding the unstable functions in the nonessential region.

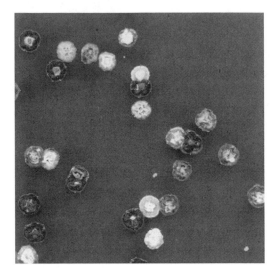

FIGURE 2 Different levels of genetic instability in *Streptomyces*. Hypervariability in *S. ambofaciens* DSM40697 is shown; no preponderant phenotype can be seen in the progeny of a pigment-defective colony generated from the WT strain (69).

Genetic instability not only leads to the loss of functions but in some cases a specific metabolic trait can be overexpressed. This is due to the amplification of the gene. Examples are the increased resistance to antibiotics, such as spectinomycin in *S. achromogenes* subsp. *rubradiris* (55), chloramphenicol in *S. coelicolor* A3(2) (41), and kanamycin in *S. griseus* and *S. rimosus* (59, 94). In contrast, the overproduction of a metabolite was rarer. Examples are the production of the antibiotic oleandomycin in *S. antibioticus* (89) and an inhibitor of alpha-amylase in *S. tendae* (65). When the positive selection for these traits was removed, these genetic regions showed a high level of instability. In *S. ambofaciens*, overexpression of *orfPS*, homologous to a polyketide synthase (PKS) gene, was associated with the amplification of the AUD205 locus, including the *orfPS* gene (2). The overexpression is correlated with the loss of production of the antibiotic spiramycin (28) and a depressed lipid content in the amplified mutant strain (103). Both phenotypes were suppressed in strains which had lost the amplification. These data strongly support the hypothesis of a metabolic drain of the precursors of the antibiotic spiramycin and lipid synthesis by the overexpressed PKS activity.

Influence of Genetic Factors on the Level of Genetic Instability

TREATMENTS THAT STIMULATE GENETIC INSTABILITY

An interesting characteristic of genetic instability is that it is inducible. Hypotheses about the possible origin of this instability are based on reports of studies where the level of instability has been altered by a variety of treatments. The spontaneous frequencies of instability can be increased by treatments as varied as exposure to UV light, culture in the presence of intercalating agents, cold storage, temperature shifts during culture, nutritional shifts, and the regeneration of protoplasts (for reviews, see references 13, 68, and 71). In the case of UV exposure and intercalating agents (well known as classical mutagens), the doses were far from the classical mutagenic conditions. The other treatments are by no means classical mutagens. These data indicate that genetic instability might be an answer to environmental stresses.

Studies undertaken in our laboratory were aimed at determining either the effect or the target of these stresses. Initially, the chromosomal or plasmid location of the unstable determinants was not known. The increased mutability of specific characters was thought to indicate that they were localized on plasmids. This hypothesis was supported by the increased mutability seen after treatment with plasmid-curing agents (26, 107). When the unstable genes were shown to be unambiguously located on the chromosome, other treatments were used to elucidate the mechanism of instability. In *S. ambofaciens*, UV light exposure, mitomycin C, and nitrous acid, known inducers of the SOS system, were shown to increase the spontaneous mutation frequencies from 0.7% to more than 30% with a high rate of survival (116). This suggested

that an SOS-like system was involved in genetic instability. Inhibitors of the DNA gyrase (topoisomerase II) were then applied to *S. ambofaciens* ATCC 23877. When the strain was grown on a medium containing either oxolinic acid or novobiocin at subinhibitory concentrations, the frequency of pigment-defective colonies increased from the basic level to almost 100%. Oxolinic acid and novobiocin interact with gyrase subunits A and B, respectively. The colonies mostly exhibited a "patchwork" phenotype. This effect was not observed in the presence of the transcriptional inhibitor rifampin or the translational inhibitor streptomycin. These experiments confirmed the involvement of an SOS-like system in genetic instability or the direct effect of DNA gyrase in the DNA rearrangements, leading to the loss of specific characters (116).

MUTATOR STATES

The involvement of RecA in genetic instability was suggested due to its role in the amplification mechanism (see "DNA amplifications" below). In order to test this hypothesis, the *recA* gene was disrupted in both *S. lividans* 66 (85) and *S. ambofaciens* (1). The spontaneous frequencies of Cml^s mutants were increased by approximately 70 times in the progeny of the *recA* *S. lividans* mutant (115). The authors suggested that the high frequency of terminal deletions might be related to unrepaired collapses of the replication fork, since RecA is implicated in repair of double-strand breaks.

Spontaneous variation in mutability was studied among wild-type subclones of *S. ambofaciens* ATCC 23877 (80). A specific aspect of the mutability of pigmentation was investigated in the formation of papillae in a wild-type background. Among the wild-type clones studied, the distribution of the pigment-negative papillae differed from the theoretical Poisson distribution, and this suggested that the colonies containing papillae were not homogeneous but polymorphic. WT clones were found to contain different numbers of papillae. Mutators at the extremity of the distribution were defined as colonies harboring more than 20 papillae per colony (Fig. 3).

In addition, the papillae were counted after different growing times on solid medium. It was found that after 14 days of growth almost all the colonies had papillae; i.e., the development of the mutator state seems to be related to the maturation process of the colony.

The mutator strains were further characterized by analyzing the frequency of pigment-defective colonies in their progeny. The frequencies continuously varied from one mutator to another within a range of values from 10 to 36.2%. The frequencies of their WT ancestors were approximately 10-fold lower. Therefore, genetic instability was increased in the mutator background.

CHROMOSOME REARRANGEMENTS

Genetic instability was historically associated with plasmid loss, DNA amplification (seen with restriction fragment length polymorphism), or chromosomal deletions visible with the PFGE technique. More recently, it has been associated with the linearity of the chromosomal DNA.

DNA Amplifications

AMPLIFIABLE UNIT OF DNA

DNA amplification is ubiquitous in bacteria (for a review, see reference 99). Mechanisms of gene amplification were first investigated in *E. coli, Salmonella typhimurium,* and *Proteus mirabilis* (reviewed in reference 100). Hot spots for DNA amplification have been identified in numerous bacterial chromosomes, and the presence of long terminal repeats in direct orientation is a characteristic of DNA regions undergoing amplification. The length rather than the nature of the repeats seems to be important for amplification.

In *Streptomyces*, DNA amplification consists of the multiple tandem repeats (several to hundreds of copies) of a basic sequence of DNA called the AUD (amplifiable unit of DNA). Hütter and Eckhardt (57) classified AUDs into two structural groups.

Type II AUDs were characterized first and are typified by the presence of repeated sequences (0.7 to 2.2 kb) flanking a unique stretch of DNA; therefore, they fit with the classical amplifiable structures found in other bacteria. Examples are seen in *S. lividans* 66 (5), *S. fradiae* (40), and *S. achromogenes* subsp. *rubradiris* (55). Moreover, in *S. lividans* 66, the AUD1 element (also called the 5.7-kb element) consists of the duplication of such a structure (93) and promotes the reproducible amplification of the AUD. Classical models for DNA amplification involved the duplication of the amplifiable unit as a rate-limiting step. Further, strains that have lost one copy of duplicated sequences no longer have the ability to amplify. The ability of AUD1 to amplify was associated with the binding of a DNA protein encoded by one of the repeated sequences (93, 117).

At first sight, the structure of type I AUDs shows a different organization and appears to impose a different amplification mechanism, as pointed out by Romero and Palacios (99). Type I AUDs are characterized by the reiteration of DNA sequences highly heterogeneous in size from a limited chromosomal region. Type I AUDs were described in *S. glaucescens* (amplifiable region, about 100 kb [45]), *S. lividans* 66 (70 kb [95]), and *S. ambofaciens* (28, 30, 69). In the last case, two main amplifiable loci, AUD6 (35 kb) and AUD90 (100 kb), generated more than 75% of the amplifications detected. The lengths of the basic DNA sequences ranged from 5.2 to more than 105 kb. These two loci are separated by approximately 400 kb on the same chromosomal arm of the linear map (Fig. 1). However, it could be shown that type I AUDs also have a repeated structure. In *S. ambofaciens*, AUD6 included two long terminal repeats (about 1 kb [8]) separated by a 4.9-kb unique intervening sequence. The repeats show a strong identity (90%) with those found in the 5.7-kb element of *S. lividans* 66. Moreover, AUD90 was investigated for the presence of repeat se-

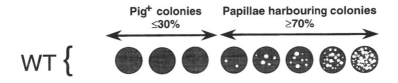

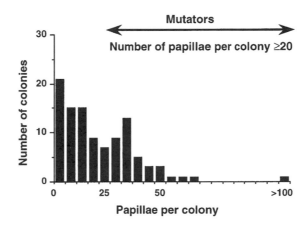

FIGURE 3 Mutator states. Mutator states are revealed by the appearance of pigment-defective papillae on the typical grey pigment of the WT colony (80a). The WT strain grown on solid medium for 14 days gives rise to a minority of fully pigmented (Pig⁺) colonies (≤30%) and to a majority of papillae-harboring colonies (≥70%) as well as fully pigment-defective colonies (about 1%; not symbolized). The colonies can also be sectored. When the number of papillae per colony was plotted against the number of colonies, the distribution differed from the theoretical Poisson distribution. Mutators at the extremity of the distribution were defined as colonies harboring more than 20 papillae per colony.

quences. The locus included different families of repeat sequences that correspond to reiterations of the functional domains homologous to a PKS gene spread over more than 100 kb (reference 2 and our unpublished results).

Finally, both types of AUD contain repeat sequences. The reproducibility of their amplification may indicate that they are processed by the same molecular mechanism. Type II AUDs could provide a more powerful template for the amplification process than the type I AUDs. In streptomycetes, DNA amplifications were characterized first in the pleiotrope mutants of *S. lividans* 66. In *S. lividans* 66, Cmlˢ mutants are generated at a fre-

quency of 1% and give rise in their progeny to Cmlˢ Arg⁻ double mutants at a frequency of 25% (4). All the Arg⁻ mutants had DNA amplifications. Similarly, in *S. ambofaciens*, amplification was correlated (frequency, 0.21) with hypervariability, the second level of genetic instability (69). Amplification may arise after several levels of genetic instability have occurred, and it may be a cyclical process (see "Hypotheses: chromosome rearrangements result from a cascade of structural instability events" below). Finally, there is no evidence to suggest a role for DNA amplification. In some cases, the phenotype conferred by the amplification can be positively selected (like hyperresistance to antibiotics), but the selective pressure remains unknown.

MODELS FOR DNA AMPLIFICATION IN *STREPTOMYCES*

Information about the molecular mechanism of DNA amplification can be deduced from the localization of the amplified DNA. The amplified DNA is mostly located at the locus of the WT AUD present as a single copy (45, 70, 95). However, extrachromosomal forms of the amplified DNA can be detected in mutants of *S. ambofaciens* (108), but in contrast to eukaryotic cells, no relocalization of DNA amplification (by means of the extrachromosomal forms) was reported in *Streptomyces*. Therefore, it seems likely that the amplification process itself takes place on the chromosome.

Models for DNA amplification mechanisms have included homologous recombination events and DNA replication. Early models involved the formation of unequal crossovers between replicated chromatids (6, 111). However, this mechanism would not explain the high copy number of identical repeats commonly seen in the *Streptomyces* species and formed within a limited number of generations. In addition, the amplified DNA was a perfect reiteration of one copy of the AUD, suggesting the involvement of a replication step.

The most convincing hypothesis was proposed by Young and Cullum (120) and was used to explain the DNA amplification seen by Petit et al. (92) in *B. subtilis*. This model takes into account the characteristics of the *Streptomyces* AUD and suggests that the first step in the amplification process is a recombination event between two repeats of the AUD (one replicated and one not yet replicated), thereby trapping the replication fork on a rolling-circle-like loop. The reassimilation of the tandem repeats on the chromosomal DNA explained the associated deletion often found immediately adjacent to the tandem repeats.

This model gives a key role to RecA. In *E. coli*, the formation of duplications is strongly dependent on *recA*; a null mutant provokes a strong decrease in the frequency of DNA duplication (7). Recent experiments involving *recA* backgrounds in *S. lividans* 66 (115) and *S. ambofaciens* (1) support the involvement of RecA at different stages of genetic instability. In *S. lividans* 66, the *recA* mutant showed a mutator phenotype (increased production of Cml^s derivates), but there was no amplification of a type II AUD (borne on a derivate of SCP2 plasmid). RecA therefore plays a role in the amplification process, at least in this plasmid model. In *S. ambofaciens*, RecA was found to be overexpressed in the amplified mutant strain NSA205, as shown by immunodetection (1). In addition, extrachromosomal multimers of the AUD are thought to result from homologous recombination between tandem repeats of the AUD and to result in the loss of the amplification (108).

By studying the chromosomal structures of the amplified strains, the molecular mechanisms underlying the amplification process can be elucidated. Amplified strains were found to have either linear or circular chromosomes. In the spontaneous Cml^s Arg$^-$ double mutants of *S. lividans* 66, the amplification process seems to produce a completely new end to one arm of the linear chromosome. These mutants have an amplified 5.7-kb element associated with the deletion of all sequences separating the AUD locus and the end of the chromosomal arm. The other end remains intact (95). In the same way, in *S. ambofaciens*, amplified strains exhibited a deletion encompassing all sequence from the chromosome end (TIR) to the AUD locus (28, 36, 70) (Fig. 4B). The chromosome of strain NSA120 (amplified for AUD90 [Fig. 1]) is linear, as shown by the protein remaining associated with the non-deleted chromosome end (36). In *B. subtilis*, where the chromosome is circular in the wild-type strain, transient amplification intermediates were found to be associated with the chromosomal DNA by only one end (92).

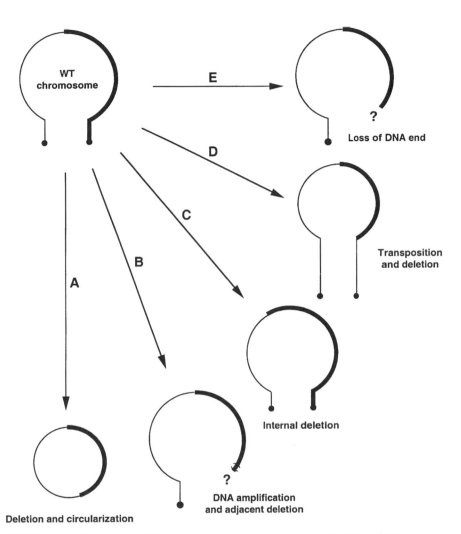

FIGURE 4 Chromosomal deletions on the linear chromosomal DNA of *Streptomyces*. The different locations of the chromosomal deletions are symbolized. The WT chromosome is represented under the racket frame model (101), where the extremities of the DNA consisting of large inverted repeats (TIRs) are associated. The length of the TIRs is symbolized by that of the parallel terminal regions. The solid circles represent the terminal proteins covalently associated with the DNA ends. (A) The TIRs and subterminal regions can be deleted, leading to circularization of the chromosome. (B) DNA amplification takes place from amplifiable loci called AUDs localized in the unstable region. Amplification is frequently associated with an adjacent deletion including all sequence separating the AUD from the chromosomal end. Thus, an unknown structure (question mark) replaces the bacterial telomere. (C) Internal deletions in one chromosomal arm were observed in *S. ambofaciens*, leading to the shortening of the TIRs. (D) Deletions were also observed associated with sister chromosomal exchanges, leading to the increase of the TIR length. (E) Loss of DNA extremities and replacement by an unknown structure (question mark) were also described in *Streptomyces*.

These data are in agreement with those reported for *Streptomyces* species (Fig. 4).

On the other hand, Volff et al. (118) and Lin and Chen (76) reported that amplification can take place on artificially circularized chromosomes of *S. lividans* 66. In this case, the circularization procedure (double-targeted recombination) leads to the formation of an AUD-like structure at the junction point. When the selection pressure for the antibiotic resistance gene inserted in the construction is applied, hyperresistant derivates that contain an amplification of the type II AUD structure can be selected for. This amplified DNA is highly unstable and can be lost when the selective pressure is removed.

The stability of the amplified DNA is rarely studied. Two effects are observed: stable amplifications over hundreds of generations and transient amplifications. These unstable amplifications are lost after a few rounds of replication (reference 33 and our unpublished data). The differences in stability may correspond to two types of structure of the amplified chromosomes (see "Hypotheses: chromosome rearrangements result from a cascade of structural instability events" below).

Large-Scale Deletions Leading to Linear or Circular Chromosomes

The loss of phenotypic characteristics has long been ascribed to deletions in the genetic determinant (57). The reports of targeted instability indicated that a limited region of the chromosome underwent deletions. This led to the idea of a region of the chromosome containing nonessential genes.

Using WT cosmid libraries or PFGE, the extent of the chromosomal deletions was investigated (15, 69, 96). In *S. ambofaciens* DSM40697, the deletion sizes ranged from 250 kb to more than 2.0 Mb and allowed us to establish a close association between genetic instability and the formation of large deletions (70). A region was defined, corresponding to approximately 25% of the total genome size, that can be deleted or amplified without lethal results (the AUDs are localized in the deletable region [Fig. 1]). A similar region was characterized in *S. lividans* 66 (33, 96), *S. glaucescens* (for a review, see reference 13), and *S. rimosus* (42).

Several types of chromosome structures result from the formation of large deletions (Fig. 4). Structural instability is associated with alterations in chromosome geometry. Different types of chromosomal structures were identified in mutant strains selected under laboratory conditions (36). In one type, deletions were internal to one chromosomal arm, leaving a linear chromosomal DNA. The deletion termini mapped within the terminal repeat sequences, leaving a TIR as small as 100 kb in one mutant strain. Another class of mutants had a deletion encompassing the entire TIR on both chromosomal arms. In each of the four cases analyzed, the deletion was approximately 2,000 kb. The chromosomes of these strains were shown to be circularized by a junction fragment, which was homologous to sequences originating from internal regions of both arms. Such spontaneous circularization of the chromosome was also described in *S. lividans* 66 (77, 96).

The mechanism by which these deletions occurred was investigated by analyzing the sequences involved at the deletion termini. In *S. glaucescens*, no long repeats were detected, but only microhomologies consisting of short direct or inverted repeats (13). Therefore, illegitimate recombination was assumed to be responsible for these DNA rearrangements. Short repeats are prone to strand slippage during replication. In *S. glaucescens*, these sequences showed a nucleotide similarity with the *att* site of the integrative element pIJ408 (14). This suggested that a site-specific recombinase was involved in the recombination event, but with an incorrect interaction site recognized. Other deletion events on a *Streptomyces* phage (109) or plasmid (86) were shown to occur between very short repeats. Similarly, Pandza et al. (91) recently reported the integration of plasmid pPZG101 on the chromosome of *S. rimosus* via a 4-bp identical sequence.

The superposition of the unstable region within terminal regions of the chromosome suggested that the terminal deletions might occur by additional mechanisms: DNA degradation from one or both DNA extremities or the formation of a double-strand break in the dispensable region (12, 23, 115). However, the internally deleted linear chromosomes described in *S. ambofaciens* (38) seem to rule out the hypothesis involving DNA degradation in that case.

Chromosome Arm Replacement

Homologous recombination is involved in numerous cases of chromosome rearrangement in bacteria. In *S. typhimurium*, recombinations among the *rrn* loci generate high frequencies of duplications and deletions (10^{-3} to 10^{-4}). Inversions are also frequent among the seven *rrn* loci of *E. coli* and *S. typhimurium* (78) (see chapter 13). The genetic plasticity seen in culture contrasts with the stability (conservation) of the genetic organization seen in enteric bacteria (67). This indicates that large-scale rearrangements are strongly selected against during evolution. However, this is not the case for *Salmonella typhi*, where an impressive plasticity of the genome is observed: inversions and transpositions are common, but deletions and duplications are not (78). These changes do not on the whole alter the distances of genes from the origin of replication (gene dosage) or dramatically modify the relative positions of the replication terminus and the origin of replication.

In a recent paper (60), the importance of homologous recombination between the *rrn* loci in the production of variable genome geometries in *Brucella* was emphasized. The variability in chromosome size and number among biovars of *Brucella suis* can be explained by recombination events between the *rrn* loci.

Homologous recombination is not only involved in the amplification process but also in a new class of DNA rearrangement characterized in *S. ambofaciens*, chromosome arm replacement (39) (Fig. 4D). Two mutant strains harbored a large deletion (several hundred kilobases) of one chromosomal arm fused to several hundred kilobases originating from the second, intact chromosomal arm. These two mutants were isolated independently but correspond to reciprocal events. These recombination events leave the chromosomal DNA linear, but the size of the TIR is greatly altered; it corresponds to the size of the translocated region (480 and 850 kb). The recombination events take place within a duplicated sequence (identical sequence, over 800 bp) but with an inverted orientation on each chromosomal arm. The duplicated sequences are 480 and 850 kb distant from the left and right ends, respectively (Fig. 1). A putative alternative sigma factor is encoded in the duplicated sequences of the *has* locus (39).

The simplest hypothesis to explain these rearrangements is a sister chromatid exchange. This would involve ectopic recombination events between the two copies of the duplicated sequence. Alternatively, a mechanism involving repair of a double-strand break occurring at a *has* locus could account for it. Therefore, the recombinogenic end could invade the homologue and initiate a replication fork. This replication process could proceed until the end of the linear DNA and duplicate the terminal sequence from the *has* locus to the chromosome end. This mechanism was described as copy break duplication in *Saccharomyces cerevisiae* (83). In the first hypothesis, the rearrangement would involve interchromosomal interactions, while the second does not determine whether the interaction is intra- or interchromosomal.

HYPOTHESES: CHROMOSOME REARRANGEMENTS RESULT FROM A CASCADE OF STRUCTURAL INSTABILITY EVENTS

It has been suggested by several groups that genetic instability could result from a cascade of molecular events leading to the loss of phenotypic traits (4, 5, 30, 33, 69). This hypothesis was supported by the phenotypic description of successive levels of instability (e.g., hypervariability following basic instability in

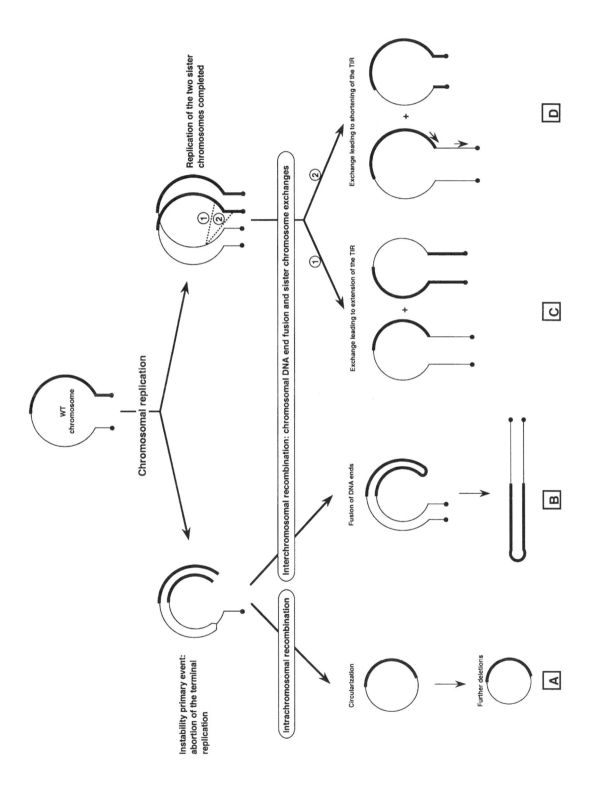

S. ambofaciens [69]). At the molecular level, these successive phenotypic stages were correlated with polar deletion or amplification events (70).

After the discovery of the linearity of the chromosomal DNA in *Streptomyces* (77), a link was thought to exist between the genetic instability seen and the presence of chromosome termini. However, recent studies have ruled out a simple correlation between the loss of DNA termini (bacterial telomeres) and structural instability (39). The current picture of instability appears to be a succession of rearrangements generated by different molecular mechanisms. The hypothesis depicted in Fig. 5 suggests that termination of replication could be the source of chromosomal instability. In addition, the putative synapsis that could result from interaction between the TIRs might favor terminal recombination. Alternatively, the primary event might also correspond to the loss of terminal protein and exonuclease attack of the DNA ends.

Thus, the terminal region of a chromosome is an area of genetic instability whatever its structure, interrupted or covalent (i.e., linear and circular, respectively). This unstable region probably corresponds to the area where the termination of chromosomal replication occurs (37, 68). In the termination process, replication forks slow down and stop at specific *Ter* sites in the *E. coli* and *B. subtilis* chromosomes (48). The impediment and arrest of the replication forks are prone to trigger illegitimate as well as homologous recombination (12). In addition, Louarn et al. (79) found a peak of homologous recombination (RecA dependent) in the terminus of replication of *E. coli*. A recent review showed that in numerous bacteria the region containing the replication terminus is prone to a high degree of polymorphism (66). Little is known about the termination of bacterial linear replicons, but specific sequences (*Ter*-like sites) might slow down the replication forks on each arm, close to the DNA ends, as suggested by Chen et al. (23). These sequences, if they exist, could be spread over several hundred kilobases in the preterminal regions, as in *E. coli* (48). Therefore, recombination (illegitimate or homologous) induced by the termination process might trigger genetic instability.

The circularization of the chromosome would stabilize the structure if DNA termini are directly and exclusively involved in genetic instability. However, in all species tested so far circularized chromosomes (occurring spontaneously or by targeted recombination) show

---

FIGURE 5 Hypothetical flow chart of the cascade of molecular events involved in genetic instability. Structural instability would proceed by two different recombination pathways: intramolecular and intermolecular interactions (sister chromosome exchange). Both recombination events could take place at the chromosomal replication in the terminus region. A primary event (terminal protein loss, DNA degradation, or collapse of the replication fork) would lead to the loss of a DNA extremity. The creation of this reactive extremity would result in either circularization of the chromosome (A) or DNA end fusion between the two newly replicated chromatids (B). This latter form would enter an equivalent to the breakage-fusion-bridge cycle described in eukaryotes by McClintock (81). The fusion of DNA ends would be a good explanation for the unknown nature of chromosomes that have lost one DNA extremity and kept the other one intact (Fig. 4). DNA amplification could take place either at the end of the replicated chromatids by a Young and Cullum mechanism (120) or on the circularized or fused molecules. Secondly, interchromosomal interactions could explain the variation of the TIR length. If the recombination event (homologous or illegitimate) takes place between two regions specific to each chromosomal arm (C, event 1), then it results in the exchange of chromosomal arms, with extension of the TIR. On the other hand, recombination between two regions, one in the TIR and the other one specific to a chromosomal arm (D, event 2), would result, on the right replicon, in the shortening of the TIR accompanied by an internal deletion. The reciprocal could remain undetectable, since no deletion would result from this event, but a duplication of an internal TIR sequence is produced (arrows). When segregated, this structure might also trigger an instability cascade by homologous recombination that could be initiated between the large duplicated areas.

an even higher level of instability. In *S. lividans* 66, circular chromosomes were found to segregate 20 times more chloramphenicol-sensitive mutants (harboring chromosome rearrangements) than the wild-type strains. Amplification of 30- to 60-kb AUDs was shown to occur at the circularization point (118). In the same way, Lin and Chen (76) showed that when the selective pressure (i.e., an antibiotic resistance gene is included in the inserted construction) for circularization is removed, instability is increased, suggesting that terminal sequences would be important for stabilization of the chromosome.

In *S. ambofaciens* we assessed the phenotypic stability of spontaneous circular DNA mutants. These mutants contain deletions which may be greater than 2,000 kb. They produce progeny exhibiting nonparental traits at frequencies exceeding that of the WT strain by 10 times (36). Additional deletions were seen in the unstable region, resulting in a total deletion size of more than 2,300 kb (the largest studied). The chromosome structure remained circular. These data confirmed that circularity of the chromosome does not correspond to stability. In *S. ambofaciens*, the chromosome might never reach a state of genetic stability, and deletions occur until an essential gene is lost, leading to cell death (Fig. 5A). Whatever the nature of the primary event leading to the loss of one DNA end, the final result would be the loss of the second chromosome end and the circularization of the molecule (intrachromosomal recombination [Fig. 5A]). Further instability might result from an abnormal termination process or partitioning on chromosomes that have lost termination signals, as noted in *E. coli* (47).

The occurrence of fused chromosomes is supported by the study of chromosomes that have lost one DNA extremity (35). This is the case for amplified as well as nonamplified deleted chromosomes (Fig. 4B and 4E). Initially, it was suggested that a new type of extremity might have replaced the protein-associated DNA end. However, molecular investigations revealed that the apparent terminal DNA fragment consisted of the fusion in inverted orientation of two deleted chromosomal-arm extremities. We speculate that the fused chromosomes might result from interchromosomal recombination between the newly replicated broken chromosomes (Fig. 5B).

In the same way, it seems unlikely that the replacement of the natural end of the chromosomal DNA by tandem amplification can stabilize the chromosome structure. The loss of the DNA end is thought to make the chromosome more unstable and to trigger a cascade of events. If the amplified sequences correspond to a new end, this DNA end might make it sensitive to exonucleases. Two recombination events could rescue the chromosome from total degradation: circularization accompanied by the loss of the second chromosomal end (Fig. 5A) or the fusion of two newly replicated amplified chromosomes (Fig. 5B). Unstable amplifications might correspond to chromosomes not rescued by the fusion of the DNA termini and subject to DNA degradation.

Fused chromosomes would be involved in a breakage-fusion-bridge cycle analogous to that described by B. McClintock in maize (81). Therefore, when the daughter nucleoids are separated, DNA breakage could occur randomly in the unstable region and the sequence would be either lost or fused again after DNA replication. Recently, a mitotic-like apparatus responsible for the movement of the origin of replication was found in a dividing cell of *B. subtilis* (119). Therefore, fused chromosomes which harbor two replication origins might break while being pulled toward opposite ends of dividing cells. This might explain the increased instability of the fused chromosomes.

Homologous recombination was involved in two cases of chromosomal arm replacement in *S. ambofaciens* DSM40697 (39). These rearrangements do not alter the general chromosomal structure (Fig. 5D); the DNA remains linear with the typical invertron structure. This stability might explain why these rearrangements would not be selected against and might even be selected for under laboratory conditions. However, such alterations could be eliminated under natural conditions if the large deletion removed genes required for facing environmental stresses. The internal deletion, shortening the TIR (Fig. 4), could also be interpreted as interchromosomal interaction (Fig. 5D). In such a hypothesis, recombinations (homologous or illegitimate) could take place between an internal locus within the TIR and a specific region on one chromosomal arm. This event would generate two reciprocal recombinant chromosomes exhibiting (i) the shortening of the TIR and its accompanying internal deletion or (ii) the increase of the TIR and the creation of a large duplicated area. The second recombinant chromosome may remain undetected, since there is no loss of information. A chromosome of this nature with a large duplicated region, although not yet observed, might be the substrate of successive homologous recombination events leading to structural instability. These rearrangements could lead to the formation of an unbalanced chromosome (alteration of the distance between *oriC* and the replication termini) prone to DNA rearrangement (Fig. 5D). Thus, one replication fork might have to wait for the completion of the replication process on the other chromosomal arm, leading to a recombinogenic site.

CHROMOSOME REARRANGEMENTS AND GENE EXPRESSION

The unstable region was for a long time thought to possess few genes, as the method of classical mutant selection failed to find many. However, DNA rearrangements associated with genetic instability indicated that genes should be present in this region. The latest update of the *S. coelicolor* A3(2) map shows that an increasing number of genes are being mapped to this area compared to the rest of the chromosome (97). The gene density might reach that of the rest of the genetic map. Under laboratory conditions, these genes can be removed, as they are mostly involved in morphological differentiation, secondary

metabolism, and the synthesis of extracellular enzymes.

Two-dimensional gel electrophoresis of total cellular protein was used in our laboratory, in collaboration with C. J. Thompson's group, to show that the deletion and amplification of hundreds of kilobases at the ends of the chromosomal DNA were associated with important alterations in the global protein expression pattern (29). This supports the hypothesis that the unstable region regulates or expresses numerous genes. This could also be explained by global alterations to nucleoid structure inducing gross changes to gene expression.

Genes may be directly identified by the phenotype accompanying their deletion or amplification (57, 61, 68). Genes have also been identified in the amplifiable unit of DNA, by either sequence analysis or, less frequently, the phenotype conferred by the amplification. Functions encoded in the AUD show considerable diversity: DNA binding protein, putative chitinase and fibronectin genes in the *S. lividans* 66 5.7-kb AUD (93, 117), chloramphenicol resistance genes in *S. lividans* 1326 (32), a PKS gene in *S. ambofaciens* (2), and antibiotic resistance determinants (spectinomycin resistance in *S. achromogenes* subsp. *rubradiris* [55], and oxytetracycline resistance in *S. rimosus* [42]).

Genes have also been identified by analysis of the deletion endpoints. For example, the deletion hot spot AUD6 of *S. ambofaciens* was shown to contain the *spa2* and *spaR* genes (31, 104). *spa2* encodes a putative regulator of the stationary phase, assumed from its homology to *rspA* of *E. coli* (56). The *S. coelicolor* A3(2) *spa2* homologue shows considerable differences in sequence and localization, being some 1,000 kb closer to the chromosome end than in *S. ambofaciens* (105). The PKS gene included in AUD90 (length, about 140 kb) was found by analysis of the spiramycin-deficient phenotype of the amplified strains (28). Overexpression of this gene is assumed to exhaust the precursors of the spiramycin biosynthetic pathway (2). The role of the product of this new PKS is unknown.

Recently, the discovery of duplicated genes, homologous to alternative sigma factors (*hasR/L*) in *S. ambofaciens* (39), corroborated the assumption that functions encoded in the deletable region could be involved in responses to environmental changes (physiological adaptation to growing conditions or to competition with microorganisms). This idea is supported by the fact that all of the *Streptomyces* species isolated from soil possess an intact terminal region (with two DNA ends). The conditions of growth in soil may therefore select for the possession of an intact terminal region.

All these data raise the question of the nature of the pressure that led to the construction and maintenance of this region that is dispensable under laboratory conditions. Essential functions could have been excluded from this region, which is prone to DNA rearrangements triggered by the processes of termination of replication or replacement of DNA sequences (e.g., conjugational or transformant DNA).

CONCLUDING REMARKS: SOME SPECULATIONS ABOUT THE EVOLUTIONARY IMPLICATIONS OF GENETIC INSTABILITY

The invertron structure is a common trait found in all of the linear replicons of *Streptomyces*. However, a considerable degree of polymorphism has been noticed at the intraspecific and interspecific levels. This polymorphism affects the length and the sequence of the TIR, suggesting that the invertron structure and, more specifically, the presence of inverted repeats is selected for. The invertron structure could be involved in the folding of the nucleoid in the racket frame (101) or the synapse model.

Genetic instability is responsible for alteration of the chromosome structure in *Streptomyces*: conversion of the linear structure into a circularized chromosome, amplification of DNA sequence, and interchromosomal exchanges. All of these DNA rearrangements are associated with the formation of large deletions.

The rearrangements that could have the most significant consequences for the evolution of the chromosome structure should confer a relative stability. Circularized chromosomes show increased instability compared to linear chromosomes, and the loss of large areas of their DNA would be selected against. In contrast, the formation of internal deletions, whatever the molecular mechanism (illegitimate or homologous recombination), leaves the chromosomal DNA linear. In the same way, interchromosomal recombination resulting from homologous recombination keeps the chromosomal DNA linear. In both cases, a considerable variability in the length of the TIR is generated. Internal deletions are associated with the shortening of the TIRs, while the intermolecular events expand the TIR up to 850 kb (39).

Homologous recombination events should be frequent between the TIRs of the same chromosome or between neoreplicated daughter chromatids (several hundred kilobases of sequence homology). This mechanism, gene conversion, could be responsible for the absence of any restriction polymorphism between the TIRs, as seen in *S. ambofaciens* (72). The exchanges of the terminal regions could be due to the structure which is suspected to keep the DNA ends together in vivo. Increased recombination frequencies could lead to sustained homogeneity of the chromosomal ends. This high recombination frequency may be due to the structure formed between the TIRs (such as synaptic structure) and/or to the replication termini, which are known to induce recombination. The only exchange events that have detectable consequences occur between loci external to the TIR (ectopic recombination between duplicated genes or dispersed copies of insertion elements). The substrate of these recombination events could be duplicated genes (such as *hasR* and *hasL* in *S. ambofaciens* [39]) or insertion elements, such as in *S. lividans* 66 [77]).

This selective pressure for the presence of TIRs could also be responsible for the rapid divergence seen at the intra- or interspecific levels. For example, no DNA homology was found between the ends of the TIRs in two strains of *S. ambofaciens* or between the TIRs of the phylogenetically closely related *Streptomyces* species, *S. lividans* 66, *S. coelicolor* A3(2), and *S. ambofaciens* (110). In contrast, the genetic organization of the rest of the chromosome showed a high degree of similarity among the same group of strains. This suggests that the evolutionary mechanism affecting the terminal region could differ from that affecting the rest of the chromosome.

The amplification mechanism could also have evolutionary effects. The repetitive structure of numerous gene clusters for secondary-metabolite synthesis (e.g., PKS genes [reviewed in reference 52]) could reflect tandem amplification of ancestral genes.

The formation of new terminal sequences could occur by an interaction with plasmid DNA (Fig. 6). An example has been reported for *S. rimosus* that resulted in the formation of hybrid molecules: plasmid prime with a chromosomal end and chromosomal DNA with a plasmid end (91). If selective pressures could act on these structures, it could lead to homogeneity of the sequences (Fig. 6). In *S. am-*

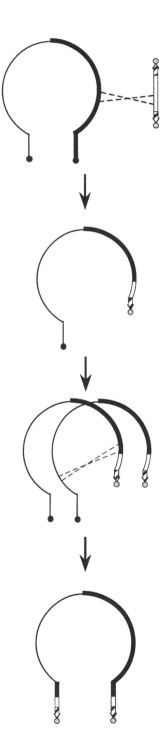

FIGURE 6 Evolutionary implications of genetic instability. Interactions between linear plasmids and chromosomes leading to the generation of hybrid chromosomes were recently reported in *Streptomyces* (91). This mechanism results in acquisition of new terminal information on the chromosomal DNA. This exchange could result from homologous or illegitimate recombination pathways. The hybrids could be unstable and undergo recombinational exchanges between newly replicated sister chromosomes, as reported in *S. ambofaciens*. This event would result in the homogeneity of the terminal sequence and thus introduce a dramatic change in the terminal sequences of this species. The symbols are as for other figures for chromosomal DNA. Plasmid DNA (open box) possesses an invertron structure with TIRs (hatched arrows) and terminal proteins (shaded circles). Broken lines symbolize the recombination event.

bofaciens, an interchromosomal recombination event caused the formation of a new invertron. A similar mechanism could explain the formation of the actual *S. lividans* 66 chromosome. The replacement of the terminal sequences of the ancestral linear chromosome of *S. lividans* by sequences of the linear plasmid SLP2 would have led to a hybrid chromosome similar to that described for *S. rimosus*. Analysis of the structure of the TIR supports this hypothesis, since only a part of the TIR (16 kb) is homologous to the linear plasmid SLP2. This could be explained if a second recombination event (translocation) occurred, leading to the ends becoming homogeneous. Comparison of the sequences of the TIRs in closely related species or strains could test this hypothesis.

REFERENCES

1. **Aigle, B., A. C. Holl, J. Angulo, P. Leblond, and B. Decaris.** 1997. Characterization of two *Streptomyces ambofaciens recA* mutants; identification of the RecA protein by immunoblotting. *FEMS Microbiol. Lett.* **149:**181–187.

2. **Aigle, B., D. Schneider, C. Morilhat, D. Vandewiele, A. Dary, A. C. Holl, J. M. Simonet, and B. Decaris.** 1996. An amplifiable and deletable locus of *Streptomyces ambofaciens* RP181110 contains a very large gene homologous to polyketide synthase genes. *Microbiology* **142:**2815–2824.

3. **Allardet-Servent, A., S. Michaux-Charanchon, E. Jumas-Bilak, L. Karayan, and M. Ramuz.** 1993. Presence of one linear and one circular chromosome in the *Agrobacterium tumefaciens* C58 genome. *J. Bacteriol.* **175:**7869–7874.

4. **Altenbuchner, J., and J. Cullum.** 1984. DNA amplification and unstable arginine gene in *Streptomyces lividans* 66. *Mol. Gen. Genet.* **195:**134–138.

5. **Altenbuchner, J., and J. Cullum.** 1985. Structure of an amplifiable DNA sequence in *Streptomyces lividans* 66. *Mol. Gen. Genet.* **201:**192–197.

6. **Anderson, R. P., and J. R. Roth.** 1981. Spontaneous tandem genetic duplications in *Salmonella typhimurium* arise by unequal recombination between ribosomal RNA (*rrn*) cistrons. *Proc. Natl. Acad. Sci. USA* **78:**3113–3117.

7. **Anderson, R. P., and J. R. Roth.** 1977. Tandem genetic duplications in phage and bacteria. *Annu. Rev. Microbiol.* **31:**473–505.

8. **Aubert, M., E. Weber, D. Schneider, J. M. Simonet, and B. Decaris.** 1993. Primary structure analysis of a duplicated region in the amplifiable AUD6 locus of *Streptomyces ambofaciens* DSM40697. *FEMS Microbiol. Lett.* **113:**49–56.

9. **Beijerinck, M. W.** 1913. Ueber Schröter und Cohn's Lakmusmicrococcus. *Folia Microbiol.* **2:**185–200.

10. **Berger, F., G. Fischer, A. Kyriacou, B. Decaris, and P. Leblond.** 1996. Mapping of the ribosomal operons on the linear chromosomal DNA of *Streptomyces ambofaciens* DSM40697. *FEMS Microbiol. Lett.* **143:**167–173.

11. **Bibb, M. J., P. R. Findlay, and M. W. Johnson.** 1984. The relationship between base composition and codon usage in bacterial genes and its use for the simple and reliable identification of protein coding sequences. *Gene* **30:**157–166.

12. **Bierne, H., and B. Michel.** 1994. When replication forks stop. *Mol. Microbiol.* **13:**17–23.

13. **Birch, A., A. Häusler, and R. Hütter.** 1990. Genome rearrangement and genetic instability in *Streptomyces* sp. *J. Bacteriol.* **172:**4138–4142.

14. **Birch, A., A. Häusler, C. Rüttener, and R. Hütter.** 1991. Chromosomal deletion and rearrangement in *Streptomyces glaucescens. J. Bacteriol.* **173:**3531–3538.

15. **Birch, A., A. Häusler, M. Vögtli, W. Krek, and R. Hütter.** 1989. Extremely large chromosomal deletions are intimately involved in genetic instability and genomic rearrangements in *Streptomyces glaucescens. Mol. Gen. Genet.* **217:**447–458.

16. **Brewer, B. J.** 1988. When polymerases collide: replication and the transcriptional organization of the *E. coli* chromosome. *Cell* **53:**679–686.

17. **Calcutt, M. J., and F. J. Schmidt.** 1992. Conserved gene arrangement in the origin region of the *Streptomyces coelicolor* chromosome. *J. Bacteriol.* **174:**3220–3226.

18. **Casjens, S., and W. M. Huang.** 1993. Linear chromosomal physical and genetic map of *Borrelia burgdorferi*, the Lyme disease agent. *Mol. Microbiol.* **8:**967–987.

19. **Casjens, S., M. Murphy, M. DeLange, R. Sampson, R. van Vugt, and W. H. Huang.** 1997. Telomeres of the linear chromosomes of Lyme disease spirochetes: nucleotide sequence and possible exchange with linear plasmid telomeres. *Mol. Microbiol.* **26:**581–596.

20. **Chang, P.-C, and S. N. Cohen.** 1994. Bidirectional replication from an internal origin in a linear *Streptomyces* plasmid. *Science* **265:**952–954.

21. **Chater, K. F.** 1993. Genetics of differentiation in *Streptomyces. Annu. Rev. Microbiol.* **47:**685–713.

22. **Chater, K. F., and D. A. Hopwood.** 1984. *Streptomyces* genetics, p. 229–286. *In* M. Good-

fellow, M. Mordarski, and S. T. Williams (ed.), *The Biology of the Actinomycetes*. Academic Press, London, United Kingdom.

23. **Chen, C. W., Y.-S. Lin, Y.-L. Yang, W.-Y. Lin, H.-W Change, H. M. Kieser, and D. A. Hopwood.** 1994. The linear chromosomes of *Streptomyces*: structure and dynamics. *Actinomycetologica* **8:**103–112.

24. **Chen, C. W., T. W. Yu, Y.-S. Lin, H. M. Kieser, and D. A. Hopwood.** 1993. The conjugative plasmid SLP2 of *Streptomyces lividans* is a 50 kb linear molecule. *Mol. Microbiol.* **7:**925–932.

25. **Coyne, V. E., K. Usdin, and R. Kirby.** 1984. The effect of inhibitors of DNA repair on the genetic instability of *Streptomyces cattleya*. *J. Gen. Microbiol.* **130:**887–892.

26. **Crameri, R., J. E. Davies, and R. Hütter.** 1986. Plasmid curing and generation of mutations induced with ethidium bromide in streptomycetes. *J. Gen. Microbiol.* **132:**819–824.

27. **Crespi, M. E., E. Messens, A. B. Caplan, M. van Montagu, and J. Desomer.** 1992. Fasciation induction by the phytopathogen *Rhodoccus fascians* depends upon a linear plasmid encoding a cytokinin synthase gene. *EMBO J.* **11:**795–804.

28. **Dary, A., N. Bourget, N. Girard, J.-M. Simonet, and B. Decaris.** 1992. Amplification of a particular DNA sequence in *Streptomyces ambofaciens* RP181110 reversibly prevents spiramycin production. *Res. Microbiol.* **143:**99–112.

29. **Dary, A., P. Kaiser, N. Bourget, C. J. Thompson, J.-M. Simonet, and B. Decaris.** 1993. Large genomic rearrangements of the unstable region of *Streptomyces ambofaciens* are associated with major changes in global gene expression. *Mol. Microbiol.* **10:**759–769.

30. **Demuyter, P., P. Leblond, B. Decaris, and J. M. Simonet.** 1988. Characterization of two families of spontaneously amplifiable units of DNA in *Streptomyces ambofaciens*. *J. Gen. Microbiol.* **134:**2001–2007.

31. **Demuyter, P., D. Schneider, P. Leblond, J.-M. Simonet, and B. Decaris.** 1991. A chromosomal region as a hotspot for multiple rearrangements associated with genetic instability in *Streptomyces ambofaciens* DSM40697. *J. Gen. Microbiol.* **137:**491–499.

32. **Dittrich, W., M. Betzler, and H. Schrempf.** 1991. An amplifiable and deletable chloramphenicol-resistance determinant of *Streptomyces lividans* 1326 encodes a putative transmembrane protein. *Mol. Microbiol.* **5:**2789–2797.

33. **Dyson, P., and P. Schrempf.** 1987. Genetic instability and DNA amplification in *Streptomyces lividans* 66. *J. Bacteriol.* **169:**4796–4803.

34. **Ferdows, M. S., and A. G. Barbour.** 1989. Megabase-sized linear DNA in the bacterium *Borrelia burgdorferi* the Lyme disease agent. *Proc. Natl. Acad. Sci. USA* **86:**5869–5973.

35. **Fischer, G.** 1998. Ph.D. thesis. Université Henri Poincaré, Nancy, France.

36. **Fischer, G., B. Decaris, and P. Leblond.** 1997. Occurrence of deletions associated with genetic instability in *Streptomyces ambofaciens* is independent of the linearity of the chromosomal DNA. *J. Bacteriol.* **179:**4553–4558.

37. **Fischer, G., A. C. Holl, J. N. Volff, D. Vandewiele, B. Decaris, and P. Leblond.** 1998. Replication of the linear chromosomal DNA from the centrally located *oriC* of *Streptomyces ambofaciens* revealed by PFGE gene dosage analysis. *Res. Microbiol.* **149:**203–210.

38. **Fischer, G., A. Kyriacou, B. Decaris, and P. Leblond.** 1997. Genetic instability and its possible evolutionary implications on the chromosomal structure of *Streptomyces*. *Biochimie* **79:** 555–558.

39. **Fischer, G., T. Wenner, B. Decaris, and P. Leblond.** 1998. Chromosomal arm replacement generates a high level of intraspecific polymorphism in the terminal inverted repeats of the linear chromosomal DNA of *Streptomyces ambofaciens*. *Proc. Natl. Acad. Sci. USA* **95:**14296–14301.

40. **Fishman, S. E., P. R. Rosteck, and C. L. Hershberger.** 1985. A 2.2-kilobase repeated DNA segment is associated with DNA amplification in *Streptomyces fradiae*. *J. Bacteriol.* **161:** 199–206.

41. **Flett, F., and J. Cullum.** 1987. DNA deletions in spontaneous chloramphenicol-sensitive mutants of *Streptomyces coelicolor* A3(2) and *Streptomyces lividans*. *Mol. Gen. Genet.* **207:**499–502.

42. **Gravius, B., T. Bezmalinovic, D. Hranueli, and J. Cullum.** 1993. Genetic instability and strain degeneration in *Streptomyces rimosus*. *Appl. Environ. Microbiol.* **59:**2220–2228.

43. **Gravius, B., D. Glocker, J. Pigac, K. Pandza, D. Hranueli, and J. Cullum.** 1994. The 387 kb linear plasmid pPZG101 of *Streptomyces rimosus* and its interactions with the chromosome. *Microbiology* **140:**2271–2277.

44. **Hanafusa, T., and H. Kinashi.** 1992. The structure of an integrated copy of the giant linear plasmid SCP1 in the chromosome of *Streptomyces coelicolor* 2612. *Mol. Gen. Genet.* **231:**363–368.

45. **Häusler, A., A. Birch, W. Krek, J. Piret, and R. Hütter.** 1989. Heterogeneous genomic amplification in *Streptomyces glaucescens*: structure, location and junction sequence analysis. *Mol. Gen. Genet.* **217:**437–446.

46. **Hayakawa, T., T. Tanaka, K. Sakaguchi, N. Otake, and H. Yonehara.** 1979. A linear plasmid-like DNA in *Streptomyces* sp. producing lan-

kacidin group antibiotics. *J. Gen. Appl. Microbiol.* **25:**255–260.

47. **Henson, J. M., and P. L. Kuempel.** 1985. Deletion of the terminus region (340 kilobase pairs of DNA) from the chromosome of *Escherichia coli. Proc. Natl. Acad. Sci. USA* **82:**3766–3770.

48. **Hill, T. M.** 1992. Arrest of bacterial DNA replication. *Annu. Rev. Microbiol.* **46:**603–633.

49. **Hinnebusch, J., and K. Tilly.** 1993. Linear plasmids and chromosomes in bacteria. *Mol. Microbiol.* **10:**917–922.

50. **Hirochika, H., K. Nakamura, and K. Sakaguchi.** 1985. A linear DNA plasmid from *Streptomyces rochei* with an inverted repetition of 614 base pairs. *EMBO J.* **3:**761–766.

51. **Hirochika, H., and K. Sakaguchi.** 1982. Analysis of linear plasmids isolated from *Streptomyces*: association to protein with the ends of the plasmid DNA. *Plasmid* **7:**59–65.

52. **Hopwood, D. A.** 1997. Genetic contributions to understanding polyketide synthases. *Chem. Rev.* **97:**2465–2497.

53. **Hopwood, D. A., T. Kieser, D. J. Lydiate, and M. J. Bibb.** 1986. *Streptomyces* plasmids: their biology and use as cloning vector. *In* S. W. Queener and E. Day (ed.), *The Bacteria*, vol. IX. *Antibiotic-Producing Streptomyces.* Academic Press, New York, N.Y.

54. **Hopwood, D. A., and H. M. Wright.** 1976. Interactions of the plasmid SCP1 with the chromosomes of *Streptomyces coelicolor* A3(2), p. 607–619. *In* K. D. McDonald (ed.), *Second International Symposium on the Genetics of Industrial Microorganisms.* Academic Press, London, United Kingdom.

55. **Hornemann, U., J. C. Otto, G. G. Hoffman, and A. C. Bertinuson.** 1987. Spectinomycin resistance and associated DNA amplification in *Streptomyces achromogenes* subsp. *rubradiris. J. Bacteriol.* **169:**2360–2366.

56. **Huisman, G. W., and R. Kolter.** 1994. Sensing starvation: a homoserine lactone-dependent signaling pathway in *Escherichia coli. Science* **265:**537–539.

57. **Hütter, R., and T. Eckhardt.** 1988. Genetic manipulation, p. 89–184. *In* M. Goodfellow, S. T. Williams, and M. Mordarski (ed.), *Actinomycetes in Biotechnology.* Academic Press, London, United Kingdom.

58. **Iismaa, T. P., and R. G. Wake.** 1987. The normal replication terminus of the *Bacillus subtilis* chromosome, *terC*, is dispensable for vegetative growth and sporulation. *J. Mol. Biol.* **195:**299–310.

59. **Ishikawa, J., Y. Koyama, S. Mizumo, and K. Hotta.** 1988. Mechanism of increased kana-

mycin-resistance generated by protoplast regeneration of *Streptomyces griseus.* II. Mutational gene alteration and gene amplification. *J. Antibiot.* **41:**104–112.

60. **Jumas-Bilak, E., S. Michaux-Charanchon, G. Bourg, D. O'Callaghan, and M. Ramuz.** 1988. Differences in chromosome number and genome rearrangements in the genus *Brucella. Mol. Microbiol.* **27:**99–107.

61. **Kessler, A., W. Dittrich, M. Betzler, and H. Schrempf.** 1989. Cloning and analysis of a deletable tetracycline-resistance determinant of *Streptomyces lividans* 1326. *Mol. Microbiol.* **3:**1103–1109.

62. **Kieser, H. M., T. Kieser, and D. A. Hopwood.** 1992. A combined genetic and physical map of the *Streptomyces coelicolor* A3(2) chromosome. *J. Bacteriol.* **174:**5496–5507

63. **Kinashi, H., M. Shimaji, and A. Sakai.** 1987. Giant linear plasmids in *Streptomyces* which code for antibiotic biosynthesis genes. *Nature* **328:**454–456.

64. **Kinashi, H., and M. Shimaji-Murayama.** 1991. Physical characterization of SCP1, a giant linear plasmid from *Streptomyces coelicolor. J. Bacteriol.* **73:**1523–1529.

65. **Koller, K. P., and G. Riess.** 1989. Heterologous expression of the alpha-amylase inhibitor gene cloned from an amplified genomic sequence in *Streptomyces tendae. J. Bacteriol.* **171:**4953–4957.

66. **Kolsto, A. B.** 1997. Dynamic bacterial genome organization. *Mol. Microbiol.* **21:**241–248.

67. **Krawiec, S., and M. Riley.** 1990. Organization of the bacterial chromosome. *Microbiol. Rev.* **54:**502–539.

68. **Leblond, P., and B. Decaris.** 1994. New insights into the genetic instability of *Streptomyces. FEMS Microbiol. Lett.* **123:**225–232.

69. **Leblond, P., P. Demuyter, L. Moutier, M. Laakel, B. Decaris, and J. M. Simonet.** 1989. Hypervariability, a new phenomenon of genetic instability related to DNA amplification in *Streptomyces ambofaciens. J. Bacteriol.* **171:**419–423.

70. **Leblond, P., P. Demuyter, J. M. Simonet, and B. Decaris.** 1991. Genetic instability and associated genome plasticity in *Streptomyces ambofaciens*: pulsed-field gel electrophoresis evidence for large DNA alterations in a limited genomic region. *J. Bacteriol.* **173:**4229–4233.

71. **Leblond, P., P. Demuyter, J. M. Simonet, and B. Decaris.** 1990. Genetic instability and hypervariability in *Streptomyces ambofaciens*: towards an understanding of a mechanism of genome plasticity. *Mol. Microbiol.* **4:**707–714.

72. **Leblond, P., G. Fischer, F. X. Francou, F. Berger, M. Guérineau, and B. Decaris.** 1996.

The unstable region of *Streptomyces ambofaciens* includes 210 kb terminal inverted repeats flanking the extremities of the linear chromosomal DNA. *Mol. Microbiol.* **19**:261–271.

73. **Leblond, P., F. X. Francou, J.-M. Simonet, and B. Decaris.** 1990. Pulsed-field gel electrophoresis analysis of the genome of *Streptomyces ambofaciens* strains. *FEMS Microbiol. Lett.* **72**:79–88.

74. **Leblond, P., M. Redenbach, and J. Cullum.** 1993. Physical map of the *Streptomyces lividans* 66 genome and comparison with that of the related strain *Streptomyces coelicolor* A3(2). *J. Bacteriol.* **175**:3422–3429.

75. **Lezhava, A., T. Mizukami, T. Kajitani, D. Kameoka, M. Redenbach, H. Shinkawa, O. Nimi, and H. Kinashi.** 1995. Physical map of the linear chromosome of *Streptomyces griseus.* *J. Bacteriol.* **177**:6492–6498.

76. **Lin, Y. S., and C. W. Chen.** 1997. Instability of artificially circularized chromosomes of *Streptomyces lividans. Mol. Microbiol.* **26**:709–719.

77. **Lin, Y. S., H. M. Kieser, D. A. Hopwood, and C. W. Chen.** 1993. The chromosomal DNA of *Streptomyces lividans* 66 is linear. *Mol. Microbiol.* **10**:923–933.

78. **Liu, S.-L., and K. E. Sanderson.** 1996. Highly plastic chromosomal organization in *Salmonella typhi. Proc. Natl. Acad. Sci. USA* **93**:10303–10308.

79. **Louarn, J.-M., J. Louarn, V. François, and J.-C. Patte.** 1991. Analysis and possible role of hyperrecombination in the termination region of the *Escherichia coli* chromosome. *J. Bacteriol.* **173**:5097–5104.

80. **Martin, P., A. Dary, and B. Decaris.** 1998. Generation of a genetic polymorphism in clonal populations of the bacterium *Streptomyces ambofaciens*: characterization of different mutator states. *Mutat. Res.* **421**:73–82.

80a. **Martin, P., A. Dary, and B. Decaris.** Unpublished data.

81. **McClintock, B.** 1951. Chromosome organization and genetic expression. *Cold Spring Harbor Symp. Quant. Biol.* **16**:13–47.

82. **Michaux, S., J. Paillisson, M. J. Carles-Nurit, G. Bourg, A. Allardet-Servent, and M. Ramuz.** 1993. Presence of two independent chromosomes in the *Brucella melitensis* 16 M genome. *J. Bacteriol.* **175**:701–705.

83. **Morrow, D. M., C. Connelly, and P. Hieter.** 1997. "Break copy" duplication: a model for chromosome fragment formation in *Saccharomyces cerevisiae. Genetics* **147**:371–382.

84. **Musialowski, M. S., F. Flett, G. B. Scott, G. Hobbs, C. Smith, and S. G. Oliver.** 1994. Functional evidence that the principal DNA replication origin of the *Streptomyces coelicolor* chromosome is close to the *dnaA-gyrB* region. *J. Bacteriol.* **176**:5123–5125.

85. **Muth, G., D. Frese, A. Kleber, and W. Wohlleben.** 1997. Mutational analysis of the *Streptomyces lividans recA* gene suggests that only mutants with residual activity remain viable. *Mol. Gen. Genet.* **255**:420–428.

86. **Nakano, M. M., H. Ogawara, and T. Sekiya.** 1984. Recombinations between short direct repeats in *Streptomyces lavendulae* plasmid DNA. *J. Bacteriol.* **157**:658–660.

87. **Okami, Y., and K. Hotta.** 1988. Search and discovery of new antibiotics, p. 33–67. *In* M. Goodfellow, S. T. Williams, and M. Mordarski (ed.), *Actinomycetes in Biotechnology.* Academic Press, London, United Kingdom.

88. **Omura, S., H. Ikeda, and C. Kitao.** 1979. The detection of a plasmid in *Streptomyces ambofaciens* KA-1028 and its possible involvement in spiramycin production. *J. Antibiot.* **32**:1058–1060.

89. **Orlova, V. A., and V. N. Danilenko.** 1983. Multiplication of DNA fragment in *Streptomyces antibioticus* producing oleandomycin. *Antibiotiki* **28**:163–173.

90. **Pandza, K., G. Pfalzer, J. Cullum, and D. Hranueli.** 1997. Physical mapping shows that the unstable oxytetracycline gene cluster of *Streptomyces rimosus* lies close to the ends of the linear chromosome. *Microbiology* **143**:1493–1501.

91. **Pandza, S., G. Biukovic, A. Paravic, A. Dadbin, J. Cullum, and D. Hranueli.** 1998. Recombination between the linear plasmid pPZG101 and the linear chromosome of *Streptomyces rimosus* can lead to exchange of ends. *Mol. Microbiol.* **28**:1165–1176.

92. **Petit, M.-A., J. M. Mesas, P. Noirot, F. Morel-Delville, and S. D. Ehrlich.** 1992. Induction of DNA amplification in the *Bacillus subtilis* chromosome. *EMBO J.* **11**:1317–1326.

93. **Piendl, W., C. Eichenseer, P. Viel, J. Altenbuchner, and J. Cullum.** 1994. Analysis of putative DNA amplification genes in the element AUD1 of *Streptomyces lividans* 66. *Mol. Gen. Genet.* **244**:439–443.

94. **Pothekin, Y. A., and V. N. Danilenko.** 1985. The determinant of kanamycin resistance of *Streptomyces rimosus*: amplification in the chromosome and reversed genetic instability. *Mol. Biol.* **19**:805–817. (In Russian. English translation, 672–683).

95. **Rauland, U., I. Glocker, M. Redenbach, and J. Cullum.** 1995. DNA amplifications and deletions in *Streptomyces lividans* 66 and the loss of one end of the linear chromosome. *Mol. Gen. Genet.* **246**:37–44.

96. **Redenbach, M., F. Flett, W. Piendl, I. Glocker, U. Rauland, O. Wafzig, R. Kliem, P. Leblond, and J. Cullum.** 1993. The *Streptomyces lividans* 66 chromosome contains a 1 MB deletogenic region flanked by two amplifiable regions. *Mol. Gen. Genet.* **241:**255–262.

97. **Redenbach, M., H. M. Kieser, D. Denapaite, A Eichner, J. Cullum, and D. A. Hopwood.** 1996. A set of ordered cosmids and a detailed genetic and physical map for the 8 Mb *Streptomyces coelicolor* A3(2) chromosome. *Mol. Microbiol.* **21:**77–96.

98. **Redshaw, P. A., P. A. McCann, M. A. Pentella, and B. M. Pogell.** 1979. Simultaneous loss of multiple differentiated functions in aerial mycelium-negative isolates of streptomycetes. *J. Bacteriol.* **137:**891–899.

99. **Romero, D., and R. Palacios.** 1997. Gene amplification and genomic plasticity in prokaryotes. *Annu. Rev. Genet.* **31:**91–111.

100. **Roth, J. R., N. Benson, T. Galitski, K. Haack, J. G. Lawrence, and L. Miesel.** 1996. Rearrangements of the bacterial chromosome: formation and applications, p. 2256–2276. *In* F. C. Neidhardt, R. Curtiss III, J. L. Ingraham, E. C. C. Lin, K. B. Low, B. Magasanik, W. S. Reznikoff, M. Riley, M. Schaechter, and H. E. Umbarger (ed.), *Escherichia coli and Salmonella typhimurium: Cellular and Molecular Biology*, 2nd ed. ASM Press, Washington, D.C.

101. **Sakaguchi, K.** 1990. Invertrons: a class of structurally and functionally related genetic elements that includes linear DNA plasmids, transposable elements, and genomes of Adenotype viruses. *Microbiol. Rev.* **54:**66–74.

102. **Salas, M.** 1991. Protein-priming of DNA replication. *Annu. Rev. Biochem.* **60:**39–71.

103. **Schauner, C., A. Dary, A. Lebrihi, P. Leblond, B. Decaris, and P. Germain.** Modulation of lipid metabolism and spiramycin biosynthesis in *Streptomyces ambofaciens* unstable mutants. *Appl. Environ. Microbiol.*, in press.

104. **Schneider, D., B. Aigle, P. Leblond, J. M. Simonet, and B. Decaris.** 1993. Analysis of genome instability in *Streptomyces ambofaciens*. *J. Gen. Microbiol.* **139:**2559–2567.

105. **Schneider, D., C. J. Bruton, and K. F. Chater.** 1996. Characterization of *spaA*, a *Streptomyces coelicolor* gene homologous to a gene involved in sensing starvation in *Escherichia coli*. *Gene* **177:**243–251.

106. **Schrempf, H.** 1985. Genetic instability: amplification, deletion, and rearrangement within the *Streptomyces* DNA, p. 436–440. *In* L. Leive (ed.), *Microbiology–1985*. American Society for Microbiology, Washington, D.C.

107. **Schrempf, H.** 1982. Plasmid loss and changes within the chromosomal DNA of *Streptomyces reticuli*. *J. Bacteriol.* **151:**701–707.

108. **Simonet, B., A. Dary, J.-M. Simonet, and B. Decaris.** 1992. Characterization of a family of multimeric CCC molecules of amplified chromosomal DNA in *Streptomyces ambofaciens* DSM40697. *FEMS Microbiol. Lett.* **78:**25–32.

109. **Sinclair, R. R., and M. J. Bibb.** 1988. The repressor gene (c) of *Streptomyces* temperate phage ΦC31: nucleotide sequence analysis and functional cloning. *Mol. Gen. Genet.* **213:**269–277.

110. **Stackebrandt, E., W. Liesack, and D. Witt.** 1992. Ribosomal RNA and rDNA sequence analyses. *Gene* **115:**255–260.

111. **Sturtevant, A. H.** 1925. The effects of unequal crossing over at the Bar locus in *Drosophila*. *Genetics* **10:**117–147.

112. **Suwanto, A., and S. Kaplan.** 1989. Physical and genetic mapping of the *Rhodobacter sphaeroides* 2.4.1 genome: presence of two unique circular chromosomes. *J. Bacteriol.* **171:**5850–5859.

113. **Umezawa, H.** 1988. Low-molecular-weight enzyme inhibitors and immunomodifiers, p. 285–325. *In* M. Goodfellow, S. T. Williams, and M. Mordarski (ed.), *Actinomycetes in Biotechnology*. Academic Press, London, United Kingdom.

114. **Vivian, A., and D. A. Hopwood.** 1973. Genetic control of fertility in *Streptomyces coelicolor* A3(2): new kind of donor strains. *J. Gen. Microbiol.* **76:**147–162.

115. **Volff, J. N., and J. Altenbuchner.** 1997. Influence of disruption of the *recA* gene on genetic instability and genome rearrangement in *Streptomyces lividans*. *J. Bacteriol.* **179:**2440–2445.

116. **Volff, J. N., D. Vandwiele, J.-M. Simonet, and B. Decaris.** 1993. Stimulation of genetic instability in *Streptomyces ambofaciens* ATCC23877 by antibiotics that interact with DNA gyrase. *J. Gen. Microbiol.* **139:**2551–2558.

117. **Volff, J.-N., C. Eichenseer, P. Viell, W. Piendl, and J. Altenbuchner.** 1996. Nucleotide sequence and role in DNA amplification of the direct repeats composing the amplifiable element AUD1 of *Streptomyces lividans* 66. *Mol. Microbiol.* **21:**1037–1047.

118. **Volff, J.-N., P. Viell, and J. Altenbuchner.** 1997. Artificial circularization of the chromosome with concomitant deletion of its terminal inverted repeats enhances genetic instability and genome rearrangements in *Streptomyces lividans*. *Mol. Gen. Genet.* **253:**753–760.

119. **Webb, C. D., A. Teleman, S. Gordon, A. Straight, A. Belmont, D. Chi-Hong Lin, A. D. Grossman, A. Wright, and R. Los-**

ick. 1997. Bipolar localization of the replication origin regions of chromosomes in vegetative and sporulating cells of *B. subtilis. Cell* **88:**667–674.

120. **Young M., and J. Cullum.** 1987. A plausible mechanism for large-scale chromosomal DNA amplification in streptomycetes. *FEBS Lett.* **212:**10–14.

121. **Zakrzewska-Czerwinska, J., and H. Schrempf.** 1992. Characterization of an autonomously replicating region from the *Streptomyces lividans* chromosome. *J. Bacteriol.* **174:**2688–2693.

122. **Zotchev, S. B., L. I. Soldatova, A. V. Orekhov, and H. Schrempf.** 1992. Characterization of a linear extrachromosomal DNA element (pBL1) isolated after interspecific mating between *Streptomyces bambergiensis* and *S. lividans. Res. Microbiol.* **143:**839–845.

123. **Zuerner, R. L., J.-L. Herrmann, and I. Saint-Girons.** 1993. Comparison of genetic maps for two *Leptospira interrogans* serovars provides evidence for two chromosomes and interspecies heterogeneity. *J. Bacteriol.* **175:**5445–5451.

GENOMIC FLUX: GENOME EVOLUTION BY GENE LOSS AND ACQUISITION

Jeffrey G. Lawrence and John R. Roth

15

Genome evolution is the process by which the content and organization of a species' genetic information changes over time. This process involves four sorts of changes: (i) point mutations and gene conversion events gradually alter internal information; (ii) rearrangements (e.g., inversions, translocations, plasmid integration, and transpositions) alter chromosome topology with little change in information content; (iii) deletions cause irreversible loss of information; and (iv) insertions of foreign material can add novel information to a genome. Although the first two processes can create new genes, they act very slowly. Gene loss and acquisition are genomic changes that can radically and rapidly increase fitness or alter some aspect of lifestyle.

Most thought on genome evolution has focused on how the slow sequence changes can cause divergence of gene functions. This is understandable because available data suggest that horizontal genetic transfer has been a minor contributor to the evolution of eukaryotic lineages (with notable exceptions, such as the

introduction of mitochondria and chloroplasts). In bacteria, however, both genetics and genome analysis provide extensive evidence for gene loss and horizontal genetic transfer. Analyses of these data suggest that gene loss and acquisition are likely to be the primary mechanisms by which bacteria adapt genetically to novel environments and by which bacterial populations diverge and form separate, evolutionarily distinct species. We suggest that bacterial adaptation and speciation are determined predominantly by acquisition of selectively valuable genes (by horizontal transfer) and by loss of weakly contributing genes (by mutation, deletion, and drift from the population) during periods of relaxed selection.

We propose that a limitation of genome expansion couples the rates of gene acquisition and loss. Genome size may be limited in part by population-based factors that limit the ability of cells to selectively maintain information; some limitation may also be imposed by physiological considerations. The balance between selective gene acquisition and secondarily imposed gene loss implies that addition of a foreign gene increases the probability of loss of some resident function of lower selective value. The interaction of these factors, we

Jeffrey G. Lawrence, Department of Biological Sciences, University of Pittsburgh, Pittsburgh, PA 15260.   *John R. Roth*, Department of Biology, University of Utah, Salt Lake City, UT 84112.

Organization of the Prokaryotic Genome, Edited by Robert L. Charlebois,
© 1999 American Society for Microbiology, Washington, D.C.

suggest, drives the divergence of bacterial types.

DYNAMICS OF GENE LOSS

Existence of a Gene Implies a Function

Traditional bacterial genetics allows identification of gene function by correlating mutant growth phenotypes with biochemical defects. One can demonstrate the functional importance of many DNA sequences by observing the consequences of their disruption. In contrast, genomic analyses identify genes solely as open reading frames, with possible similarity to genes of known function but without a direct tie to either phenotype or biochemical defect.

In identifying a gene by its sequence, rather than by mutations and phenotypes, one assumes implicitly that the very presence of a gene implies that it must confer a selectable function. That is, the gene could only have remained in the population if the encoded function is important; that is, mutants lacking the function show reduced fitness and are removed from the population by natural selection. This evolutionary argument implies that a mutant phenotype (perhaps difficult to demonstrate) will result if the gene is disrupted. In principle, a few genes might be encountered which either have just been introduced or have escaped selection, and the process of elimination has not yet run its course; data supporting such cases will be presented below. An important question arises when one tries to define the function of a gene (or determine whether it is one of the rare nonfunctional examples). That is, how important must a function be to assure the maintenance of its gene? How large a fitness contribution must a gene confer to remain in a genome?

A Spectrum of Fitness Contributions

All bacterial genes are not equally important. Functions performed by bacterial cells are diverse, and while some are essential for life under all conditions (e.g., RNA polymerase), others may provide a benefit only under certain circumstances (e.g., TMAO reductase). In this way, one may sort bacterial genes into broad classes that reflect their average importance to the cell (Table 1). Mutations in genes making essential or very important contributions to the cell will be strongly counterselected in bacterial populations, since these mutants cannot compete effectively against otherwise uncompromised conspecific individuals. However, mutations in genes that do not contribute to cell fitness (e.g., selfish genes on transposons) will not be counterselected, since these neutral mutants do not put their bearers at a selective disadvantage.

Between these extremes lies the gray area of mutations that have subtle effects on fitness; we include two extreme classes of genes. Some genes may make a minimal contribution to fitness under all growth conditions; others may make a large contribution to fitness but do so only in a rare subset of environments. For either class of gene, the average selection coefficient is low. The fate of mutations in the most weakly selected genes is governed by a complex interaction of natural selection, random genetic drift, population size, population subdivision, and genetic exchange within the species.

For any one species, a different fraction of genes may make up each category in Table 1. In small genomes (e.g., that of *Mycoplasma genitalium*), a large fraction of genes are likely to be essential (42), while a smaller fraction of genes are likely to be essential in prokaryotes with larger genomes (e.g., *Escherichia coli*). Regardless of the distribution, some genes in any genome will fall at the bottom of the list and make a minimal contribution to cellular fitness (i.e., they are nearly neutral). These genes will be at greatest risk for functional loss by mutation, deletion, and genetic drift. This class of genes may comprise a large portion of genomic information.

In *E. coli*, few of the 4,286 protein-coding genes are essential. Isolation of temperature-sensitive lethal mutations suggests that only ~200 genes are essential on rich medium (55).

TABLE 1 Fitness contributions of bacterial genes

Class	Fitness contribution	Physiological role	Consequence of null mutation
Top	Large	Essential	Lethality
High	Large	Very important	Strong impairment
Middle	Moderate	Important	Obvious impairment
Low (I)	Very low	Minimally useful in all conditions	Subtle impairment; hard to detect experimentally
Low (II)	Very low	Important in rare conditions	Strong impairment in some environments
Neutral	None	None	None

This estimate of the minimal number of essential genes in the *E. coli* chromosome (even when adjusted for failure to detect some genes by this method) is congruent with theoretical estimates of minimal gene number (256 genes) based on comparisons of several bacterial genomes (42). Moreover, a similar number of genes are detected by mutations that cause a nutritional requirement for growth. Together, these rather important gene classes constitute approximately 10% of the *E. coli* genome; the remaining 90% of genes must make smaller contributions to fitness or be needed only under particular conditions. This conclusion is supported by the analysis of *E. coli* mutants described above and by the observation that the genomes of free-living bacteria vary greatly in size. In addition, experimental approaches have revealed that lesions in a large proportion of genes in the yeast *Saccharomyces cerevisiae* have only minimal effects on fitness (65).

A Minimal Fitness Contribution Is Required for Gene Maintenance

While a complete description of the processes governing the selective maintenance of bacterial genes is beyond the scope of this paper, some general guidelines can be described. If a gene is to avoid loss by mutation and genetic drift, it must provide some minimal average selective benefit to the cell; we call that value s. (This is the maintenance threshold in Fig. 1.) If mutation rates are low, genes with small fitness contributions can be maintained in a population. If mutation rates are high, a stronger selective coefficient would be required to maintain a gene in a population of the same size. Therefore, s is proportional to the mutation rate, μ ($s \propto \mu$). As the mutation rate increases, null mutations in a gene under weak selection are more likely to drift to fixation, since defective alleles are created more rapidly than they can be removed by selection. As population size decreases, loss by drift becomes more likely and more genes become effectively neutral (49, 50). In these smaller populations, a larger selective value is required to maintain a gene in the population. That is, a gene must make a stronger contribution to fitness to be selectively maintained. Therefore, s is inversely proportional to the effective population size, N_e ($s \propto \mu/N_e$). The dynamics of how selection acts on mutant alleles is influenced by the rate of intraspecific recombination. As the recombination rate increases, selection can more effectively remove the steady accumulation of detrimental alleles from a population, and the species can avoid to some extent the fitness decline mandated by Muller's ratchet (41). As a result, the minimum fitness contribution required for selective maintenance in a haploid genome decreases as the recombination rate, r, increases:

$$s \propto \mu/rN_e \qquad (1)$$

Here, recombination allows selection to act more efficiently on individual alleles by plac-

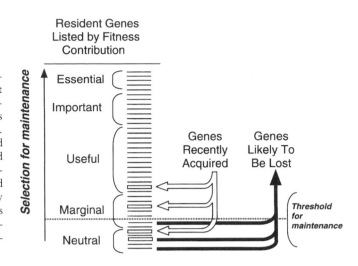

FIGURE 1 Genome evolution by genomic flux. A genome is depicted as a set of genes (lines and boxes) ranked by average selection coefficient; classes of genes discussed in Table 1 are noted on the left. Genes inherited vertically are represented by solid lines; foreign genes are indicated by open boxes. Genes below the threshold for maintenance cannot be maintained by natural selection and will be lost by mutation and deletion. Acquired genes introduced by horizontal transfer will ultimately be lost if they fail to confer sufficient fitness.

ing them in more genetic contexts and minimizing their ability to persist by association with valuable genes.

Thus, a gene must make a larger fitness contribution to be maintained as the mutation rate increases, as the population size decreases, or as the recombination rate decreases. Conversely, with lower mutation rates, bigger populations, or more recombination, genes making extremely subtle contributions to cellular fitness can maintain their positions. This relationship leads to two conclusions: (i) genes failing to make a minimal contribution to cellular fitness will be eliminated from the genome, and (ii) barring any change in mutation rate, recombination rate, or population size, the genome will have an upper size limit at which all resident genes can persist in a population of genomes by counterselection of less-fit mutants. Additional genes cannot be maintained by natural selection, and those genes making the smallest contribution to the organism's fitness will be lost.

Thus, the minimal selective value needed for gene maintenance is a variable dependent on several factors. These dependencies make it difficult to estimate the absolute values for the minimum required fitness contribution. Eukaryotic sexual species typically have high recombination frequencies, small population sizes, and mutation rates similar to or lower than those of bacteria. For such species, the estimate is made that a new mutation with no effect on fitness has a probability of $1/N$ of drifting to fixation, with N being the population size. Therefore, in the absence of selection, every neutral allele in the gene pool has an equal probability of displacing all other alleles and drifting to fixation in the population. To prevent loss by null mutations and drift, a gene must provide a selection value great enough to oppose this drift, placing the required fitness contribution in the range of $1/N$.

For bacteria, which have much larger population sizes, the above estimate would suggest that an extremely small fitness contribution would be sufficient to allow maintenance in the population. In bacteria, mutation rate is likely to be a more prominent dictator of loss than drift; opposing this would require a fitness contribution in the realm of the likelihood of mutational loss (10^{-5} per gene per generation), which is probably larger than $1/N$. Despite these reasonable calculations, the minimal fitness contribution could become much higher if many beneficial genes compete with each other to maintain a position in a genome of limited size. This possibility emerges from the model presented below.

Regardless of the numerical evaluation of the minimal fitness contribution, if genes are sorted in order of decreasing average fitness contribution (Table 1), some must necessarily fall at the end of the list and be prone to loss by mutation and drift. It is these genes that are always at risk for loss, especially when conditions change or newly acquired genes are inserted above them on the list.

Evidence for Gene Loss

Genome sequence comparisons among enteric bacteria reveal genes that have been lost from certain lineages while being maintained in others. For example, the *phoA* gene, encoding alkaline phosphatase, has been lost from the *Salmonella* lineage but has been maintained in the genomes of virtually all other enteric bacteria (16). Such cases demonstrate that genes conferring a sufficient selectable phenotype in one ecological context may fail to provide such a benefit in another context and therefore be subject to loss by mutation and genetic drift.

DYNAMICS OF GENE ACQUISITION

Horizontal Transfer of Genetic Information

DNA may be introduced into bacterial genomes by many processes, including conjugation, bacteriophage-mediated transduction, and transformation. The general term *recombination* typically refers to the introduction of DNA from a conspecific cell, whereas *horizontal genetic transfer* or *lateral genetic transfer* refers to interspecific exchanges; we will use these definitions in the discussion that follows.

The insertion of a foreign gene into a bacterial genome does not guarantee its survival. Rather, the gene must confer a sufficiently large selective advantage to the cell to avoid loss by mutation and genetic drift; above, we have described this minimal fitness contribution (s). The fate of an introduced DNA sequence can be predicted by the model developed above for the loss of DNA sequences under weak selection.

One may infer that the vast majority of sequences introduced by horizontal transfer would fail to make a minimal contribution and would be lost. Several factors may explain the failure to make a contribution. (i) The introduced DNA does not encode a product. (ii) The acquired genes are not expressed in the new host. (iii) The acquired genes are expressed and could contribute to a valuable function, but they cannot provide that function without the help of other genes that were not cotransferred (see "Genomic Flux and the Evolution of Gene Clusters" below). (iv) The acquired gene produces a functional protein, but this function does not increase the fitness of the new host cell (either the cell already possessed the function, or the activity is not useful). Therefore, the introduction and incorporation of foreign DNA per se does not ensure its persistence in the recipient gene pool. We must distinguish between horizontal transfer of genetic information (the physical process of incorporating foreign DNA into one of a cell's replicons) and horizontal transfer of useful phenotypic information (foreign DNA sequences that can be maintained for long periods by selection for the encoded functions). Although cursory inspection of genomic sequences may reveal genes that were introduced by horizontal transfer (38, 47, 69), such analyses alone do not reveal whether these sequences are providing a useful function (i.e., have been maintained for long periods by selection). The acquisition of sequence that may or may not increase fitness is diagrammed (with loss) in Fig. 1.

Horizontal Transfer of Useful Phenotypic Information

Horizontally acquired DNA can confer a selectable new function only if the genes are expressed and all of the new genes required for that function have been cotransferred. If the new function makes a sufficiently large contribution to fitness ($>s$, the threshold for maintenance), then the genes providing for this function may be maintained within the new host genome and escape loss from the population by mutation and genetic drift. In this way, the horizontal transfer of genetic information is successful. That is, long-term maintenance of the acquired DNA occurs only when that DNA provides phenotypic information that is useful to the recipient cell.

Some phenotypic capabilities may provide a selective advantage only to hosts growing in certain environmental regimes. Similarly, hosts growing in a particular ecological niche may find only certain kinds of functions useful. Therefore, although DNA may be transferred nonspecifically among a broad range of organisms, it persists in only a small subset of fortuitous cases, where the transferred DNA imparts a function of value to the host organism in its traditional, or a newly available, ecological context. Only in these rare cases does the horizontal-transfer event contribute to genome evolution (ignoring any mechanistic consequences of recombination). Below we describe methods for assessing the frequency of successful gene acquisition and the fraction of modern genomes that has been acquired by horizontal transfer of useful phenotypic information.

Evidence for Gene Acquisition

The evidence for gene acquisition in many organisms has been detailed elsewhere (64). In bacteria, horizontally transferred genes frequently confer phenotypes that are characteristic of particular taxonomic groups. Bacterial taxonomists assign a new isolate to a particular species by scoring possession of particular phenotypes that are present in one lineage but not in another. Examples of group-specific traits among enteric bacteria are lactose utilization in *E. coli* and citrate transport in *Salmonella enterica*. *Salmonella*'s ability to synthesize cobalamin and use it to support propanediol degradation forms the basis of a test diagnostic for this bacterium (52). These species-specific abilities are frequently ones whose genes appear to have been acquired horizontally (based on nucleotide composition and codon usage bias) (37). These data suggest that horizontal transfer is an important aspect of the genomic changes that have caused the divergence of phylogenetic groups. Large-scale surveys which assess the numbers and ages of acquired genes in bacterial genomes by DNA sequence analysis are discussed below.

Limits to Genome Expansion

The gene acquisition process outlined above implies that a genome might continuously accumulate genes that impart a minimal selective coefficient to the cell. However, it is clear that finite populations of cells cannot maintain an infinite number of genes. There is no evidence that bacterial genomes have been continuously increasing over evolutionary time. Rather, genome size is constrained by the population-genetic considerations outlined above. As detailed in equation 1, the minimal fitness contribution required to maintain a gene can be described as a function of the mutation rate, recombination rate, and population size ($s \propto \mu/rN_e$). If the selective benefit of an allele falls short of this level, it may be lost from the population by genetic drift. When applied to many genes in a genome, these forces limit genome expansion, because the minimum selective value required for each gene increases with the total number of genes under selection (G, the informative genome size).

The effect of genome size on the minimum selective value is a bit more difficult to explain but can be visualized in the following way. The overall fitness of a genome is a complex function of the combined fitness contributions of the many individual genes. As the number of genes increases, the target for deleterious mutations increases. Each deleterious mutation causes a fitness decrease for the organism and for alleles of other genes carried by that organism. Natural selection can reduce the frequency of deleterious mutations by favoring cells that do not carry deleterious mutations. However, as a genome increases in size, selection becomes less efficient at removing deleterious mutations from any one gene, since there are many more genes with potentially deleterious mutations.

First, consider a genome of sufficient size that, on average, one mutation occurs per genome per generation. Assuming that the number of individuals in the population exceeds the number of genes in a chromosome, every cell will acquire a potentially deleterious mutation every generation. If the recombination rate is sufficiently high, such mutations may be removed and Muller's ratchet may be avoided. As the number of genes increases, this selective cleansing becomes more difficult, until the number of mutations occurring every generation is more than can be counterselected. Eventually, null mutations will completely eliminate some genes from the population, specifically, those that made the smallest contribution to the organism's fitness. Thus, as genome size increases, the component genes effectively compete with one another for maintenance in the genome. The consequence of these effects is that maintenance of a gene in a large genome requires a higher selective value than maintenance of the same gene as part of a smaller genome.

The number of genes that can be simultaneously maintained by selection is limited by these population factors, which therefore limit genome expansion. The relationship between genome size (G) and these population factors is

$$G \propto rN_e/\mu \qquad (2)$$

Genome size, G, can increase only if recombination rates increase (increasing the efficiency with which selection can favor nonmutant alleles of each gene), population sizes increase (to decrease the rate of genetic drift), or mutation rates decrease (to minimize gene damage). The coupled gain and loss process is diagrammed in Fig. 1.

Empirical evidence supports this limitation on genome size. Despite high rates of horizontal genetic transfer among enteric bacterial species, the genomes of *E. coli*, *S. enterica*, and related organisms are notably uniform in size (2, 3, 45).

Acquisition of Gene Clusters

For a horizontally transferred gene to provide a new, selectively valuable function, all other genes required for that function must be present or cointroduced into the new host genome and be appropriately expressed. Without simultaneous transfer of all required genes, a single acquired gene cannot provide a selective benefit and the gene will not remain in the new population.

In bacteria, genes for nonessential (but useful) metabolic functions are often found in operons or in closely linked clusters of independently transcribed genes. These operons and clusters frequently include all the genes needed to provide a selectable function. We have proposed that evolutionary formation of these clusters (selfish operons) can occur by stepwise aggregation of unlinked genes and that it is driven by selection for increased efficiency of horizontal transfer among prokaryotic genomes (36). Evidence for this model is detailed later.

DYNAMICS OF GENOMIC FLUX

Competition between Genes for Maintenance in a Genome

To this point, we have considered gene loss and acquisition as separate processes. Although both loss and acquisition strongly influence genome evolution, we suggest that the two processes are synergistic due to the limits on genome expansion. As detailed above, the minimal selective contribution required for gene maintenance, s, is a function of the mutation rate, the recombination rate, and the population size (equation 1). The number of genes, G, that can be maintained by selection is also a function of these parameters (equation 2). If mutation rate, recombination rate, and population size remain constant, it is clear that the minimum selective contribution required for maintenance of each individual gene is a function of the genome size:

$$s_i \propto \mu G/rN_e \qquad (3)$$

That is, as genome size approaches the maximum number of genes maintainable by natural selection, each gene must make a greater contribution to cellular fitness in order to persist. If a genome is small relative to the maximum maintainable size, a gene making only a small contribution to cellular fitness can be maintained, since mutations in this gene can be effectively counterselected. However, as the number of genes under simultaneous selection increases, mutations in this same gene may no longer be effectively counterselected.

Thus, genes in a genome are competing with each other for maintenance in the face of mutation, deletion, and genetic drift. Consequently, an increase in genome size caused by a valuable horizontally acquired gene may increase the likelihood that some less valuable gene (probably of totally unrelated function) will be lost. Since genome size is limited by selection, each gene within a genome competes with the new invading genes, and with other genes, for maintenance in the genome.

The competition among genes for selective maintenance complicates the process of gene acquisition and loss. Because of this competition, the selective hierarchy of genes outlined above (Table 1) changes when a gene for a new function is acquired. Consider genes for a weakly selected function (*wsf*) poised at the threshold of maintenance by natural selection, s_i. If the cell acquires genes for a novel function that provides a selective advantage greater than s_i, the *wsf* genes will be lost from the genome, because the minimum selective contribution required for maintenance has increased and the value provided by *wsf* is no longer sufficient for maintenance (equation 3). In this way, once genome size has reached its upper limit, the process of gene acquisition can indirectly cause loss of less valuable genes. Genomes that are continually acquiring useful functions probably persist at this upper size limit as genes are continually added and lost.

Acquired genes may also cause selective loss of previously resident sequences if the old functions conflict with the introduced ones. That is, an introduced function may disrupt metabolism or cellular processes and be unable to contribute maximally until some resident functions are eliminated by mutations. Thus, the innovations provided by horizontally acquired genes may drive selective elimination of some previous functions from the host genome. In this way, the deletion of ancestral sequences effectively increases the fitness impact of the acquired sequences.

Genomic Flux and Ecological Differentiation

Gene acquisition, coupled with gene loss, is a dynamic process by which genetic material flows in and out of bacterial genomes. As noted above, long-term stability of transferred DNA must be associated with a selectable phenotype. Hence, the positively selected gain of valuable genes by a genome will be associated with loss of other genes, resulting in a constantly changing portfolio of phenotypic capabilities. The process of gene acquisition and loss will serve to create descendent populations that are genetically and phenotypically different from ancestral populations. This process is the inevitable outcome of the horizontal transfer of phenotypic information; the valuable acquired sequences will be maintained by selection, and insufficiently valuable native sequences will be displaced (Fig. 1).

Genomic flux can alter the phenotypic character of the recipient organism even when the acquired sequences encode functions that are highly similar to functions encoded by native sequences (e.g., *E. coli* maintains the horizontally acquired *argI* and *gapA* genes although each has a homologue performing a similar function). Unless the new and old versions have some sort of functional difference, natural selection cannot counterselect mutations in both sets of genes. An acquired sequence that provides precisely the same function as a native gene, and makes an identical, redundant contribution to cellular fitness, will have an equal chance of surviving mutational loss. This predicts that we may expect to see occasional examples of old functions being superseded by functionally equivalent genes acquired by horizontal transfer.

One can envision two outcomes of gene acquisition. First, the phenotypic alteration may allow the population to exploit its current ecological niche more effectively. This refines an existing species but does not create an ecologically distinct group of organisms. Alternatively, the acquired information may broaden the ecological niche of the descendent population, allowing it to exploit or survive new aspects of its surroundings. When this happens, the descendent population has novel capabilities that distinguish it from the ancestral population. This process of niche alteration defines bacterial speciation, whereby a group of organisms evolves and forms a distinct evolutionary lineage (67, 70).

The genotypic divergence of speciating lineages is increased by the inevitable loss of genes that accompanies gene acquisition. The losses make it unlikely that gene acquisition will simply broaden the ecological niche; rather, they mandate a qualitative change in niche definition. Lineages that prosper due to acquired traits will differ from the parent strain both by the newly acquired properties and by the gene losses that necessarily occur. As multiple genotypic differences that distinguish newly diverging species accumulate, the infrequent recombination among them will be less likely to cause coalescence of the lineages. To complete the speciation process, neutral point mutations lead to sequence differences between the diverging lineages; this reduces the homologous-recombination rate and imposes reproductive isolation. This drop in recombination is imposed by the sequence specificity of the recombination and mismatch repair (Mut) systems (53, 61, 68, 71).

Genomic flux makes bacterial divergence a natural outcome of the exploration of novel environments. The divergence of nascent species accelerates due to the combination of horizontal transfer and concomitant mutational loss of information. We think speciation would be less likely to occur (or would occur much more slowly) if it depended only on accumulation of internal point mutations. However, it is difficult to compare the two processes. It is unclear how frequently useful information is transferred or how efficiently novel metabolic capabilities can evolve by point-mutational change (internal gene duplication and divergence). It is also unclear how effectively internal gene formation could compete with acquisition of equivalent functions by horizontal transfer. Only by quantifying these two processes (discussed below) can we compare their importance. Below we provide evidence that in enteric bacteria, the magnitude of genomic flux is so large that it probably dominates the process of adding new functions to a genome.

Limits to Applicability of Genomic Flux in Speciation

The genomic-flux model may be less important for some bacterial lineages. The genomes of enteric bacteria show properties that fit well with predictions of the model. These enteric bacteria have large (5 Mb) genomes that encode many nonessential (but presumably useful) functions; such modestly important functions are subject to loss and acquisition. Differences in these nonessential functions distinguish enteric bacterial species, allowing each lineage to effectively exploit a different ecological niche.

The genomic flux paradigm may apply less well to organisms that do not experience high rates of horizontal transfer or those unlikely to derive any selective benefit from acquired genes. For example, bacterial endosymbionts have little access to horizontal genetic transfer or intraspecific recombination and persist in the relatively constant environment of the host. One might expect that their smaller population sizes and lower recombination rates would serve to reduce the number of genes the organisms could maintain (equation 3). Not surprisingly, phylogenetic analyses show that the genomes of these organisms are becoming smaller (39). Genome size reduction is more dramatic in obligate endosymbionts, although these inferences are confounded by possible gene transfer from the endosymbiont genome to the host genome. In this lineage, acquisition of novel metabolic capabilities is unlikely to allow differentiation of this lineage into related, but unexplored, ecological niches. The minimal fitness contribution required to maintain a gene in such organisms would be expected to be high.

MEASURING GENOMIC FLUX

Assessing Gene Loss and Acquisition

To assess the contribution of genomic flux to genome evolution and speciation, one must measure rates of gene loss and acquisition. To do this, one must identify foreign genes and determine when each was acquired (and how many other genes were lost) since the divergence of sibling species from a common ancestor. One must then determine how long each acquired sequence has persisted in the genome. By comparing the complete genome sequences of two closely related organisms, one can easily identify the sequences that are unique to each species; these sequences have either been gained unilaterally by one taxon or lost unilaterally from the other. The next problem is to assess the arrival times of the foreign genes.

Complete genome sequences allow one to count genes unique to each member of a closely related species pair. *M. genitalium* has a 580-kb genome (19), while the closely related species *Mycoplasma pneumoniae* possesses an 816-kb genome (20). Although each gene in the *M. genitalium* genome has a homologue in the *M. pneumoniae* genome, the *M. pneumoniae* genome contains 209 genes not found in *M. genitalium* (21). Similarly, the genome of the sulfate-reducing archeon *Archaeoglobus* sp. (29) is 25% larger than that of the strict methanogen *Methanococcus janaschii* (11). One could speculate that the additional genes found in *Archaeoglobus* allow its heterotrophic lifestyle; such genes would not confer a selective advantage on autotrophs like *M. janaschii*.

These genome comparisons demonstrate that gene loss and acquisition contributed to the divergence of these bacterial species. However, these species pairs are difficult to analyze further because we lack robust phylogenies describing their relationships to each other and to similar bacteria. Without these phylogenies, we cannot easily identify which genes were lost by one species (and would be present in a closely related outgroup taxon) and which sequences were gained (and would be absent from other members of the phylogenetic group). More importantly, genome comparisons do not tell us whether an acquired foreign gene is providing a function that increases the fitness of the organism. This information would be provided if we knew how long each gene had persisted in the genome. A long persistence time would suggest that the sequence is valuable; that is, its mutant alleles are being removed from the population by selection.

An alternative tactic is to identify genes of foreign origin and determine how long ago each entered the genome. This method relies on intrinsic sequence characteristics rather than genome comparisons. From the entry times of foreign genes, one can estimate the age structure of the gene population and the overall rate of gene acquisition. If genomes are not increasing in size, one can conclude that the overall gene acquisition rate is accompanied by an approximately equal rate of gene loss. The lost genes would include some old ancestral genes and some of the recently added genes that failed to confer sufficient selective value.

The assumption of a constant genome size seems reasonable in cases where the members of a robust phylogeny possess similar genome sizes. In these cases one can compare the rate of information influx to the rate at which information is introduced by point-mutational change and thereby assess the relative contributions of these processes to genome evolution. We have explored this general strategy by examining the genomes of *Salmonella* and *E. coli*.

Enteric bacteria have a robust phylogeny (33) and do not vary substantially in genome size (45). Moreover, the various members of this clade inhabit diverse ecological niches—

they exploit soil and water regimes and the digestive tracts of insects, fish, amphibians, reptiles, birds, and mammals and show path ogenic lifestyles. Furthermore, they have obvious phenotypic characteristics that distinguish one species from another. The two enteric species *E. coli* and *S. enterica* are clearly distinguishable, and each is the closest major relative of the other. The sequence of the *E. coli* genome is available (4), and that of the *Salmonella* genome will be available shortly. Considerable sequence information is already available for *Salmonella*. Therefore, we can identify genes unique to *E. coli* and *S. enterica*, identify the foreign-looking genes, and analyze the age structure of each genome, making the assumption of no genome expansion. The genomic-flux model predicts that the differences between these taxa (including metabolic differences or degree of pathogenicity) arose by the process of gene loss and acquisition. The selfish operon model (see below) predicts that multigene phenotypes acquired by horizontal transfer will be found in gene clusters and operons.

Identification of Foreign Genes Acquired by *E. coli*

Genes that are native to a genome show characteristic and species-specific nucleotide compositions, dinucleotide fingerprints (28), and codon usage biases (56, 57). These patterns emerge when genes experience the same directional mutation pressures for a long time (62, 63). The rate of mutations that convert an AT (or TA) pair to a GC (or CG) base pair is not necessarily the same as the rate of the reverse substitutions, GC (or CG) to AT (or TA). The relative rates of the two substitution types reflect intracellular deoxynucleoside triphosphate pools, the error frequency of DNA polymerases, and the error-specific effectiveness of mismatch repair systems. Organism-specific differences in these functions dictate the relative rates of AT and GC pair interconversion and establish the characteristic differences in nucleotide composition (percent G+C).

Codon usage bias measures the translational preferences among synonymous codons (some codons are translated more quickly, more efficiently, or more faithfully than others). These preferences reflect the relative abundance of different cognate tRNAs (23, 24) and the nature of the tRNA modifications that allow for discrimination among synonymous codons (29a). These factors modulate the effect of directional mutation pressures and are reflected in the relative percent G+C contents of the three codon positions in protein-coding sequences (43).

The effect of directional mutation pressure on position-specific nucleotide composition is shown in Fig. 2A. Muto and Osawa (43) correlated the overall nucleotide compositions of various bacterial genomes with the nucleotide composition of each of the three codon positions in coding sequences. They showed that nucleotides experiencing very weak selection for function (e.g., third codon positions, at which most substitutions fail to alter the encoded amino acid) show a dramatic increase in percent G+C content as genome percent G+C nucleotide composition increases. In contrast, nucleotides under strong natural selection (e.g., the second codon position, whose substitution alters the character of the encoded amino acid) show more modest changes. The behavior of the first position is intermediate, since a few changes there are synonymous. Although the original data of Muto and Osawa show substantial variance from a strict linear relationship between positional and overall percent G+C contents, their study was done at a time when rather little sequence data was available. When large sets of genes or whole genome sequences are added to the analysis, the relationships appear quite robust (linear equations are provided in Lawrence and Ochman [31]).

Directional mutation pressures are evident, in that every long-time resident gene of a genome shows an extremely predictable and characteristic nucleotide composition for the first, second, and third codon positions (Fig. 2A). In contrast, nonnative genes are identi-

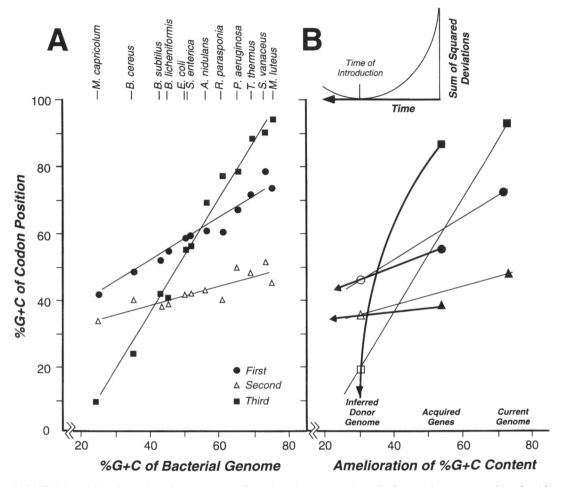

FIGURE 2 (A) Relationships between overall nucleotide composition of a bacterial genome and nucleotide compositions of the three codon positions (first to third); after Muto and Osawa (43) and Lawrence and Ochman (31). The organisms providing the data are shown at the top. The data for *E. coli* and *S. enterica* were calculated from 100 and 25% of the genome sequences, respectively, after known horizontally transferred sequences were removed. (B) Process of amelioration used to infer the time of introduction of acquired genes (31). The acquired genes (shaded symbols) are atypical for the genome (solid symbols) in which they are found. The codon position-specific nucleotide compositions of acquired genes are back-ameliorated (equation 4) until the minimum deviation (by least-squares analysis) from the Muto and Osawa relationships (open symbols) are obtained. The heavy lines indicate the codon position-specific nucleotide compositions during back-amelioration. The arrows indicate the calculated back-amelioration process used to estimate the elapsed time since the sequence showed the pattern of the donor. The inset graph shows the deviation of the curves from the Muto and Osawa relationships as a function of time rather than overall percent G+C.

fiably different, because they still show the sequence patterns imposed by the directional mutation pressures of their donor organisms. For example, a gene from *Bacillus cereus*, with 28% G+C at the third codon position, would be readily detectable as unusual in a back-

ground of *E. coli* genes with 58% G+C content at this position. Moreover, the codon usage bias and dinucleotide frequencies of *B. cereus* genes would be strikingly different from the major patterns found in the *E. coli* genome. These patterns have been used by many workers to identify foreign genes in partially sequenced bacterial genomes (28, 31, 38, 47, 69).

Lawrence and Ochman (32) have used these criteria to identify all of the horizontally transferred genes in the complete *E. coli* genome sequence. They used the nucleotide composition of codon positions, patterns of codon usage bias, and dinucleotide frequencies to determine that 15% of the *E. coli* genome (755 of 4,288 genes) was made up of genes identifiably introduced by horizontal transfer. This figure agrees well with previous estimates based on subsets of *E. coli* genes (38, 69). It should be noted that these methods provide a minimum estimate of the numbers of acquired genes, since they would not detect foreign genes donated by an organism with directional mutation pressures similar to those of *E. coli*. The 755 foreign genes (547.8 kb) were introduced into the *E. coli* genome in at least 234 lateral-transfer events since this species diverged from the *Salmonella* lineage 100 million years ago.

The large number of acquired genes supports the magnitude of horizontal transfer but does not reveal how many of the acquired genes contribute a useful function. Some of these genes, like those found on mobile genetic elements and prophages, may have been introduced recently into the *E. coli* genome and may not contribute to the fitness of the organism. To assess the role of gene acquisition in selective divergence, we must determine how many of these foreign genes have remained in the genome for sufficient time to assure us that they are maintained by selection. That is, we must estimate the entry time of genes that have arrived since the divergence of the *E. coli* and *Salmonella* lineages ~100 million years ago (40, 48).

Estimating the Introduction Time of Acquired Genes

The same characteristics of genes that suggest a foreign origin can help assess their time of introduction. As described above (see also Fig. 2), all genes that have been long-term residents of a bacterial genome experience the same directional mutation pressures and consequently evolve to exhibit relatively uniform patterns of nucleotide composition, codon usage bias, and dinucleotide frequencies. These genes are readily identified as foreign when encountered, following horizontal transfer, in a recipient genome which experiences different directional mutation pressures.

As these foreign genes are selectively maintained in the new genome, they accumulate substitutions that occur under the new set of directional mutation pressures. Over time, their nucleotide compositions will evolve to resemble patterns reflected by native genes (31). During this period of amelioration, however, the nucleotide composition of ameliorating genes will reflect a combination of directional mutation pressures: those imparted by their donor organism and the modifications imposed since entry into the new host genome. A gene caught in the act of amelioration is between the equilibrium states of the donor and recipient genomes and will show a sequence pattern that is unlike that of either donor or recipient (or any other well-ameliorated genome). The unique properties of nonequilibrium genes not only show their foreign origin but also permit an estimation of how long the amelioration process that has brought them from the well-ameliorated donor condition to their present (nonequilibrium) intermediate state has been under way. One can estimate the elapsed time since transfer by assessing the degree to which patterns deviate from the Muto and Osawa relationships (31).

The degree to which the compositional patterns of ameliorating genes depart from the relationships defined by typical genomes allows quantification of the time these genes have experienced the directional mutation pressures of their recipient genome. Lawrence and Ochman (31) described the rates at which the nucleotide composition of each codon position will change over time. At any moment, the amelioration rate for each codon position can be expressed as a function of the overall substitution rate, R (a function of the synonymous and nonsynonymous substitution rates), the current nucleotide composition of the horizontally transferred sequence, GC^{HT}, the nucleotide composition of the recipient genome, GC^{Native}, and the transition/transversion ratio (IV):

$$\Delta GC^{HT} = [(IV + 1/2)/(IV \quad (4)$$
$$+ 1)] \cdot R \cdot (GC^{Native} - GC^{HT})$$

Three equations, each derived from equation 4, describe the rate of amelioration for each codon position. If the positional nucleotide composition of an acquired gene deviates from the Muto and Osawa equilibrium relationships, each codon position may have been ameliorating, as described in equation 4. These amelioration equations can be used to determine the length of time that would be required to cause the observed deviation of the three codon positions from Muto and Osawa equilibria. By computational "back-amelioration" of the sequence, one can estimate the introduction time of an acquired gene. This is diagrammed in Fig. 2B. Normal amelioration in this example is proceeding leftward toward the host pattern.

The Age Structure of the *E. coli* Genome

Each of the 750 foreign genes in the *E. coli* genome was subjected to amelioration analysis (32) to estimate the time at which each entered the genome. Introduction times ranged from 0 to 100 million years ago. As seen in Fig. 3, about a third (225) of the foreign genes do not show any sign of amelioration, since their codon-positional percent G+C contents conform to the Muto and Osawa relationships expected for a sequence of their overall percent G+C content. These genes were likely introduced very recently and still reflect the directional mutation pressures of their donor genomes. Without direct genetic evidence, we cannot assess whether any of these newly acquired genes influences the fitness of *E. coli* and will be maintained in the genome.

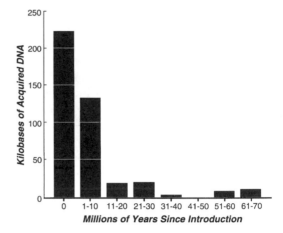

FIGURE 3 The distribution of times of introduction for horizontally acquired genes in *E. coli*; after Lawrence and Ochman (32). All foreign genes that could be ameliorated successfully are included. Roughly one-third of the foreign genes are not included because their positional percent G+C contents did not converge to fit the Muto and Osawa relationships (see the text).

Another third of the foreign genes have been impossible to date by the amelioration method. They are abnormal with respect to the Muto and Osawa relationships, but the back-amelioration process does not converge on a pattern that fits these relationships, so they are not represented in Fig. 3. Since this group includes transposable elements, prophages, and other selfish elements, we suspect that the odd nature of their sequences may result from their having moved repeatedly from one genome to another with long periods in which their sequences ameliorated without selection. Thus, two-thirds of the foreign genes are either newly acquired or from selfish elements; neither class of genes is likely to impart a selectable phenotype and contribute to species divergence. We expect that these genes will ultimately be deleted from the chromosome; selfish elements will continue to propagate and to persist, but older copies will eventually be removed (34, 44).

Another third of the acquired sequences shows signs of amelioration since entering the *E. coli* genome (Fig. 3). In these genes, mutations that abolish gene function appear to have been counterselected (the reading frames are not interrupted by nonsense codons), while mutations that alter nucleotide composition but do not abolish function have accumulated. These data suggest that these acquired genes have improved the average fitness of *E. coli* and have been maintained by natural selection. The age distribution of these useful foreign genes shows that fairly recent additions are more common than later arrivals, suggesting that many useful genes are held by selection for considerable time but are ultimately lost because of insufficient selective value. We assume that the rate of introduction of DNA by horizontal transfer has remained constant since the divergence of the *E. coli* and *Salmonella* lineages, and the vast majority of horizontally acquired genes have failed to confer a sufficiently useful function to become permanent residents.

After correcting for this inevitable deletion of ameliorated foreign selected genes, we estimate the rate of horizontal transfer of genes conferring selectable phenotypes (horizontal transfer of phenotypic information) into *E. coli* to be ~16 kb/million years (32). The age structure of this group of genes suggests that despite being held by selection for some time, most of these genes are ultimately eliminated (a small proportion of genes has ameliorated beyond our ability to detect them, but this process is much slower than the apparent deletion rate). As noted above, we assume that this rate of acquisition is balanced by a comparable rate of deletion of ancestral genes that could no longer be maintained by natural selection. The overall rate of introduction of DNA is much higher (in excess of 64 kb/million years) and includes sequences which fail to make a contribution to the fitness of the cell and have remained only long enough to be detected in the genome of *E. coli* K-12. We suspect that this vastly underestimates the true flux of total sequence. Our interpretation of the genome structure is diagrammed in Fig. 4.

The identities of the horizontally acquired sequences support the genomic-flux model of speciation. Horizontally acquired genes encode many of the functions by which *E. coli* and *Salmonella* differ. These include *E. coli* genes for lactose utilization (*lac*), phosphonate utilization (*phn*), iron citrate transport (*fec*), and tryptophan degradation (*tna*) and *Salmonella* genes for cobalamin biosynthesis (*cob*), propanediol degradation (*pdu*), citrate utilization (*tct*), and host invasion (*spa*). Other functions present in only one taxon (like alkaline phosphatase [PhoA] in *E. coli* [16]) are unique because the corresponding genes were deleted from the other taxon. We know of no phenotypic property that discriminates between these taxa and is not correlated with a gene loss or acquisition event. No taxonomically useful function has been identified in either taxon that has arisen by internal duplication and divergence of an ancestral gene to allow evolution of a new function by point mutation.

Genomic Flux versus Point–Mutational Change

The overall rate of horizontal transfer of selectively valuable phenotypic information (16 kb/million years) is lower than the total rate of sequence introduction, since much information does not contribute to fitness. Based on the substitution rates of *E. coli* genes (56), scattered point mutations are estimated to alter the information of the *E. coli* genome at a rate equivalent to 22 kb/million years (31). However, unlike the overall rate of DNA introduction by horizontal transfer, very few of these changes are likely to contribute to cellular fitness; the bulk (~90%) are at synonymous codon positions and cause no change in the nature of the encoded protein. Therefore, while gene acquisition and point mutations both introduce change into bacterial genomes, the qualitative nature of the information they furnish is quite different. Acquired sequences must provide a selectable function in order to be maintained by natural selection and show signs of amelioration; very few point muta-

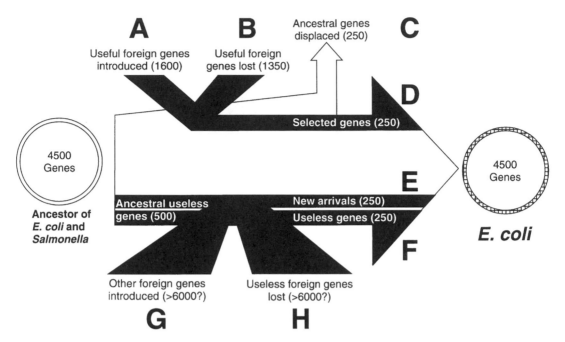

FIGURE 4 Schematic representation of the effects of genomic flux on the *E. coli* genome. The events depicted occurred since divergence of *E. coli* and *Salmonella* from the common ancestor. The useful gene flux (ABD) depicted above the major arrow describes genes that are kept in the genome long enough to show measurable amelioration. Although large numbers of potentially useful genes enter the chromosome (A) and may be maintained for some time, most of these are ultimately lost (B); however, about 250 remain (D). The number of selectively maintained foreign genes (D) is matched by a loss of ancestral genes (C). The flux below the major arrow (EFGH) represents sequences that provide no selective advantage. Large amounts of DNA with no value are likely introduced into the chromosome (G), but the vast majority are removed by deletion (H). About 500 of these genes are still in the genome but will eventually be removed by mutation and drift; these include those known to be useless (F), like mobile genetic elements, and those merely too recently arrived to have been subject to deletion (E). The newly arrived genes have not been tested by selection, but very few are likely to remain. As shown by the black bars, the ancestral chromosome also contained a portion of selectively useless genes that would have been deleted.

tions are likely to improve cellular fitness. For this reason, we maintain that new genes contributing to the long-term evolution of bacterial species are more likely to have been obtained by horizontal transfer than by internal duplication and divergence (point mutations).

Salmonella and Its Divergence from *E. coli*

Analysis of the *E. coli* and *Salmonella* genomes by using the principles outlined above allows several conclusions regarding the divergence of *Salmonella* and *E. coli* from their common ancestor. Many of these predictions will be

better tested when the complete sequence of the *Salmonella* genome is available. We predict that 15 to 30% of each genome will comprise sequences absent from the other taxon; DNA hybridization data supports this conclusion. Whole-genome DNA-DNA hybridization studies showed that the *E. coli* and *S. enterica* genomes are 45% "related" (7–10). This estimate reflects two processes: (i) more than half of each genome is comprised of shared sequences that are ~85% identical (48, 56), and (ii) the remaining portion of each genome (between 25 and 45%) is unique (8, 10). As detailed above, the sequences unique to each genome include both acquired sequences and genes that have been lost from the other species' genome (Fig. 4).

Lateral genetic transfer contributed many of the features used to identify strains of *Salmonella*. A notable example is B$_{12}$ metabolism, the basis of the Rambach test (52) for *Salmonella* identification, which scores the ability to synthesize cobalamin (*cob*) and perform cobalamin-dependent degradation of propanediol (*pdu*). *Salmonella* acquired these functions in a single horizontal-transfer event that added a block of over 40 contiguous genes, nearly 1% of the *Salmonella* genome. These abilities are shared by virtually every isolate of *S. enterica* (35). Amelioration analysis suggests that the genes were acquired 71 million years ago (31), after the divergence of the *E. coli* and *Salmonella* lineages (~100 million years ago) but before the radiation of the salmonellae (~50 million years ago). The phylogenetic distribution of these functions supports the amelioration method for estimating transfer time.

The gene arrangement within the acquired sequence block (*pdu-cob*) supports the importance of horizontal transfer. This block includes about 40 genes that act together to provide a single selectable phenotype—the ability to degrade a carbon source and synthesize the cofactor needed for this degradation. The block of genes includes four independently transcribed units: the *pdu* operon for degradation of propanediol (15 to 20 genes), the *pduF* gene for importing propanediol, the *cob* operon for synthesis of B$_{12}$ (20 genes), and the *pocR* gene for regulation of all three transcripts (54). There is no known reason that these transcription units need to be close together in order to perform their functions. The fact that all were acquired in a single transfer event suggests why they happen to be close together—their proximity in some donor organism made it possible for *Salmonella* to acquire a complex metabolic capability and a selectable phenotype. Below, we will propose that such gene clusters and operons are formed in a process driven by the horizontal-transfer process discussed above.

A notable feature of the *Salmonella* lineage is its widespread exploitation of pathogenic lifestyles. Strains of *S. enterica* are pathogens of many organisms, including humans, mice, cows, reptiles, amphibians, and poultry (22). Many genes facilitating this pathogenic lifestyle are found in clusters known as pathogenicity islands. Analysis of these regions in *Salmonella* has indicated that many essential virulence factors, including those for attachment (*Salmonella* pathogenicity island [SPI-1]), invasion (SPI-2), and macrophage survival (SPI-3), are found on segments of DNA acquired by horizontal transfer (46, 58). The occurrence of horizontally acquired pathogenicity islands in many pathogens (1) suggests that the genomic-flux model may provide a general framework for describing the evolution of many bacterial lineages. A more complete evaluation of the impact of genomic flux in *Salmonella* evolution can be performed when the genomic sequence becomes available.

Examination of available *Salmonella* sequence data suggests that the rate of foreign-sequence acquisition has been similar to that estimated (see above) for *E. coli*. However, this equality of acquisition rates need not be true and may not be seen for all pairs of sister species. One species may have continued to live in a manner similar to that of the common ancestor, while the sister species diverged to explore some novel environmental situation. The exploratory species may have acquired more foreign genes that gave it the capabilities needed to support its divergence, while the conservative species did not experience significant selection for acquisition of new functions.

If *Salmonella* acquires useful foreign sequence at the rate of 16 kb per million years estimated for *E. coli*, we assume (as we did for *E. coli*) that it will lose ancestral sequence at a similar rate. However, gain and loss of sequences will occur independently in the two lineages, and their genomes will diverge to the extent that they gain and lose different information. By this analysis alone (without comparison to an outgroup taxon), we cannot estimate directly how much sequence was lost unilaterally from the *Salmonella* or *E. coli* lineage or how much ancestral sequence was lost from both genomes independently. However, even without this information, we can use the above considerations to predict what will be seen when the two genomes are compared after 100 million years of divergence.

The general expectations are diagrammed in Fig. 5. Assuming similar events in each lineage and a random loss of ancestral sequences, we predict that the roughly 4,000 genes under selection in both *Salmonella* and *E. coli* will prove to be distributed in the following way. The two genomes will each contain roughly 3,000 shared genes inherited vertically from their common ancestor. Each will contain an added 250 ancestral genes that have been lost

from the other species by deletion. Each genome will contain about 750 foreign genes, a different set in each organism, that were acquired by horizontal transfer; of these, about 250 are under selection and contributing to the fitness of the organism. The rest contribute no valuable function: they are either new arrivals that have yet to be eliminated (250) or they are selfish elements that are decaying without selection (250). These expectations are generally borne out by the available sequence data and by DNA hybridization tests done many years ago (7, 10).

IMPACT OF GENOMIC FLUX

Mechanisms of Gene Acquisition

The high rate of introduction of DNA into the *E. coli* genome, in excess of 64 kb/million

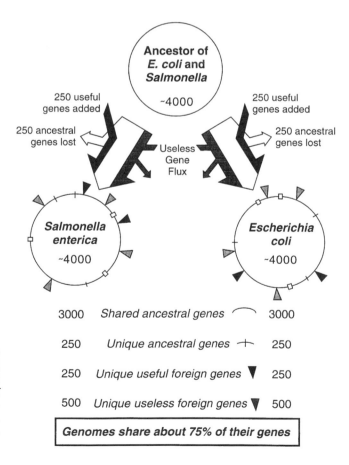

FIGURE 5 Divergence of *Salmonella* and *E. coli* genomes. Based on the model presented in the text, the likely events and final genomic consequences of this act of speciation are portrayed. On genomes, triangles indicate acquired foreign material. Open boxes indicate lost ancestral material.

years, requires plausible mechanisms for the introduction and stable chromosomal incorporation of heterologous genes. It may seem unlikely that sequences could be transferred in natural environments at the rates inferred here. Several facets of this process should be appreciated. First, even an extremely low rate of heterologous recombination will ensure a high rate of gene acquisition by a species if the population size is large and the transferred gene provides a selective advantage. This combination allows extremely rare transfer events to introduce genes that will rise to high frequency in the population. Second, genes may initially be introduced into the cytoplasm on independent replicons (plasmids or bacteriophages) that can provide a long period of replication and selective maintenance prior to integration into the bacterial chromosome. Lastly, transposons are site-specific recombination mechanisms that may mediate insertion of foreign genes; these mechanisms circumvent the need for homologous recombination with its demand for close sequence similarity.

There is evidence for transposon- or bacteriophage-mediated acquisition of foreign genes in the *E. coli* chromosome (32). First, many acquired genes lie adjacent to tRNA genes; bacteriophages are known to use tRNA genes as chromosomal integration sites (14, 25, 51), and many acquired gene blocks adjacent to tRNA genes include homologues of bacteriophage genes. We suspect that other acquired sequences adjacent to tRNA genes may be remnants of prophages from which identifiable bacteriophage genes have been deleted; far too many more foreign genes are found adjacent to tRNA genes than would be expected at random.

A significant fraction (68%) of the insertion sequences found in the *E. coli* chromosome are associated with horizontally acquired genes, often lying at the boundary between native and acquired sequences (32). A foreign segment would initially be flanked by direct-order copies of the insertion sequence (IS) element if an introduced circular DNA fragment were integrated by replicative transposition. (If replicative transposition occurs between a linear fragment and the chromosome, a second exchange would be required—transposition, recombination, or an illegitimate event—to recircularize the recipient chromosome ([59]). The IS element mediating the transposition probably resided on the *E. coli* chromosome and not on the acquired fragment, since IS elements of each class within *E. coli* are nearly identical in DNA sequence.

One might argue that the association of IS elements with acquired genes is not due to a role in integration but rather reflects the dispensability of foreign sequences, which allows them to be used as targets with minimal selective consequences. This alternative would predict that all IS elements would be found more often with foreign sequences. The mechanistic involvement in incorporation seems more likely to us because certain IS elements show a much higher likelihood of association than others (six of seven for IS2 but zero of three for IS186). A notable example of apparent IS-mediated gene acquisition is *E. coli*'s G+C-rich *argI* region at min 7, which is flanked by IS1 elements.

Genomic Flux and Speciation

Gene loss and acquisition represent a powerful mechanism by which genomes adjust to major changes in organism lifestyle. Such adjustments are implicit in bacterial speciation. Speciation is the process by which an ancestral population of organisms—defined by their exploitation of a particular ecological niche—evolves to form two populations, each exploiting a separate ecological niche and recombining more often within their own group than with the sister species. As detailed above, genomic flux appears to have played a big role in the speciation event by which *E. coli* and *S. enterica* diverged. The distribution of these organisms in natural environments (18, 66) and their distinctive behavior in clinical situations (22) suggest that they enjoy significantly different lifestyles and inhabit substantially different ecological niches.

We postulate that such diversification into ecologically distinct organisms would not be feasible (in the 100-million-year period) if the genomes of these organisms diverged solely by point mutation or internal rearrangement. We estimate (see above) that about 80% of the genomes of *E. coli* and *Salmonella* are derived from their common ancestor and that each species possesses about 1,000 genes that are absent from the other's genome. The observed differences appear to be the result of unilateral gene acquisition and unilateral gene loss. None of the features now used to distinguish these organisms taxonomically can be attributed to an accumulation of point mutations and internal functional divergence. Rather, gene acquisition (and attendant gene loss) has provided each organism with distinct selectable phenotypes.

Evolution of Metabolic Novelty

Even if horizontal transfer of phenotypic information is a major factor in bacterial evolution, metabolic novelties must ultimately evolve by stepwise mutational changes that cause functional alteration of preexisting genes. We argue that, when the possibility of horizontal transfer exists, the slow process of functional modification cannot provide for efficient competitive exploitation of an environment. We suggest that the obviously homologous genes with distinct functions seen in a single bacterial genome (apparent paralogues) probably arose initially in distinct lineages as the sole member of the gene family (orthologues). After the orthologues diverged in different genomes, they were brought into a common lineage by horizontal transfer. Thus, we suggest that most of the apparent paralogues seen in bacterial genomes are actually horizontally transferred orthologues. After the microbial world came to include a wide variety of highly adapted genes, assorting these information bits by horizontal transfer became much more likely than reinvention of functions by internal duplication and divergence. In this sense, we suspect there will be few examples of true paralogues in bacterial genomes.

The evolution of metabolic novelty by duplication and divergence is almost always a slow, stepwise, and inherently inefficient process. For example, the evolution of a catabolic pathway for utilization of a hexose may require the following:

1. alteration of binding specificities for several enzymes to accommodate binding of new substrates and intermediates

2. alteration of binding sites of these enzymes to minimize binding of their original substrates

3. alteration of the active sites of these enzymes to allow them to perform their catalytic functions efficiently and effectively on the new substrates and intermediates

4. alteration of release activities of the new products to optimize enzyme turnover

5. alteration of regulatory proteins to discriminate between old and new substrates

6. alteration of regulatory interactions to allow gene expression under potentially different growth conditions

7. coordinate evolution of multiple enzymes to accommodate new substrates and intermediates

We maintain that all of these evolutionary steps are required to provide a function that allows efficient, competitive exploitation of novel environments. Organisms at any stage in the evolutionary process outlined above will not effectively compete with those that have acquired a preformed highly evolved gene complex which performs the same task.

We propose that when horizontal acquisition is possible and appropriate information modules preexist, internal evolution of metabolic novelty (reinvention of the wheel by duplication and divergence) cannot occur in the context of competition. When bacterial speciation entails the competitive invasion of an ecological niche, the evolution of metabolic novelty cannot be correlated with bacterial speciation. The analysis of *E. coli* and *S. enterica* detailed above supports this hypothesis; none of the characteristics that discriminate between these taxa can be attributed to the evolution of metabolic novelty by intragenomic duplication and divergence. Rather, all of these differences, which we posit were intimately involved in the divergence of the lineages and adaptation to their individual niches, can be attributed to gene loss and gene acquisition. All novel phenotypes were conferred by horizontally transferred genes. Therefore, genes for the diverse metabolic pathways have formed slowly at earlier times and not during the course of competitive invasion of novel ecological niches. Regardless of the relative rates of these two processes, it is clear that gene loss and acquisition have facilitated exploration of novel environments and allowed more rapid divergence of bacterial types in competitive situations.

Genomic Flux and the Evolution of Gene Clusters

The genomic-flux model predicts that horizontal gene transfer mediates the acquisition of novel phenotypic capabilities. If multiple gene products are required to confer a selectable phenotype, the phenotype cannot be transmitted unless all of the required genes are simultaneously mobilized. A subset of the necessary genes will not provide a selectable phenotype and cannot be maintained by natural selection. Therefore, the applicability of the genomic-flux model beyond simple functions requiring the product of only a single gene is contingent upon the physical clustering of genes that provide for a single function.

Bacterial genomes are notable for having clusters of cotranscribed genes, usually contributing to a single selectable function (26). These clusters include genes from a variety of gene families that must have been brought together after their initial formation. As described previously (36) and outlined below, we believe that formation of these clusters was driven by the selective advantage conferred on the clustered alleles themselves (in comparison to the same genes in an unclustered state). The clustered alleles are fitter in a global sense be-

cause they can transfer more widely and (like transposable elements) have a selfish advantage. This advantage need not be associated with any physiological improvement of host phenotype. We propose that the prevalence of gene clusters stands as evidence that genomic flux has historically been a primary contributor to genome evolution.

The cotranscription of many clustered genes (operons) is a further refinement of this process and has allowed subsequent coregulation of gene clusters whose products participate in a single metabolic process. Since the elucidation of gene clusters in the 1950s (15) and that of operon structure in the 1960s (26, 27), four theories have been offered to explain the evolution of bacterial gene clusters; these are reviewed in reference 36. The natal theory proposes that gene clusters resulted from the tandem duplication and divergence of parental genes that occurred during initial evolution of metabolic pathways. While this process may explain clusters of homologous genes (e.g., mammalian globin gene clusters), bacterial operons typically contain genes whose products belong to distinct gene families (e.g., kinases, methylases, and dehydrogenases) and were likely assembled from previously existing unlinked genes.

The Fisher theory postulates that coadapted genes—those whose products have been selected to work together efficiently—will appear to be more tightly linked than expected, since recombination may disrupt particularly advantageous combinations of coadapted alleles (17). Consider two genes, each with two alleles, A and a at one locus and B and b at another locus. If natural selection favors organisms bearing a coadapted combination of alleles at these loci (AB or ab) and counterselects organisms bearing more poorly cooperating combinations (Ab or aB), linkage disequilibrium will occur among the alleles at these loci, simply because the good combinations confer greater fitness. Apparent disequilibrium among alleles would be expected, even for genes on different chromosomes, because the alternative combinations have lower fitness. This model was extended (5, 60) to explain the origins of clusters of genes in haploid organisms, especially in bacteriophage genomes (6, 12, 13). It was postulated that selection favors the assembly of genes into clusters, so that unfavorable recombination between coadapted genes occurs less frequently. This model is restricted to the assembly of operons bearing coadapted alleles (probably encoding products that interact physically) in a freely recombining, variable population. The model is not supported by the fact that organisms with the highest recombination rates (obligatory sexual species) show little evidence of clustering related genes. Conversely, the largely asexual bacterial lineages show a strong tendency to cluster genes with related functions.

The coregulation model (probably the most widely accepted) postulates that genes are found in operons because coordinate expression of the constituent genes is beneficial to the cell. While the benefits of coregulation may contribute to selection for maintenance of a gene cluster, this selection cannot drive formation of the clusters, since coregulation cannot provide a strong selective advantage for the intermediate states that must exist prior to transcriptional fusion. Moreover, coregulation is seen for many bacterial genes (e.g., the *E. coli arg*, *nad*, *pur*, and *pyr* genes) that are not assembled into operons, making it clear that clustering is not essential to coregulation. We have proposed a different model to account for the evolution of operons.

The selfish operon model posits that gene clusters were assembled as a natural consequence of frequent horizontal transfer and that they serve to increase the fitness of the constituent genes by facilitating their distribution to a wide variety of genomes (36). As detailed below, this model provides a way of selecting for progressive clustering of genes and for the maintenance of gene clusters once formed.

Genomic Flux Facilitates Gene Clustering

The assembly of genes into clusters must entail a series of intermediate steps, during which not all of the members of the gene clusters are physically close. More precise positioning of genes in cotranscribable groups is likely to require additional steps. The most effective means of juxtaposing genes (deletion of intervening genetic material) will not be useful when the intervening material is selectively valuable. For this reason, we discount coregulation as a plausible selective force to drive the evolution of operons.

Horizontal transfer of genes to new genomes provides a selective force that is not subject to the above limitations (36). As detailed above, the introduction of a foreign DNA sequence to a genome does not ensure its persistence. Sequences that do not contribute to fitness are subject to deletion. Therefore, when a large fragment is introgressed, those functions contributing a valuable phenotype will be selectively maintained while intervening noncontributing material will be deleted. In this way, transferred genes that act together to confer a valuable phenotype will be maintained and brought closer together. Although deletion of intervening genes was not possible in the context of the donor genome (where the deletions were deleterious), deletion is inevitable in the context of a new recipient genome. Therefore, any loosely linked collection of cooperating genes, just close enough to be transferred between organisms, will become progressively more tightly linked with time in the new host. As the cluster of related genes becomes tighter, it will transfer more efficiently to new hosts. It should be noted that the base sequence of genes in a cluster may be identical before and after clustering has occurred. Thus, clustering is not expected to provide any phenotypic improvement to the host (compared to the same genes in an unclustered state) but rather provides a benefit to the clustered genes themselves (wider distribution). We termed this the selfish operon model because we consider physical proximity a selfish property of the constituent genes, allowing more frequent transfer to naive genomes. The process for evolution of gene clusters is diagrammed in Fig. 6.

Evolutionary advantage of clustered selectable alleles

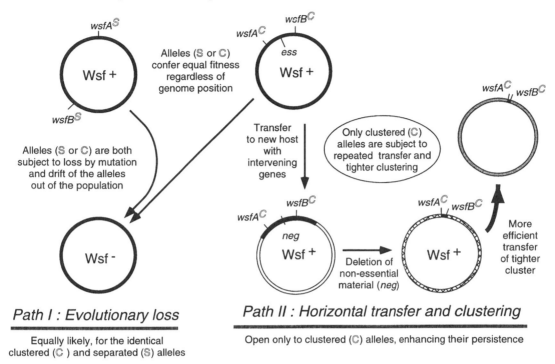

FIGURE 6 Mechanism for clustering of genes by horizontal transfer. The diagram compares the fates of identical alleles of two genes, *wsf*, that together confer a weakly selectable function. In one organism these genes are clustered (C), and in the other they are separated (S). Both sets are equally subject to loss by mutation and drift. However, clustered *wsf* genes can spread horizontally to new genomes. Following transfer, genes (*ess*) that were essential in the donor become nonessential (*neg*) in the new species. While the selected *wsf* genes are selectively maintained, the *neg* genes are deleted. This tightens the *wsf* cluster and enhances the likelihood of its further transfer.

Tighter Gene Clustering and Cotranscription Facilitate Horizontal Transfer

As described above, assembly of related genes into clusters is driven because the clustered state improves the efficiency with which those genes spread between and among species. As gene clusters become tighter, their ability to be mobilized increases, making the model described here operate with ever-increasing efficiency. All known mechanisms of gene transfer increase in efficiency with decreasing size of the fragment that must be transferred. Genes contributing to a single function have a very low probability of cotransfer if they are dispersed on a bacterial chromosome. How-

ever, once even loosely aggregated genes have been assembled into a cluster following a single horizontal transfer (see above), these genes have a higher probability of successful cotransfer (and cooperative provision of a phenotype) to new genomes.

Cotranscription of gene clusters (organization of gene clusters into operons) facilitates horizontal transfer by minimizing the number of promoters that must function in the new context. Gene clusters may frequently be transferred into recipient cells with a transcription apparatus different from that of the donor; no phenotype can be conferred unless all promoters function in the new host. This problem is minimized as the number of required promoters is reduced and is eliminated if a block of cotranscribable genes integrates near a host promoter. Furthermore, operons of translationally coupled genes gain the additional advantage of not requiring de novo translation start signals in the new host. Therefore, operons of cotranscribed, translationally coupled genes are highly portable packages of genetic information that function in the widest variety of organisms. For this reason, the cotranscription can be considered an additional selfish property of the constituent genes that extends the benefits accrued by proximity. Once an operon has formed, for purely selfish reasons, a regulatory mechanism may provide additional benefits; in this way, coregulation may contribute to selection for the maintenance of a gene cluster but it cannot provide selection for the initial assembly, or cotranscription, of that cluster.

Summary

We have outlined a model for the evolution of bacterial genomes through the synergistic processes of gene acquisition and gene loss. From analysis of the *E. coli* genome, we estimate that genes are lost and gained at a rate of 16 kb/million years. The information gained by this process allows exploration of novel ecological niches under competitive conditions; the coupled loss of genes cuts the new organism off from its old lifestyle and contributes to its divergence from the ancestral population. We suggest that this process has driven the organization of bacterial genes into clusters and cotranscribed operons, which allow a single horizontally transferred fragment to confer novel phenotypic capabilities on recipient cells. The organization of genes into operons reflects the important role in bacterial evolution and speciation played by genomic flux—the development of bacterial genomes by gene loss and gene acquisition.

REFERENCES

1. **Barinaga, M.** 1996. A shared strategy for virulence. *Science* **272:**1261–1263.

2. **Bergthorsson, U., and H. Ochman.** 1995. Heterogeneity of genome sizes among natural isolates of *Escherichia coli. J. Bacteriol.* **177:**5784–5789.

3. **Bergthorsson, U., and H. Ochman.** 1998. Distribution of chromosome length variation in natural isolates of *Escherichia coli. Mol. Biol. Evol.* **15:**6–16.

4. **Blattner, F. R., G. Plunkett III, C. A. Bloch, N. T. Perna, V. Burland, M. Riley, J. Collado-Vides, J. D. Glasner, C. K. Rode, G. F. Mayhew, J. Gregor, N. W. Davis, H. A. Kirkpatrick, M. A. Goeden, D. J. Rose, B. Mau, and Y. Shao.** 1997. The complete genome sequence of Escherichia coli K-12. *Science* **277:**1453–1474.

5. **Bodmer, W. F., and P. A. Parsons.** 1962. Linkage and recombination in evolution. *Adv. Genet.* **11:**1–100.

6. **Botstein, D.** 1980. A theory of modular evolution for bacteriophages. *Ann. N.Y. Acad. Sci.* **354:**484–491.

7. **Brenner, D. J., and D. B. Cowie.** 1968. Thermal stability of *Escherichia coli-Salmonella typhimurium* deoxyribonucleic acid duplexes. *J. Bacteriol.* **95:**2258–2262.

8. **Brenner, D. J., and S. Falkow.** 1971. Molecular relationships among members of the enterobacteriaceae. *Adv. Genet.* **16:**81–118.

9. **Brenner, D. J., G. R. Fanning, K. E. Johnson, R. V. Citarella, and S. Falkow.** 1969. Polynucleotide sequence relationships among members of the *Enterobacteriaceae. J. Bacteriol.* **98:** 637–650.

10. **Brenner, D. J., G. R. Fanning, F. J. Skerman, and S. Falkow.** 1972. Polynucleotide sequence divergence among strains of *Escherichia coli* and closely related organisms. *J. Bacteriol.* **109:** 953–965.

11. **Bult, C. J., O. White, G. J. Olsen, L. Zhou, R. D. Fleischmann, G. G. Sutton, J. A. Blake, L. M. FitzGerald, R. A. Clayton, J. D. Gocayne, A. R. Kerlavage, B. A. Dougherty, J.-F. Tomb, M. D. Adams, C. I. Reich, R. Overbeek, E. F. Kirkness, K. G. Weinstock, J. M. Merrick, A. Glodek, J. L. Scott, N. S. M. Geoghagen, J. F. Weidman, J. L. Fuhrmann, D. Nguyen, T. R. Utterback, J. M. Kelley, J. D. Peterson, P. W. Sadow, M. C. Hanna, M. D. Cotton, K. M. Roberts, M. A. Hurst, B. P. Kaine, M. Borodovsky, H.-P. Klenk, C. M. Fraser, H. O. Smith, C. R. Woese, and J. C. Venter.** 1996. Complete genome sequence of the methanogenic archaeon, *Methanococcus jannaschii. Science* **273:** 1058–1073.

12. **Campbell, A., and D. Botstein.** 1983. Evolution of lambdoid phages, p. 365–380. *In* R. W. Hendrix, J. W. Roberts, F. W. Stahl, and R. A. Weisberg (ed.), *Lambda II*. Cold Spring Harbor Laboratory, Cold Spring Harbor, N.Y.

13. **Casjens, S., G. Hatfull, and R. Hendrix.** 1992. Evolution of dsDNA tailed-bacteriophage genomes. *Virology* **3:**383–397.

14. **Cheetham, B. F., and M. E. Katz.** 1995. A role for bacteriophages in the evolution and transfer of bacterial virulence determinants. *Mol. Microbiol.* **18:**201–208.

15. **Demerec, M., and P. Hartman.** 1959. Complex loci in microorganisms. *Annu. Rev. Microbiol.* **13:**377–406.

16. **DuBose, R. F., and D. L. Hartl.** 1990. The molecular evolution of alkaline phosphatase: correlating variation among enteric bacteria to experimental manipulations of the protein. *Mol. Biol. Evol.* **7:**547–577.

17. **Fisher, R. A.** 1930. *The Genetical Theory of Natural Selection*. Oxford University Press, Oxford, United Kingdom.

18. **Foltz, V. D.** 1969. *Salmonella* ecology. *J. Am. Oil Chem. Soc.* **46:**222–224.

19. **Fraser, C. M., J. D. Gocayne, O. White, M. D. Adams, R. A. Clayton, R. D. Fleischmann, C. J. Bult, A. R. Kerlavage, G. Sutton, J. M. Kelley, J. L. Fritchman, J. F. Weidman, K. V. Small, M. Sandusky, J. L. Fuhrmann, D. T. Nguyen, T. R. Utterback, D. M. Saudek, C. A. Phillips, J. M. Merrick, J.-F. Tomb, B. A. Dougherty, K. F. Bott, P.-C. Hu, T. S. Lucier, S. N. Peterson, H. O. Smith, C. A. I. Hutchison, and J. C. Venter.** 1995. The minimal gene complement of *Mycoplasma genitalium*. *Science* **270:**397–403.

20. **Himmelreich, R., H. Hilbert, H. Plagens, E. Pirkl, B. C. Li, and R. Herrmann.** 1996. Complete sequence analysis of the genome of the bacterium *Mycoplasma pneumoniae*. *Nucleic Acids Res.* **24:**4420–4449.

21. **Himmelreich, R., H. Plagens, H. Hilbert, B. Reiner, and R. Herrmann.** 1996. Comparative analysis of the genomes of the bacteria *Mycoplasma pneumoniae* and *Mycoplasma genitalium*. *Nucleic Acids Res.* **25:**701–712.

22. **Hoff, G. L., and D. M. Hoff.** 1984. Salmonella and Arizona, p. 69–82. *In* G. L. Hoff, F. L. Frye and E. R. Jacobson (ed.), *Diseases of Amphibians and Reptiles*. Plenum Press, New York, N.Y.

23. **Ikemura, T.** 1980. The frequency of codon usage in *E. coli* genes: correlation with abundance of cognate tRNA, p. 519–523. *In* S. Osawa, H. Ozeki, H. Uchida, and T. Yura (ed.), *Genetics and Evolution of RNA Polymerase, tRNA and Ribosomes*. University of Tokyo Press, Tokyo, Japan.

24. **Ikemura, T.** 1982. Correlation between the abundance of yeast transfer RNAs and the occurrence of the respective codons in protein genes. Differences in synonymous codon choice patterns of yeast and Escherichia coli with reference to the abundance of isoaccepting transfer RNAs. *J. Mol. Biol.* **158:**573–597.

25. **Inouye, S., M. G. Sunshine, E. W. Six, and M. Inouye.** 1991. Retrophage phi R73: an *E. coli* phage that contains a retroelement and integrates into a tRNA gene. *Science* **252:**969–971.

26. **Jacob, F., and J. Monod.** 1962. On the regulation of gene activity. *Cold Spring Harbor Symp. Quant. Biol.* **26:**193–211.

27. **Jacob, F., D. Perrin, C. Sanchez, and J. Monod.** 1960. L'opéron: groupe de gènes à expression coordonée par un opérateur. *C. R. Acad. Sci.* **250:**1727–1729.

28. **Karlin, S., and C. Burge.** 1995. Dinucleotide relative abundance extremes: a genomic signature. *Trends Genet.* **11:**283–290.

29. **Klenk, H. P., R. A. Clayton, J. F. Tomb, O. White, K. E. Nelson, K. A. Ketchum, R. J. Dodson, M. Gwinn, E. K. Hickey, J. D. Peterson, D. L. Richardson, A. R. Kerlavage, D. E. Graham, N. C. Kyrpides, R. D. Fleischmann, J. Quackenbush, N. H. Lee, G. G. Sutton, S. Gill, E. F. Kirkness, B. A. Dougherty, K. McKenney, M. D. Adams, B. Loftus, and J. C. Venter.** 1997. The complete genome sequence of the hyperthermophilic, sulphate-reducing archaeon *Archaeoglobus fulgidus*. *Nature* **390:**364–370.

29a. **Lawrence, J.** Unpublished results.

30. **Lawrence, J. G.** 1997. Selfish operons and speciation by gene transfer. *Trends Microbiol.* **5:**355–359.

31. **Lawrence, J. G., and H. Ochman.** 1997. Amelioration of bacterial genomes: rates of change and exchange. *J. Mol. Evol.* **44:**383–397.

32. **Lawrence, J. G., and H. Ochman.** 1998. Molecular archaeology of the *Escherichia coli* genome. *Proc. Natl. Acad. Sci. USA* **95:**9413–9417.

33. **Lawrence, J. G., H. Ochman, and D. L. Hartl.** 1991. Molecular and evolutionary relationships among enteric bacteria. *J. Gen. Microbiol.* **137:**1911–1921.

34. **Lawrence, J. G., H. Ochman, and D. L. Hartl.** 1992. The evolution of insertion sequences within enteric bacteria. *Genetics* **131:**9–20.

35. **Lawrence, J. G., and J. R. Roth.** 1996. Evolution of coenzyme B12 among enteric bacteria: evidence for loss and reacquisition of a multigene complex. *Genetics* **142:**11–24.

36. **Lawrence, J. G., and J. R. Roth.** 1996. Selfish operons: horizontal transfer may drive the evolution of gene clusters. *Genetics* **143:**1843–1860.

37. **Lawrence, J. G., and J. R. Roth.** 1997. Roles of horizontal transfer in bacterial evolution. *In* M. Syvanen and C. Kado (ed.), *Horizontal Gene Transfer.* Chapman and Hall, London, United Kingdom.

38. **Medigue, C., T. Rouxel, P. Vigier, A. Henaut, and A. Danchin.** 1991. Evidence for horizontal gene transfer in *Escherichia coli* speciation. *J. Mol. Biol.* **222:**851–856.

39. **Moran, N. A.** 1996. Accelerated evolution and Muller's rachet in endosymbiotic bacteria. *Proc. Natl. Acad. Sci. USA* **93:**2873–2878.

40. **Moran, N. A., M. A. Munson, P. Baumann, and H. Ishikawa.** 1993. A molecular clock in endosymbiotic bacteria is calibrated using insect hosts. *Proc. R. Soc. Lond. B* **253:**167–171.

41. **Muller, H.** 1932. Some genetic aspects of sex. *Amer. Nat.* **66:**118–138.

42. **Mushegian, A. R., and E. V. Koonin.** 1996. A minimal gene set for cellular life derived by comparison of complete bacterial genomes. *Proc. Natl. Acad. Sci. USA* **93:**10268–10273.

43. **Muto, A., and S. Osawa.** 1987. The guanine and cytosine content of genomic DNA and bacterial evolution. *Proc. Natl. Acad. Sci. USA* **84:**166–169.

44. **Naas, T., M. Blot, W. M. Fitch, and W. Arber.** 1994. Insertion sequence-related genetic variation in resting *Escherichia coli. Genetics* **136:**721–730.

45. **Ochman, H., and U. Bergthorsson.** 1995. Genome evolution in enteric bacteria. *Curr. Opin. Genet. Dev.* **5:**734–738.

46. **Ochman, H., and E. A. Groisman.** 1996. Distribution of pathogenicity islands in *Salmonella* spp. *Infect. Immun.* **64:**5410–5412.

47. **Ochman, H., and J. G. Lawrence.** 1996. Phylogenetics and the amelioration of bacterial genomes, p. 2627–2637. *In* F. C. Neidhardt, R. Curtiss III, J. L. Ingraham, E. C. C. Lin, K. B. Low, B. Magasanik, W. S. Reznikoff, M. Riley, M. Schaechter, and H. E. Umbarger (ed.), Escherichia coli *and* Salmonella typhimurium: *Cellular and Molecular Biology,* 2nd ed. American Society for Microbiology, Washington, D.C.

48. **Ochman, H., and A. C. Wilson.** 1988. Evolution in bacteria: evidence for a universal substitution rate in cellular genomes. *J. Mol. Evol.* **26:**74–86.

49. **Ohta, T.** 1973. Slightly deleterious mutant substitutions in evolution. *Nature* **264:**96–98.

50. **Ohta, T.** 1976. Role of very slightly deleterious mutations in molecular evolution and polymorphism. *Theor. Popul. Biol.* **10:**254–275.

51. **Pierson, L. S. D., and M. L. Kahn.** 1987. Integration of satellite bacteriophage P4 in *Escherichia coli.* DNA sequences of the phage and host regions involved in site-specific recombination. *J. Mol. Biol.* **196:**487–496.

52. **Rambach, A.** 1990. New plate medium for facilitated differentiation of *Salmonella* spp. from *Proteus* spp. and other enteric bacteria. *Appl. Environ. Microbiol.* **56:**301–303.

53. **Rayssiguier, C., D. S. Thaler, and M. Radman.** 1989. The barrier to recombination between *Escherichia coli* and *Salmonella typhimurium* is disrupted in mismatch-repair mutants. *Nature* **342:**396–401.

54. **Roth, J. R., J. G. Lawrence, and T. A. Bobik.** 1996. Cobalamin (coenzyme B12): synthesis and biological significance. *Annu. Rev. Microbiol.* **50:**137–181.

55. **Schmid, M. B., N. Kapur, D. R. Isaacson, P. Lindroos, and C. Sharpe.** 1989. Genetic analysis of temperature-sensitive lethal mutants of Salmonella typhimurium. *Genetics* **123:**625–633.

56. **Sharp, P. M.** 1991. Determinants of DNA sequence divergence between *Escherichia coli* and *Salmonella typhimurium*: codon usage, map position, and concerted evolution. *J. Mol. Evol.* **33:** 23–33.

57. **Sharp, P. M., and W.-H. Li.** 1987. The codon adaptation index—a measure of directional synonymous codon usage bias, and its potential applications. *Nucleic Acids Res.* **15:**1281–1295.

58. **Shea, J. E., M. Hensel, C. Gleeson, and D. W. Holden.** 1996. Identification of a virulence locus encoding a second type III secretion system in *Salmonella typhimurium. Proc. Natl. Acad. Sci. USA* **93:**2593–2597.

59. **Sonti, R. V., D. H. Keating, and J. R. Roth.** 1992. Lethal transposition of Mud phages in Rec⁻ strains of *Salmonella typhimurium. Genetics* **133:**17–28.

60. **Stahl, F. W., and N. E. Murray.** 1966. The evolution of gene clusters and genetic circularity in microorganisms. *Genetics* **53:**569–576.

61. **Stambuk, S., and M. Radman.** 1998. Mechanism and control of interspecies recombination in *Escherichia coli.* I. Mismatch repair, methylation, recombination and replication functions. *Genetics* **150:**533–542.

62. **Sueoka, N.** 1988. Directional mutation pressure and neutral molecular evolution. *Proc. Natl. Acad. Sci. USA* **85:**2653–2657.

63. **Sueoka, N.** 1992. Directional mutation pressure, selective constraints, and genetic equilibria. *J. Mol. Evol.* **34:**95–114.

64. **Syvanen, M., and C. Kado.** 1998. *Horizontal Gene Transfer.* Chapman & Hall, Ltd., London, England.

65. **Thatcher, J. W., J. M. Shaw, and W. J. Dickinson.** 1998. Marginal fitness contributions of nonessential genes in yeast. *Proc. Natl. Acad. Sci. USA* **95:**253–257.

66. **Thomason, B. M., J. W. Biddle, and W. B. Cherry.** 1975. Detection of salmonellae in the environment. *Appl. Microbiol.* **30:**764–767.

67. **Van Valen, L.** 1976. Ecological species, multi-species, and oaks. *Taxon* **25**:223–239.

68. **Vulic, M., F. Dionisio, F. Taddei, and M. Radman.** 1997. Molecular keys to speciation: DNA polymorphism and the control of genetic exchange in Enterobacteria. *Proc. Natl. Acad. Sci. USA* **94**:9763–9767.

69. **Whittam, T. S., and S. Ake.** 1992. Genetic polymorphisms and recombination in natural populations of *Escherichia coli*, p. 223–246. *In* N. Takahata and A. G. Clark (ed.), *Mechanisms of Molecular Evolution*. Japan Scientific Society Press, Tokyo, Japan.

70. **Wiley, E. O.** 1978. The evolutionary species concept reconsidered. *Syst. Zool.* **27**:17–26.

71. **Zahrt, T. C., and S. Malot.** 1997. Barriers to recombination between closely related bacteria: MutS and RecBCD inhibit recombination between *Salmonella typhimurium* and *Salmonella typhi*. *Proc. Natl. Acad. Sci. USA* **94**:9786–9791.

GENE TRANSFER IN
ESCHERICHIA COLI

Roger Milkman

16

Recombination includes both the rearrangement of the genetic material in an individual genome (46) and *gene transfer*, the incorporation of exogenous genetic material into an individual genome. While rearrangement and gene transfer share some molecular processes, their functions are different in most basic ways.

The transfer of genetic material into a genome may take the form of replacement of a similar component (homologous recombination). Alternatively, nonhomologous recombination may involve either the replacement of an unrelated component or incorporation without replacement.

Gene transfer in bacteria may be vertical, within a line of descent from parent to offspring, or horizontal, between two nonlineal individuals. While horizontal gene transfer is possible between two immediate descendants of a common ancestor, its detectability and potential impact are significantly greater between individuals whose common ancestry is not recent.

Horizontal gene transfer between members of a single species is far more common than

that between individuals at a greater phylogenetic (not necessarily physical) distance. Nevertheless, the rarity of phylogenetically distant gene transfer is occasionally compensated for by an increased likelihood of its retention. This is important, because the retention of any given replacement (even a favorable replacement) is highly improbable, and only *retained* replacements (those that are not lost by selection or random genetic drift) count in evolution. As will be detailed later, an unusually great selective advantage can confer a vast increase in the probability of retention.

Bacterial recombination differs from the basic sexual metazoan paradigm, in whose simplest form a diploid genome is rearranged (by homologous crossing over) and divided into transitory haploid parts in gametes, which combine to form new diploid individuals of dual ancestry. In bacteria, gene transfer is not linked to reproduction, and it is not frequent, so that bacterial population structure is basically clonal (77). This limits evolutionary trajectories. The accumulation of the progressive improvements that drive natural selection is limited when better genotypes must be assembled from the mutations occurring only within a single line of descent. This limitation is relieved by horizontal gene transfer, gener-

Roger Milkman, Department of Biological Sciences, The University of Iowa, Iowa City, IA 52242-1324.

Organization of the Prokaryotic Genome, Edited by Robert L. Charlebois,
© 1999 American Society for Microbiology, Washington, D.C.

ally within the species but, in an important minority of cases, from beyond as well (38, 45, 71, 92).

In bacteria there are three major categories of gene transfer mechanisms: conjugation, transduction, and transformation. All operate in *Escherichia coli*. Conjugation (3, 25, 31, 44) typically involves the specific binding of a donor cell's pilus to a recipient's cell surface. Single-stranded DNA generated by the rolling-circle mechanism passes from the donor to the recipient cell (the precise physical path is not yet established), and a complementary strand forms on the incoming template. The DNA is now in either of two forms, circular (as in a conjugational plasmid) or linear (a chromosomal segment beginning at an origin of transfer). It is known to be double-stranded before incorporation into the recipient chromosome because of the evident action of restriction endonucleases (which in general attack only double-stranded DNA) on donor DNA (49, 52, 53, 60). While conjugation is ordinarily the most species specific of the three gene transfer mechanisms, experimental constructs have resulted in conjugational DNA transfer from *E. coli* to *Saccharomyces cerevisiae* (30, 92).

Transduction is mediated either by the filling of a phage head by a random fragment of double-stranded donor DNA instead of a similar-sized phage genome concatemer (generalized transduction [48], e.g., via bacteriophage P1) or by the inclusion of flanking donor DNA with a phage genome as it is excised from a host chromosome (specialized transduction [88], e.g., via bacteriophage λ). Species specificity stems from the host range of the phage, which varies with phage species and strain.

Transformation (29, 45) is in general the least specific mechanism of gene transfer, although numerous bacterial species do use recognition sequences to admit conspecific DNA preferentially. While many species conduct a lively internal trade in *natural transformation*, in which healthy cells extrude DNA (45, 92), the evidently rare exchange of DNA across great phylogenetic distances is clearly also of great importance in evolution. The wide-range transmission of drug resistance is of course known to be mediated by plasmids. Other nonhomologous genes, such as new restriction-modification system components (7, 8, 72) and O-antigen structure determinants (36), are often seen on plasmids, and there are practical reasons why their transfer by plasmids should be common: first, host specificity can be quite low; second, plasmids are not vulnerable to exonucleases; and third, the relatively small size of the transmitted fragment reduces the amount of excess baggage, which is especially apt to be counterproductive when acquired from a phylogenetically distant host. Once such DNA has been incorporated into a new genome, it can of course be relayed throughout the species by the resident intraspecific mechanisms (presumably resulting in the transmission and homologous recombination of extensive flanking DNA).

It is worth noting that, in addition to the well-known mechanisms of plasmid transfer by conjugation and by transformation, there are other less well-defined but clearly demonstrated processes, such as transformation by cell contact (92, p. 160–161) and transformation in aquatic habitats (92, p. 164–166). Plasmids have been observed to move by both routes, in *Escherichia* as well as numerous other bacteria. The implication of plasmids in lateral transfer to *E. coli* and the demonstration of *E. coli*'s ability to display natural competence for plasmid transformation (10, 92) argue for the (nonexclusive) feasibility of this specific pathway.

Recent years have seen the expansion of the subject of natural transformation. The discovery of notable properties in the diverse environmental theaters in which it takes place (92), the analysis of its component processes (45), and a variety of small experiments with potentially big implications (10, 26, 66, 81) suggest that gene transfer over great phylogenetic distances is frequent enough to be significant whenever the transferred genes are highly advantageous to the recipient.

GENOMIC ORGANIZATION

The organization of the *E. coli* chromosome can be seen in three perspectives: the arrangement of genes and basic chromosomal functions in an individual genome, the genetic and structural variation among strains of the species, and the dynamics of DNA exchange. An example of the first aspect is now available in admirable detail, clarity, and completeness for *E. coli* K-12 strain MG1655 (6, 14). The second has been available for some time in the results of detailed samples of a large number of wild isolates, including the standard set of 72 *E. coli* Reference (ECOR) strains (64), and a variety of other strains, benign (75, 77, 89) and pathogenic (1, 76); it is revealed by multilocus enzyme electrophoresis (MLEE) (15, 64, 74, 76, 90), restriction fragment length polymorphism (RFLP) (57), and comparative sequencing (13, 15, 23, 24, 51, 76, 90). Here, the presence of three major groups, as well as several sets of two to four nearly identical strains, among the 72 independent ECOR isolates (which were chosen to maximize diversity) suggests the presence of some vast clones among the millimol (6×10^{20}) or so of *E. coli* cells in the world. A view of the dynamics of DNA exchange is just now developing, and it will be the primary focus of this chapter, although the complete details are not yet definitive. The interpretation of existing empirical information rests on some simple principles of population genetics.

The basic clonal structure of *E. coli*, first proposed as a dynamic model in 1951 (4), appears to stem from the occasional (in evolutionary time) selection of a very rare mutation, or a combination of mutations; current views also allow the possibility of a recombinational replacement. The entire genome is evidently carried to high frequency by hitchhiking with the selected motivating allele (53, 54). The original periodic selection model (4) explicitly excluded recombination; later, this exclusion was not regarded as a necessary element. In any event, recombination frequency currently appears to be too low, except in two hypervariable regions (see "The Hypervariable

Regions" below) to scramble the chromosomes before the clones reach substantial size. The motivating allele must have been produced by a very rare event, as opposed to a single nucleotide substitution such as $T \rightarrow A$, which would occur with a frequency of perhaps 10^{-10} per nucleotide per generation (22). The appearance of an individual new allele, moreover, is in itself generally a highly transitory phenomenon, since the probability of its retention is ordinarily vanishingly small. A brief consideration of the underlying dynamics will be useful, and this leads first to the basic population genetic parameters of selection.

FITNESS AND RETENTION

A standard assumption is that the number of individuals in a species remains constant over long periods of time. However, the number of individuals sharing a certain property may increase progressively as other groups become smaller. For any group of individuals, the ratio of N_1, the number of individuals present after a unit time period, to N_0, the initial number of individuals, is called *fitness*, (w). Thus, many fitness values can be found within a species. When the number of individuals in a group does not change, its fitness, w_i, has the absolute value of 1. If the overall number of individuals in the species is constant, the mean fitness, $\overline{w}$, equals 1. Note that fitness is a rate; specifically, it is a multiplier of the number of individuals per unit time. In the present context, a convenient time unit for this rate is the generation, whose real-time equivalent, taken here as 0.005 year, is based on a frequently used rough estimate of the average number of divisions per year as 200 for *E. coli* in nature (reference 73; see also reference 65, p. 1652, and reference 90, p. 2716). The generation is compatible, as are various real units of time, with the modeling of each of four rate processes that must be considered in the discussion to follow: change in numbers, origin of large clones, nucleotide sequence divergence, and recombination. However, it is *necessary* for the fifth process: random genetic drift. Random genetic drift is a random-sampling process that is conceived to operate once per generation in sexual organisms, whose reproduction is assumed to be a random variable with binomial variance (18, p. 101). A good deal of population genetics theory rests on this assumption. In bacteria, whose growth and chromosomal replication is continuous between cell divisions, there seems to be only one stage in which random genetic drift can take place, namely, cell division, since for a constant population size, an average of one daughter cell must survive, and the individual possibilities are 2, 1, and 0. (In contrast, the cumulative effect of a series of instantaneous random minute changes in growth rate during a generation would be essentially zero.) This point is made because it supports a relationship between fitness and the retention of a new mutation or replacement (17, p. 184–186; 42, p. 47–48) upon which much that follows rests.

The time can be estimated for the rise of an allele to a given frequency in nature. Individual components of the species, for example, E. coli cells carrying a particular trp allele, may have a fitness value greater than 1; they therefore increase in number as generations pass. The choice of $w_i = 1.00001$ generation^{-1} as an example may seem modest, even trivial, but such a fitness could carry a population from 10^6 cells to 10^{18} cells in about 2.8 million generations, or (again at 200 generations per year) 14,000 years, which is a short time for a species that has existed for well over 100 million years. This comes from a calculation based on $(w_i)^t = N_t/N_0$; $1.00001^t = 10^{12}$. Taking $\log_{10}$ of both sides, $t \log_{10} 1.00001 = 12$, and $t = 12/\log_{10} 1.00001 \cong 2.8 \times 10^6$.

A rate of 1.00001/generation means an advantage of 10^{-5} per generation compared to the average member of the species. The difference between any fitness and 1 is called the selection coefficient: it may be positive or negative. The selection coefficient $s_i \equiv w_i - 1$, and here $s_i = 1.00001 - 1.00000 = 0.00001 = 10^{-5}$. The value s of the selection coefficient, which, like fitness, is a rate and

here shares the time unit generation^{-1}, is of critical importance in the retention of a new allele. For any selectively advantageous allele, the probability of being retained (instead of being lost due to random genetic drift) is only about twice the selection coefficient (17, 18, 42). Crow and Kimura use the symbol u for this "probability of ultimate survival" (18, p. 421). In the present case, $u \cong 2s = 2 \times 10^{-5}$. For neutral alleles (neither better nor worse than their common homologues) or deleterious alleles (at a selective disadvantage), the situation is far worse. Also, there are certain unusual cases where genes confer an extremely high selective advantage, such as an antibiotic resistance gene in the presence of the antibiotic, where s could easily be 10 or 100 initially. In fact, that is far more than necessary, for the probability u cannot exceed 1; when $s = 0.5$, $u = 0.6$, and from a similar calculation, when $s = 1$, $u = 0.8$. These values are obtained from the iterative solution of the equation $u_t = 1 - e^{-wu_{t-1}}$ (reference 18, p. 423, where c is used in place of w). Thus, a mutant allele or a recombinational replacement resulting in a doubled reproductive rate will have a probability of retention approaching 1. Clearly, u can vary over a vast range. In considering favorable genes only, this range is at least 4×10^4; with neutral genes also considered, the range extends to at least 10^9.

Retention is important both in the formation of species-wide clones, which will now be discussed, and in the effective rate of recombinational replacement (see below). First, taking $s = 10^{-3}$ as an example of a higher selection coefficient, $u \cong 2 \times 10^{-3}$. A nucleotide substitution with a selective advantage of 10^{-3} would occur and be retained $10^{-10} \times 2 \times 10^{-3} \times 6 \times 10^{20} = 1.2 \times 10^8$, or 120 million, times per generation in the species, forming tiny local clones of no species-wide significance. Even if the same selection coefficient, $s = 10^{-3}$, were conferred by a specific combination of two substitutions, with neither being favorable alone, the combination would occur and be retained at a rate of $10^{-20} \times 2 \times 10^{-3} \times 6 \times 10^{20} = 1.2 \times 10^{-2}$ per generation, about once every 80 generations somewhere in the world, or 2.5 times per year. Thus, it appears that either a combination of three nearly simultaneous nucleotide substitutions or a rare favorable recombinational event (perhaps originating in another species) would have to be responsible in most cases for the origin of a motivating allele leading to the formation of a species-wide clone. Although any given combination of three nucleotide substitutions would appear to be too improbable, the likely existence of many possible combinations makes this event a reasonable mechanism for the origin of a vast clone. Finally, it should be noted that the selective advantage of the motivating allele should be effectively independent of circumstances, including its own frequency (up to, say, 0.1, after which it begins to compete significantly with itself). Also, as Peter Reeves (71) illustrates, specific very large clones may dominate certain environments.

A more limited process of clone formation will be seen to result from negatively frequency-dependent selection (NFDS) operating on an imported gene whose initial probability of retention is near 1 (see below).

SPECIES TREES AND GENE TREES; CLONAL FRAMES AND CLONAL SEGMENTS

It is useful to note that a clone is ordinarily defined as a group of individuals whose entire genome is descended from a single common ancestor (2). Thus, mutation does not compromise membership in a clone, but a recombinational replacement does, if it has originated outside the clone. The replacement's ancestry and therefore phylogenetic relationships will be different from those of its unreplaced neighbor. So gene transfer would result in new local phylogenies (52, 58). In 1986 this was the decisive criterion used by Dykhuizen and Green (24) to establish the fact that natural gene transfer in *E. coli* had taken place to an evolutionarily significant extent. Dykhuizen and Green sequenced *gnd* in a set of strains that had previously been sequenced in the *trp*

region (54). The critical observation was a striking difference in the trees constructed for the two data sets. This demonstration that phylogeny is not genome wide in *E. coli* made it evident that recombination had introduced replacements with different ancestries. The distinction between *gene trees* and *population* or *species trees* was expressed by Avise (5): "A gene tree is the phylogeny of a particular gene or stretch of DNA. . . . a population tree must in some sense represent a compilation of genealogies for many genes." Moreover, a *clonal sweep* was understood to involve the rise to high frequency of any chromosomal stretch, from an entire genome to a gene or less. This was an important conceptual link between the allele frequency of classical population genetics on one hand and the genome on the other. Previously, the high rate of recombination in eukaryotic model organisms had made typical individual genes appear to evolve independently, while in bacteria their evident clonality had made the genome the working unit of phylogeny.

The 72 ECOR strains include three major groups of 25, 16, and 15 strains, called the A, B_1, and B_2 groups, respectively (32, 64) (Fig. 1). The remaining 16 strains are divided into smaller, deeply branched groups. Each major group is characterized by a common *clonal frame*, which is derived from a single ancestral chromosome. Each member of the group shares a large sample of this ancestral relic, distributed over the chromosome so that most members share both a common phenotype in a set of restriction digests of a given PCR fragment and a common electrophoretic mobility for each protein; but the similarities are not always shared by all the group members, and these exceptions vary from case to case. The collective results of RFLP and MLEE surveys lead to the interpretation that each major group's clonal frame is dotted with recombinational replacements, which vary in position and come from sources outside the clone. These replacements, which often are seen to share a similarity in a varying smaller group of strains, are called *clonal segments*: they have an

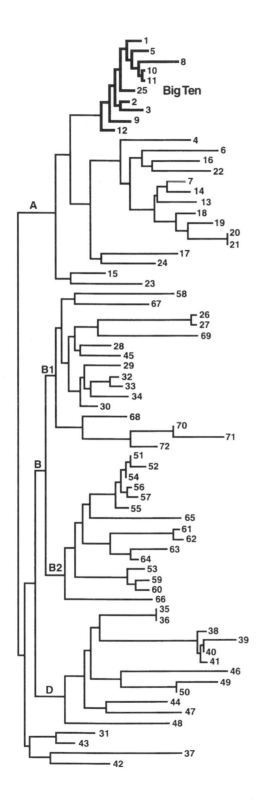

apparent local clonal affinity which may or may not be identified with known strains outside the group in question. Each of the three groups, then, is like a clone in sharing a clonal frame descended from a single common ancestor but unlike a clone in having numerous local exceptions to that common ancestry, stemming from recombination. To describe this pattern of genome variation, the term *meroclone* (i.e., "partial clone") has been used (51, 56).

The clonal segments are so named because they are inferred to be shared by members of a clone nested within a meroclone. The nesting might be clearer if the phenogram in Fig. 1 excluded the two hypervariable regions, describing instead only the 85% of the chromosome in which variation is restricted to the clonal frames and the sporadic clonal segments.

Two evident levels of tighter relationship between strains exist. First, the Big Ten ECOR strains (Fig. 1) and their close relative, K-12, evince very little variation in numerous RFLP analyses over most of the chromosome. Moreover, in a 12.7-kb stretch in which 40 strains were comparatively sequenced (51), the four sequenced members of this group (K-12 and ECOR 1, 8, and 12) differ by an average of 3 nucleotides per 10 kb in pairwise comparisons, but in the hypervariable region, they vary dramatically. And second, a number of small groups meet the following criteria: very little within-group variation over the entire chromosome and a single O-antigen type (see below). The members of each of these tight groups have diverged so recently from a common ancestor that the introduction of a recombinational replacement to some, but not all, of the group members is rare indeed. If K-12 and the 72 ECOR strains are representative

FIGURE 1 Phenogram of the 72 ECOR strains (after Herzer et al. [32]). The Big Ten group comprises the 10 ECOR strains (branches in boldface) and their close relative, K-12 (not shown).

of *E. coli* as a whole, the groups of nearly identical strains reflect the existence of large (~10^{19}-member) clones in the species that may be genome wide, or nearly so.

ABRIDGEMENT OF INCOMING DNA

The clonal frame-clonal segment pattern is seen on a finer scale in the results of comparative sequencing, where the size of the clonal segments is often on the order of 1 kb (51, 58), raising a question about the nature of the recombinational processes that are responsible. Specifically, in a single recombinational event, how small can the stretches of incorporated donor DNA be?

The answer to this question has come from experimental crosses among K-12 and several wild ECOR strains (49, 53, 55, 60). This approach is based on the expectation that widespread natural polymorphism exists in restriction-modification systems among the different strains of any given bacterial species and on the considerable DNA sequence polymorphism among the ECOR strains. Both features would obviously have interfered with the analysis of the macromolecular biochemistry of recombination; fortunately, that remarkable analysis has employed experiments conducted within a single genetic strain, specifically, *E. coli* K-12 (6, 19, 83). K-12 has literally thousands of genetically marked derivatives, but they all belong to the same strain.

The main hypothesis to be tested here, however, was that endogenous restriction endonucleases in the recipient cell would often cut up unprotected donor DNA before its incorporation into the recipient chromosome. The sequence differences between the strains used range from 1 to 4%, permitting an adequate level of resolution when contiguous or closely spaced 1,500-bp PCR fragments were amplified from the genomic DNA of each progeny strain from a cross. As a means of exploiting these sequence differences as genetic markers, each progeny PCR fragment was digested with a commercial restriction endonuclease previously found to distinguish the donor and recipient forms of each fragment at one or more restriction sites (whose locations were known or inferred from the K-12 sequences).

The initial experiments (49) involved P1 transduction of K-12 *trpA33* by ECOR 47. Eighteen progeny were analyzed; seven had multiple donor fragments, and the largest extents of donor DNA (each was a single fragment, as it turned out) were 25 and 20 kb, which contrasted with the 80 kb normally carried by P1 phage. Two other crosses, involving different donors, produced comparable results but lower proportions of multiple fragments. The two ECOR 47 → K-12 progeny with 25- and 20-kb donor fragments were crossed back to the original recipient, whose known restriction endonucleases were all coded near 99 min, far from the selected marker's location at 28 min. It was thus expected that the backcross donor DNA would be protected against the recipients' restriction endonucleases, and the results were consistent with this expectation. Twenty-eight of the thirty backcross progeny contained the entire original donor DNA fragment present in the respective backcross donor, and the extent of the backcross donor's DNA actually incorporated was likely much greater than the marked DNA that it included. The observed effects, fragmentation (presumably by restriction endonucleases) and shortening (presumably by exonucleases), were later described together as *abridgement* (55, 60), with the fragmentation taken to be a direct effect, and the shortening primarily a subsequent indirect effect, of restriction endonuclease activity.

The study then focused on conjugation, with comparable results. Moreover, reciprocal crosses also demonstrated that DNA mismatch in this range (1 to 4%) did not play a major role in the observed abridgement, since the results of crosses in the two directions were often extremely different while the degrees of mismatch between donor and recipient were essentially identical (60). The role of mismatch was investigated because of its effects on recombination under different circumstances (62, 63, 87, 91).

To illustrate the patterns of incorporation, Tables 1 and 2 summarize the results of conjugational crosses corresponding to the described transductions and back transductions, respectively. The same basic strains were used, with derivatives of ECOR 47 (now an Hfr with the origin of transfer at 31 min) as the donor and K-12 W3110 trpA33 (now also carrying rpoB to counterselect for rifampin resistance) as the recipient. The distribution of ECOR 47 and K-12 DNA is given for the series of PCR fragments, ranging in position from 1,297 to 1,335 kb previously described for the transduction crosses (49), plus six more spaced at greater distances. The abridgement evident in the original ECOR 47 → K-12 conjugational cross (Table 1) contrasts with the results of the backcross to K-12 (Table 2) and also with the results of a reciprocal (K-

12 → ECOR 47) conjugation (Table 3). In order to permit selection on minimal medium for trp+ in this reciprocal cross, as in the other crosses, a short section including the 59-kb region from 1,276 to 1,335 kb (and thus the respective trpA alleles) had been switched between K-12 and ECOR 47. This switch left the extent of mismatched DNA (local and genome wide) essentially equivalent to that of the original cross. Reciprocal crosses involving ECOR 47 and ECOR 72, as well as K-12 and ECOR 72, also showed direction-dependent differences (55, 60). Reciprocal crosses, backcrosses, and other serial crosses were made possible by the use of an F plasmid that establishes a dynamic equilibrium between free and integrated (with origin of transfer always at the Broca 7 region near 31 min) states (3, 25, 31, 34, 35, 47, 60, 61, 82).

It seems reasonable to expect that in most bacteria, chromosomal incorporation pat-

TABLE 1 HAZ: ECOR 47 → K12 W3110 trpA33 transconjugants selected on minimal medium[a]

No. of donor Fragments	Position, min → Location, kb →	22 / 1033	24 / 1115	25 / 1184	27.5 / 1276	28.3 / 1297	28.3 / 1305	28.3 / 1306	28.3 / 1308	28.3 / 1310	28.3 / 1311	28.3 / 1312	28.3 / 1314	28.3 / *	28.3 / 1316	28.3 / 1320	28.3 / 1327	28.3 / 1328	28.3 / 1330	28.3 / 1335	29.3 / 1362	29.8 / 1385
(1)	HAZ-12	-	-	-	D	D	D	D	D	D	D	D	D	M	D	D	D	D	D	D	-	-
	4	-	-	-	-	D	D	D	D	D	D	D	D	M	D	D	D	D	D	D	-	-
	8	-	-	-	-	D	D	D	D	D	D	D	D	M	D	D	D	D	D	D	-	-
	11	-	-	-	-	-	-	-	-	-	D	D	D	M	D	D	D	D	D	D	D	-
	21	-	-	-	-	D	D	D	D	D	D	D	D	M	D	D	D	-	-	-	-	-
	3	-	-	-	-	-	D	D	D	D	D	D	D	M	D	D	D	D	D	D	-	-
	7	-	-	-	-	D	D	D	D	D	D	D	D	M	D	D	D	-	-	-	-	-
	16	-	-	-	-	-	-	-	-	-	D	D	D	M	D	D	D	D	D	D	-	-
	13	-	-	-	-	-	-	-	-	-	-	-	-	M	D	D	D	D	D	D	-	-
	25	-	-	-	-	-	-	-	-	-	-	-	-	M	D	D	D	D	D	D	-	-
	29	-	-	-	-	-	D	D	D	D	D	D	D	M	D	-	-	-	-	-	-	-
	22	-	-	-	-	-	D	D	D	D	D	D	D	M	D	-	-	-	-	-	-	-
	6	-	-	-	-	-	-	D	D	D	D	D	D	M	-	-	-	-	-	-	-	-
	14	-	-	-	-	-	-	-	-	-	-	-	D	M	D	D	-	-	-	-	-	-
	24	-	-	-	-	-	-	-	-	-	-	-	-	M	D	D	-	-	-	-	-	-
	15	-	-	-	-	-	-	-	-	-	D	D	D	M	D	-	-	-	-	-	-	-
	26	-	-	-	-	-	-	-	-	-	-	-	-	M	D	-	-	-	-	-	-	-
(2)	27	-	-	-	-	-	D	D	D	D	D	D	D	M	-	D	D	D	D	D	-	-
	2	-	-	-	D	-	-	-	-	-	-	-	D	M	-	-	-	-	-	-	-	-
	28	-	-	-	-	-	D	D	-	D	D	D	D	M	D	D	D	-	-	-	-	-
	17	-	-	-	-	-	-	-	D	-/d	D	D	D	M	D	D	-	-	-	-	-	-
	18	-	-	-	-	-	-	-	D	-/d	D	D	D	M	D	D	-	-	-	-	-	-
	1	-	-	-	-	-	-	-	-	d/-	D	-	D	M	-	-	-	-	-	-	-	-
(3)	19	D	D	-	-	-	D	D	D	D	D	D	D	M	D	D	D	D	D	d/-	D	-
	30	D	D	-	D	-	-	-	-	-	-	D	D	M	D	D	-	-	-	-	-	-
	23	-	-	-	-	-	-	-	-	-	-	-	D	M	-	D	-	-	-/d	-	-	-
(4)	20	-	-	D	-	-	-	-	-	-	-	-	-	M	D	-/d/-	d/-	-	-	-	-	-
	10	-	-	-	-	-	D	D	D	D	D	-	D	M	d/-	-/d	-	-	D	-	-	-
(5)	9	-	D	D	-	-	-	D	D	D	D	-	-	M	D	-/d	D	D	D	d-d	-	-

[a] Locations of donor DNA (D) and recipient DNA (−) determined by single restriction digests of specific 1,500-bp PCR fragments. Mosaic distributions are indicated by "d" and "−" according to the respective patterns within the PCR fragment. Asterisk and M, selected marker.

TABLE 2 JAZ: (ECOR 47 → K12W3110trpA33 transconjugant HAZ-12) → K12 W3110 trpA33 back-transconjugants[a]

PCR Fragment	1	1	1	1	1	1	1	1	1	1	1	1	*	1	1	1	1	1	1	1	1
Location, kb	115	12	2	27	30	30	30	31	31	31	32	34		36	39	37	38	30	35	32	35
[DONOR: HAZ-12	-	-	D	D	D	D	D	D	D	D	D	D	M	D	D	D	D	D	D	-	-]
	@																				
JAZ: 28	-	D	D	D	D	D	D	D	D	D	D	M	D	D	D	D	D	D	D	D	D
JAZ: 1	-	-	-	-	-	D	D	D	D	D	D	M	D	D	D	D	D	D	D	-	
JAZ: 1	-	-	-	-	-	-	-	-	-	-	D	M	D	D	D	D	D	D	D	-	

[a] D, original donor DNA determined in progeny of initial cross (Table 1). @, number of backcross progeny strains with indicated donor DNA pattern. For other symbols, see note *a* to Table 1.

terns following gene transfer will be similar, since restriction-modification systems are widespread (7, 8, 12, 20, 67, 80) and since their part in the effective defense of a species against bacteriophage requires that they be highly polymorphic (and thus likely to fragment incoming conspecific DNA as well as phage DNA). A monomorphic restriction-modification system would be ineffectual against the descendants of any lucky infecting phage that produced progeny, whose perpetual protection would leave the entire species vulnerable (20, 23, 41, 68).

The incorporation pattern of donor DNA fragments into a recipient chromosome in bacteria constitutes another important recombinational difference between bacteria and sexual metazoa. The further apart two genes are in a metazoan linear chromosome, the greater the frequency of crossing over between them, and thus the lower the correlation of transmission of particular alleles at the two loci. This soon leads to the independent allelic transmission of all but the most closely linked genes; gene genealogies are increasingly likely to differ with increasing chromosomal distance. In bacteria, the clonal frame is likely to be quite discontinuous, but the gene trees will be uniform over the frame's entire extent, though different in the replacement fragments (clonal segments).

Clearly, this pattern of variation has one formal similarity to that caused by mutation. Lines of descent differ from one another in the sets of mutations they have acquired since diverging from their most recent common an-

TABLE 3 Conjugation: K12 → ECOR 47[a]

Min	90 +	22	24	25	27.5	----------------28----------------								*	-----29-----						-30
Kb	4180	1033	1115	1184	1276	1297	1305	1306	1308	1310	1311	1312	1314	1315	1316	1319	1327	1328	1330	1335	...
@																					
2	C	D	D	D	D	D	D	D	D	D	D	D	D	M	D	D	D	D	D	D	D D
1	C	D	D	D	D	D	D	D	D	D	D	D	D	M	D	D	D	D	D	D	- -
1	C	-	-	D	D	D	D	D	D	D	D	D	D	M	D	D	D	D	D	D	D D
4	C	-	-	D	D	D	D	D	D	D	D	D	D	M	D	D	D	D	D	D	- -
1	C	-	-	D	D	D	D	D	D	D	D	D	D	M	d/-	-	-	-	-	-	- -
5	C	-	-	-	D	D	D	D	D	D	D	D	D	M	D	D	D	D	D	D	D D
1	C	-	-	-	D	D	D	D	D	D	D	D	D	M	D	D	D	D	D	D	D -
12	C	-	-	-	D	D	D	D	D	D	D	D	D	M	D	D	D	D	D	D	- -
1	C	-	-	-	D	D	D	D	-	-	-	-	-	M	D	D	D	-	-	-	- -
1	C	-	-	-	-	D	D	D	D	D	D	D	D	M	D	D	D	D	D	D	D D
1	C	-	-	-	-	D	D	D	D	D	D	D	D	M	D	D	D	D	D	D	- -

[a] Conjugational cross reciprocal to that in Table 1. D, donor DNA; −, recipient DNA; @, number of backcross progeny strains with indicated donor DNA pattern; d/−, PCR fragment has donor DNA at left and recipient DNA at right. * and M denote selected marker (*trpA*[+]); + and C denote counterselected marker, *rpoB* (*rif*).

cestor (42). Similarly, lines of descent differ from one another in the sets of recombinational replacements they have acquired since diverging from their most recent common ancestor (52, 58). Of course, in any given strain, while mutation is a random process emanating from a basically common source, gene transfer emanates from an importantly diverse set of sources. In any event, clonal diversification can result from recombination as well as mutation (27), and the rates of both processes, whose kinetics are (to a first approximation) similar functions of time, are important to the genetic structure of the species. Recognizing this, and given the availability of an estimate of the nucleotide substitution rate in *E. coli* (22), Guttman and Dykhuizen (27) took a first step toward the systematic estimation of the rate of recombination by searching for local differences in gene genealogies among a set of 12 strains: K–12 and 11 ECOR strains. They compared 1-kb sequences from each of four genes spaced over a 1.9-min (87.5 kb in this case) region (Fig. 2). (The map positions have recently been recalibrated to accord with the K–12 MG1655 genome sequence [11, 14, 37].) Three changes of phylogeny were noted, each attributed to a single gene transfer border. In principle, this approach, applied to a larger, more diverse set of strains and comparing continuous 100-kb DNA sequences for each strain, should lead to a reasonable estimate of the gene transfer rate and a sample of the length and form (including discontinuity of the donor fragments), as well as revealing insertions and deletions along the way. Of critical advantage to such a venture are the increasingly rapid sequencing methods and the K–12 MG1655 (6, 14) sequence data that have become available since this work was done. The pertinent feature of the MG1655 genome sequence is a complete, continuous basis for the formulation of PCR primers with which to amplify fragments and sequence them.

THE HYPERVARIABLE REGIONS

The variation among the ECOR strains as indicated by extensive MLEE studies, as well as by somewhat fewer RFLP analyses, is consistent with the clonal frame-clonal segment model already described: members of a meroclone are clustered where they share the clonal frame; they vary locally due to unshared replacements. RFLP studies have the virtues of focusing on precisely known chromosomal regions and revealing a sample of sequence differences, but unlike MLEE surveys, they do not generally cover the entire ECOR collection. A number of the enzymes studied electrophoretically are not yet referable to a specific gene. All in all, the patterns of variation evinced by the ECOR strains are consistent over most of the chromosome. They depart from the model dramatically, however, in two local singularities in the chromosome: the hypervariable regions. One of these centers on the *hsd-mcr-mrr* restriction (and modification) region, located near 99 min. This region, de-

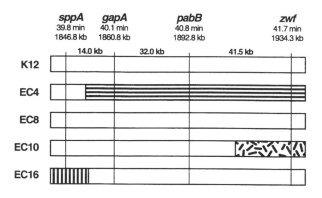

FIGURE 2 Diagram of a 2-min region of the *E. coli* chromosome in four ECOR strains. Modified (see the text) after Guttman and Dykhuizen (27). The open bars represent the common clonal background. The other symbols represent regions introduced by recombination. The locations of the breakpoints are approximate.

scribed in detail as the immigration control region (69), is hypervariable in *E. coli* (8) and in enteric bacteria as a whole (7, 69); sequencing evidence suggests the recent acquisition of the *mcrBC* (21) and *hsd* (8) genes by *E. coli*.

Across the chromosomal circle, near 45 min, is the *rfb* gene complex, which determines the structure of the O antigen (9, 33, 38, 43, 70, 71, 85). It houses the ongoing mechanism responsible for the hypervariability that extends 3 to 5 min in both directions. The O antigen covers most of the cell surface of noncapsular (i.e., "rough") *E. coli*, and other species as well. It is a highly polymorphic lipopolysaccharide: some 170 varieties of this antigen are known in *E. coli* (71), and variation is seen in numerous other species. This polymorphism is a classic case of NFDS, where the selective advantage of an allele decreases with its increasing frequency. The explanation here is that even beneficial bacteria must not be allowed to multiply out of control, and the hosts, recognizing their presence by their surface antigens, act to constrain their growth. When a novel antigen escapes detection, the cell it covers gains a big selective advantage. Its probability of retention may be near 1, like that conferred by a drug resistance gene, and far greater than most new alleles ever confer. As the new O-antigen gene rises in frequency, host antibodies appear and nullify its advantage; but for a considerable time, the gene spreads throughout the species, just ahead of the sheriff. Eventually, as the global relative frequency approaches, say 10^{-3} (this would be equivalent to an absolute frequency of 6×10^{17} cells), the advantage disappears: the gene is no better but no worse than the other determinants of O antigens. Thus, there is a continuing demand for new variants.

The clones formed by this type of diversifying selection clearly differ from the vast clones described earlier. T. S. Whittam (personal communication) has identified a large number of known O antigens in the ECOR strains. These are shared by members of some small groups of very closely related strains which also share RFLPs at the extremely variable *cpsB* locus, about 10 kb from the *rfb* genes themselves. The groups are ECOR 26 and 27 (O104); 38, 39, 40, and 41 (O7); 49 and 50 (O2); 51, 52, 54, and 55 (O25); 53 and 60 (O4); and 61 and 62 (also O2). This suggests that they may still be genome-wide clones. But in one interesting case, two of these groups (49 and 50 and 61 and 62) are quite different over most of the chromosome but share both an O antigen and a six-enzyme restriction pattern in a PCR fragment 10 kb away from *rfb*; this suggests recombination of the relay type outlined above, where flanking DNA has been transmitted along with the O antigen.

Evidently numerous new *rfb* complexes are spreading at once, and the rapid occurrence of superimposed intraspecific replacements results in a near-chaotic form of variation in the region near the selected gene complex, as opposed to the near uniformity expected of the classical motivating allele and its genome in the early stages of clone formation. Finally, NFDS limits these antigen-driven clones to a size 2 or 3 orders of magnitude less than that of the classical clones.

Unlike that of protein antigens, polysaccharide variation is based on nonhomologous genetic differences. A protein's shape, however, may be changed by a particular nonsynonymous substitution or by the insertion or deletion of some multiple of 3 nucleotides in the coding gene (52, 58), and so it is possible to generate considerable antigenic variety in the descendants of one ancestor. But the O-antigen variation results from the substitution of one type of saccharide monomer for another or a change in the attachment of a given type of monomer. This ordinarily requires a new enzyme, coded by a new nonhomologous gene, and importation is the only practical means of acquiring it.

Can lateral transfer really repeatedly fill the demand for such antigenic novelty? The answer lies in the dynamics of recombination, specifically, in the basis of the effective rate of recombination. The important parameter is the *rate of retained replacement*: the product of the rate at which a certain replacement occurs and the probability of its retention (see above). Obviously, intraspecific recombination occurs far more frequently than lateral transfer from other species. Were this not so, species would not retain their distinctive properties. On the other hand, the probability of retention, which can be a compensating factor, varies over a vast range, actually approaching 1. What quantitative advantage in growth rate would confer such a probability of retention? A doubling of growth rate for a limited period of time should easily be sufficient.

Then how long must this rapid growth rate continue? Alleles with a selective advantage will cease to be vulnerable to extinction by random genetic drift if they reach a "safe" number, which is equal to $1/s$, and the probability of doing that is about $2s$, as previously noted. At the safe number, the further increase in number, generation after generation, due to the selective advantage is essentially certain to exceed any series of random losses and thus to prevent the removal of the allele by random genetic drift. Thus, its probability of reaching an absolute frequency of 10^9 (and beyond, until or unless its selective advantage changes) is now 1. That is why the probability of retention (or ultimate survival), u, of a new mutation or replacement is equal to $2s$. As for a recombinant cell that grows on the average twice as fast as its competitors, after only 15 generations the probability is less than 10^{-5} that it will be lost if its fitness has not changed (using the iterative formula given above). After 20 doublings the progeny of any cell will number a million, which is the safe number for cells whose selective advantage is a mere 10^{-6}. Thus, the need for a high selection coefficient persists only until the initial barrier is passed.

In the hypervariable regions, the diversifying selection dominates the local scene: only the most recently separated lines remain identical. This diversification must be due to the new *rfb* complexes' relatively high frequency of occurrence and retention. It would appear that the classical motivating alleles' selective advantage is not high enough to unleash their recombinational transfer and retention at a comparable rate: otherwise, genome-wide clonal frames would not exist. The possibility of major recent changes in the rates of intraspecific gene transfer in *E. coli* seems contradicted by the presence of clonal segments.

The product of the frequency of imports, the proportion of all imports that are highly advantageous, and their probability of retention must be tiny. Nevertheless, the reported number of different O antigens in *E. coli*, the variety of restriction enzymes in *E. coli* coded in the 99-min hypervariable region (8, 69, 86), and the evidence that a considerable portion of the entire *E. coli* genome is exogenous (39, 40) (see chapter 15), as well as the increasing evidence in nature of the common occurrence of diverse forms of distant horizontal transfer (45, 92), combine to indicate that the proportion is large enough to figure in a rational model in which the ongoing importation of new genetic material is part of the recombinational dynamics of *E. coli*, and presumably of other bacteria as well. And it is possible that the proportion of highly advantageous replacements among imports is far higher than in local products (other than retransmissions of recent imports from the original recipient to other *E. coli* cells). Indeed, it is possible that a large proportion of all the highly advantageous replacements in any given *E. coli* genome originated outside the species (and that their evolution was thus predominantly independent of *E. coli*).

The impact of the importation and retention of a new gene into the *rfb* region begins on a small spatial scale, but it unleashes a cascade of retained recombination that extends 3 to 5 min in both directions. Once the novel gene has replaced a previous member of the *rfb* complex and risen in frequency, it is included in the normal intraspecific recombination events in *E. coli*, namely, conjugation and transduction. These processes distribute the new gene to other strains, together with various continuous or discontinuous lengths of flanking homologous DNA. The resulting replacements now have a radically increased probability of retention, anchored in the selected gene but of course applicable to the entire replacement, whether the donor DNA is continuous or discontinuous. The result is the hypervariability associated with this region: the donor DNA flanking the new gene may be incorporated in a recipient with a different sequence. Comparative sequencing of the Big Ten strains in this region shows variation that reflects recombination: the variation consists of stretches of DNA that are often found in more than one strain, as opposed to a scattering of individual nucleotide differences that would indicate an unusually high local mutation rate. Pairwise sequence differences range up to 7.6% in this set of strains (59), whose pairwise differences are next to nothing over the nonhypervariable ~85% of the chromosome. The variation tapers off and disappears with increasing distance from *rfb*, and RFLP studies of the Big Ten strains indicate the same pattern in the other hypervariable region near 99 min as in that near *rfb* (59). Evidently replacements from a variety of *E. coli* strains (of which ECOR strains are representative) arrive everywhere on the chromosome in these closely related 11 strains but are retained too infrequently (except in the hypervariable regions) to produce observable sequence variation among them in the short time since they diverged from their most recent common ancestor.

As a postscript to this consideration of the high effective recombination rate in the two hypervariable regions, it should be noted that Dykhuizen and Green's choice of *gnd* to compare with *trp* was wise. Fortunately, they knew that variation at the *gnd* locus was relatively high (50, 64, 77); indeed, it now appears that the chance of such a dramatic difference in local phylogenies between *trp* and a locus selected at random is rather low. In ~85% of the chromosome, the clonal frame prevails, with individual exceptions (as opposed to completely new gene trees). In contrast, *gnd* is next to the *rfb* gene complex at the center of the hypervariable region, and its variation is

pervasive and great enough, for example, to make it difficult to design PCR primers common to a large number of strains (13).

IMPACT OF SOME RECENT ADVANCES

The Tree Needs Pruning

The characterization of the hypervariable regions means that the phenogram in Fig. 1 conflates two strikingly different data sets. Thus, it would be improved by the removal of electrophoretic mobility data on enzymes coded in the hypervariable 15% of the genome. The use of MLEE began when there was no reason to believe that the genes blindly sampled varied in a degree strongly dependent on their regional location, and the major groupings of the ECOR strains have never been contradicted. A problem of less importance is that several enzyme activities on which Fig. 1 is based were never characterized beyond an operational definition (which is perfectly satisfactory in the absence of a complete genome sequence) (74); others have not yet been assigned definitively to a specific gene (15). Enough MLEE data remain, however, and more informative RFLP and sequence data (90) are obtainable as needed. Still, the hypervariable regions have greatly exaggerated the variation among closely related strains. For example, there is a good possibility that most of the branching in the Big Ten strains (Fig. 1), and even the occasional twig in the most closely related strains (see above), stems from variation in the hypervariable regions (reference 59 and my unpublished sequence data). A clearer picture of the evolution of the clonal frames is likely to emerge from a phenogram based on the "normal" 85% of the chromosome. Indeed, such a phenogram may prove to be sufficiently tree-like to test the clonality of the clonal segments. Since every clone is a *clade* (an ancestor together with all of its descendants), a retained replacement from the past should be seen in all the recipient's descendants. Outside of the hypervariable regions, strains carrying common clonal segments should form consistent clades; exceptions (where not all the descendants of a most recent common ancestor share a clonal segment) would presumably be due to a superimposed replacement in a smaller clone nested within the exceptional intermediate clone. This replacement might well be one that has restored the original clonal frame. A test of this general expectation, using restriction analysis followed by local sequencing, could demonstrate the consistency of the meroclonal model and perhaps lead to the development of a replacement clock, analogous to a nucleotide substitution clock. There are two technical points which might interfere. First, because bacterial chromosomes divide continuously during the growth cycle, some donor fragments might be incorporated into an undivided part and others into one or the other of two divided parts. This might result in two or more similar but not identical sets of clonal segments among the descendants. And second, any greater likelihood of within-group versus between-group recombination would open the door to variation in the rate of detectable replacement and complicate the formulation of a clock pattern.

Gene Shuffling

One striking example of the potential value of horizontal gene transfer is presented in a recent application of the technique of *gene shuffling* (16), which combines the in vitro self-priming assembly of random DNA fragments of a cephalosporinase gene with the introduction of the products into *E. coli,* followed by selection for moxalactamase activity. (Moxalactam is a type of cephalosporin.) Genetic variation was introduced either during error-prone amplification of a gene from one species or by the mixing and random fragmentation of homologous genes taken from *Citrobacter, Enterobacter, Klebsiella,* and *Yersinia.* Response to selection was observed in each case, but the latter was far more effective. Of course, very few of the randomly induced mutations in the first case would have been advantageous (and of those, most probably were combined with

deleterious mutations). The combination of fragments of functional genes from four diverse lines, each with a presumably long history of natural selection, appears to have provided a far better set of building blocks for a vastly improved gene. Again we are reminded that natural selection does not speculate on future benefits but demands progressive improvement.

General Additivity Results in Many Evolutionary Trajectories

There are few cases in which individual improvements appear to be so strictly additive, so progressive in their cumulative benefits, as codon usage in highly expressed genes (42, 78). One is apparent in *E. coli,* where the *gapA* gene has the highest codon adaptation index (CAI) value. The CAI is a parameter reflecting the usage frequency of codons that are generally prevalent in highly expressed *E. coli* genes (42, 79, 84). In an important and provocative paper, Guttman and Dykhuizen (28) focused on *gapA* as a potential motivating allele in a clonal sweep. They envisioned a local, rather than genome-wide, clonal sweep, in which a small region of the *E. coli* genome, limited by recombination, hitchhiked to (near) fixation as *gapA* was selected. The most striking evidence that a *gapA* allele had risen to fixation over a very short time was its remarkable near uniformity in a sample of 11 ECOR strains and K-12 (using the same sequence data referred to in reference 27 and described above), augmented by data from 6 additional ECOR strains (28). Recombination between *gapA* and loci on either side was indicated by the distinct difference evinced between *gapA* and these nearby genes with respect to the phylogenetic relationships of the same ECOR strains, recalling Dykhuizen and Green's (24) landmark demonstration of an evolutionarily significant rate of gene transfer in *E. coli.* The importance of the present paper lies in its goal and approach, rather than its conclusion; it indicates important characteristics that should be useful in identifying a motivating allele and the region that has participated in a clonal sweep. In the case of *gapA*, however, the uniformity could be due to stringent selection leading to its high CAI. Since almost (84) every replacement of any codon complementary to a scarce transfer RNA by a synonymous codon complementary to an abundant tRNA would improve fitness (in other words, since no trajectory of improvement would be preferred over any other), all *gapA* alleles in *E. coli* would be under strong selective pressure to converge and remain uniform at their codons' third nucleotides. Sequencing of additional ECOR strains may confirm and extend a pattern suggested by the original analysis, namely, that the strains exhibit their characteristic common (clonal frame) phylogeny in *gapA*.

PERSPECTIVE

It would appear that with both a body of basic information and an effective technology now in place, the next few years should see the elementary articulation of the properties of the *E. coli* species genome, as well as the answers to a number of specific questions only recently formulated, ranging from important details to overarching patterns. The platform offered by the *E. coli* K-12 MG1655 genome sequence provides an effective means of pursuing the detailed determination of recombinational replacement rates, the search for a consistent ECOR tree based on nucleotide substitution rates and replacement rates, and the identification of motivating alleles. It also opens the prospect of this new level of organization, whose basic character must lie in the quantitative details of the structure of variation over chromosomal regions and strains combined. That is what evolutionary genetic analysis can lead to. Then is there to be found an adaptive gestalt emerging from physiological, epigenetic, and ecological dynamics that can be applied to bacterial species?

ACKNOWLEDGMENTS

Work in my laboratory and preparation of this chapter have been supported in part by grants MCB-9420613 and 9728230 from the National Science Foundation.

REFERENCES

1. **Achtman, M.** 1994. Clonal spread of serogroup A meningococci: a paradigm for the analysis of microevolution in bacteria. *Mol. Microbiol.* **11:** 15–22.
2. **Allaby, M.** 1985. *The Oxford Dictionary of Natural History.* Oxford University Press, New York, N.Y.
3. **Ankenbauer, R. G.,** 1997 Reassessing forty years of genetics doctrine; retrotransfer and conjugation. *Genetics* **145:**543–549.
4. **Atwood, K. C., L. K. Schneider, and F. J. Ryan.** 1951. Selective mechanisms in bacteria. *Cold Spring Harbor Symp. Quant. Biol.* **16:**345–355.
5. **Avise, J.** 1989. Gene trees and organismal histories: a phylogenetic approach to population biology. *Evolution* **43:**1192–1208.
6. **Bachmann, B. J.** 1996. Derivations and genotypes of some mutant derivatives of *Escherichia coli* K-12, p. 2460–2488. *In* F. C. Neidhardt, R. Curtiss III, J. L. Ingraham, E. C. C. Lin, K. B. Low, B. Magasanik, W. S. Reznikoff, M. Riley, M. Schaechter, and H. E. Umbarger (ed.), *Escherichia coli and Salmonella: Cellular and Molecular Biology,* 2nd ed. American Society for Microbiology, Washington, D.C., [See p. 2470.]
7. **Barcus, V. A., and N. E. Murray.** 1995. Barriers to recombination: restriction, p. 31–58. *In* R. Bishop (ed.), *Population Genetics of Bacteria.* Cambridge University Press, Cambridge, United Kingdom.
8. **Barcus, V. A., J. B. Titheradge, and N. E. Murray.** 1995. The diversity of alleles at the *hsd* locus in natural populations of *Escherichia coli. Genetics* **140:**1187–1197.
9. **Bastin, D. A., G. Stevenson, P. K. Brown, A. Haase, and P. R. Reeves.** 1993. Repeat unit polysaccharides of bacteria: a model for polymerization resembling that of ribosomes and fatty acid synthetase, with a novel mechanism for determining chain length. *Mol. Microbiol.* **7:**725–734.
10. **Baur, B., K. Hanselmann, W. Schlimme, and B. Jenni.** 1996. Genetic transformation in freshwater: *Escherichia coli* is able to develop natural competence. *Appl. Environ. Microbiol.* **62:** 3673–3678.
11. **Berlyn, M. K. B., K. B. Low, K. E. Rudd, and M. Singer.** 1996. Linkage map of *Escherichia coli* K-12, edition 9, p. 1715–1902. *In* F. C. Neidhardt, R. Curtiss III, J. L. Ingraham, E. C. C. Lin, K. B. Low, B. Magasanik, W. S. Reznikoff, M. Riley, M. Schaechter, and H. E. Um-

barger (ed.), Escherichia coli *and* Salmonella: *Cellular and Molecular Biology*, 2nd ed. American Society for Microbiology, Washington, D.C.

12. **Bickle, T. A., and D. H. Krüger.** 1993. Biology of DNA restriction. *Microbiol. Rev.* **57:**434–450.

13. **Bisercic, M., J. Feutrier, and P. R. Reeves,** 1991. Nucleotide sequences of *gnd* genes from nine natural isolates of *Escherichia coli*: evidence of intragenic recombination as a contributing factor in the evolution of the polymorphic *gnd* locus. *J. Bacteriol.* **173:**3894–3900.

14. **Blattner, F. R. G. Plunkett III, C. A. Bloch, N. T. Perna, V. Burland, M. Riley, J. Collado-Vides, J. D. Glasner, C. K. Rode, G. F. Mayhew, J. Gregor, N. W. Davis, H. A. Kirkpatrick, M. A. Goeden, D. J. Rose, B. Mau, and Y. Shao.** 1997. The complete genome sequence of *Escherichia coli*. *Science* **277:**1453–1474.

15. **Boyd, E. F., K. Nelson, F.-S. Wang, T. S. Whittam, and R. K. Selander.** 1994. Molecular genetic basis of allelic polymorphism in malate dehydrogenase (*mdh*) in natural populations of *Escherichia coli* and *Salmonella enterica*. *Proc. Natl. Acad. Sci. USA* **91:**1280–1284.

16. **Crameri, A., S.-A. Raillard, E. Bermudez, and W. P. C. Stemmer.** 1998. DNA shuffling of a family of genes from diverse species accelerates directed evolution. *Nature* **391:**288–291.

17. **Crow, J. F.** 1986. *Basic Concepts in Population, Quantitative and Evolutionary Genetics*, p. 185–186. W. H. Freeman & Co., New York, N.Y.

18. **Crow, J. F., and M. Kimura.** 1970. *An Introduction to Population Genetics Theory*, p. 418–430. Harper and Row, New York, N.Y.

19. **Dabert, P., and G. R. Smith.** 1997. Gene replacement with linear DNA fragments in wildtype *Escherichia coli*: enhancement by chi sites. *Genetics* **145:**877–889.

20. **Daniel, A. S., F. V. Fuller-Pace, D. M. Legge, and N. E. Murray.** 1988. Distribution and diversity of *hsd* genes in *Escherichia coli* and other enteric bacteria. *J. Bacteriol.* **170:**1775–1782.

21. **Dila, D., E. Sutherland, L., Moran, B. Slatko, and E. A. Raleigh.** 1990. Genetic and sequence organization of the *mcrBC* locus of *Escherichia coli* K-12. *J. Bacteriol.* **172:**4888–4900.

22. **Drake, J. W.** 1991. A constant rate of spontaneous mutation in DNA-based microbes. *Proc. Natl. Acad. Sci. USA* **88:**7160–7164.

23. **DuBose, R. F., D. E. Dykhuizen, and D. L. Hartl.** 1988. Genetic exchange among natural isolates of bacteria: recombination within the *phoA* gene of *Escherichia coli*. *Proc. Natl. Acad. Sci. USA* **85:**7036–7040.

24. **Dykhuizen, D. E., and L. Green.** 1991. Recombination in *Escherichia coli* and the definition of biological species. *J. Bacteriol.* **173:**7257–7268.

25. **Firth, N., K., Ippen-Ihler, and R. A. Skurray.** 1996. Structure and function of the F factor and mechanism of conjugation, p. 2377–2401. *In* F. C. Neidhardt, R. Curtiss III, J. L. Ingraham, E. C. C. Lin, K. B. Low, B. Magasanik, W. S. Reznikoff, M. Riley, M. Schaechter, and H. E. Umbarger (ed.), Escherichia coli *and* Salmonella: *Cellular and Molecular Biology*, 2nd ed. American Society for Microbiology, Washington, D.C.

26. **Graupner, S., and W. Wackernagel.** 1996. Identification of multiple plasmids released from recombinant genomes of *Hansenula polymorpha* by transformation of *Escherichia coli*. *Appl. Environ. Microbiol.* **62:**1839–1841.

27. **Guttman, D. S., and D. E. Dykhuizen.** 1994. Clonal divergence in *Escherichia coli* as a result of recombination, not mutation. *Science* **266:**1380–1383.

28. **Guttman, D. S., and D. E. Dykhuizen.** 1994. Detecting selective sweeps in naturally occurring *Escherichia coli*. *Genetics* **138:**993–1003.

29. **Hanahan, D., and F. R. Bloom.** 1996. Mechanisms of DNA transformation, p. 2449–2459. *In* F. C. Neidhardt, R. Curtiss III, J. L. Ingraham, E. C. C. Lin, K. B. Low, B. Magasanik, W. S. Reznikoff, M. Riley, M. Schaechter, and H. E. Umbarger (ed.), Escherichia coli *and* Salmonella: *Cellular and Molecular Biology*, 2nd ed. American Society for Microbiology, Washington, D.C.

30. **Heinemann, J., and G. Sprague.** 1989. Bacterial conjugative plasmids mobilize DNA transfer between bacteria and yeast. *Nature* (London) **340:**205–209.

31. **Heinemann, J. A., H. E. Scott, and M. Williams.** 1996. Doing the conjugative two-step: evidence of recipient autonomy in retrotransfer. *Genetics* **143:**1425–1435.

32. **Herzer, P. J., S. Inouye, M. Inouye, and T. Whittam.** 1990. Phylogenetic distribution of branched RNA-linked multicopy single-stranded DNA among natural isolates of *Escherichia coli*. *J. Bacteriol.* **172:**6175–6181.

33. **Hobbs, M., and P. R. Reeves.** 1994. The JUMPstart sequence: a 39 bp element common to several polysaccharide gene clusters. *Mol. Microbiol.* **12:**855–856.

34. **Holloway, B. W., and K. B. Low.** 1987. F-prime and R-prime factors, p. 1145–1153. *In* F. C. Neidhardt, J. L. Ingraham, K. B. Low, B. Magasanik, M. Schaechter, and H. E. Umbarger (ed.), Escherichia coli *and* Salmonella typhimurium: *Cellular and Molecular Biology*. American Society for Microbiology, Washington, D.C.

35. **Holloway, B. W., and K. B. Low.** 1996. F-prime and R-prime factors, p. 2413–2420. *In* F. C. Neidhardt, R. Curtiss III, J. L. Ingraham, E. C. C. Lin, K. B. Low, B. Magasanik, W. S. Reznikoff, M. Riley, M. Schaechter, and H. E. Umbarger (ed.), Escherichia coli *and* Salmonella: *Cellular and Molecular Biology*, 2nd ed. American Society for Microbiology, Washington, D.C.

36. **Keenleyside, W. J., and C. Whitfield.** 1995. Lateral transfer of *rfb* genes: a mobilizable ColE1-type plasmid carries the *rfbO:54* (O:54 antigen biosynthesis) gene cluster from *Salmonella enterica* serovar Borreze. *J. Bacteriol.* **177:**5247–5253.

37. **Kröger, M., and R. Wahl.** 1997. Compilation of DNA sequences of *Escherichia coli* K12; description of the interactive databases ECD and ECDC (update 1996). *Nucleic Acids Res.* **25:**39–42.

38. **Lan, R., and P. R. Reeves.** 1996. Gene transfer is a major factor in bacterial evolution. *Mol. Biol. Evol.* **13:**47–55.

39. **Lawrence, J. G., and H. Ochman.** 1997. Amelioration of bacterial genomes: rates of change and exchange. *J. Mol. Evol.* **44:**383–397.

40. **Lawrence, J. G., and J. R. Roth.** 1996. Selfish operons: horizontal transfer may drive the evolution of gene clusters. *Genetics* **143:**1843–1860.

41. **Levin, B. R.** 1986. Restriction-modification immunity and the maintenance of genetic diversity in bacterial populations, p. 669–688. *In* S. Karlin and E. Nevo (ed.), *Evolutionary Processes and Theory*. Academic Press, New York, N.Y.

42. **Li, W.-H.** 1997. *Molecular Evolution.* Sinauer Associates, Sunderland, Mass.

43. **Liu, D., and P. R. Reeves.** 1994. Presence of different O antigen forms in three isolates of one clone of *Escherichia coli*. *Genetics* **138:**6–10.

44. **Lloyd, R. G., and C. Buckman.** 1995. Conjugational recombination in *Escherichia coli*: genetic analysis of recombinant formation in Hfr X F crosses. *Genetics* **139:**1123–1148.

45. **Lorenz, M. G., and W. Wackernagel.** 1994. Bacterial gene transfer by natural genetic transformation in the environment. *Microbiol. Rev.* **58:**563–602.

46. **Louarn, J., F. Cornet, V. Francois, J. Patte, and J. M. Louarn.** 1994. Hyperrecombination in the terminus region of the *Escherichia coli* chromosome: possible relation to nucleoid organization. *J. Bacteriol.* **176:**7524–7531.

47. **Low, K. B.** 1996. Hfr strains of *Escherichia coli* K-12, p. 2402–2405. *In* F. C. Neidhardt, R. Curtiss III, J. L. Ingraham, E. C. C. Lin, K. B. Low, B. Magasanik, W. S. Reznikoff, M. Riley, M. Schaechter, and H. E. Umbarger (ed.), Escherichia coli *and* Salmonella: *Cellular and Molecular Biology*, 2nd ed. American Society for Microbiology, Washington, D.C.

48. **Masters, M.** 1996. Generalized transduction, p. 2421–2441. *In* F. C. Neidhardt, R. Curtiss III, J. L. Ingraham, E. C. C. Lin, K. B. Low, B. Magasanik, W. S. Reznikoff, M. Riley, M. Schaechter, and H. E. Umbarger (ed.), Escherichia coli *and* Salmonella: *Cellular and Molecular Biology*, 2nd ed. American Society for Microbiology, Washington, D.C.

49. **McKane, M., and R. Milkman.** 1995. Transduction, restriction and recombination patterns in *Escherichia coli*. *Genetics* **139:**35–43.

50. **Milkman, R.** 1973. Electrophoretic variation in *Escherichia coli* from natural sources. *Science* **182:**1024–1026.

51. **Milkman, R.** 1996. Recombinational exchange among clonal populations, p. 2663–2684. *In* F. C. Neidhardt, R. Curtiss III, J. L. Ingraham, E. C. C. Lin, K. B. Low, B. Magasanik, W. S. Reznikoff, M. Riley, M. Schaechter, and H. E. Umbarger (ed.), Escherichia coli *and* Salmonella: *Cellular and Molecular Biology*, 2nd ed. American Society for Microbiology, Washington, D.C.

52. **Milkman, R.** 1997. Recombination and DNA sequence variation in *E. coli*, p. 177–189. *In* B. A. M. van der Zeijst, W. P. M. Hoekstra, J. D. A. Van Embden, and A. J. W. van Alphen (ed.), *Ecology of Pathogenic Bacteria, Molecular and Evolutionary Aspects*. North-Holland, Amsterdam, The Netherlands.

53. **Milkman, R.** 1997. Recombination and population structure in *Escherichia coli*. *Genetics* **146:**745–750.

54. **Milkman, R., and I. P. Crawford.** 1983. Clustered third-base substitutions among wild strains of *Escherichia coli*. *Science* **221:**378–380.

55. **Milkman, R., D. Cryderman, M. McKane, K. McWeeny, and E. A. Raleigh.** 1998. Evolutionary evidence for recombination among bacteria in nature: *E. coli*, p. 226–240. *In* M. Syvanen and C. Kado (ed.), *Horizontal Gene Transfer*. Chapman and Hall, London, United Kingdom.

56. **Milkman, R., and M. McKane.** 1995. DNA sequence variation and recombination in *E. coli*. p. 127–142. *In* S. Baumberg, J. P. W. Young, E. M. H. Wellington, and J. R. Saunders (ed.), *Population Genetics of Bacteria*. Cambridge University Press, Cambridge, United Kingdom.

57. **Milkman, R., and M. McKane Bridges.** 1990. Molecular evolution of the *Escherichia coli* chromosome. III. Clonal frames. *Genetics* **126:**505–517.

58. **Milkman, R., and M. McKane Bridges.** 1993. Molecular evolution of the *Escherichia coli* chromosome. IV. Sequence comparisons. *Genetics* **133:**455–468.

59. **Milkman, R., R. Melvin, E. Jaeger, and R. McBride.** Unpublished data.

60. **Milkman, R., E. A. Raleigh, M. McKane, D. Cryderman, P. Fiscus, and K. McWeeny.** Molecular evolution of the *Escherichia coli* chromosome. V. Recombination patterns among strains of diverse origin. *Genetics*, in press.

61. **Miller, J. H.** 1992. *A Short Course in Bacterial Genetics.* Cold Spring Harbor Laboratory Press, Cold Spring Harbor, N.Y.

62. **Modrich, P.** 1991. Mechanisms and biological effects of mismatch repair. *Annu. Rev. Genet.* **25:**229–253.

63. **Modrich, P., and R. Lahue.** 1996. Mismatch repair in replication fidelity, genetic recombination and cancer biology. *Annu. Rev. Biochem.* **65:** 101–133.

64. **Ochman, H., and R. K. Selander.** 1984. Standard reference strains of *Escherichia coli* from natural populations. *J. Bacteriol.* **157:**690–693.

65. **Ochman, H., and A. C. Wilson.** 1987. Evolutionary history of enteric bacteria, p. 1649–1654. *In* F. C. Neidhardt, J. L. Ingraham, K. B. Low, B. Magasanik, M. Riley, M. Schaechter, and H. E. Umbarger (ed.), Escherichia coli *and* Salmonella typhimurium: *Cellular and Molecular Biology.* American Society for Microbiology, Washington, D.C.

66. **Oh, S. H., and K. F. Chater.** 1997. Denaturation of circular or linear DNA facilitates targeted integrative transformation of *Streptomyces coelicolor* A3(2): possible relevance to other organisms. *J. Bacteriol.* **179:**122–127.

67. **Povilionis, P. I., A. A. Lubys, R. I. Vaisvila, S. T. Kulakauskas, and A. A. Janulaitis.** 1989. Investigation of methyl-cytosine specific restriction in *Escherichia coli* K-12. *Genetika* **25:** 753–755. (In Russian.)

68. **Price, C., and T. A. Bickle.** 1986. A possible role for DNA restriction in bacterial evolution. *Microbiol. Sci.* **3:**296–299.

69. **Raleigh, E. A.** 1992. Organization and function of the mcrBC genes of *E. coli* K-12. *Mol. Microbiol.* **6:**1079–1086.

70. **Reeves, P.** 1993. Evolution of *Salmonella* O-antigen variation by interspecific gene transfer on a large scale. *Trends Genet.* **9:**17–22.

71. **Reeves, P.** 1997. Specialised clones and lateral transfer in pathogens, p. 237–254. *In* B. A. M. van der Zeijst, W. P. M. Hoekstra, J. D. A. van Embden, and A. J. W. van Alphen (ed.), *Ecology of Pathogenic Bacteria, Molecular and Evolutionary Aspects.* North-Holland, Amsterdam, The Netherlands.

72. **Roberts, R. J., and D. Macelis.** 1998. REBASE—restriction enzymes and methylases. *Nucleic Acids Res.* **26:**338–350.

73. **Savageau, M. A.** 1983. *Escherichia coli* habitats, cell types and mechanisms of gene control. *Am. Nat.* **122:**732–744.

74. **Selander, R. K., D. A. Caugant, H. Ochman, J. M. Musser, and T. S. Whittam.** 1986. Methods of multilocus enzyme electrophoresis for bacterial population genetics and systematics. *Appl. Environ. Microbiol.* **51:**873–884.

75. **Selander, R. K., D. A. Caugant, and T. S. Whittam.** 1987. Genetic structure and variation in natural populations of *Escherichia coli*, p. 1625–1648. *In* F. C. Neidhardt, J. L. Ingraham, K. B. Low, B. Magasanik, M. Riley, M. Schaechter, and H. E. Umbarger (ed.), Escherichia coli *and* Salmonella typhimurium: *Cellular and Molecular Biology,* American Society for Microbiology, Washington, D.C.

76. **Selander, R. K., J. Li, and K. Nelson.** 1996. Evolutionary genetics of *Salmonella enterica.* p. 2691–2707. *In* F. C. Neidhardt, R. Curtiss III, J. L. Ingraham, E. C. C. Lin, K. B. Low, B. Magasanik, W. S. Reznikoff, M. Riley, M. Schaechter, and H. E. Umbarger (ed.), Escherichia coli *and* Salmonella: *Cellular and Molecular Biology,* 2nd ed. American Society for Microbiology, Washington, D.C. [See p. 2695.]

77. **Selander, R. K., and B. R. Levin.** 1980. Genetic diversity and structure in *Escherichia coli* populations. *Science* **210:**545–547.

78. **Sharp, P., and W.-H. Li.** 1986. An evolutionary perspective on synonymous codon usage in unicellular organisms. *J. Mol. Evol.* **24:**28–38.

79. **Sharp, P., and W.-H. Li.** 1987. The codon adaptation index—a measure of directional synonymous codon usage bias, and its potential applications. *Nucleic Acids Res.* **15:**1281–1295.

80. **Sharp, P., J. E. Kelleher, A. S. Daniel, G. M. Cowan, and N. E. Murray.** 1992. Roles of selection and recombination in the evolution of type I restriction-modification systems in enterobacteria. *Proc. Natl. Acad. Sci. USA* **89:**9836–9840.

81. **Sikorski, J., S. Graupner, M. G. Lorenz, and W. Wackernagel.** 1998. Natural transformation of *Pseudomonas stutzeri* in a non-sterile soil. *Microbiology* **144:**569–576.

82. **Singer, M., T. A. Baker, G. Schnitzler, S. M. Deischel, M. Goel, W. Dove, K. J. Jaacks, A. D. Grossman, J. W. Erickson, and C. A. Gross.** 1989. A collection of strains containing genetically linked alternating antibiotic resistance elements for genetic mapping of *Escherichia coli. Microbiol. Rev.* **53:**1–24.

83. **Smith, G. R.** 1991. Conjugational recombination in *E. coli*: myths and mechanisms. *Cell* **64:** 19–27.

84. **Smith, J. M., and N. H. Smith.** 1996. Site specific codon bias in bacteria. *Genetics* **142:** 1037–1043.

85. **Stevenson, G., K. Andrianopoulos, M. W. Hobbs, and P. R. Reeves.** 1996. Organization of the *Escherichia coli* K-12 gene cluster responsible for the extracellular polysaccharide colanic acid. *J. Bacteriol.* **178:**4885–4893.

86. **Titheradge, A. J. B., D. Ternent, and N. E. Murray.** 1996. A third family of allelic *hsd* genes in *Salmonella enterica*: sequence comparisons with related proteins identify conserved regions implicated in restriction of DNA. *Mol. Microbiol.* **22:** 437–447. (Corrigendum, **23:**851.)

87. **Vulic, M., F. Dionisio, F. Taddei, and M. Radman.** 1997. Molecular keys to speciation: DNA polymorphism and the control of genetic exchange in enterobacteria. *Proc. Natl. Acad. Sci. USA* **94:**9763–9767.

88. **Weisberg, R. A.** 1996. Specialized transduction, p. 2442–2448. *In* F. C. Neidhardt, R. Curtiss III, J. L. Ingraham, E. C. C. Lin, K. B. Low, B. Magasanik, W. S. Reznikoff, M. Riley, M. Schaechter, and H. E. Umbarger (ed.), Esch-erichia coli *and* Salmonella: *Cellular and Molecular Biology*, 2nd ed. American Society for Microbiology, Washington, D.C.

89. **Whittam, T. S., H. Ochman, and R. K. Selander.** 1983. Multilocus genetic structure in natural populations of *Escherichia coli*. *Proc. Natl. Acad. Sci. USA* **80:**1751–1755.

90. **Whittam, T. S.** 1996. Genetic variation and evolutionary processes in natural populations of *Escherichia coli*, p. 2708–2720. *In* F. C. Neidhardt, R. Curtiss III, J. L. Ingraham, E. C. C. Lin, K. B. Low, B. Magasanik, W. S. Reznikoff, M. Riley, M. Schaechter, and H. E. Umbarger (ed.), Escherichia coli *and* Salmonella: *Cellular and Molecular Biology*, 2nd ed. American Society for Microbiology, Washington, D.C.

91. **Worth, L., Jr., S. Clark, M. Radman, and P. Modrich.** 1994. Mismatch repair proteins MutS and MutL inhibit RecA-catalyzed strand transfer between diverged DNAs. *Proc. Natl. Acad. Sci. USA* **91:**3238–3241.

92. **Yin, X., and G. Stotzky.** 1997. Gene transfer among bacteria in natural environments. *Adv. Appl. Microbiol.* **45:**153–212.

GENETIC INVENTORY: *ESCHERICHIA COLI* AS A WINDOW ON ANCESTRAL PROTEINS

Bernard Labedan and Monica Riley

17

ESCHERICHIA COLI: AN INVALUABLE TOOL FOR GENOMICS STUDIES

Why *E. coli* Is Invaluable

Knowledge of the identities and functions of gene products enhances any analysis of gene and protein sequences. The delineation of groups of genes and proteins that trace back to common ancestors derives legitimacy and depth when functions are known and can be assessed for relatedness within any sequence-related group. There are two organisms whose genes and gene products have been under study for over 50 years: *Escherichia coli* and *Saccharomyces cerevisiae*. For these organisms the functions of many gene products are known, often very well known (in terms of biochemistry, physiology, cell anatomy, genetic regulation, etc.). This is due to the sustained efforts of many independent microbiologists, geneticists, and biochemists across the world, each one working on his or her gene and/or protein. The wealth of information on *E. coli* has been well summarized in the latest edition of the encyclopedic Escherichia coli *and* Salmo-

nella: *Cellular and Molecular Biology* (23), and one of the chapters compiles all of the cellular functions of genes and gene products that have been determined by the *E. coli* community (27). Such a detailed functional classification (see also reference 26) provided a valuable dimension to our project of thorough analysis of all *E. coli* genes.

Objectives

We initiated this analysis in order to understand the evolutionary history of both the genes and the proteins of present-day *E. coli* with the prospect that studying these particular entities will help to disclose the general mechanisms underlying the evolution of genes, proteins, and genomes in all organisms. Accordingly, we aimed in the following directions.

E. COLI: A TOOL FOR STUDYING GENES AND GENOME EVOLUTION

Although other mechanisms may also have played important roles, the duplication of ancestral genes followed by divergence has been suggested to be a major mechanism of molecular evolution (18, 24). Presumably, primitive cells would keep the two products of a gene duplication event if they evolved separate and useful cellular roles (15, 33). In order to check

Bernard Labedan, Institut de Génétique et Microbiologie, CNRS UMR 8621, Bâtiment 409, Université de Paris-Sud, 91405 Orsay Cedex, France.   *Monica Riley*, Marine Biological Laboratory, Woods Hole, MA 02543.

Organization of the Prokaryotic Genome, Edited by Robert L. Charlebois,
© 1999 American Society for Microbiology, Washington, D.C.

the validity of this attractive hypothesis, we have undertaken to determine the number of present-day genes putatively created by an ancestral gene duplication, since this number will give us an estimate of the importance of gene duplication in genome shaping.

E. COLI: A TOOL FOR STUDYING PROTEIN EVOLUTION

The goal in understanding protein evolution is the reconstruction of past events that have given rise to the inventory of extant proteins. Early ancestral organisms that existed before the separation of the three branches of the tree of life (defined as domains in reference 32) are believed to have possessed many of the functions of cell physiology and metabolism that are found in all living forms today. Preceding this development, at an earlier time in evolution, there were ancient cells that contained even more ancient proteins (8). The ancient proteins were the progenitors of the partially developed proteins in the last universal common ancestor, and the "ancestral proteins" were in turn the ancestors of many contemporary proteins. Products of duplication and divergence of the ancient proteins, including many enzymes and regulatory and transport entities, exist today in organisms from all three branches. By analysis of contemporary sequences we may be able to access some of the molecular mechanisms of the evolution of proteins that occurred in primordial primitive ancestral and ancient cells. Sets of sequence-related proteins in any one organism may be descendants of ancient proteins, created by duplication and divergence. Thus, the analysis of sequence-related proteins in E. coli will define sets of contemporary proteins that are products of duplication and divergence of an earlier ancestor. If divergence has not stretched relationships beyond recognition, we may be able to trace back through time to enumerate at least some of the ancient proteins. There are many complexities. One that affects reconstruction of related sets of proteins is the rearrangement of some of the proteins over time—fusion in various combinations of parts of proteins. Systematic identification of the elementary bricks that have been reassorted will help us to trace the way proteins were formed and evolved far into the past.

E. COLI: A TOOL FOR STUDYING RELATIONSHIPS BETWEEN SEQUENCE SIMILARITIES AND FUNCTIONAL SIMILARITIES

Are sequence similarities associated with functional similarities, and vice versa? Is the knowledge of gene and protein histories helpful in understanding how divergence in sequences permitted elaboration and specification of the numerous distinct and specific functions present in contemporary organisms? These are the types of questions that E. coli can help us answer.

A First Appraisal

With the final release of the complete sequence of the E. coli genome (4), it has become possible to make an appraisal of the progress we (and others [14, 31]) have made both in concepts and in the most important results obtained. In this review, we will describe (i) the main concepts we used and/or introduced; (ii) the methodological approaches we used to compare all of the E. coli proteins, in terms of similarities of sequences and similarities of function; (iii) the data we obtained in identifying some members of the putative set of ancestral genes present in the last universal common ancestor (their number, their sizes, and the functions they encode); and (iv) the data relevant to the mechanisms of protein formation and evolution.

INTRODUCING SOME BASIC CONCEPTS

Phylogeny of Organisms versus Genealogy of Proteins

Many molecular evolution studies have focused on the phylogeny of organisms. Molecular phylogeny tries to trace all the speciation events back to the last universal common ancestor. This approach is fundamental in reconstructing the tree of life and is also very useful in the context of species systematics. As to the relationships among proteins, many studies of protein sequence and structure have organized proteins into families of similar characteristics (references 14 and 31 and references therein). Connections among the protein families based on structures and primary sequences are only beginning to be made. We are outlining an approach to building genealogical trees of the descent of proteins from their ancestors by using the sequence data from fully sequenced genomes.

With the determination of the sequences of the complete genomes of several microorganisms, it has become possible to analyze the full panoply of genes and proteins of one organism. Through analysis of sequence relationships, it becomes possible to study both gene ancestry and protein evolution independently of speciation events. In other words, the data obtained using this genomics approach will enable us to group and identify the products of gene duplication events. Some duplication events occurred prior to the emergence of the last common ancestor, early in evolution, while other duplication events occurred more recently in the course of the emergence and evolution of organisms. Ultimately, comparative genomic analysis will allow us to distinguish ancient events common to all organisms from more recent events common to only a subset of organisms. The analysis of descendants of ancient proteins allows us to understand how genes that were present in the last common ancestor were formed over the course of evolution and will give indications about the identities of the ancestral proteins that were present in the last common ancestor. The two fields, molecular phylogeny of species and genomic analysis of protein families, are complementary and share several basic concepts.

Similarity and Homology

A newly determined sequence may have similarities to one or more previously determined sequences. It is important to stress the difference between similarity, which is an empirical observation that can be quantified by using various parameters (the most straightforward being the percentage of identical residues between two sequences), and homology, which is just a hypothesis of common ancestry. Indeed, sequence similarity can be due to at least three different mechanisms: homology, convergence, and gene conversion (25). Therefore, stating that two similar sequences are homologous is (i) an all-or-none condition and (ii) only the more plausible hypothesis.

Orthology and Paralogy

There are two main ways to be homologous according to the fundamental distinction initially made by Fitch (10). When homologues are found to be present in different genomes (species), they are called orthologues. Orthologous genes descend from a unique ancestral gene, and their divergence from comparable genes in different organisms is simply parallel to speciation. Therefore, they are good instruments for building phylogenetic trees of organisms. Paralogous genes, on the other hand, descend directly from copies of a gene that duplicated within an ancestral genome. These copies have diverged independently of speciation and are good candidates for understanding the course of protein evolution. (A third category, xenologous genes, complicates building species trees. They have been brought together by lateral transfer.) Nevertheless, as a first approximation, sets of proteins with similar sequences in the genome of one organism represent the products of duplication and divergence of a common ancestral protein. Some of the ancestors will be recent (the ancestors of specialized and unique functions), but many will be ancient ancestors of proteins performing similar functions in almost all existing organisms. Thus, sets of paralogous genes and their gene products can inform us of the ancient evolutionary history of macromolecules. Deep evolutionary times are accessible by sequence analysis of paralogous proteins as long as the changes to sequences over time by processes of mutation, recombination, and repair have not blurred the similarities so they cannot be discerned today over background noise.

What expectations are there regarding the functions of the descendants of a common ancestral protein? Paralogy and orthology as they have been defined by Fitch (10) have absolutely no implication about any similarity of function. In other words, even if we expect that orthologues will code for identical functions, we cannot make any assumption about the extent of divergence of function to be expected among a set of paralogous genes (see below). One view of functional relatedness is that the earliest ancestral proteins had broad specificity, catalyzing whole classes of reactions with one enzyme, and that progressively more specialized proteins with narrow specificity have been produced over time by duplication and divergence of the corresponding ancestral genes (15).

When looking at the full set of genes present in a single genome, we can distinguish two classes: those which are paralogues and those which are single (nonparalogues). We consider sets of paralogues to be descended from ancestral proteins by duplication and divergence. We consider single genes to be one of three possibilities. First, the singles could be descended from ancestral genes that never duplicated during their history, and thus the descendant gene present in *E. coli* has simply evolved by mutation in the course of unilineal descent at the same pace as speciation. Second, the single genes could be the only surviving members of earlier paralogous groups, many of which were retained in the direct descent to a bacterial genome. Third, the singles could be members of paralogous sets whose members have diverged too far for their familial relationship to be detectable today. (Lateral transfer of an unrelated foreign gene is always an additional possibility.)

Protein Family

For our purposes we consider a protein family to be a group of polypeptides, a large proportion of which descended from the same ancestor.

Module, Domain, and Motif

Protein nomenclature can be confusing. We previously defined a long segment (>100 amino acid residues) of homology within a protein as a "module" (28). A module as we define it and use it here is always far larger than a motif, which in common usage is normally composed of fewer than 20 amino acids. In our use of the term module, it is an element of sequence that has independent evolutionary history. It may correspond to the kind of functional unit, defined by protein structure analysis, that is called a domain, or a module may contain several such domains (28). A module is an element that is capable of independent existence but in some instances may be joined to another module of independent function in a multimodular protein (28). One example is the FadB protein of *E. coli*, which is composed of four enzyme entities joined together, enzyme entities that in other organisms can be separate and independent proteins. Another example is a two-component regulator protein in *E. coli*, EvgS, which contains multiple copies of both a sensor protein and the related response regulator, entities that in other cases are separate, independent proteins.

DISCOVERING SIGNIFICANT SEQUENCE SIMILARITIES AMONG *E. COLI* PROTEINS

Choosing a Methodology To Identify Homologous Proteins

To assess the homologous relationships among proteins, we adopted the following strategy, which is based in part on the experimental and theoretical works of Margaret Dayhoff and coauthors (6, 29), Stephen Altschul (1, 2), and Gaston Gonnet and coauthors (3, 11, 12).

We adopted the stochastic model of protein evolution proposed by Margaret Dayhoff and coauthors (6, 29), in which any protein will evolve by undergoing point mutation. In this stochastic model, the evolutionary distance separating two proteins deriving from a common ancestor and displaying significant sequence similarities is given in PAM units. A PAM unit is defined as the number of accepted point mutations per 100 residues separating two sequences; the term "accepted" emphasizes that only some amino acid substitutions are compatible with function and therefore survive in viable form. The frequency with which any particular pair of (mutated) amino acids occur at a given position in two properly aligned homologous proteins can be used as a PAM score to evaluate the evolutionary distance separating the two proteins. It has been shown (6) and frequently confirmed (9, 11, 12) that 250-PAM scores appear to be the best ones in sequence comparisons.

Following this fundamental pioneering work of Dayhoff and coauthors, Altschul (1) used an approach based on information theory. He showed that 30 bits of information are necessary to distinguish an alignment from chance. Accordingly, to be statistically significant, an alignment of sequences separated by a distance of 250 PAM units would need to have a length of at least 83 residues (see Table 1 in reference 1 and comments in the relevant text). Therefore, to define the significance of sequence similarities in terms of putative homology, we adopted the following two limits: any sequence alignment must extend for at least 80 residues and have a PAM distance of less than 250 PAM units.

The DARWIN program created by Gaston Gonnet and coauthors (3, 11, 12) was remarkably fitted to our theoretical approach and experimental needs. The program (available on-line [11]) is based on a probabilistic approach. Using the method of maximum likelihood, it tries to distinguish biologically meaningful similarities from those due to chance. To do that, DARWIN uses in a first step two tools which have been designed to align homologous proteins: the dynamic programming algorithm (22) is used with the PAM 250 matrix as a substitution score matrix. In this first step, the introduction of gaps is strictly regulated by penalties computed as a function of the PAM distance separating the two sequences (3). Then, in a second step, which is conceptually identical to the Altschul (2) approach but independently designed by the Gonnet group (3, 11, 12), each alignment is refined with two other tools. The first tool tries to extend the initial alignment as far as possible by using the Smith and Waterman algorithm (30), a powerful adaptation of the Needleman and Wunsch algorithm (22), which was designed for local alignment. The second tool recalculates the initial substitution score matrix with the current data set of proteins in order to find the best matrix. This optimization process is monitored by computing the variance of the PAM distance, searching for the lowest value. When this lowest value cannot be decreased further, the alignment is registered as optimal for the two proteins studied.

Moreover, DARWIN was designed to make an exhaustive matching of all members of an entire protein database. In other words, when a set of proteins has been conveniently formatted, it is possible to obtain, in one step, the optimized alignments for all of the putative homologues present within this set of proteins. The entire contents of the Swissprot database can be treated by this approach (12).

Numbering the *E. coli* Paralogous Proteins

We have adapted this DARWIN approach to the study of the *E. coli* proteins. Since DARWIN (which is written in the Maple programming language designed for a symbolic algebra system) is fully programmable, we have adapted several of the preexisting procedures and added several new ones to get in one step the full set of matches between *E. coli* proteins separated by less than 250 PAM units and having alignments extending for at least 80 residues. Renaud de Rosa, a Ph.D. student at the Université de Paris-Sud, designed the set of commands (7).

When we applied this strategy to the 3,996 *E. coli* sequences longer than 79 amino acids, we obtained 11,160 matches. These matches, separated by less than 250 PAM units, correspond to the alignments of 2,629 proteins that have sequence similarities extending for at least 80 residues. Since both length and PAM requirements are conservative, there is a high probability that these similar proteins are homologous. Moreover, according to the Fitch nomenclature (10), virtually all similar proteins may be defined as paralogues, although we cannot exclude the possibility that some of them are xenologues, foreign genes which entered the ancestral genome of *E. coli* at various times by horizontal transfer (20) followed by acclimatization to the genome (see chapter 15). Conversely, proteins which do not have a paralogous partner within the set of contemporary *E. coli* sequences, the 1,367 proteins we call singles, may in fact be paralogues which either lost their evolutionarily related kin at some period in the history of the *E. coli* genome or have diverged too far for the relationship to be detected.

This first result confirmed our previous results (16, 17, 27, 28), which indicated that the *E. coli* genome contains a high proportion of paralogous genes (65.8% of the total set of sequences longer than 79 amino acids). This high proportion strongly suggests that the mechanism of gene duplication has been one of the main forces helping to shape the evolution of all proteins, and notably, proteins present in the last common ancestor of all life.

Figure 1 shows the distribution of the PAM distances for these 11,160 matches. As can be seen, the majority of the matches correspond to fairly distant proteins, with a mean at around 160 PAM units. The fact that many proteins have maintained a PAM distance between 120 and 200 could suggest the following:

1. Contrary to what may have been feared, the divergence of many proteins did not proceed so far as to be lost through the saturation of repeated substitutions of amino acids, where so many replacements have occurred that no measure of distance between two proteins can be made (21). Although this saturation phenomenon may have blurred some events (the relationships among some singles may be in this class), nevertheless we are able to see many ancient evolutionary relationships through protein sequence comparison.

2. The distribution of distance between related proteins is unimodal (Fig. 1). One interpretation of the broad peak from 120 to 200 PAM units is that a majority of the descendants' ancestral sequences started to diverge at about the same time in the distant evolutionary past and have undergone evolutionary change over time to about the same extent. The number of pairs with low PAM values (more closely related) is much lower than the number in the peak range. Duplication seems to have been relatively rare in more recent times. This suggests that genome plasticity has not always been present, conferring the same level of freedom over long periods of time, but rather that there were windows in the evolutionary history where some *E. coli* ancestors could accommodate gene redundancy and divergence to new functions more easily than their decendants can today.

3. An alternative but not exclusive interpretation would be that in many cases functional constraints have hindered further divergence among related proteins. Although in large families of paralogues we see great variation in the level of divergence among members of the family—even to the point where there are no detectable sequence similarities between the most diverged sequences—the phenomenon of functional constraint may have been important for certain classes of proteins. The degree of functional constraint varies widely among different kinds of proteins, as reflected in the range of sizes of the groups of sequence-related paralogues. The sizes of paralogous groups range from large, over 100 members, to the smallest, pairs. Members of large groups evidently have not diverged very far from their common ancestors; at least they have retained sufficient similarity to enable sequence relatedness to be detected today among members of the family. Most large groups are either transport proteins or regulators. Members of small groups, such as pairs

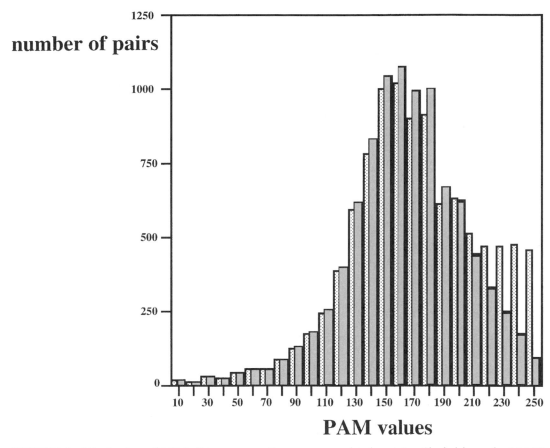

FIGURE 1 Distribution of PAM distances separating sequence-related proteins. Shaded bars, the 11,160 DARWIN matches separated by less than 250 PAM units and having alignments extending over at least 80 residues; solid bars, the 9,375 matches required to have identities of at least 20%.

and possibly also the singles, evidently have diverged and particularized to such an extent that it is not possible with current tools to detect relationships with other proteins that share the common ancestor. Many of the proteins in these small groups are enzymes.

Figure 1 also shows that the distribution of the PAM distances (with a mean of 169 PAM units) descends toward zero on the left, but there is an excess of pairs at large PAM distances, and therefore no descent toward a limiting PAM value on the high side. We observed that when we looked at the correlation between the two parameters used to measure the relatedness of proteins in each match, i.e., the PAM distance and the percentage of iden-

tical residues, the correlation was excellent at low PAM values and became progressively looser when the PAM distance increased. If we redraw this distribution leaving aside the matches having identity values below 20% (the remainder is a set of 9,375 pairs), we obtain a more symmetrical histogram with a mean PAM distance of 158 (Fig. 1). More than 95% of the excess pairs (1,785 pairs) correspond to distantly related members of the largest families (see below), with a mean PAM distance of 226.6.

Since the 2,629 paralogous proteins are engaged in at least 11,160 matches, gene duplication must have occurred frequently and repeatedly for some genes. Each sequence-related family must derive from a unique ancestral gene. Therefore, numbering the families is operatively equivalent to counting the ancestral genes.

Before we can sort paralogous *E. coli* proteins into families, we must deal with the phenomenon of modules as presented above. A few *E. coli* proteins have been rearranged and have fused since their initial duplication and divergence, and the separate elements within complex proteins need to be identified and separated. Defining the modules helped us to better understand mechanisms of protein evolution.

Defining the Module

When looking at the different matches present in the DARWIN output, we observed that there were at least three types of alignment. These are summarized in Fig. 2.

Figure 2A shows the first case, where the alignment extends over the entire length of both proteins. This is the most frequent case, and it is independent of the protein size.

Figure 2B shows a more complex case, where a shorter protein aligns its entire length along a segment of a longer protein. For example, proteins f, g, and h behave like proteins a, b, and c of Fig. 2A but align with only the second half of proteins d and e. In a previous paper (28), we named the separate regions of alignment present in the larger proteins modules, and called the proteins containing more than one module multimodular. Although modules are found fused in a multimodular protein, a defining characteristic is that they are known to have independent existence as simple proteins either in the same organism or in other organisms (13). Thus, the second half

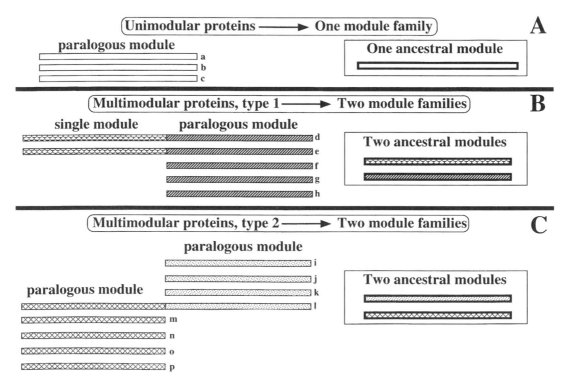

FIGURE 2 Schematic view of the different module arrangements found for the 11,160 DARWIN matches. The natures of the family, of the present-day modules, and of the ancestor are indicated for the three cases A, B, and C. "Paralogous" and "single" refer to modules with paralogous (duplication and divergence of the two copies) or nonparalogous (no duplication and unilineal descent) behaviors, respectively, during their evolutionary histories.

of proteins d and e and the complete proteins f, g, and h correspond to paralogous modules which seem likely to derive from one ancestral module of about the same size. Since proteins d and e align along their entire lengths, one may further imagine that at some time one copy deriving from this ancestral module fused with another single module which was descended from a separate, nonduplicated ancestor, giving the progenitor of the present-day proteins d and e. Therefore, this group of proteins corresponds to two independent families of modules that derive from two ancestral modules, one with paralogous behavior that duplicated more than once and generated a family of proteins, the other with unilineal behavior that did not duplicate but generated a single line of progeny proteins by unilineal descent. Following this defining approach, unimodular proteins, such as that shown in Fig. 2A, would derive from one ancestral module that had paralogous behavior. Figure 2C shows another example of a multimodular protein, but this time the fused protein l corresponds to the union of two independent and unrelated modules, both of which had simple paralogous behavior. One module corresponds to the first half of l and the four short proteins m, n, o, and p, and the other module corresponds to the second half of protein l and the three short proteins i, j, and k. One can see that if we did not make the distinction in terms of modules we would have put two unrelated protein families without any evolutionary relationship in the same group.

Numbering the Modules

We have tried to detect and label all the modules present in paralogous proteins of E. coli. This must be an underestimate of all the putative modules present in E. coli proteins, since not all will have single-module paralogues in the E. coli genome. In future work, delineation of modules in other genomes will in time lead to a listing of all the independently evolved modules of currently known contempororay proteins. In E. coli many of the paralogous modules were clearly distinguishable by inspection of the DARWIN output. Although it was possible to analyze the modular makeup of E. coli proteins by hand, to work toward a reproducible procedure, it was necessary to develop an automatic approach. An automatic program was developed to detect straightforward modular structure in many of the matches and to assign them a suffix signifying the location of the module in the protein. The names of modules located in the N-terminal part of the protein were given the suffix _1. The names of modules located in either the middle or the C-terminal part of the protein were given the suffix _2. Modules located further into the C-terminal part of large proteins were given a suffix with the next higher number. In some cases the modular structures had ambiguous boundaries and were too complex to be assessed directly by the automatic approach. These cases, amounting to some 10% of the 11,160 matches, required a one-by-one manual examination with reiterative checks of module boundaries in relation to all earlier module assignments.

Figure 3 shows the distribution of the lengths of the 3,717 paralogous modules (including the 1,173 unimodular paralogous proteins). There is a small group centered at around 95 amino acid residues, and the large majority is centered around 220 residues.

The size relationships tell us that in early times ancestral genes were coding for proteins of sizes similar to those of contemporary proteins. Moreover, Table 1 compares module lengths with the total lengths of paralogous proteins and with the total lengths of the single proteins (proteins that are not members of paralogous groups, evidently inherited simply by unilineal descent). The single nonparalogues (mean length, 229.5) have about the same size as the main class of modules (mean length, 220), whereas paralogues (mean length, 381) are close to twice the size of the modules, suggesting that many present-day paralogues are the products of at least one fusion event between two ancestral modules.

Grouping Modules into Families

As previously stated, once we have determined the total number of paralogous mod-

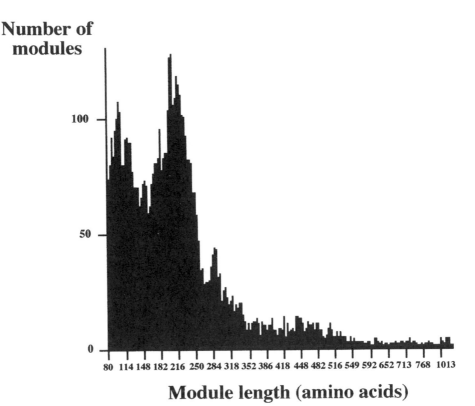

FIGURE 3 Distribution of module lengths.

ules present in *E. coli*, we can assemble them into sequence-related families in order to find out how many ancestral genes were at the origin of the present-day proteins of this bacterium. To do that we used simple automatic programs designed in our respective laboratories (7, 28). The most efficient, which is also the most recent, was written by Renaud de Rosa in the Caml language. It allows us to assemble families of sequence-related modules transitively and to sort the families by size (number of members) and then by the alphabetical order of gene names, which are themselves sorted inside each family (7).

This program grouped the 3,717 modules into some 873 families, suggesting that the 3,717 segments of homology which form the 2,629 paralogous proteins are descended from 873 ancestral genes. The families are formed by including any module that is homologous

to at least one member of the family. Not all modules are related to all members of the family. However, the groups are independent of one another in the sense that no member of one family is also a member of another family. In this way, we are assembling modules whose

TABLE 1 Mean lengths of the different kinds of proteins and modules

Protein or module	Total no.	Mean length (residues)
Total proteins	3,996	329.0
Singles[a]	1,367	229.5
Paralogues	2,629	381.0
Paralogous modules	3,717	220.0

[a] Proteins that are inherited simply by unilineal descent at the same pace as speciation or that are members of earlier paralogous groups which lost their evolutionarily related kin at some period in the history of the *E. coli* genome.

relationships can be quite distant but nevertheless perceptible above background noise.

We believe each module family corresponds to an ancestral protein or, in some cases, even farther back, to an ancient protein ancestor. Subsequent fusions of some modules in different lines of descent have given rise to multimodular proteins with different arrangements in different organisms. A comparative study years ago of the *trp* operons in different microoganisms showed that different genes for enzymes of tryptophan biosynthesis have been fused in different organisms (5). TrpA and -B are fused in yeast but not in *E. coli*. TrpD and -G are fused in *E. coli* but not in *Serratia marcescens*, and so forth.

To inquire into the descent of the proteins of a family from their ancestral sequence, we have studied further the history of individual families. For each family larger than three members, we used other DARWIN procedures (MulAlignment and ProbAncestral) to make a multiple alignment of each module sequence and to calculate the probabilistic ancestral sequence (PAS) simultaneously (11). The PAS is a sequence of probability vectors, one for each amino acid position. It is computed with a bottom-up approach, using the distance tree (see below) to compute an intermediate PAS for each node up to the selected root of the tree. The evolutionary tree is computed by the DARWIN procedure Phylotree, which is based on a least-squares approach using the estimated PAM distances between pairs of sequences. Since the deduced evolutionary distance between nodes is weighted by computing the variance of the distance, these distance trees are approximations to maximum-likelihood trees (see references 11 and 12 for additional details of the method, as well as a new DARWIN manual under construction at the Internet address http://cbrg.inf.ethz.ch/personal/hallett/drive/drive.html). Moreover, although these trees are unrooted, a weighted centroid of the tree is calculated as being the virtual root (i.e., the location of the PAS). Since we are looking at the history of *E. coli* proteins, we propose

calling these evolutionary trees genealogical trees of *E. coli* modules. Let us look at a few examples of families.

A HYDRATASE FAMILY

Figure 4 shows the genealogical tree and module alignment of a four-member family. These paralogous modules belong to two isoenzymes (aconitate hydratases 1 and 2), a subunit (LeuC) of a functionally similar enzyme, the 3-isopropylmalate dehydratase, and the product of one unknown open reading frame (b0771). This genealogical tree may be seen as a good example of the topology of small families.

Several interesting features specific to this family must be emphasized. (i) Surprisingly, the two isoenzymes are very distant. As shown in the inset in Fig. 4, their similarities of sequence are too low for our criteria of homology. They are found to belong to the same group only because both aconitate hydratases display significant sequence similarities to LeuC and b0771. We will see in more detail later this general rule of correlation between sequence similarities and function similarities. (ii) The module of this family is about the size of the entire LeuC protein. The corresponding segment of homology is located on the N-terminal side in aconitate hydratase 1, and the same thing is true for b0771, but it is located on the C-terminal side in aconitate hydratase 2. This striking difference in location suggests, at least in this case, that module location has no effect on protein activity. It also confirms that the evolutionary histories of the two aconitate hydratases are significantly different. (iii) The long segment present in the N-terminal side of aconitate hydratase 2 has no known homologue either in *E. coli* or in the whole protein database. It is a nonparalogous module of independent origin, seemingly descended unilinearly without duplication other than speciation. On the other hand, we can detect on the C-terminal side of aconitate hydratase 1 a short module homologue to the corresponding segment present in the second subunit (LeuC) of the 3-isopropylmalate de-

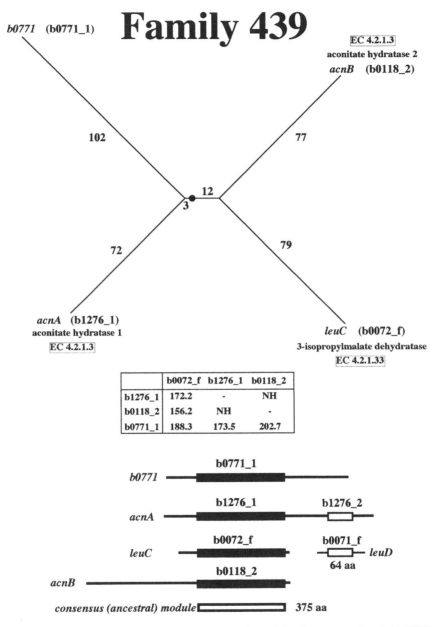

FIGURE 4 Family 439: genealogical tree and module alignment. The DARWIN PhyloTree procedure was used to reconstruct the genealogical tree from the alignment of modules found in proteins b0072, b0118, b0771, and b1276. The matrix of the PAM distances separating the modules of these different proteins is shown in an inset. NH, nonhomologous; aa, amino acids.

hydratase. This module is only 64 amino acids long and thus has not been detected in our exhaustive search, where we put the lower limit at an alignment length of 80 residues. This emphasizes how our present figures represent an underestimation of the exhaustive listing of the homologous modules. Thus, it seems that during their evolutionary history the two similar enzymes aconitate hydratase 1 and 3-isopropylmalate dehydratase have followed two different paths: in one case, two independent modules remain as separate subunits (encoded by two adjacent genes inside the *leu* operon) which form a multimeric active complex. In the other case, a similar function is performed by a monomer which corresponds to the fusion of the two ancestral modules. Note also that b0771 is apparently too short to contain this entire 64-amino-acid module, and this may predict an absence of functionality. This first example shows how the study of the genealogy of a small family like family 439 can give us important information on several aspects of protein evolution.

A FAMILY OF KINASES OF SUGAR

Figure 5 shows the genealogical tree of a larger family. There are six kinases of various sugars and two unknown open reading frames. This family is rather homogeneous in terms of the functions and sizes of the proteins. The topology of this tree is rather typical of many midsize families, having both a center with poor resolution and distances that are probably underestimated due to the saturation phenomenon (21). However, this tree strongly suggests that the ancestral module which was at the origin of this family, and which extends along the entire length of these eight present-day proteins, was already a sugar kinase with a size of about 500 amino acids. Moreover, it must be emphasized that other paralogous kinases of sugar having different sizes are present in other families, showing that different evolutionary histories have produced this kind of functional protein.

A MIXED FAMILY OF REPRESSORS AND PERIPLASMIC BINDING PROTEINS

Figure 6 shows an example of the genealogical tree of a mixed family. There are evolutionary relationships between repressors of type LacI, which group on the left side of the tree, and several periplasmic binding proteins (PBP) of transport systems, which group on the right of the tree. This is the case, for example, with the pair made up of RbsB, which participates in the transport of ribose, and RbsR, the repressor of the ribose operon. An explanation, already proposed by Mauzy and Hermodson (19), might be that differentiation of a protein specifically recognizing ribose gave rise to a ribose-specific repressor on one hand and a ribose-specific transport protein on the other hand. More generally, the ancestor, which is probably located at the connection point of the two subtrees, could have been a protein able to bind to ribose and which, after gene duplication and divergence of the copies, gained progressively different functionalities and different cellular locations. The addition of a signal sequence could have been a crucial step for such a transfer from cytoplasm to periplasm.

THE CASE OF THE PORINS

Figure 7 shows an interesting example of two interconnected module families. Three porins present in the outer membrane, PhoE, OmpF, and OmpC, are very similar. One open reading frame, b1377, appears to also be evolutionarily very close to these porins. These four proteins appear to be made up of two modules: an N-terminal module which is a homologue of the open reading frame b1964 and a C-terminal module which is a homologue of the open reading frame b1966. These correspond to families 500 and 481, respectively. Moreover, there is another, shorter open reading frame, b1472, which belongs to family 500. The two proteins, which help to define the two modules of the porins, are encoded by neighboring genes which are separated by only a short open reading frame, b1965, en-

Family 523

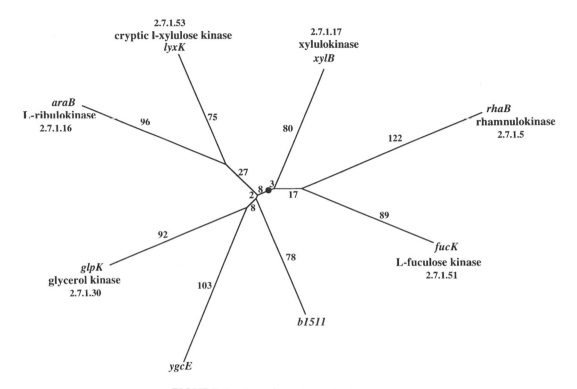

FIGURE 5 Genealogical tree for family 523.

coding a putative 69-amino-acid peptide. This peptide is highly similar to the segments separating the two modules of the porins and porin-like protein. Therefore, two hypotheses are possible: (i) the ancestral modules (corresponding in size to b1964, b1965, and b1966) have fused to give the present-day porins; (ii) alternatively, the ancestor was of the size of the present-day porins and duplicated several times to produce these proteins. Then, one of the copies was interrupted twice to give the three genes b1964, b1965, and b1966.

RELATIONSHIPS BETWEEN SEQUENCE SIMILARITIES AND FUNCTION SIMILARITIES

Sequence Similarities versus Function Similarities

We have observed that virtually all of the proteins in any one group with significant sequence similarities (whose function is known) have similar functions (references 17 and 27 and unpublished data). This reinforces the idea that when an unknown open reading frame is highly similar to several proteins which code for similar functions, it may be assigned a related function with good probability. This could be done, for example, for the sequence b1377 in the case of the porins (Fig. 7) or for YgcE and b1511 in the case of the kinases of sugar (Fig. 5). This correlation also strongly suggests that the original progenitor sequence of such a homogeneous family coded originally for a function similar to that of its present-day descendants.

Nearly half of all the proteins of *E. coli* have at least one partner with a similar sequence and function. Some sets of similar proteins are quite large, over 100 members (unpublished data). The persistence of so many groups with

Family 549

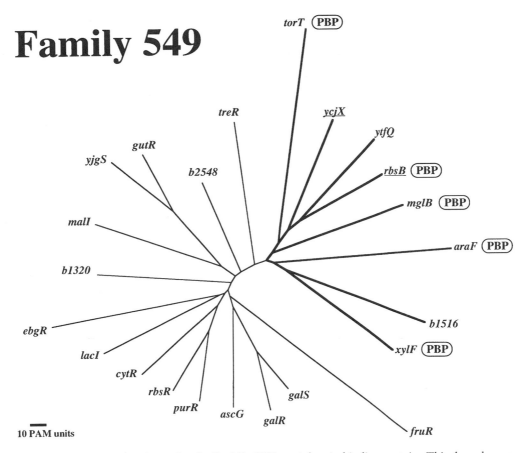

FIGURE 6 Genealogical tree for family 549. PBP, periplasmic binding protein. Thin branches, repressors of type LacI; thick branches, PBP. The PBP with significant sequence similarities to one or several repressors are underlined.

Families 481/500

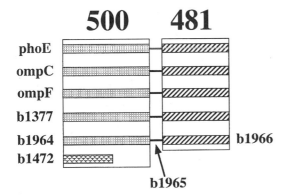

FIGURE 7 Module alignment for families 481 and 500.

similar sequences and functions to the present time underscores the importance of the mechanism of duplication and divergence as an evolutionary device.

However, there are a few exceptions, cases of paralogous proteins having apparently completely different functions. Several examples may be found in biosynthetic pathways, such as the pair MetB and MetC. These proteins are in the pathway of synthesis of methionine and are related, with a PAM value of 145. MetB encodes cystathionine gamma-synthase and MetC encodes cystathionine beta-lyase. The similarity in this case resides in substrate specificity, not in the reaction catalyzed.

HisA and HisF are two unrelated enzymes which catalyze steps 4 and 6 of the histidine biosynthesis pathway, respectively. HisA is the *N*-(5′-phospho-1-ribosyl-formimino)-5-amino-1-(5′-phosphoribosyl)-4-imidazole carboxamide isomerase, whereas HisF is imidazole glycerol phosphate synthase. Although the similarity is weak (PAM = 236), the two protein sequences are related over their whole lengths. It may be that recognition of the imidazole derivatives is the common feature of HisA and HisF, although the enzymes catalyze nonidentical reactions.

A similar case occurs with the pair PurT-PurK, two unrelated enzymes which catalyze step 3 (5′-phosphoribosyl-glycinamide formyltransferase 2) and step 6 (the CO_2-fixing subunit of 5′-phosphoribosyl-aminoimidazole carboxylase) of purine biosynthesis, respectively. PurT and PurK are related over most of their lengths, with a PAM value of 174. Although their substrates share a 5′-phosphoriboxyl group, the reactions catalyzed are different.

In all three of these unusual cases, a weak sequence similarity of enzymes evidently relates to some element of substrate similarity, not to a similarity of reaction mechanisms.

However, these cases are rare. Most sets of proteins of *E. coli* with similar sequences are proteins that carry out similar actions, whether those actions are enzymatic catalysis, transport of metabolites by similar mechanisms, or regulation of gene expression by similar mechanisms (reference 28 and unpublished data).

Function Similarities versus Sequence Similarities

If we look now at the relationship of sequence and function from the opposite point of view, in many cases, we see proteins which have identical or similar functions but have no detectable sequence similarities. We have already mentioned several cases, such as the two aconitate hydratases (family 439 [Fig. 4]) or the two phosphoribosylglycinamide formyltransferases. This absence of correlation between similarities of function and similarities of sequence is strikingly clear in the case of some instances of isoenzymes. In a previous work, we pre-

sented a list of 76 pairs of isoenzymes known at that time in *E. coli* (17, 27). The majority, 43 pairs, were found to have sequence similarities, but as many as 33 were poorly related or not at all (with PAM distances of over 250). It seems likely that the latter group have evolved by convergence to ultimately display identical or similar functions with dissimilar sequences. Although we understand the mechanism of gene duplication, we do not know today what the molecular mechanisms are which allow this formation of analogous proteins by functional convergence. Alternatively, these analogous proteins may have been invented independently to accomplish the same task. Furthermore, we cannot exclude the possibility that some of the unrelated isoenzymes have been acquired by lateral transfer from another organism.

Function Similarities: Insight on Protein Evolution

Defining a family of paralogous modules is helpful in reconstructing the putative ancestral sequence of the progenitor and the subsequent evolutionary history of the family. Analysis of all paralogous families of all organisms based on maximum likelihood could lead to a picture of the gene content of the last universal common ancestor. It appears that many of these ancestral genes were of the size of the modules, i.e., not so different from unimodular present-day proteins. Moreover, it appears that the last universal common ancestor had already developed several genes which code for similar functions that appear not to be evolutionarily related. What can we learn from such data to understand how protein function evolved?

Let us see, for example, the case of the decarboxylases, starting with the two arginine decarboxylases, Adi and SpeA. These two isoenzymes are apparently analogous: they have no discernible sequence similarities and different features. SpeA is an anabolic enzyme (the first step in the biosynthesis of spermidine from arginine) located in the periplasm, and its gene expression is induced by growth in minimal medium at neutral pH and repressed by putrescine and spermidine. Adi is a catabolic enzyme located in the cytoplasm, and its gene expression is induced by growth in acid medium and anaerobiosis. Adi is thought to regulate internal pH by consuming proteins. SpeA is homologous to LysA (family 280), the diaminopimelic decarboxylase. Adi is part of family 513, which contains two pairs of paralogous isoenzymes, SpeC-SpeF (ornithine decarboxylase, constitutive and inducible, respectively) and CadA-LdcC (lysine decarboxylase, inducible and constitutive, respectively). Moreover, there is another pair of paralogous isoenzymes, GadA-GadB (glutamate decarboxylase), not related today to the others. This is summarized in Fig. 8, where we propose that these three families of paralogues issued from three ancestral modules of different sizes. The three ancestral modules may have been present in the last universal common ancestor,

along with the ancestors of the four different single modules, which appear to have evolved separately and independently into four other present-day decarboxylases, namely, HemE, PyrF, Psd, and SpeD. What we do not yet understand is how these seven ancestral modules were formed. Were they independent creations de novo of different sorts of functional decarboxylases, or did they come from a single very ancient decarboxylase with very broad specificity, and divergence has proceeded to the point that we can no longer detect common parentage?

FOR THE FUTURE

Analysis of all available genomic sequence data is far from complete, but already at this stage we can see the outline of the process of creation of the multiplicity of proteins known today. We suppose that a first level of variation on the ancient proteins was created by dupli-

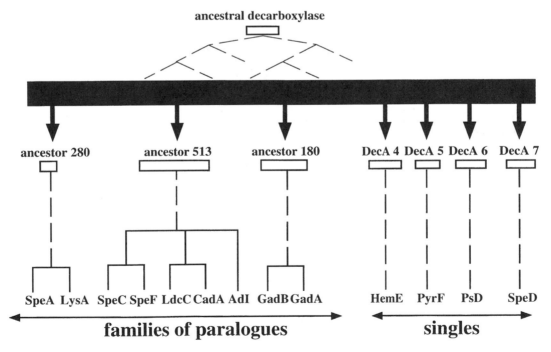

FIGURE 8 Schematic view comparing evolutionary histories of similar sequences and similar functions. Broken lines indicate putative lineages of descent of each putative ancestor of either the ancestor of one family of paralogous modules or unique modules. Solid lines indicate schematic trees of the determined families of paralogous modules.

cation and divergence within the ancient organisms. Ultimately, the last common ancestor of the three domains of the tree of life inherited the products of this variation. Further variation occurred by further divergence in descendant organisms created by repeated speciation events. The unit of functional protein, the module, which had evolved independently, fused with other modules in various lines of descent to create additional variation.

With the many whole-genome-sequencing projects under way today and planned for the future, there will be an abundance of data to be analyzed further along these lines, with the ambitious aim of eventually reconstructing the paths of evolution of all proteins.

REFERENCES

1. **Altschul, S. F.** 1991. Amino acid substitution matrices from an information theoretic perspective. *J. Mol. Biol.* **219:**555–565.
2. **Altschul, S. F.** 1993. A protein alignment scoring system sensitive at all evolutionary distances. *J. Mol. Evol.* **36:**290–300.
3. **Benner, S. A., M. A. Cohen, and G. H. Gonnet.** 1993. Empirical and structural models for insertions and deletions in the divergent evolution of proteins. *J. Mol. Biol.* **229:**1065–1082.
4. **Blattner, F. R., G. Plunkett III, C. A. Bloch, N. T. Perna, V. Burland, M. Riley, J. Collado-Vides, J. D. Glasner, C. K. Rode, G. F. Mayhew, J. Gregor, N. W. Davis, H. A. Kirkpatrick, M. A. Goeden, D. J. Rose, B. Mau, and Y. Shao.** 1997. The complete genome sequence of *Escherichia coli* K-12. *Science* **277:**1453–1474.
5. **Crawford, I. P.** 1975. Gene rearrangements in the evolution of the tryptophan synthetic pathway. *Bacteriol. Rev.* **39:**87–120.
6. **Dayhoff, M. O., R. M. Schwartz, and B. C. Orcutt.** 1978. A model for evolutionary change, p. 345–352. *In* M. O. Dayhoff (ed.), *Atlas of Protein Sequence and Structure*, vol. 5, suppl. 3. National Biomedical Research Foundation, Washington, D.C.
7. **De Rosa, R., and B. Labedan.** 1998. The evolutionary relationships between the two bacteria *Escherichia coli* and *Haemophilus influenzae* and their putative last common ancestor. *Mol. Biol. Evol.* **15:**17–27.
8. **Doolittle, R. F.** 1981. Similar amino acid sequences: chance or common ancestry? *Science* **214:**149–159.
9. **Feng, D. F., M. S. Johnson, and R. F. Doolittle.** 1985. Aligning amino acid sequences: comparison of commonly used methods. *J. Mol. Evol.* **21:**112–125.
10. **Fitch, W. D.** 1970. Distinguishing homologous from analogous proteins. *Syst. Zool.* **19:**99–113.
11. **Gonnet, G., and M. Hallett.** 29 August 1997, posting date. [Online.] *The DARWIN Manual.* http://cbrg.inf.ethz.ch/. [25 May 1999, last date accessed.]
12. **Gonnet, G. H., M. A. Cohen, and S. A. Benner.** 1992. Exhaustive matching of the entire protein sequence database. *Science* **256:**1443–1445.
13. **Guigo, R., I. Muchnik, and T. F. Smith.** 1996. Reconstruction of ancient molecular phylogeny. *Mol. Phylogenet. Evol.* **6:**189–213.
14. **Henikoff, S., E. A. Greene, S. Pietrokovski, P. Bork, T. K. Attwood, and L. Hood.** 1997. Gene families: the taxonomy of protein paralogs and chimeras. *Science* **278:**609–614.
15. **Jensen, R.** 1976. Enzyme recruitment in evolution of new function. *Annu. Rev. Microbiol.* **30:**409–425.
16. **Labedan, B., and M. Riley.** 1995. Widespread protein sequence similarities: origins of *Escherichia coli* genes. *J. Bacteriol.* **177:**1585–1588.
17. **Labedan, B., and M. Riley.** 1995. Gene products of *Escherichia coli*: sequence comparisons and common ancestries. *Mol. Biol. Evol.* **12:**980–987.
18. **Lewis, E. B.** 1951. Pseudoallelism and gene evolution. *Cold Spring Harbor Symp. Quant. Biol.* **16:**159–174.
19. **Mauzy, C. A., and M. A. Hermodson.** 1992. Structural homology between rbs repressor and ribose binding protein implies functional similarity. *Protein Sci.* **1:**843–849.
20. **Médigue, C., T. Rouxel, P. Vigier, A. Henaut, and A. Danchin.** 1991. Evidence for horizontal gene transfer in *Escherichia coli* speciation. *J. Mol. Biol.* **222:**851–856.
21. **Meyer, T. E., M. A. Cusanovich, and M. D. Kamen.** 1986. Evidence against use of bacterial amino acid sequence data for construction of all-inclusive phylogenetic trees. *Proc. Natl. Acad. Sci. USA* **83:**217–220.
22. **Needleman, S. B., and C. D. Wunsch.** 1970. A general method applicable to the search for similarities in the amino acid sequence of two proteins. *J. Mol. Biol.* **48:**443–453.
23. **Neidhardt, F. C., R. Curtiss III, J. Ingraham, E. C. C. Lin, K. B. Low, B. Magasanik, W. S. Reznikoff, M. Riley, M. Schaechter, and H. E. Umbarger (ed.).** 1996. Escherichia coli *and* Salmonella: *Cellular and Molecular Biology*, 2nd ed. ASM Press, Washington, D.C.

24. **Ohno, S.** 1970. *Evolution by Gene Duplication.* Springer-Verlag, New York, N.Y.

25. **Patterson, C.** 1988. Homology in classical and molecular biology. *Mol. Biol. Evol.* **5:**603–625.

26. **Riley, M.** 1993. Functions of the gene products of *Escherichia coli. Microbiol. Rev.* **57:**862–952.

27. **Riley, M., and B. Labedan.** 1996. *E. coli* gene products: physiological functions and common ancestries, p. 2118–2202. *In* F. C. Neidhardt, R. Curtiss III, J. L. Ingraham, E. C. C. Lin, K. B. Low, B. Magasanik, W. S. Reznikoff, M. Riley, M. Schaechter, and H. E. Umbarger (ed.), Escherichia coli *and* Salmonella: *Cellular and Molecular Biology*, 2nd ed. ASM Press, Washington, D.C.

28. **Riley, M., and B. Labedan.** 1997. Protein evolution viewed through *Escherichia coli* protein sequences: introducing the notion of a structural segment of homology, the module. *J. Mol. Biol.* **268:**857–868.

29. **Schwartz, R. M., and M. O. Dayhoff.** 1978. Matrices for detecting distant relationships, p. 353–358. *In* M. O. Dayhoff (ed.), *Atlas of Protein Sequence and Structure*, vol. 5, suppl. 3. National Biomedical Research Foundation, Washington, D.C.

30. **Smith, T. F., and M. S. Waterman.** 1981. Identification of common molecular subsequences. *J. Mol. Biol.* **147:**195–197.

31. **Tatusov, R. L., E. V. Koonin, and D. J. Lipman.** 1997. A genomic perspective on protein families. *Science* **278:**631–637.

32. **Woese, C. R., O. Kandler, and M. L. Wheelis.** 1990. Towards a natural system of organisms: proposal for the domains Archaea, Bacteria, and Eucarya. *Proc. Natl. Acad. Sci. USA* **87:**4576–4579.

33. **Ycas, M.** 1974. On earlier states of the biochemical system. *J. Theor. Biol.* **44:**145–160.

PROTEOME APPROACH TO THE IDENTIFICATION OF CELLULAR *ESCHERICHIA COLI* PROTEINS

Amanda S. Nouwens, Femia G. Hopwood, Mathew Traini, Keith L. Williams, and Bradley J. Walsh

18

In the ever-changing environment, prokaryotes, like all living organisms, must adapt or respond to change in order to survive and proliferate. The importance of prokaryotes in the environment and their impact on humans have caused an enormous amount of time and money to be invested in research to understand the functions and organization of their cellular parts. In the past, research has usually focused on specific genes or gene products, and it was not thought possible to be more holistic. Now, with expression studies and proteomics it is possible to view globally gene activity and the resultant protein makeup of an organism. *Escherichia coli* has been an obvious initial study for proteomics, as it is one of the best-understood organisms and its genome was one of the first sequenced (4, 12, 32, 53). Indeed, one of the first organisms to be separated by two-dimensional polyacrylamide gel electrophoresis (2-D PAGE) was *E. coli* (44). Here we review the technologies that make proteomics a genuinely new approach to protein science, and we reflect on the emergence of global protein studies that allow proteins separated on gels to be identified, thus enabling analysis of protein fluxes and networks.

PROTEOMICS

The word *proteome* describes the protein products expressed by the genome. Simply speaking, it is a new word to describe an old concept: that of functional genomics and the systematic analysis of proteins and their modifications in whole organisms, as well as in tissues or organelles. The application of some proteome technologies, such as 2-D PAGE, was pioneered approximately 20 years ago by researchers such as Garrels (19, 20), Görg (23, 65), Celis (7, 8), O'Farrell (44), and Van-Bogelen and Neidhardt (42, 43, 62). In the past, proteome projects were arduous, as the technology was not ready for preparative protein array and true parallel protein identifications were not possible. Thus, progress could be slow and frustrating. Advances in this emerging area led to the coining of the new word proteome to describe the field (67), the invention of core technologies, and their application to a variety of biological problems (15, 16, 22, 23, 27, 66). The novelty of the approach is that it is now possible to conduct

Amanda S. Nouwens, Femia G. Hopwood, Mathew Traini, Keith L. Williams, and Bradley J. Walsh, Australian Proteome Analysis Facility (APAF), Macquarie University, Sydney, NSW 2109, Australia.

Organization of the Prokaryotic Genome, Edited by Robert L. Charlebois,
© 1999 American Society for Microbiology, Washington, D.C.

parallel studies of many proteins separated by highly reproducible and sensitive techniques.

THE COMPLEMENTARY ROLES OF PROTEOME AND GENOME STUDIES

Proteomics is a means of studying an organism's gene expression on a global scale. Compared to genomics, proteomics provides information on the end products of gene expression and, in essence, is an examination of the "tools" an organism uses to survive and proliferate in an environment. Genomics provides the total information base or capacity of an organism to survive in terms of the potential gene products, open reading frames, and organization of the genes. Genomic projects may tell us which genes are turned on or off in response to stimuli, but proteomics indicates the dynamic or static status of the gene products (proteins).

Often genomic information is compared with that from other organisms to identify homologous genes, thereby assigning potential functions to genes or open reading frames which are currently unknown. Genome-sequencing projects, such as the recently completed *E. coli* project, have identified numerous potential open reading frames with as-yet-unassigned gene functions (http://www.genetics.wisc.edu:80/index.html). Proteome studies can confirm that these open reading frames encode functional genes through the identification of the gene products. It is particularly challenging to identify genes encoding small proteins, for example, those comprising less than 150 amino acids, from DNA sequence alone (49).

Other reasons for the importance of examining the output of the gene as well as the gene itself include the following.

1. There is a lack of direct correspondence between mRNA levels and those of proteins. So far, this comparison has only been conducted in eukaryotes (2, 24). Transcription rates do not adequately reflect the protein status of the cell, and assessment of gene expression based solely on mRNA levels may, at times, be misleading. The turnover of mRNA and protein, as well as rates of synthesis for both, are important factors in a holistic assessment of the cell biology of any organism.

2. Co- and posttranslational modifications are important for protein structure, function, and stability. Common modifications include N- or C-terminal truncation, phosphorylation, glycosylation, and deamidation (reviewed in references 21 and 63). The significance of these modifications is often underestimated. While genome studies can indicate likely sites of modification, they cannot give insight into the actual modification status of the protein. Initial studies of *E. coli* suggest that each gene is represented on average by two protein isoforms on 2-D gels. This indicates that even in prokaryotes, processing and modification of gene products often occur (see "Bioinformatics" below).

3. Protein levels reflect a dynamic balance between synthesis and turnover. In the complicated web of biochemical pathways and differing protein half-lives, protein turnover cannot be deduced from genome sequencing but instead requires both static and dynamic global analysis, such as that obtained with proteome studies.

4. Overexpression or knockout of a particular gene has a global impact on protein levels which cannot be anticipated through knowledge of the genome alone (28). Using proteome technology, and in particular 2-D PAGE, a better understanding of the effects of a single-gene change at a global level can be gained.

5. Identification of gene products may uncover a section of genomic DNA that contains a gene not previously annotated or detected (reviewed in reference 3). Sequencing errors can mask protein coding regions, but evidence in the form of actual cellular proteins can highlight these mistakes (14).

PROTEOME DISPLAYS

At the heart of proteome studies is the need to resolve as many of the expressed proteins from the chosen organism as possible. The

2-D PAGE technique is the best available for this purpose, as it can simultaneously separate thousands of proteins. The resulting array consists of purified proteins appearing as spots.

E. coli, with a genome size of 4.63 Mb, is proposed to have approximately 4,288 gene products (4). One-dimensional (1-D) separations, such as sodium dodecyl sulfate (SDS)-PAGE (separating proteins by apparent molecular mass) or isoelectric focusing (separating proteins based on charge), are impractical for resolving thousands of proteins simultaneously. Attempts to employ sequential liquid chromatography separations with various parameters, such as size and charge, are proving successful (13), but such separations still fail to resolve the numbers of proteins that can be separated with 2-D gels.

The 2-D PAGE technique provides protein separation and purification based on the combination of two parameters.

1. Separation by net charge on the protein, using isoelectric focusing. Isoelectric focusing separates proteins in a pH gradient by the application of an electric current until the proteins reach a stationary point at which their net charge is zero. The pH at which the net charge of a protein is zero is referred to as its isoelectric point (pI).

2. Separation by apparent molecular mass. The proteins, once separated according to charge, are separated by their apparent molecular masses by methods based on standard PAGE techniques. The second dimension can be carried out either horizontally or vertically. The isoelectric focusing gel strip is loaded across the top of an SDS-PAGE gel, and proteins are electrophoretically moved into the second dimension. The gel can then be stained in a fashion similar to that used with 1-D PAGE gels, with silver or other sensitive stains (e.g., fluorescence based) for analytical gels and Coomassie blue stains for preparative gels. If desired, the gel can be blotted to a membrane (such as nitrocellulose or polyvinylidene difluoride) to enable further work, such as Western blotting or Edman protein sequencing, to be carried out (38).

The position of a protein in a 2-D gel can be related to physical properties of the protein, and one can make predictions as to where a protein will migrate on a 2-D gel, assuming no posttranslational modifications. For example, Color Plate 1 shows the predicted distribution of all *E. coli* proteins on a 2-D gel.

Many researchers have used pH gradients that were formed by combining soluble carrier ampholytes (CA) for isoelectric focusing. While excellent maps are possible with this technology (Fig. 1) (60), it is difficult to make highly reproducible maps, especially when attempting preparative loads. Proteins themselves are amphoteric molecules, and they affect the soluble pH gradient established with carrier ampholytes. The method is also inad-

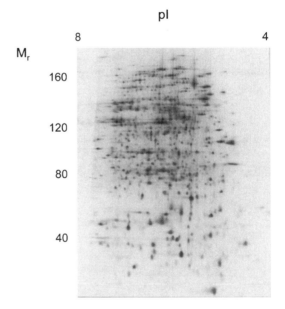

FIGURE 1 2-D PAGE reference map for *E. coli* K-12 strain W3110 created with CA technology. Proteins were separated by four to eight CA in the first dimension and 11.5% acrylamide in the second dimension. The convention for pI orientation with CA is often the reverse of that used by IPG. Molecular weight estimates were made with protein spots of known molecular mass (deduced from sequence). Reproduced from ftp://ncbi.nlm.nih.gov/repository/ECO2DBASE/F1nogrid.tif.

equate for resolving proteins with extremely high or low pI values, even when only analytical amounts are applied. Today, there is renewed interest in isoelectric focusing with immobilized pH gradients (IPGs), which are commercially available, have greater stability at extreme pH regions, and, because the pH gradient is chemically immobilized, allow reproducibility of 2-D gels (6, 10). In addition, a key advantage of 2-D PAGE with IPG strips is compatibility with both analytical (microgram) and preparative (milligram) amounts of material. A direct comparison between analytical and preparative gels can be made.

Conventional 2-D gels are approximately 20 cm square and allow the resolution of thousands of spots on a single gel. Gels of this size, however, can be cumbersome to make, and the process to obtain an end product is laborious and time-consuming. Minigels, with a protein resolution area of 7 by 6 cm, are much easier to handle. A 2-D gel, using the minigel format, can be completed in a single day (from sample preparation to stained gel).

For maximum resolution and protein separation on 2-D gels, preparation of the sample is crucial, as this determines what proteins enter and are resolved and visualized on the gel. Sample preparation must include (i) extraction of the proteins from the biological sample, (ii) disruption of any interaction among the proteins, (iii) keeping the proteins in solution during the separation process, (iv) prevention of any nonbiological modifications to the proteins, and (v) compatibility with 2-D gel electrophoresis systems (47). If these conditions are not met, the result can be spots in the wrong place on the 2-D gel (such as when protein-protein interactions are not disrupted) or missing spots (when the protein is not solubilized and so does not enter the gel). The final interpretation of the protein spots (such as pI and apparent molecular mass), and therefore the proteome, may be flawed and inconclusive.

Challenging problems for 2-D gel electrophoresis include:

1. detecting low-abundance proteins which may be "swamped" by abundant proteins that resolve in a similar location on the 2-D gel
2. solubilizing highly hydrophobic proteins (48)
3. resolving, on a conventional 2-D gel, mixtures that contain proteins with extremely low or high pI values

It is becoming apparent that the concept of using a single gel (even a large-format gel) to display the complete proteome for an organism, even a so-called "simple" prokaryote, may not be achievable due to the number of proteins that must be resolved. To circumvent this problem, prefractionation techniques are required. These techniques currently include the following.

1. Sequential extractions (39), in which various solubilizing reagents are used to separate or extract proteins based on their hydrophobicities (Fig. 2).
2. Narrow-range pH gradients (11, 50, 57), in which the distance normally traversed by 3 to 7 pH units is instead covered by 0.5 to 1 unit. This results in "stretching" the pH gradient to resolve proteins with pI values that are too close to be seen as individual spots with the conventional 4 to 7 or 3 to 10 pH ranges.
3. Subcellular fractionation. This technique is more appropriate for eukaryotic systems, in which proteins are first separated based on their cellular or organelle locations (e.g., cytosolic, nuclear, or membrane associated), often by gradient centrifugation, prior to 2-D PAGE. There has been limited application of this method to *E. coli* to separate membrane proteins prior to 2-D PAGE (1, 35, 52).

A complexity introduced by prefractionation techniques is the generation of many 2-D gels from one proteome, resulting from the inability to display a complete proteome on a single gel (Fig. 3). However, the proteome can be displayed by combining gel images, us-

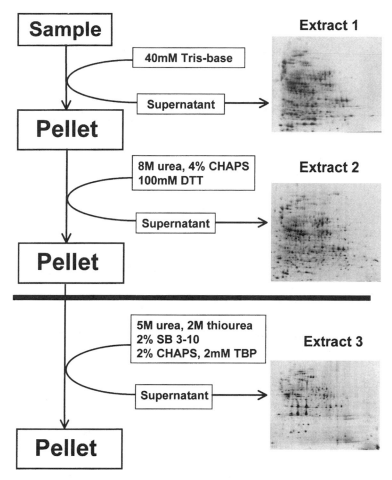

FIGURE 2 Outline of the steps involved in a sequential extraction procedure. *E. coli* proteins are fractionated by their hydrophobicity by using different reducing agents, chaotropes, and surfactants. The proteins are then separated by 2-D PAGE. The result is several simplified patterns of protein spots, making image analysis and identification less complicated. The technique also extracts proteins not usually obtained in a one-step procedure. Modified from Molloy et al. (39). CHAPS, 3-[(3-cholamidopropyl)dimethylammonio]-1-propanesulfonate; DTT, dithiothreitol; SB 3-10, sulfobetaine 3-10; TBP, tributyl phosphine.

ing software such as Melanie II or PDQuest (Bio-Rad Laboratories, Hercules, Calif.). Overlapping landmark spots can be used to align the gels, and the result is a composite or cyber map with a complete proteome display.

PROTEIN IDENTIFICATION

After arraying samples by 2-D PAGE, the next goal in proteome studies is the identification

of the expressed proteins. This allows the establishment of reference maps for proteome comparisons. Advances in methods for solubilization and separation by 2-D PAGE have shifted the need for technological developments from the initial protein separation to postgel analysis.

Early studies to identify proteins from 2-D gels, such as those conducted by Phillips et al.

pI

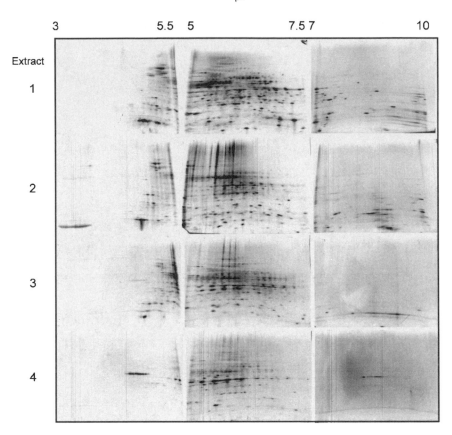

FIGURE 3 Use of narrow-range pH gradients and differential solubility to display the *E. coli* K-12 strain W3110 proteome more easily. Using a four-step sequential extraction procedure (the solutions detailed in Fig. 3 plus a final solubilization with 2% amidosulfobetaine 3-10 instead of sulfobetaine 3-10), proteins were separated by the Bio-Rad minigel format on three IPG strips of overlapping pI ranges. The second-dimension gels were Bio-Rad 10 to 20% ReadyGels, which fractionate in the range of 10 to 100 kDa. Note the well-spread-out pattern in the horizontal plane (different pH) and the different proteins solubilized in the vertical plane (different extraction conditions).

(46) and Bloch et al. (5), were based on co-migrating purified proteins with unknowns. Identifications were possible, but they relied on the contribution of purified material by other investigators. The laborious task of isolating individual proteins and characterizing them meant the supply of purified proteins was quickly exhausted. In addition, the use of previously purified proteins as an identification tool for 2-D PAGE defeats one of the key advantages of the technique: the simultaneous preparation of many proteins in a pure form suitable for chemical identification.

Alternatively, identification and detection of proteins have been achieved with antibodies, as illustrated by Celis (8), who has identified proteins from human bladder cancer by this approach (http://biobase.dk/cgi-bin/celis). However, the technique still requires antibodies to be available for the proteins of interest.

For large-scale proteome projects, a rapid method of chemically identifying proteins is needed. Fundamental approaches to protein identification have not changed much over the past decade, e.g., N-terminal sequencing, molecular weight, protein pI, amino acid analysis and peptide mapping. However, enormous technological improvements have been made in the chemistry and equipment and the manner in which samples are treated during analysis. These advances (discussed in detail below) are providing rapid throughput of samples.

Peptide Mapping

As each protein has a specific and unique sequence, cleavage of the protein into fragments (either chemically or enzymatically) will result in unique peptide maps being generated. The proteins can then be identified based on the molecular masses of their peptide fragments. Originally, a peptide map was produced by cleaving the protein of choice and separating the peptide fragments based on size with an SDS-PAGE gel (9). The banding pattern was unique to the protein and could be compared to a purified protein treated in the same fashion to confirm identity. Improved methods are based on greater resolution of peptide masses by mass spectrometry (MS) (37). Proteins are cleaved as previously discussed, and peptide masses are obtained via matrix-assisted laser desorption ionization time of flight (MALDI-TOF) MS or liquid chromatography-MS. New-generation MALDI-TOF and electrospray ionization-TOF instruments allow determination of isotopic masses with extremely high mass accuracy (accurate to less than 50 ppm). The data (often referred to as the peptide mass fingerprint [PMF]), once generated, can then be compared to a database containing the theoretical mass peaks for "database-digested" proteins, and a protein identity can be assigned. In a small genome, such as that of E. coli, as few as three peptides may identify a protein with a high degree of confidence. If a pure protein has been used (e.g., one separated by 2-D PAGE), other masses which do not match the predicted PMF can be interrogated to look for modifications by using database tools such as FindMod, which is part of the ExPASy server (http://www.expasy.org.au).

N- and C-Terminal Tagging

Obtaining a tag (3 to 6 residues from the N or C terminus of a protein, is an informative and reliable method for the identification of proteins. This is particularly useful when PMF indicates a protein with a pI or molecular weight very different from that estimated from a 2-D gel (as is often the case with truncated proteins). As there are 20 amino acids available for each position in a protein sequence, the number of combinations in which they can be arranged is enormous. A four-amino-acid peptide has 20^4 combinations, meaning the likelihood that two unrelated proteins have similar N- or C-terminal sequences is very small. However, this assumes that the N and C termini have randomly arranged amino acids. This is not true, especially for N termini, where the same initial sequence can often occur in different proteins. For E. coli, N-terminal tags of four residues specify a protein uniquely in 57% of cases, while C-terminal tetrapeptide tags are close to being unique (92%) (70). Those proteins that are identical at their C or N termini can usually be distinguished by using other properties, such as molecular weight and pI—parameters that are obtainable from a 2-D gel (70). The above-mentioned techniques (PMF and N- and C-terminal tags) are ideal for identifying proteins from prokaryotes, such as E. coli, for which the complete genome sequence is available.

Amino Acid Analysis

While proteins can be identified by their amino acid compositions (68), this technique is best used as a definitive quantitative tool rather than an identification tool. It provides the total amount (in picomoles) of each amino acid present in a protein. In addition to quantitation, it provides a check on purity, which is indicated by the correct amino acid composition.

The techniques described above, when taken in conjunction, allow confident, positive identification of the chosen protein spot from a 2-D gel. New technologies discussed below, based on automation and robotization, are allowing high throughput of samples, providing rapid, accurate identification of proteins.

AUTOMATION FOR HIGH THROUGHPUT

Although the methods discussed above are informative and allow accurate protein identification, the time and resources required to complete a proteome project based on such methods, if conducted manually, would be astronomical. Thus, to improve efficiency in postgel analyses, automation and robotization are required. Automation and robotization not only reduce analysis time but also remove human error and contaminants (such as keratin).

Automation can be achieved in a number of areas:

1. 2-D PAGE staining—automated staining procedures

2. imaging and excising of protein spots by robotics

3. automated enzymatic digests for PMF

4. automated interpretation of mass spectrum data

5. automated database searching

A recent project initiated in the Australian Proteome Analysis Facility has involved automation and integration of techniques to allow high-throughput identification of samples derived from *E. coli*. Briefly, whole *E. coli* (K-12) lysates were separated by 2-D PAGE. From these gels, 1,085 spots were analyzed by MALDI-TOF MS. A prototype robot (ARRM-214; Advanced Rapid Robotic Manufacturing, Kent Town, Australia), capable of cutting 1.5-mm-diameter spots from gel or polyvinylidene difluoride membrane, was used to allow fast excision, thereby eliminating user error and sample contamination. After spot excision, enzymatic trypsin digestion was carried out with a Multiprobe 104 (Packard Instrument Co., Downers Grove, Ill.). Manual handling was only required for moving the microtiter plate from the robot to a 37°C shaker to allow digestion. The peptides and then the matrix sample were automatically de-

livered to a sample plate for MALDI-TOF MS (Fig. 4).

Using the combination of automated technologies listed above, 565 proteins were confidently identified and a further 55 were putatively identified in a space of 8 weeks. On a master map (Color Plate 2), 324 spots were identified, of which 195 were unique proteins. This indicates a ratio of 1.7 proteins per gene. Spots not identified were due to low confidence in the scores obtained by the software program Rapid Automated Interrogation of Databases (RAID) (see below) and/or poor mass spectra. Failed identifications can be examined in detail by expert operators to improve the success rate.

BIOINFORMATICS

Proteome studies generate vast amounts of data. The ability to store, retrieve, and systematically organize this information is a large challenge for bioinformatics. A single automated MALDI-TOF MS is capable of acquiring spectra from approximately 300 samples per day. Although the generation of these data is rapid, the data themselves are complex, making interpretation and analysis difficult and highly time-consuming. To overcome this problem, RAID was developed (56). This program sorts through the data produced by MALDI-TOF MS and can submit it to PMF engines across the Internet and record the results. The program is flexible, allowing user-defined parameters to be added, such as pI, molecular weight, and mass tolerance (Fig. 5).

CHARACTERIZATION OF PROTEIN MODIFICATIONS

A fully functional protein is not necessarily the result of only mRNA translation. Often, proteins undergo co- and posttranslational modifications. The above-mentioned *E. coli* pilot study demonstrated that for the genes studied there were 1.7 proteins produced per gene. The occurrence of protein modifications is still being unraveled, as illustrated by the recent discovery that *E. coli* phosphorylates tyrosine residues (17). Protein modifications

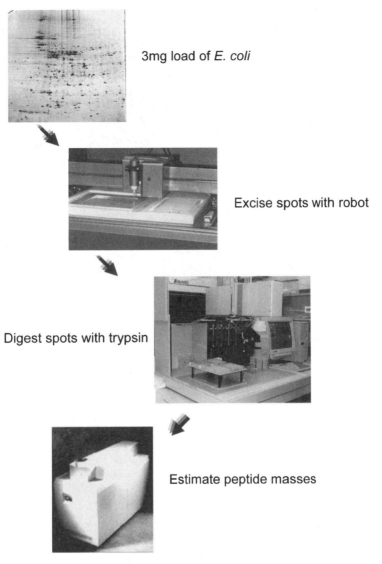

3mg load of *E. coli*

Excise spots with robot

Digest spots with trypsin

Estimate peptide masses

FIGURE 4 Flow chart showing equipment used in the process of preparing proteins separated by 2-D PAGE for PMF.

contribute to protein stability, function, protection from enzymes, signaling, cell-cell interactions, and recognition (reviewed in reference 63). Most information on protein modifications has been gained from studies of eukaryotic organisms. In fact, up until the mid-1970s, glycosylation of proteins was believed to occur only in eukaryotic organisms. The roles of protein modifications in pro-

karyotes appear to be similar to those for eukaryotes. However, the biosynthesis of these modifications and target sites can be different. For example, proteins from eukaryotes are typically phosphorylated on the hydroxyl amino acids: tyrosine, serine and threonine (21). Prokaryotes phosphorylate carboxyl amino acids and histidine, although histidine phosphorylation has now been discovered in

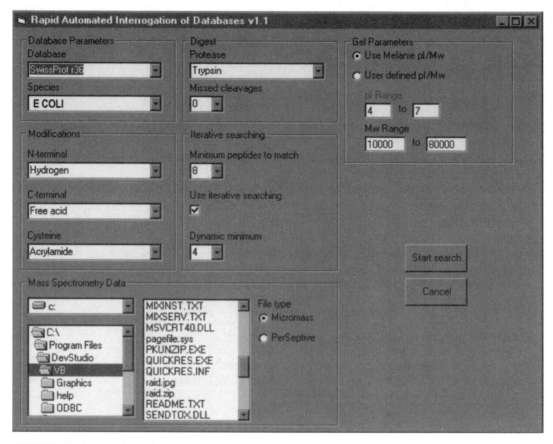

FIGURE 5 Screen shot of RAID. The program first retrieves monoisotopic peak lists and filters out contaminating masses (such as trypsin). It then automatically submits the data to the MS–Fit program (http://prospector.ucsf.edu/msfit). The program can also extract pI/*Mr* information from the Melanie II imaging software that is used to create gel images like that in Fig. 6.

eukaryotes (30, 36). Despite our still-limited knowledge about how, why, and when proteins are modified, some general rules for predicting modifications to amino acids are available for both eukaryotes and prokaryotes (71). Typical modifications that can be detected by the resulting increase or decrease in mass are shown in Table 1.

The very nature of 2-D PAGE means that physical properties, such as pI and apparent molecular mass, can be derived directly from the gel. Some indication of protein modifications may also be present in the gels. For example, trains of spots lying horizontally across the gel may suggest charge variation (e.g., due to phosphorylation or deamidation), and changes in apparent molecular mass and charge are often indicative of glycosylation (34). The application of proteome studies and associated techniques is allowing faster analysis of protein. Traditional affinity purification techniques are unable to resolve protein isoforms. In contrast, many protein modifications result in changes to pI and apparent mass, and thus the different isoforms can be separated by 2-D PAGE (21).

PROTEIN FLUXES AND NETWORKS

In living organisms it is unlikely that at any one time all genes in the genome will be ex-

TABLE 1 Average and monoisotopic mass values for typical posttranslational modifications used by the FindMod program (http://expasy.proteome.org.au/sprot/findmod/findmod_masses.html).

Posttranslational modification	Abbreviation	Mass value (Da)	
		Monoisotopic	Average
Acetylation	ACET	42.0106	42.0373
Amidation	AMID	−0.9840	−0.9847
Biotin	BIOT	226.0776	226.2934
C Mannosylation	CMAN	162.052823	162.1424
Deamidation	DEAM	0.9840	0.9847
N-acyl diglyceride cysteine (tripalmitate)	DIAC	788.7258	789.3202
Dimethylation	DIMETH	28.0314	28.0538
FAD	FAD	783.1415	783.542
Farnesylation	FARN	204.1878	204.3556
Formylation	FORM	27.9949	28.0104
Geranyl-geranyl	GERA	272.2504	272.4741
Gamma-carboxyglutamic acid	GGLU	43.98983	44.0098
O-GlcNac	GLCN	203.0794	203.1950
Hydroxylation	HYDR	15.9949	15.9994
Lipoyl	LIPY	188.033	188.3027
Methylation	METH	14.0157	14.0269
Myristoylation	MYRI	210.1984	210.3598
Palmitoylation	PALM	238.2297	238.4136
Phosphorylation	PHOS	79.9663	79.9799
Pyridoxal phosphate	PLP	229.014	229.129
Phosphopantetheine	PPAN	339.078	339.3234
Pyrrolidone carboxylic acid	PYRR	−17.0266	−17.0306
Sulfatation	SULF	79.9568	80.0642
Trimethylation	TRIMETH	42.0471	42.0807

pressed (64). Genomic studies and even mRNA display analyses cannot tell us which proteins are present. The application of proteomics, however, can provide a global view—a "snapshot" of protein expression and how it changes with varying cellular processes and environmental conditions. It has been conjectured that the total percentage of the genome expressed is dependent on the genomic complexity of the organism, meaning that in bacterial systems, a higher proportion of the genome is predicted to be expressed at any one time (28). Bacteria, like all organisms, can change gene expression by either gene induction and repression or increases and decreases in protein synthesis in response to stimuli. By comparing 2-D maps from an organism under different conditions (e.g., heat shock or starvation), a global analysis of protein expression under the different conditions

can be obtained. This process is termed protein differential display.

Proteome technology with radiolabelled- or nonradioactive-isotope (e.g., ^{15}N) substrates is an ideal tool for investigating differential expression, synthesis of individual proteins, and steady-state levels (29, 31, 33, 41). Through the use of radiolabelled substrates (typically amino acids), newly synthesized proteins can be distinguished from steady-state proteins, in turn allowing identification of protein networks. A simultaneous qualitative and quantitative measurement of newly synthesized proteins, as well as rates of protein synthesis, can be obtained. Microbes are the only group for which in vivo radiolabelling techniques are facile. Proteomic studies with radiolabelled substrates have been applied to E. coli to examine, among other things, heat shock proteins, the effect of phosphate limitation, and protein expression under anaerobiosis (54, 58, 59, 61).

Another useful approach to studying protein networks is to use gene deletions and/or overexpression of specific genes. The 2-D PAGE technique can display the networks of proteins that are related to the gene product of the deleted gene as an increase or decrease in expression of other proteins. This has been elegantly demonstrated by Sankar et al., who used 2-D PAGE to visualize overexpression of an *E. coli* gene in a low-copy-number plasmid vector (51).

SUMMARY OF PROTEOME STUDIES OF *E. COLI*

The differential display of proteins by 2-D PAGE has long been used with *E. coli* to monitor response to stimuli. Gage and Neidhardt (18) showed the induction of approximately 50 proteins in *E. coli* in response to 2,4-dinitrophenol. Likewise, heat shock proteins have been well studied over the years, and they are now regarded as universal stress response proteins in all cells (25, 26, 60).

Neidhardt and Savageau (43) recognized the importance of studying intracellular processes and molecular interactions, and the lack of methods to do so. Typically, enzyme regulation has been examined through activity assays, although these are not without fault. Besides being time-consuming, these methods lack accuracy due to in vitro degradation and inactivation, the presence of unknown inhibitors, and an inability to measure small changes in expression levels. More importantly there are many proteins for which no assays exist. Indeed, 40% of the *E. coli* genome codes for proteins of unknown function (4). Proteome studies offer a way to overcome these problems.

The work by the pioneers of molecular biology has left us with a detailed understanding of many metabolic pathways in *E. coli* (43). After the development of 2-D PAGE, a gene-protein database for *E. coli* was initiated (62). This database, maintained and coordinated by VanBogelen and Neidhardt (60), is a catalogue of *E. coli* proteins, including the levels and conditions under which each protein is expressed. The database is now subdivided into three sections: the Genome Expression Map, the Response/Regulation Map, and the Cell Architecture Map. The goal of the Genome Expression Map is to link every protein-encoding gene to a spot on a 2-D gel, while the Response/Regulation Map is a catalogue of the conditions under which each of the open reading frames is expressed and how expression is regulated (60). In turn, this information will tell us the regulatory networks to which the proteins belong. The final section, the Cell Architecture Map, has been designed to map the locations, abundance, and intermolecular arrangements of the *E. coli* proteins. The *E. coli* gene-protein database has been highly successful in cataloguing proteins and protein expression. However, the database is limited to mostly soluble proteins and has been based on CA technology for the first dimension of 2-D gels, thus limiting interlaboratory comparisons and postseparation identification of proteins.

A collaborative effort between the Clinical Chemistry Laboratory, Geneva University Hospital, Geneva, Switzerland, and the Australian Proteome Analysis Facility at Macquarie University, Sydney, Australia, has begun an IPG gel-based proteome project for *E. coli* K-12. The goal of this project is to create reference maps of *E. coli* proteins that can be compared and aligned with the original CA maps (Color Plate 3). Since recent work by Nawrocki et al. (40) has demonstrated that there is poor correlation between spot migration patterns on CA gels when compared to IPGs, it is necessary to identify proteins to allow comparisons to be made. This was initiated by Pasquali et al. (45), and IPGs were used to separate the same strain of *E. coli* as Neidhardt's group used. Gel matching, amino acid composition, and N-terminal sequencing identified a total of 153 proteins. This has now been extended to 231 proteins (55), and the work is continuing with the application of PMF (56).

PROTEOMIC CHALLENGES

The 2-D PAGE technique is still unparalleled in its ability to simultaneously resolve thousands of proteins, but the hard work is not over yet. Further advances in protein solubilization and detection methods are required to resolve in gel all proteins from a proteome, in particular, those that are membrane associated or of high hydrophobicity. However, the bulk of future work lies in postgel analysis: improving identification methods by decreasing analysis time, better integration of the process of protein separation-detection-analysis-identification, as well as improving methods for identification of co- and posttranslational modifications that allow rapid throughput. In addition, challenges lie ahead in developing good strategies for systematic storage and archiving of the vast amounts of data that will be generated from both large- and small-scale proteome studies.

ACKNOWLEDGMENTS

This research has been facilitated by access to the Australian Proteome Analysis Facility established under the Australian Government's Major National Research Facilities Program.

We thank several members of our laboratory for contributing figures for this chapter: Andrew Gooley, Ben Herbert, Mark Molloy, Keli Ou, and Marc Wilkins.

REFERENCES

1. **Ames, G. F.-L., and K. Nikaido.** 1976. Two-dimensional gel electrophoresis of membrane proteins. *Biochemistry* **15:**616–623.

2. **Anderson, L., and J. Seilhamer.** 1997. A comparison of selected mRNA and protein abundances in human liver. *Electrophoresis* **18:**533–537.

3. **Bairoch, A.** 1997. Proteome databases. *In* M. R. Wilkins, K. L. Williams, R. D. Appel, and D. F. Hochstrasser (ed.), *Proteome Research: New Frontiers in Functional Genomics.* Springer-Verlag, Berlin, Germany.

4. **Blattner, F. R., G. Plunkett, C. A. Bloch, N. T. Perna, V. Burland, M. Riley, J. Collado-Vides, J. D. Glasner, C. K. Rode, G. F. Mayhew, J. Gregor, N. W. Davis, H. A. Kirkpatrick, M. A. Goeden, D. J. Rose, B. Mau, and Y. Shao.** 1997. The complete genome sequence of *Escherichia coli* K-12. *Science* **277:**1453–1474.

5. **Bloch, P. L., T. A. Phillips, and F. C. Neidhardt.** 1980. Protein identifications on O'Farrell two-dimensional gels: locations of 81 *Escherichia coli* proteins. *J. Bacteriol.* **141:**1409–1420.

6. **Blomberg, A., L. Blomberg, J. Norbeck, S. J. Fey, P. M. Larsen, M. Larsen, P. Roepstorff, H. Degand, M. Boutry, A. Posch, and A. Görg.** 1995. Interlaboratory reproducibility of yeast protein patterns analyzed by immobilized pH gradient two-dimensional gel electrophoresis. *Electrophoresis* **16:**1935–1945.

7. **Bravo, R., and J. E. Celis.** 1980. A search for differential polypeptide synthesis throughout the cell cycle of HeLa cells. *J. Cell Biol.* **84:**795–802.

8. **Celis, J. E., M. Ostergaard, N. A. Jensen, I. Gromova, H. H. Rasmussen, and P. Gromov.** 1998. Human and mouse proteomic databases: novel resources in the protein universe. *FEBS Lett.* **430:**64–72.

9. **Cleveland, D.** 1983. Peptide mapping in one dimension by limited proteolysis of sodium dodecyl sulfate-solubilized proteins. *Methods Enzymol.* **96:**222–229.

10. **Corbett, J. M., M. J. Dunn, A. Posch, and A. Görg.** 1994. Positional reproducibility of protein spots in two-dimensional polyacrylamide gel electrophoresis using immobilised pH gradient isoelectric focusing in the first dimension: an interlaboratory comparison. *Electrophoresis* **15:**1205–1211.

11. **Cordwell S. J., D. J. Basseal, B. Bjellqvist, D. C. Shaw, and I. Humphery-Smith.** 1997. Characterization of basic proteins from *Spiroplasma melliferum* using novel immobilised pH gradients. *Electrophoresis* **18:**1393–1398.

12. **Daniels, D. L., G. Plunkett III, V. Burland, and F. R. Blattner.** 1992. Analysis of the *Escherichia coli* genome: DNA sequence of the region from 84.5 to 86.5 minutes. *Science* **257:**771–778.

13. **Ducret, A., I. Van Oostveen, J. K. Eng, J. R. Yates III, and R. Aebersold.** 1998. High throughput protein characterization by automated reverse-phase chromatography/electrospray tandem mass spectrometry. *Protein Sci.* **7:**706–719.

14. **Dujon, B.** 1996. The yeast genome project: what did we learn? *Trends Genet.* **12:**263–270.

15. **Dunn, M. J., J. M. Corbett, and C. H. Wheeler.** 1997. HSC-2DPAGE and the two-dimensional gel electrophoresis database of dog heart proteins. *Electrophoresis* **18:**2795–2802.

16. **Fey, S. J., A. Nawrocki, M. R. Larsen, A. Görg, P. Roepstorff, G. N. Skews, R. Williams, and P. M. Larsen.** 1997. Proteome analysis of *Saccharomyces cerevisiae*: a methodological outline. *Electrophoresis* **18:**1361–1372.

17. **Freestone, P., M. Trinei, S. C. Clarke, T. Nyström, and V. Norris.** 1998. Tyrosine phosphorylation in *Escherichia coli. J. Mol. Biol.* **279:** 1045–1051.

18. **Gage, D. J., and F. C. Neidhardt.** 1993. Adaptation of *Escherichia coli* to the uncoupler of oxidative phosphorylation 2,4-dinitrophenol. *J. Bacteriol.* **175:**7105–7108.

19. **Garrels, J. I., and W. Gibson.** 1976. Identification and characterization of multiple forms of actin. *Cell* **9:**793–805.

20. **Garrels, J. I., C. S. McLaughlin, J. R. Warner, B. Futcher, G. I. Latter, R. Kobayashi, B. Schwender, T. Volpe, D. S. Anderson, R. Mesquita-Fuentes, and W. E. Payne.** 1997. Proteome studies of *Saccharomyces cerevisiae*: identification and characterization of abundant proteins. *Electrophoresis* **18:**1347–1360.

21. **Gooley, A. A., and N. H. Packer.** 1997. The importance of protein co- and post-translational modifications in proteome projects. *In* M. R. Wilkins, K. L. Williams, R. D. Appel, and D. F. Hochstrasser (ed.), *Proteome Research: New Frontiers in Functional Genomics.* Springer-Verlag, Berlin, Germany.

22. **Gooley, A. A., and K. L. Williams.** 1997. How to find, identify and quantitate the sugars on proteins. *Nature* **385:**557–559.

23. **Görg, A., G. Boguth, C. Obermaier, and W. Weiss.** 1998. Two-dimensional electrophoresis of proteins in an immobilized pH 4–12 gradient. *Electrophoresis* **19:**1516–1519.

24. **Haynes, P. A., S. P. Gygi, D. Figeys, and R. Aebersold.** 1998. Proteome analysis: biological assay or data archive. *Electrophoresis* **19:** 1862–1871.

25. **Hecker, M., W. Schumann, and U. Völker.** 1996. Heat-shock and general stress response in *Bacillus subtilis. Mol. Microbiol.* **19:**417–428.

26. **Herendeen, S. L., R. A. VanBogelen, and F. C. Neidhardt.** 1979. Levels of major proteins of *Escherichia coli* during growth at different temperatures. *J. Bacteriol.* **139:**185–194.

27. **Hochstrasser, D. F., S. Frutiger, M. R. Wilkins, G. Hughes, and J.-C. Sanchez.** 1997. Elevation of apolipoprotein E in the CSF of cattle affected by BSE. *FEBS Lett.* **416:**161–163.

28. **Humphery-Smith, I., S. J. Cordwell, and W. P. Blackstock.** 1997. Proteome research: complementarity and limitations with respect to the RNA and DNA worlds. *Electrophoresis* **18:** 1217–1242.

29. **Jones P. G., R. A. VanBogelen, and F. C. Neidhardt.** 1987. Induction of proteins in response to low temperature in *Escherichia coli. J. Bacteriol.* **169:**2092–2095.

30. **Kennelly, P. J., and M. Potts.** 1996. Fancy meeting you here! A fresh look at "Prokaryotic" protein phosphorylation. *J. Bacteriol.* **178:**4759–4764.

31. **Knopf, U. C., A. Sommer, J. Kenny, and R. R. Traut.** 1975. A new two-dimensional gel electrophoresis system for the analysis of complex protein mixtures: application to the ribosome of *E. coli. Mol. Biol. Rep.* **2:**35–40.

32. **Kohara, Y., K. Akiyama, and K. Isono.** 1987. The physical map of the whole *E. coli* chromosome: application of a new strategy for rapid analysis and sorting of a large genomic library. *Cell* **50:**495–508.

33. **Lambert, L. A., K. Abshire, D. Blankenhorn, and J. L. Slonczewski.** 1997. Proteins induced in *Escherichia coli* by benzoic acid. *J. Bacteriol.* **179:**7595–7599.

34. **Link, A. J., L. G. Hays, E. B. Carmack, and J. R. Yates III.** 1997. Identifying the major proteome components of *Haemophilis influenzae* type-strain NCTC 8143. *Electrophoresis* **18:** 1314–1334.

35. **Link, A. J., K. Robison, and G. M. Church.** 1997. Comparing the predicted and observed properties of proteins encoded in the genome of *Escherichia coli* K-12. *Electrophoresis* **18:**1259–1313.

36. **Loomis, W. F., G. Shaulsky, and N. Wang.** 1997. Histidine kinases in signal transduction pathways of eukaryotes. *J. Cell Sci.* **110:**1141–1145.

37. **Mann, M., P. Hojrup, and P. Roepstorff.** 1993. Use of mass spectrometric molecular weight information to identify proteins in sequence databases. *Biol. Mass Spectrom.* **22:**338–345.

38. **Matsudaira, P.** 1987. Sequence from picomole quantities of proteins electroblotted onto polyvinylidene difluoride membranes. *J. Biol. Chem.* **262:**10035–10038.

39. **Molloy, M. P., B. R. Herbert, B. J. Walsh, M. I. Tyler, M. Traini, J.-C. Sanchez, D. F. Hochstrasser, K. L. Williams, and A. A. Gooley.** 1998. Extraction of membrane proteins by differential solubilization for separation using two-dimensional gel electrophoresis. *Electrophoresis* **19:**837–844.

40. **Nawrocki, A., M. R. Larsen, A. V. Podtelejnikov, O. N. Jensen, M. Mann, P. Roepstorff, A. Görg, S. J. Fey, and P. M. Larsen.** 1998. Correlation of acidic and basic carrier ampholyte and immobilized pH gradient two-dimensional gel electrophoresis patterns based on mass spectrometric protein identification. *Electrophoresis* **19:**1024–1035.

41. **Neidhardt, F. C., P. L. Bloch, S. Pedersen, and S. Reeh.** 1977. Chemical measurement of

steady-state levels of ten aminoacyl-transfer ribonucleic acid synthetases in *Escherichia coli*. *J. Bacteriol.* **129**:378–387.

42. **Neidhardt, F. C., T. A. Phillips, R. A. VanBogelen, M. W. Smith, Y. Georgalis, and A. R. Subramanian.** 1981. Identity of the B56.5 protein, the A-protein, and the *groE* gene product of *Escherichia coli*. *J. Bacteriol.* **145**:513–520.

43. **Neidhart, F. C., and M. A. Savageau.** 1996. Regulation beyond the operon, p. 1311–1324. *In* F. C. Neidhart, R. Curtiss III, J. L. Ingraham, E. C. C. Lin, K. B. Low, B. Magasanik, W. S. Reznikoff, M. Riley, M. Schaechter, and H. E. Umbarger (ed.), Escherichia coli *and* Salmonella: *Cellular and Molecular Biology*, 2nd ed. American Society for Microbiology, Washington, D.C.

44. **O'Farrell, P. H.** 1975. High resolution two-dimensional gel electrophoresis of proteins. *J. Biol. Chem.* **250**:4007–4021.

45. **Pasquali, C., S. Frutiger, M. R. Wilkins, G. J. Hughes, R. D. Appel, A. Bairoch, D. Schaller, J.-C. Sanchez, and D. F. Hochstrasser.** 1996. Two-dimensional gel electrophoresis of *Escherichia coli* homogenates: the *Escherichia coli* SWISS-2DPAGE database. *Electrophoresis* **17**:547–555.

46. **Phillips, T. A., P. L. Bloch, and F. C. Neidhardt.** 1980. Protein identification on O'Farrell two-dimensional gels: location of 55 additional *Escherichia coli* proteins. *J. Bacteriol.* **144**:1024–1033.

47. **Rabilloud, T.** 1996. Solubilization of proteins for electrophoretic analyses. *Electrophoresis* **17**:813–829.

48. **Rabilloud, T., C. Adessi, A. Giraudel, and J. Lunardi.** 1997. Improvement of the solubilization of proteins in two-dimensional electrophoresis with immobilized pH gradients. *Electrophoresis* **18**:307–316.

49. **Rudd, K. E., I. Humphery-Smith, V. C. Wasinger, and A. Bairoch.** 1998. Low molecular weight proteins: a challenge for postgenomic research. *Electrophoresis* **19**:536–544.

50. **Sanchez, J.-C., V. Rouge, M. Pisteur, F. Ravier, L. Tonella, M. Moosmayer, M. Wilkins, and D. F. Hochstrasser.** 1997. Improved and simplified in-gel sample application using reswelling of dry immobilized pH gradients. *Electrophoresis* **18**:324–327.

51. **Sankar, P., M. E. Hutton, R. A. VanBogelen, R. L. Clark, and F. C. Neidhardt.** 1993. Expression analysis of cloned chromosomal segments of *Escherichia coli*. *J. Bacteriol.* **175**:5145–5152.

52. **Sato, T., K. Ito, and T. Yura.** 1977. Membrane proteins of *Escherichia coli* K-12:

two-dimensional polyacrylamide gel electrophoresis of inner and outer membranes. *Euro. J. Biochem.* **78**:557–567.

53. **Smith, C. L., J. G. Econome, A. Schutt, S. Klco, and C. R. Cantor.** 1987. A physical map of the *Escherichia coli* K12 genome. *Science* **236**: 1448–1453.

54. **Smith, M. W., and F. C. Neidhardt.** 1983. Proteins induced by anaerobiosis in *Escherichia coli*. *J. Bacteriol.* **154**:336–343.

55. **Tonella, L., B. J. Walsh, J.-C. Sanchez, K. Ou, M. R. Wilkins, M. Tyler, S. Frutiger, A. A. Gooley, I. Pescaru, R. D. Appel, J. X. Yan, A. Bairoch, C. Hoogland, F. S. Morch, G. J. Hughes, K. L. Williams, and D. F. Hochstrasser.** 1998. '98 *Escherichia coli* SWISS-2DPAGE database update. *Electrophoresis* **19**: 1960–1971.

56. **Traini, M., A. A. Gooley, K. Ou, M. R. Wilkins, L. Tonella, J.-C. Sanchez, D. F. Hochstrasser, and K. L. Williams.** 1998. Towards an automated approach for protein identification in proteome projects. *Electrophoresis* **19**: 1941–1949.

57. **Urquhart, B. L., T. E. Atsalos, D. Roach, D. J. Basseal, B. Bjellqvist, W. L. Britton, and I. Humphery-Smith.** 1998. 'Proteomic contigs' of *Mycobacterium tuberculosis* and *Mycobacterium bovis* (BCG) using novel immobilised pH gradients. *Electrophoresis* **18**:1384–1392.

58. **VanBogelen, R. A., P. M. Kelley, and F. C. Neidhardt.** 1987. Differential induction of heat shock, SOS, and oxidation stress regulons and accumulation of nucleotides in *Escherichia coli*. *J. Bacteriol.* **169**:26–32.

59. **VanBogelen, R. A., and F. C. Neidhardt.** 1990. Ribosomes as sensors of heat and cold shock in *Escherichia coli*. *Proc. Natl. Acad. Sci. USA* **87**:5589–5593.

60. **VanBogelen, R. A., K. Z. Abshire, A. Pertsemlidis, R. L. Clark, and F. C. Neidhardt.** 1996. Gene-protein database of *Escherichia coli* K-12, edition 6. *In* F. C. Neidhardt, R. Curtiss III, J. L. Ingraham, E. C. C. Lin, K. B. Low, B. Magasanik, W. S. Reznikoff, M. Riley, M. Schaechter, and H. E. Umbarger (ed.), Escherichia coli *and* Salmonella: *Cellular and Molecular Biology,* 2nd ed. American Society for Microbiology, Washington, D.C.

61. **VanBogelen, R. A., E. R. Olson, B. L. Wanner, and F. C. Neidhardt.** 1996. Global analysis of proteins synthesized during phosphorus restriction in *Escherichia coli*. *J. Bacteriol.* **178**:4344–4366.

62. **VanBogelen, R. A., K. Z. Abshire, B. Moldover, E. R. Olson, and F. C. Neidhardt.** 1997. *Escherichia coli* proteome analysis using the

gene-protein database. *Electrophoresis* **18**:1243–1251.

63. **Varki, A.** 1993. Biological roles of oligosaccharides: all of the theories are correct. *Glycobiology* **3**:97–130.

64. **Wasinger, V. C., S. J. Cordwell, A. Cerpa-Poljak, J. X. Yan, A. A. Gooley, M. R. Wilkins, M. W. Duncan, R. Harris, K. L. Williams, and I. Humphery-Smith.** 1995. Progress with gene-product mapping of the Mollicutes: *Mycoplasma genitalium. Electrophoresis* **16**:1090–1094.

65. **Westermeier, R., W. Postel, J. Weser, and A. Görg.** 1983. High-resolution two-dimensional electrophoresis with isoelectric focusing in immobilized pH gradients. *J. Biochem. Biophys. Methods* **8**:321–330.

66. **Wilkins, M., K. Ou, R. D. Appel, J.-C. Sanchez, J. X. Yan, O. Golaz, V. Farnsworth, P. Cartier, D. F. Hochstrasser, K. L. Williams, and A. A. Gooley.** 1996. Rapid protein identification using N-terminal "sequence tag" and amino acid analysis. *Biochem. Biophys. Res. Commun.* **221**:609–613.

67. **Wilkins, M. R., J.-C. Sanchez, A. A. Gooley, R. D. Appel, I. Humphery-Smith, D. F. Hochstrasser, and K. L. Williams.** 1995. Progress with proteome projects: Why all proteins expressed by a genome should be identified and how to do it. *Biotechnol. Genet. Eng. Rev.* **13**:19–50.

68. **Wilkins, M. R., C. Pasquali, R. D. Appel, K. Ou, O. Golaz, J.-C. Sanchez, J. X. Yan, A. A. Gooley, G. Hughes, I. Humphery-Smith, K. L. Williams, and D. F. Hochstrasser.** 1996. From proteins to proteomes: large scale protein identification by two-dimensional electrophoresis and amino acid analysis. *Bio/Technology* **14**:61–65.

69. **Wilkins, M. R., and A. A. Gooley.** 1997. Protein identification in proteome projects. *In* M. R. Wilkins, K. L. Williams, R. D. Appel, and D. F. Hochstrasser (ed.), *Proteome Research: New Frontiers in Functional Genomics.* Springer-Verlag, Berlin, Germany.

70. **Wilkins, M. R., E. Gasteiger, L. Tonella, K. Ou, M. Tyler, J.-C. Sanchez, A. A. Gooley, B. J. Walsh, A. Bairoch, R. D. Appel, K. L. Williams, and D. F. Hochstrasser.** 1998. Protein identification with N and C-terminal sequence tags in proteome projects. *J. Mol. Biol.* **278**:599–608.

71. **Wilkins, M. R., E. Gasteiger, A. A. Gooley, B. R. Herbert, M. Molloy, P. A. Binz, K. Ou, J.-C. Sanchez, A. Bairoch, K. L. Williams, and D. F. Hochstrasser.** Large-scale mass spectrometric discovery of protein post-translational modifications. (Submitted for publication).

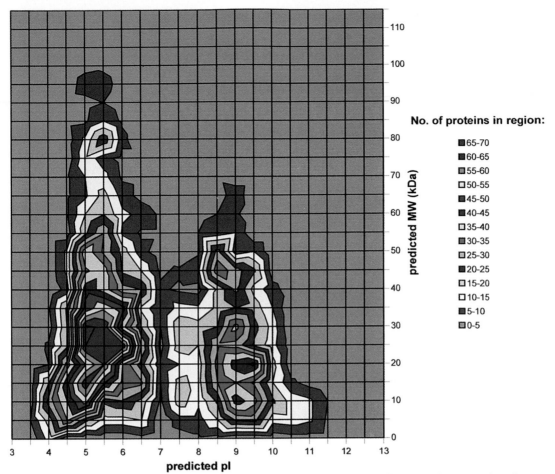

No. of proteins in region:

- ■ 65-70
- ■ 60-65
- ■ 55-60
- □ 50-55
- ■ 45-50
- ■ 40-45
- □ 35-40
- ■ 30-35
- □ 25-30
- ■ 20-25
- □ 15-20
- □ 10-15
- ■ 5-10
- ■ 0-5

COLOR PLATE 1 Contour map showing the predicted distribution of cellular *E. coli* proteins found in any pI and mass region on a 2-D gel. SwissProt release 34 was used to construct the map from approximately 3,500 proteins. Proteins larger than 115 kDa are not shown. Reprinted from *Proteome Research: New Frontiers in Functional Genomics* (69) with permission of the publisher.

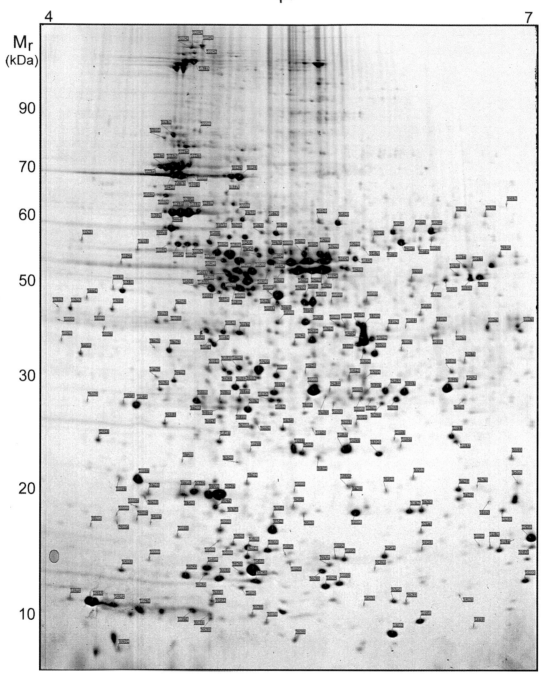

COLOR PLATE 2 2-D PAGE reference map for *E. coli* K-12 strain W3110. A 3-mg load of *E. coli* was solubilized with extract 3 (Fig. 3) without sample fractionation. The proteins were focused in the first dimension on a pH 4 to 7 IPG strip and separated on an 8 to 18% gradient gel. The proteins were stained with colloidal Coomassie brilliant blue G-250. SwissProt accession numbers were assigned to proteins that have been identified through PMF. Peptide mass estimates between 800 and 2,600 Da were obtained with MALDI-TOF MS to 50 ppm.

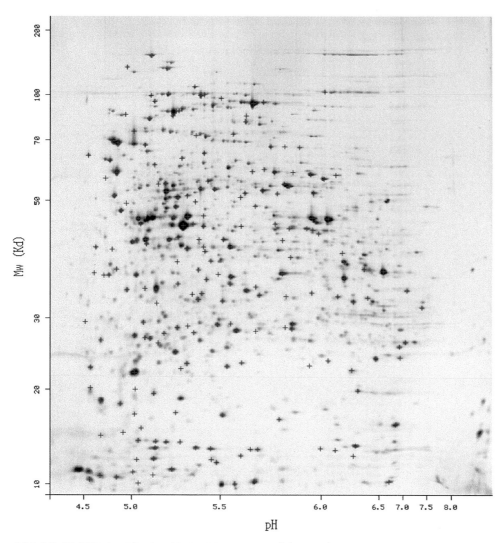

COLOR PLATE 3 The SWISS-2D PAGE image of the *E. coli* K-12 strain W3110 map. Proteins were separated by pI 3 to 10 IPG strips in the first dimension and then by an 8 to 18% acrylamide gel in the second dimension. Proteins that have been identified are indicated by red crosses. The total number of proteins identified is 206. Reproduced from http://expasy.proteome.org.au/cgi-bin/ map2/def?ECOLI (55).

GENOME-WIDE STRATEGIES FOR STUDYING GENE FUNCTION BY USING MODEL SYSTEMS

Reginald K. Storms

19

The complete DNA sequences of 19 genomes (1, 3, 5, 6, 10, 11, 14, 17–19, 25, 34, 39, 41–43, 71, 74, 75) are now publicly accessible in databases like GenBank at the National Center for Biotechnology Information (http://www.ncbi.nlm.nih.gov/). Of these 19 organisms, two prokaryotes, the gram-positive bacterium *Bacillus subtilis* (43) and the gram-negative bacterium *Escherichia coli* (5), as well as the unicellular eukaryote *Saccharomyces cerevisiae* (25), have been used extensively for over 40 years as model systems for basic research. The completion of the sequencing of these three genomes marked the official end of the sequencing phase of research on these popular model organisms and marked the beginning of broad genome-wide or functional genomics approaches.

Although *E. coli, B. subtilis*, and *S. cerevisiae* are among the most thoroughly studied genetic systems, less than 30% of their respective genes had been defined experimentally prior to the initiation of their respective genome-sequencing projects (5, 25, 43). With the completion of these DNA-sequencing pro-

jects, it has been relatively easy to identify essentially all of their genes, because the genomes of these organisms are information rich and do not contain many introns. DNA sequence analysis identified about 4,300, 4,100, and 6,200 potential protein-coding genes in *E. coli, B. subtilis*, and *S. cerevisiae*, respectively. Clearly one impressive outcome from genome sequencing, even for these intensively studied model organisms, was the large number of new genes and gene products identified.

Protein data bank searches with algorithms like FASTA (56, 57) and BLAST (2) identified at least one significant counterpart with something known about its function for about 60% of the potential protein-coding genes in *E. coli, B. subtilis*, and *S. cerevisiae*. For all three of these organisms, computer-assisted database searching therefore increased the number of genes to which some functional information could be assigned about twofold. Also significant was the discovery that functions could not be assigned to about 40% of the genes in each of these intensively studied organisms. The genes of unknown function could be assigned to two arbitrary classes, the "orphan genes," whose products were not significantly similar to any other proteins in publicly accessible databases (about 30% for all three or-

Reginald K. Storms, Centre for Structural and Functional Genomics, Department of Biology, Concordia University, Montreal, Quebec, Canada H3G 1M8.

Organization of the Prokaryotic Genome, Edited by Robert L. Charlebois,
© 1999 American Society for Microbiology, Washington, D.C.

ganisms) and genes that code for proteins that were similar to proteins or predicted proteins of unknown function from other organisms (about 10% for all three organisms). Understanding the biological roles of these two classes of genes is a major challenge for researchers working with these model organisms.

At first glance, knowing something about the function of about 60% of the potential genes in these three organisms suggests that we are well on our way to understanding in intricate detail the workings of these model systems. However, knowing that a potential gene, identified as an open reading frame (ORF) in a DNA sequence, shows significant similarity to other kinases, phosphatases, a previously characterized transcription factor, an enzyme of intermediate metabolism, or a specific structural protein is far from completely understanding its biological and biochemical function. Knowledge of gene function is a continuum, from learning its product's primary amino acid sequence, to a complete understanding of its role in a biological process, and finally, a detailed understanding of that biological process at the level of the organism under investigation.

Often additional information (a clue or clues) is necessary before we have significant insight into the function of a gene identified by DNA sequence analysis. Information about the regulation of a gene, including at the transcriptional, translational, and posttranslational levels, is critical. Knowledge of a protein's intracellular location and what proteins it interacts with are also important factors in determining biological relevance. Thus, just as the sequence of a genome cannot define the workings of even the simplest of organisms, the assignment of a potential function to all the proteins encoded by a genome is only a useful and important additional step along a continuum that leads to a detailed understanding of that organism.

Genome sequences are valuable resources that not only complement traditional biochemical and genetic research but can also be used to develop new approaches for advancing biological understanding. For the first time, it is now possible to develop global or genome-wide approaches for assessing gene function (known as "functional genomics"). Because of their broad scope, these approaches will play increasingly important roles. Developing a more detailed understanding of an organism's entire gene set will be the goal of functional genomics; however, of significant near-term interest is the large number to genes to which functions have not been assigned. This is particularly true for the orphan genes, which may define a large portion of the uniqueness associated with an individual biological species, genus, etc. The objective of this chapter is to outline how complete genome sequences promote the development of broad genome-wide or functional-genomics approaches for investigating individual genes to entire gene sets.

GENOME SEQUENCES AND THE IDENTIFICATION OF NEW GENES

The large number of new genes identified by DNA sequence analysis showed that over 40 years of intensive investigation by a large research community had focused on only a small portion of the genetic potential of the model organisms, *E. coli, B. subtilis,* and *S. cerevisiae.* One of the first indications that traditional approaches were unable to fully tap the genetic potential of an organism came from a study reported by David Kaback and colleagues (38). They isolated 32 temperature-sensitive mutations that mapped to chromosome I (CHI) of *S. cerevisiae.* The fact that these mutations included multiple, independent alleles of just three genes strongly suggested that a very small portion of the genes on CHI were essential. Subsequent analysis of the complete CHI gene set (reference 8 and unpublished results) revealed that only 13 of its 103 potential protein-coding genes were essential for growth on rich medium at 30°C. Although most yeast genes are expressed (52, 76, 80, 81, 86), less than 20% of them are essential (7, 8, 24, 38, 58). Furthermore, significantly less than 50% produced an informative phenotype when inactivated by mutation (7, 24, 58, 72, 73, 79). Similar results have been obtained with bacterial systems. Itaya estimated the number of indispensable chromosomal loci in *B. subtilis* by determining whether 79 randomly inactivated genetic loci were essential for growth (35). He found that only six of the loci examined were essential for colony formation on rich medium. The author used these results to estimate that the indispensable genetic material would occupy less than 562 kbp, a mere 14% of the *B. subtilis* genome. It is interesting that this estimate closely corresponds to the size of the smallest sequenced bacterial genome, that of *Mycoplasma genitalium* (18).

Classical approaches had focused on a small portion of the *S. cerevisiae*, *B. subtilis*, and *E. coli* genes. At least within the laboratory setting, many genes were either "unimportant" or played subtle roles that were difficult to observe as a distinct phenotype compared with "wild-type" laboratory strains. But why are most genes, at least under the physiological conditions tested in the laboratory, apparently functionally unimportant?

First, as stated by Steven Oliver, "many [yeast] genes may be required to deal with challenges that are never encountered in the laboratory, but which are common in the rotting fig or the brewery" (55). The demonstration of roles for many genes, therefore, awaits the examination of functional relevance under a myriad of physiological conditions. These should be designed to test the biochemical and genetic demands imposed during natural evolution rather than in the controlled environment of the research laboratory.

A second factor that probably contributes to the small number of apparently important genes is functional redundancy. Many important functions may be performed by the protein products of more than one gene. Functionally related genes could arise in two ways, gene duplication (resulting in sets of parologous genes) or convergent evolution. For functionally related genes that arose by gene

duplication, residual similarities in amino acid sequences may exist. One indicator of the level of redundancy within a given genome can thus be assessed (once the genome is sequenced) by estimating the number of sets of two or more genes related by sequence. An indication that functional redundancy plays an important role is the fact that the predicted protein products of almost one-third of all yeast genes are highly similar (i.e., with BLAST scores above 200) to the product of at least one additional gene elsewhere in the genome. Approximately 900 "simple gene pairs" (BLAST score > 200), where each gene is the other's only highly homologous counterpart in the yeast genome, have been identified (85). In addition, many larger families of highly similar genes also exist (32, 85).

In their comparison of sequenced *E. coli* genes (59), Riley and Labedan found that 971 of 1,862 proteins (52%) in the SwissProt database (version 28) had significant similarity to at least one other *E. coli* sequence. When the functional relatedness of parologous proteins of known function was classified, they found that 587 of 599 proteins performed related or similar functions. Thus, since a large portion of the *E. coli* proteins were related and performed similar functions, gene redundancy is also an important feature of the *E. coli* genome. Similarly almost one-half of the *B. subtilis* genome consists of related simple gene pairs and larger gene families. Since related simple gene pairs account for a large portion of each of these genomes, functional overlap could contribute to the small proportion of these organisms' genes that are found to be important when subjected to systematic functional analysis.

In *S. cerevisiae*, about 42% of all the highly homologous simple gene pairs are included within extended regions of gene synteny. The arrangement of these syntenic regions suggests that the present-day haploid genome evolved from an ancient genome duplication event that occurred about 10^8 years ago (85). In neither *E. coli* nor *B. subtilis* is there evidence to suggest that whole-genome duplication played an important role in the generation of parologous gene families. Rather, the duplication and divergence of individual genes seem responsible for generating most of the evolutionarily related gene sets. The maintenance of these parologous gene sets suggests that each member provides a selective advantage that in many cases awaits discovery.

GENOME SEQUENCES FACILITATE GENETIC AND BIOCHEMICAL ANALYSIS

The power of genetically tractable model systems is that they facilitate the analysis of gene function at the level of the intact organism. Particularly important is the ease with which genetic manipulations can be performed with these organisms (49, 67–69). For example, it is often desirable to precisely mutate a specific gene or set of genes. The molecular analysis and genetic manipulation of these organisms are made dramatically easier by the availability of complete genome sequences. This is particularly true for *S. cerevisiae*, because of the ease with which targeted gene replacement (63) can be performed with this unicellular eukaryote.

Complete Genome Sequences Allow Accurate and Efficient Targeted Gene Replacement

Functional genomics in yeast will be used as an example of how gene replacement can be applied, since its methods are relatively advanced and the technology could be invented or adapted to work with other systems, including bacteria. Indeed, methods and strains suitable for generating precise mutations or gene deletions in the genomes of *B. subtilis* (36) and *E. coli* (45) have recently been developed.

Targeted gene replacement, which can be used to genetically alter organisms ranging from bacterial to mouse and human cell lines, was initially developed with yeast (63). Yeast can be transformed at a high frequency (22, 23), and integrative transformation occurs at homologous sequences far more efficiently

than at nonhomologous sequences. Strains harboring targeted gene replacements, or knockouts, can therefore be generated very efficiently with yeast. These attributes have been exploited to develop an elegant PCR-based (50) strategy for replacing genomic sequences (4). This DNA replacement method can be used to replace any contiguous genomic sequence (from a few base pairs to 3,000 bp or more in length) with essentially any other PCR-amplified DNA sequence that contains a selectable marker.

Briefly outlined here is a PCR-based method (4) presently used for replacing yeast genome sequences with a selectable cassette. First, a selectable marker (cassette) is amplified with two long primers with at least 40-nucleotide-long targeting sequences on their 5′ ends and about 18 nucleotides of homology to the selectable cassette on their 3′ ends. Amplification of the selectable-cassette DNA is performed with these long oligonucleotides. The 5′-terminal 40 nucleotides of each oligonucleotide provide genome targeting sequences because they are designed to be identical to two 40-nucleotide regions that flank the genomic region to be replaced. PCR products are targeted to the regions of homology and by homologous recombination replace the intervening information. Cassette sequences with homology to sequences other than the desired targeting sites (defined by the two 40-nucleotide-long targeting sequences at the 5′ ends of the amplification oligonucleotides) can also promote integrative transformation. It is therefore desirable to use a cassette that does not harbor any sequences present in the genome of the strain to be genetically altered by transformation. Cassettes harboring *S. cerevisiae* sequences, such as a genetic marker for selection, can also be efficiently targeted for gene replacement if all sequences with homology to the cassette have been deleted from the recipient strain's genome. A selectable marker that is now widely used confers dominant G418 (Geneticin) resistance and does not harbor any yeast sequences. This cassette (82) contains the coding region of the *kanR* gene (aminoglycoside phosphotransferase) from the *E. coli* transposon Tn*903* fused to transcriptional and translational control sequences of the *TEF* gene from the filamentous fungus *Ashbya gossypii*. *TEF* encodes the translation elongation factor 1-alpha.

The G418 resistance cassette is amplified as described above. The PCR amplification product can be transformed into any haploid or diploid yeast strain, where it will replace the yeast genome sequences flanked by the 40-nucleotide targeting sequences previously described. Yeast transformation is performed by standard procedures (23) with about 1 μg of linear PCR product. Transformants, selected on rich-medium plates containing 200 μg of G418 (Gibco BRL)/ml, appear after 3 to 5 days of growth at a frequency that is 4 to 5 orders of magnitude less than the frequency obtained with autonomously replicating plasmids (23).

Putative gene replacement mutants can be confirmed by PCR analysis of the targeted locus with genomic DNA isolated from G418-resistant transformants. To accomplish this, one reaction is performed with oligonucleotides complementary to sequences just upstream and downstream of the targeted ORF. Additional PCRs can be done to verify that the desired mutant locus is present. These can utilize the same upstream and downstream oligonucleotides in combination with appropriate oligonucleotides that hybridize to sequences found within the resistance cassette. In about 80% of the G418-resistant transformants, the cassette has correctly replaced the targeted ORF, although the frequency can vary from 10 to 100% depending upon the strain, locus, and oligonucleotides (30, 70, 79, 82).

In various adaptations of this method, PCR-generated replacement DNAs are used for constructing strains harboring simple point mutations or large deletions, making reporter gene constructs for any gene, inserting epitope or fluorescent tags in any protein, and studying the regulatory information controlling gene expression (53).

Strategies for the Analysis of Essential Genes

Almost 15% of the genes in yeast are essential for survival. Since "dead" is neither a very useful nor an interesting phenotype and heterozygous mutants may fail to elicit a detectable phenotype, methods that facilitate the functional analysis of essential genes have been and are being developed. Traditionally, essential genes have been studied by using conditional mutants isolated by screening populations of mutagenized cells for a phenotype that is temperature dependent. This approach has been extremely valuable; however, it cannot be easily adapted to study specific genes. Furthermore, mutagenesis often generates multiple mutations, making the genetic analysis of mutant phenotypes difficult. Temperature shifts also result in the induction of a battery of stress response genes, potentially making it difficult to identify the specific effects associated with inactivating the gene product under investigation. Techniques that allow conditionally expressed derivatives of specific genes to be constructed and studied without accompanying effects associated with additional genetic or physiological changes are therefore particularly desirable for the systematic analysis of gene function. Completely sequenced genomes allow accurate insertion of regulatable promoters to control the expression of any gene.

Several systems developed for generating gene-specific conditional mutants can be employed to analyze any gene in the yeast genome. Most of these systems use regulatable promoters (60, 61), and for yeast these are often based on the highly regulated *GAL1-GAL10* promoter region (64). The *GAL1-GAL10* promoter region is induced 10^3-fold by galactose (28, 29). Usually, regulatable promoters have been used to construct plasmid-borne copies of the essential gene of interest. This approach requires the construction of an appropriate host strain with the gene of interest deleted from its genome and a conditionally expressed derivative of the same gene harbored on a yeast plasmid vector.

Lafontaine and Tollervey recently used the galactose promoter to develop a system (44) that can be used for the systematic characterization of essential genes. Their system employs a modification of the PCR-based DNA replacement method described above; however, rather than replacing a gene's coding region, it replaces the upstream control region with a module containing the highly regulated GAL10 promoter. This promoter cassette can replace the promoter of essentially any yeast gene, thereby generating derivatives expressed in a galactose-dependent fashion. A cassette that includes an epitope tag is also available. This epitope-tagged GAL10 regulatory cassette generates fusion proteins which can be used for immunolocalization or immunopurification and/or for following the effects of protein depletion without antibodies directed against the targeted gene's protein product. Systems like the one developed by Lafontaine and Tollervey (44) allow conditionally expressed mutant alleles of essential genes to be constructed in haploid strains.

Although widely used for studies requiring conditionally expressed genes, galactose-inducible promoters are also strongly repressed by glucose. GAL promoter-based expression therefore requires growth without glucose (29, 62), conditions that alter expression of a large portion of the yeast genome (12). To address this, highly regulatable promoters that are not affected by growth conditions have recently been developed (21, 51). Transcription in these systems is driven by a promoter consisting of the tetracycline operator (tetO) from the E. coli transposon Tn10 fused to the TATA and leader region of a yeast gene. Expression of these hybrid promoters is dependent upon a chimeric protein that includes the tetracycline-inducible repressor (TetR protein) from the Tn10-encoded tetracycline resistance operon fused to a eukaryotic transcriptional activator.

Yeast tetO-based promoters are expressed only in the absence of tetracycline because the TetR protein DNA binding domain is inactivated when it is associated with tetracycline. These tetracycline-repressible promoters are highly regulated (700-fold or greater) and have several advantages over galactose-inducible promoters and traditional temperature-sensitive mutants. Foremost, tetO promoter systems do not require changes in nutrient composition or temperature to modulate gene expression. Northern analysis of a tetO-driven version of the yeast G1-cyclin gene CLN1 showed that mRNA expression was turned off and mRNA disappeared about 30 min after the addition of tetracycline (21). tetO-tetR systems therefore provide a simple method for turning gene expression on and off. Regulatable promoter replacement cassettes in combination with the availability of complete genome sequences will be important resources for studying gene function in yeasts and other eukaryotic systems.

GENOME SEQUENCES FACILITATE LARGE-SCALE METHODS FOR CHARACTERIZING ORF FUNCTION

Even for the well-characterized model organisms whose genomes have been sequenced, very little is known about the biological functions of most ORFs. Additional clues about a gene can therefore provide important insight into its function. Completely sequenced genomes enable researchers to use approaches that screen the entire genome for genes that are important under specific growth conditions or for adaptation to the alteration of some physiological parameter. Three experimental approaches that can be used to examine the entire gene set for informational clues about gene function are briefly described below. The first of these approaches is made dramatically easier by the availability of completely sequenced genomes. The second and third approaches require access to the complete sequence of the organism's genome.

Transposon Tagging for Analyzing Gene Expression, Gene Function, and Protein Localization

One of the first global or genome-wide strategies for studying gene function in yeast was reported by Burns and colleagues (7). They mutagenized an *S. cerevisiae* genomic DNA library harboring inserts with an average size of about 3 kbp with a mini-Tn*3*::*lacZ*::*LEU2* by using modifications of a previously developed Tn*3* insertion procedure (65). The Tn*3*::*lacZ*::*LEU2*-mutagenized library of yeast sequences was then used to transform a diploid yeast strain and generate about 2×10^4 independent transformants. Homologous recombination between the mutagenized inserts and the yeast genome results in each transformant being heterozygous for one of the 3-kb genomic clones disrupted by transposon mutagenesis. Heterozygous disruption mutants are identified by selection for Leu$^+$ transformants. Since yeast has about one gene per 2 kbp (25, 26), about one-sixth of the transformants should harbor the *lacZ* portion (which lacks its own translational start and control sequences) in the correct reading frame and orientation for its expression. This is an overestimate, since it assumes that all inserts are targeted to ORFs and not sequences between ORFs. Yeast ORF sequences do, however, occupy over 80% of the yeast genome.

Gene disruption libraries created with this approach can be used in several ways to gain insights into yeast gene function. These include the assessment of (i) gene expression by monitoring β-galactosidase activity by simple endpoint- or indicator plate-based enzyme assays, (ii) the subcellular localization of many gene products by indirect immunofluorescence with antibodies against β-galactosidase, and (iii) mutant phenotypes resulting from insertional mutagenesis of the genes in haploids derived by sporulation of the original diploid transformants.

Analysis of β-galactosidase expression in the diploid insertion strains enabled Burns et al. (7) to estimate that 80 to 86% of yeast genes were expressed during vegetative growth. By following β-galactosidase expression during sporulation, they estimated that the yeast genome encodes about 100 meiotically induced genes. Immunolocalization studies found that 245 different LacZ fusion proteins were localized to discrete locations in the cell. Analysis of the haploid products derived by sporulating 59 independent Tn*3*::*lacZ*::*LEU2* diploids revealed that about 15% of the gene disruptions resulted in inviable meiotic products and a further 25% displayed a mutant phenotype. The initial analyses of Tn*3*::*lacZ*::*LEU2* insertion mutants have been expanded, and the complete results can be found at the Yale Genome Analysis Center (http://ycmi.med.yale.edu/YGAC/home.html).

This technology can also be applied to address specific aspects of yeast biology, for example, to identify genes that are important for the yeast cell surface (47). Lussier et al. screened a population of Tn*3*::*lacZ*::*LEU2* disruptants for altered sensitivity to calcofluor white (these mutants are enriched in strains with cell surface defects). The identity of the mutated genes was revealed by retrieval into *E. coli*, single-pass sequencing of a short portion of the disrupted gene, and finally, searching for matching sequences in the yeast genome sequence. Eighty-two genes were identified, and 65 of these were found to exhibit additional phenotypes that could be attributed to cell surface defects. The utility and power of this analysis are reflected in the fact that it identified both genes already known to affect the cell surface and new genes that, based on sequence similarity, would be predicted to be important for the biochemical composition of the yeast cell surface.

Methods for Screening Entire Gene Sets for Functionally Important Genetic Entities

Genome-wide approaches that utilize sequenced genomes to assess gene function are being developed. One such approach is useful for screening the entire genome for genes that are important for the organism's fitness under specific physiological conditions. To accomplish this objective, an international consortium is systematically constructing a set of over 6,000 mutant *S. cerevisiae* strains (each strain has a mutation in a different gene, and the collection represents essentially all the protein-coding genes). A strategy developed in Ron Davis's laboratory at Stanford University and disruption DNA supplied by Anna Astromoff, Dan Shoemaker, and Elizabeth Winzeler (http://sequence-www.stanford.edu/group/yeast_deletion_project/deletions3.html) are being used. Each mutant strain will be deleted for one of yeast's approximately 6,000 predicted genes. Using the PCR-based gene replacement method (4) outlined above, each ORF is being replaced by a selectable marker and a unique sequence oligonucleotide, or "bar code." The null mutation associated with each gene replacement strain will therefore include a unique DNA tag (bar code) to identify the individual mutant genes and the strains that carry them (70). Each gene replacement strain's associated bar code has been engineered so that it is flanked by two primer binding sites that are common to all the gene replacements. All 6,000 bar codes can therefore be PCR amplified by a single pair of primers.

Using the complete set of bar coded mutants, the entire set of nonessential yeast genes can be screened for genes that are important for a specific function in a single experiment. To accomplish this, the bar coded mutants are pooled and grown under control conditions and an experimental condition. The experimental condition could involve some treatment, for example, exposure to a cell wall or DNA synthesis inhibitor, with a DNA-damaging agent or some physiological change. DNA samples isolated from the control and treated populations can be used as templates for asymmetrical PCR to amplify the bar codes. In one possible scenario, two different fluorescent tags could be used to label the bar codes amplified from the control and treated cultures, respectively. The resulting two sets of probes would then be hybridized to an array of unique DNA sequences (representing all the bar coded tags in the mutant strains) immobilized on an appropriate support. Bar codes that show significantly reduced hybridization signals after treatment would identify genes that are important for survival under the conditions tested. This strategy and the technology to perform such experiments have

been developed and tested in a pilot study (70). In the pilot study, equal numbers of cells of a total of 11 bar coded adenine and tryptophan auxotrophs were pooled and subjected to competitive growth in complete synthetic medium, medium lacking adenine, and medium lacking tryptophan. PCR-generated probes made from genomic DNA isolated from each culture were used as templates to probe the bar codes representing each of the inactivated genes. The signal intensities representing hybridization to the bar codes for the individual adenine and tryptophan auxotrophs became progressively weaker as the pooled strains were grown over 24 h in media without adenine and tryptophan, respectively.

This bar coded collection of 6,000 reference strains can be used by *S. cerevisiae* researchers specializing in the study of a particular gene, cellular process, or class of genes to devise assays and genetic screens for their individual research interests. It will also be valuable for the systematic functional analysis of the entire set of nonessential genes. Isogenic haploid and diploid disruption strains representing the entire set of nonessential *S. cerevisiae* genes should be available in 2000. These bar coded yeast disruption strains are being made available at Research Genetics (http://www.resgen.com/). It should be possible to construct similar sets of bar coded gene disruption strains for any genetically tractable organism amenable to PCR-based targeted gene replacement.

Limitations on the utility of the disruption set include (i) the need to deal with those genes that are members of functionally redundant or overlapping gene sets and (ii) the fact that bar coded strains cannot be used to address the functional importance of essential genes except as heterozygous diploids.

Assessing Global Gene Expression with Transcript-Specific DNA Arrays

A second genome-wide approach utilizes the completed genome sequence to generate filter-based DNA arrays (http://www.resgen.com/), DNA microarrays (66), or oligonucleotide microarrays (9) to assess global gene expression. DNA arrays (with long DNA sequences usually generated by PCR amplification) and oligonucleotide arrays (with chemically synthesized DNA from 20 to 100 bases in length) have many applications. One such application is to monitor expression of the entire genome in response to physiological and/or genetic change. Probes derived from total mRNA are hybridized to filter-based DNA arrays, DNA microarrays, or oligonucleotide microarrays. All three types of arrays immobilize relatively long DNA sequences complementary to entire ORFs or short oligonucleotides complementary to portions of specific ORFs to a solid support. Any organism for which the entire genome has been sequenced and the protein-coding regions have been identified can have its entire gene set represented on a DNA array.

Nylon membrane-based yeast ORF DNA arrays are available from Research Genetics Inc. (http://www.resgen.com/). They provide a yeast ORF index that has 6,144 individual ORFs from yeast spotted onto two nylon membrane filters. Each ORF DNA sequence has been individually amplified by PCR and spotted to the membranes. DNA microarrays, pioneered by DeRisi and colleagues (12), also utilize PCR-amplified ORF DNA; however, the ORF DNAs are attached to glass supports. Glass supported arrays allow more than one fluorescently labelled probe to be used for a single hybridization experiment. Hybridization signals are detected and analyzed by fluorescent scanning confocal microscopy.

The use of DNA microarrays to study yeast gene regulation is illustrated in a recent publication by DeRisi and colleagues (12). They followed the changes in gene expression that occurred during the switch from anaerobic (fermentation) to aerobic (respiration) metabolism. They also examined how genetic manipulation of a transcriptional corepressor, *TUP1*, and a transcriptional activator, *YAP1*, affected expression of the entire yeast gene set. These results and their analysis can be searched

on the Internet at http://cmgm.stanford.edu/pbrown/. A list of parts and instructions for building a DNA arrayer are also included at the website.

Oligonucleotide microarrays have been developed by several companies, including Affymetrix (Santa Clara, Calif.), Brax (Cambridge, United Kingdom), Nanogen (San Diego, Calif.), Protogene Laboratories (Palo Alto, Calif.), and the German Cancer Institute (Heidelberg, Germany). Affymetrix Inc. (http://www.affymetrix.com/research.html) manufactures DNA microarrays by using technologies similar to those used for producing computer chips (9, 15, 16, 46). Using their photolithographically based method, Affymetrix Inc. can synthesize more than 100,000 unique-sequence oligonucleotides on a silicon surface to generate a "DNA chip." Each oligonucleotide is placed on the chip so that its location is known and it does not overlap with any of the other oligonucleotides. A mixture of fluorescently labelled sequences (these can be derived from mRNA or DNA) can then be hybridized to the DNA chip. Relative levels of individual species in a complex mixture of fluorescent DNA- or RNA-derived probes are determined by analyzing the intensity of fluorescence emanating from each oligonucleotide patch. The Affymetrix system measures fluorescence from the individual oligonucleotide patches with a laser-based reader and image-processing software. The resulting data is processed with custom bioinformatics software.

Affymetrix manufactures a yeast oligonucleotide microarray set. Included in this set are four arrays (gene chips) that contain more than 260,000 specifically chosen oligonucleotides. Their yeast gene chip set has about 20 pairs of 25-base oligonucleotides specific for each of the 6,100 yeast genes. Each pair includes a perfect-match oligonucleotide and a single-base-mismatch oligonucleotide. Hybridization to the Affymetrix gene chip set can be used to directly measure mRNA levels in a highly parallel approach (84). Oligonucleotide DNA microarrays were used by Wodicka and colleagues to compare the levels of mRNA expression by *S. cerevisiae* grown in rich and minimal media. Almost 90% of all yeast genes were transcribed at detectable levels under both growth conditions. Furthermore, although many genes predicted to be differentially expressed under these two conditions behaved as expected, large differences were observed for genes that had not been expected or previously tested. There are many other applications of oligonucleotide chips and similar technologies, such as monitoring gene expression in a strain with a mutant transcription factor or assessing the expression of downstream genes regulated by a signalling pathway. A recent review (48) describes the different types of DNA and oligonucleotide arrays and lists several companies that are producing them.

GENOME SEQUENCES FACILITATE ANALYSIS OF PROTEIN-PROTEIN AND PROTEIN-NUCLEIC ACID INTERACTIONS

Establishing whether a protein interacts with another protein or proteins often provides information that can be used to predict, or to expand our understanding of, its function. Similarly, identifying and defining DNA-protein and RNA-protein interactions provides important functional links between nucleic acid sequences and specific proteins. Knowing these interactions will also provide information that can be used to identify the regulatory networks that define the basic cell. Genome sequences significantly enhance our ability to identify and define these macromolecular interactions.

Assessing Protein–Protein Interactions on a Global Scale

Determining the biological importance of ORFs whose function remains unknown represents a formidable task. The two-hybrid screen is a system for studying protein-protein interactions (13). It is an in vivo method for studying a protein's function by identifying the other proteins it interacts with and for de-

fining the nature of these interactions. The two-hybrid system as initially developed takes advantage of the functional independence of the DNA binding and transcriptional activation domains of the *S. cerevisiae* protein Gal4p. Hybrid genes are constructed by using each of these two domains. One hybrid encodes the DNA binding domain of Gal4p fused to a known protein (the bait). Typically, the second hybrid gene consists of the Gal4p transcription activation domain fused to protein sequences encoded by a library of DNA sequences. Usually, but not necessarily, the library and bait sequences are derived from the same organism. A functional transcriptional activator is formed by binding between the bait and a protein (the prey) encoded by a library plasmid. A reporter gene and/or selectable gene with expression controlled by a Gal4p binding site is used to screen and/or select for protein interactions.

Analysis of protein-protein interactions with the two-hybrid screen requires an extensive investment of time and effort to identify the interactions of even a single protein. The two-hybrid screen has, however, been adapted by Fromont-Racine, Rain, and Legrain so that exhaustive screens for protein-protein interactions can be performed at the level of entire genomes. This adaptation of the two-hybrid method can detect protein interactions reliably and rapidly (20). To facilitate this, the authors developed tools, including an exhaustive genomic DNA library and a selective two-hybrid procedure for efficiently screening the entire yeast genome for interactors with a bait protein. They illustrated the potential of their modified two-hybrid method by examining spliceosomal protein interactions. Known spliceosomal proteins were used as bait to identify interacting proteins (prey). These prey were in turn used as bait for a second round of screens. Repeating the procedure several times enabled them to establish a network of protein interactions. The authors claim that by using their system it will be possible to proceed from proteins of known function as anchor points to an "interaction map of the yeast proteome."

Identification of Proteins and Characterization of Protein Complexes by Mass Spectroscopy

Two-hybrid screens often identify fortuitous interactions; therefore, although a systematic interaction map of the yeast proteome based on two-hybrid analysis will provide important information, verification and further elucidation by alternative methods will be important. Mass spectroscopy is an independent method that can be used to identify the individual proteins and peptides in complexes. Furthermore, it can estimate the relative amounts of different peptides in a protein complex.

During the last decade technological advances have made it possible to measure the molecular masses of intact proteins by mass spectrometry. Perhaps the most powerful of these methods is matrix-assisted laser desorption ionization-mass spectroscopy (MALDI-MS) (33). In MALDI-MS, laser light is used to desorb proteins and peptides from a matrix. It allows the masses of proteins up to 300 kDa to be determined with 99.9% accuracy (40). This method is also very sensitive and allows mass to be determined with only picomoles of protein.

Simultaneously, advances in protein- and DNA-sequencing technology are resulting in an exponential increase in the number of sequences deposited in databases. It is now possible to combine mass spectrometric data on proteins and protein complexes with data on the predicted masses of proteins derived from the analysis of genome sequences. Searching protein databases with the molecular mass of a protein found by mass spectrometry is an easy and often sufficient approach to unambiguously identify a protein. For increased specificity, a partial mass spectrometric peptide map of the protein or proteins of interest can also be determined (33). Knowing the masses of just a few proteolytic peptides usually allows protein databases to be successfully searched for a unique match. As DNA and protein sequence databases grow, protein identification by using mass spectrometric data should become an increasingly important method for identifying and characterizing individual proteins and protein complexes. Mann and colleagues identified several ligands for profilin I and profilin II from mouse brain extracts. These interacting proteins, including dynamin I, clathrin, synapsin, Rho-associated coiled-coil kinase, the Rac-associated protein NAP1, and a member of the NSF/sec18 family, were all identified by MALDI-MS (83).

Mass spectrometric analysis combined with estimated protein masses based on the sequenced yeast genome has also been used to analyze and identify components of the anaphase-promoting complex, or cyclosome, of S. cerevisiae (87). Using mass spectrometric data obtained for the protein subunits and for peptide fragments of selected proteins, it was possible to identify at least 12 different subunits in the purified anaphase-promoting complex.

Identification and Analysis of RNA-Protein and DNA-Protein Interactions

Gold et al. (27) and Tuerk and Gold (78) proposed using the SELEX (systematic evolution of ligands by exponential enrichment) technique to generate protein-nucleic acid linkage maps. The first SELEX experiment was performed with a pool of about 65,000 sequences representing all possible sequences of the 8-nucleotide loop portion of a T4 mRNA. The loop is required for the binding of bacteriophage T4 gene 43 protein to the mRNA sequence. The best protein 43 binding sequences were identified by iterated partitioning of the pool by binding to protein 43. To achieve this, the pool of potential binding loops was flanked by fixed sequences. Following enrichment by protein 43 binding, the enriched sequences were amplified by PCR. The binding and amplification procedures were then repeated. In the end, two sequences able to bind 43 protein were purified from the pool of random sequences (78). One sequence had the wild-type loop, and the second was a quadruple mutant.

Since successes have been obtained with the SELEX technology when the in vitro selection process is started with pools of 10^{14} to 10^{15} unique oligonucleotides (77), it is at least potentially feasible to select sequences from within an organism's entire genome. For example, pools of genomic clones with about 2×10^7 E. coli sequences, less than 10^8 yeast sequences, and about 10^{10} human sequences can represent the entire complexity of the E. coli, yeast, and human genomes, respectively. Using the SELEX technique with a SELEX genomic library of E. coli, several DNA binding sites for the regulatory protein MetJ were identified within the E. coli genome (27).

CONCLUDING REMARKS

Elucidating the biological importance of genes has traditionally been carried out by approaches where individual laboratories pursued their particular biological interests. Often this approach has proceeded from a mutant phenotype to protein function. New methods for assessing gene expression, protein-protein interactions, and protein-nucleic acid interactions mean that it is now possible to use genome-wide approaches to study gene function. These global approaches should significantly accelerate progress toward a detailed understanding of the model organisms B. subtilis, E. coli, and S. cerevisiae.

How will functional genomics contribute to the goal of attaining a complete understanding of these model organisms? Will it replace traditional research focused on one or a few genes? First, global genome-wide strategies cannot fully replace the traditional approaches where individual laboratories study a few genes as they pursue focused biological problems (31, 37, 54). Genome-wide approaches will, however, supply a plethora of additional clues for solving specific biological questions. These broad-based methods will also increasingly be used to pursue more focused biological questions. Finally, genome-wide methods will generate exponentially increasing amounts of biological data. Effective systems of data storage, management, and analysis must keep pace with data generation so that the data and the tools necessary to apply these resources to individual research programs are accessible to all researchers.

ACKNOWLEDGMENT

This work was supported by an NSERC of Canada operating grant to R.K.S.

REFERENCES

1. **Alm, R. A., L.-S. L. Ling, D. T. Moir, B. L. King, E. D. Brown, P. C. Doig, D. R. Smith, B. Noonan, B. C. Guild, B. L. deJonge, G. Carmel, P. J. Tummino, A. Caruso, M. Uria-Nickelsen, D. M. Mills, C. Ives, R. Gibson, D. Merberg, S. D. Mills, O. Jiang, D. E. Taylor, G. F. Vovis, and T. J. Trust.** 1999. Genomic sequence comparison of two unrelated isolates of the human gastric pathogen *Helicobacter pylori*. *Nature* **397:**176–180.
2. **Altschul, S. F., T. L. Madden, A. A. Schäfer, J. Zhang, Z. Zhang, W. Miller, and D. J. Lipman.** 1997. Gapped BLAST and PSI-BLAST: a new generation of protein database search programs. *Nucleic Acids Res.* **25:**3389–3402.
3. **Andersson, S. G., A. Zomorodipour, J. O. Andersson, T. Sicheritz-Ponten, U. C. Alsmark, R. M. Podowski, A. K. Naslund, A. S. Eriksson, H. H. Winkler, and C. G. Kurland.** 1998. The genome sequence of *Rickettsia prowazekii* and the origin of mitochondria. *Nature* **396:**133–140.
4. **Baudin, A., O. Ozier-Kalogeropoulos, A. Denouel, F. Lacroute, and C. Cullin.** 1993. A simple and efficient method for direct gene deletion in *Saccharomyces cerevisiae*. *Nucleic Acids Res.* **21:**3329–3330.
5. **Blattner, F. R., G. Plunkett III, C. A. Bloch, N. T. Perna, V. Burland, M. Riley, J. Collado-Vides, J. D. Glasner, C. K. Rode, G. F. Mayhew, J. Gregor, N. W. Davis, H. A. Kirkpatrick, M. A. Goeden, D. J. Rose, B. Mau, and Y. Shao.** 1997. The complete genome sequence of *Escherichia coli* K-12. *Science* **277:**1453–1474.
6. **Bult, C. J., O. White, G. J. Olsen, L. Zhou, R. D. Fleischmann, G. G. Sutton, J. A. Blake, L. M. FitzGerald, R. A. Clayton, J. D. Gocayne, A. R. Kerlavage, B. A. Dougherty, J. F. Tomb, M. D. Adams, C. I. Reich, R. Overbeek, E. F. Kirkness, K. G. Weinstock, J. M. Merrick, A. Glodek, J. L.**

Scott, N. S. M. Geoghagen, J. F. Weidman, J. L. Fuhrmann, D. Nguyen, T. R. Utterback, J. M. Kelly, J. D. Peterson, P. W. Sadow, M. C. Hanna, M. D. Cotton, K. M. Roberts, M. A. Hurst, B. P. Kaine, M. Borodovsky, H.-P. Klenk, C. M. Fraser, H. O. Smith, C. R. Woese, and J. C. Venter. 1996. Complete genome sequence of the methanogenic archaeon, *Methanococcus jannaschii*. *Science* **273**: 1058–1073.

7. Burns, N., B. Grimwade, P. B. Ross-Macdonald, E. Y. Choi, K. Finberg, G. S. Roeder, and M. Snyder. 1994. Large-scale analysis of gene expression, protein localization, and gene disruption in *Saccharomyces cerevisiae*. *Genes Dev.* **8**:1087–1105.

8. Bussey, H., D. B. Kaback, W. Zhong, D. T. Vo, M. W. Clark, N. Fortin, J. Hall, B. F. Ouellette, T. Keng, A. B. Barton, Y. Su, C. J. Davies, and R. K. Storms. 1995. The nucleotide sequence of chromosome I from *Saccharomyces cerevisiae*. *Proc. Natl. Acad. Sci. USA* **92**: 3809–3813.

9. Chee, M., R. Yang, E. Hubbell, A. Berno, X. C. Huang, D. Stern, J. Winkler, D. J. Lockhart, M. S. Morris, and S. P. Fodor. 1996. Accessing genetic information with high-density DNA arrays. *Science* **274**:610–614.

10. Cole, S. T., R. Brosch, J. Parkhill, T. Garnier, C. Churcher, D. Harris, S. V. Gordon, K. Eiglmeier, S. Gas, C. E. Barry III, F. Tekaia, K. Badcock, D. Basham, D. Brown, T. Chillingworth, R. Connor, R. Davies, K. Devlin, T. Feltwell, S. Gentles, N. Hamlin, S. Holroyd, T. Hornsby, K. Jagels, A. Krogh, J. McLean, S. Moule, L. Murphy, S. Oliver, J. Osborne, M. A. Quail, M. A. Rajandream, J. Rogers, S. Rutter, K. Seeger, S. Skelton, S. Squares, R. Squares, J. E. Sulston, K. Taylor, S. Whitehead, and B. G. Barrell. 1998. Deciphering the biology of *Mycobacterium tuberculosis* from the complete genome sequence. *Nature* **393**:537–544.

11. Deckert, G., P. V. Warren, T. Gaasterland, W. G. Young, A. L. Lenox, D. E. Graham, R. Overbeek, M. A. Snead, M. Keller, M. Aujay, R. Huber, R. A. Feldman, J. M. Short, G. J. Olson, and R. V. Swanson. 1998. The complete genome of the hyperthermophilic bacterium *Aquifex aeolicus*. *Nature* **392**: 353–358.

12. DeRisi, J. L., V. R. Iyer, and P. O. Brown. 1997. Exploring the metabolic and genetic control of gene expression on a genomic scale. *Science* **278**:680–686.

13. Fields, S., and O. Song. 1989. A novel genetic system to detect protein-protein interactions. *Nature* **340**:245–246.

14. Fleischmann, R. D., M. D. Adams, O. White, R. A. Clayton, E. F. Kirkness, A. R. Kerlavage, C. J. Bult, J.-F. Tomb, B. A. Dougherty, J. M. Merrick, K. McKenney, G. Sutton, W. FitzHugh, C. A. Fields, J. D. Gocayne, J. D. Scott, R. Shirley, L.-I. Liu, A. Glodek, J. M. Kelley, J. F. Weidman, C. A. Phillips, T. Spriggs, E. Hedblom, M. D. Cotton, T. R. Utterback, M. C. Hanna, D. T. Nguyen, D. M. Saudek, R. C. Brandon, L. D. Fine, J. L. Fritchman, J. L. Fuhrmann, N. S. M. Geoghagen, C. L. Gnehm, L. A. McDonald, K. V. Small, C. M. Fraser, H. O. Smith, and J. C. Venter. 1995. Whole-genome random sequencing and assembly of *Haemophilus influenzae* Rd. *Science* **269**:496–512.

15. Fodor, S. P., R. P. Rava, X. C. Huang, A. C. Pease, C. P. Holmes, and C. L. Adams. 1993. Multiplexed biochemical assays with biological chips. *Nature* **364**:555–556.

16. Fodor, S. P. A., J. L. Read, M. C. Pirrung, L. Stryer, A. T. Lu, and D. Solas. 1991. Light-directed, spatially addressable parallel chemical synthesis. *Science* **251**:767–773.

17. Fraser, C. M., S. Casjens, W. M. Huang, G. G. Sutton, R. A. Clayton, R. Lathigra, O. White, K. A. Ketchum, R. Dodson, E. K. Hickey, M. Gwinn, B. Dougherty, J.-F. Tomb, R. D. Fleischmann, D. Richardson, J. Peterson, A. R. Kerlavage, J. Quackenbush, S. Salzberg, M. Hanson, R. van Vugt, N. Palmer, M. D. Adams, J. D. Gocayne, J. Weidman, T. Utterback, L. Watthey, L. McDonald, P. Artiach, C. Bowman, S. Garland, C. Fujii, M. D. Cotton, K. K. Horst, K. Roberts, B. Hatch, H. O. Smith, and J. C. Venter. 1997. Genomic sequence of a Lyme disease spirochete, *Borrelia burgdorferi*. *Nature* **390**:580–586.

18. Fraser, C. M., J. D. Gocayne, O. White, M. D. Adams, R. A. Clayton, R. D. Fleischmann, C. J. Bult, A. R. Kerlavage, G. Sutton, J. M. Kelley, J. L. Fritchman, J. F. Weidman, K. V. Small, M. Sandusky, J. L. Fuhrmann, D. T. Nguyen, T. R. Utterback, D. M. Saudek, C. A. Phillips, J. M. Merrick, J.-F. Tomb, B. A. Dougherty, K. F. Bott, P.-C. Hu, T. S. Lucier, S. N. Peterson, H. O. Smith, C. A. Hutchison III, and J. C. Venter. 1995. The minimal gene complement of *Mycoplasma genitalium*. *Science* **270**:397–403.

19. Fraser, C. M., S. J. Norris, G. M. Weinstock, O. White, G. G. Sutton, R. Dodson, M. Gwinn, E. K. Hickey, R. Clayton, K. A. Ketchum, E. Sodergren, J. M. Hardham, M. P. McLeod, S. Salzberg, J. Peterson, H. Khalak, D. Richardson, J. K. Howell, M.

Chidambaram, T. Utterback, L. McDonald, P. Artiach, C. Bowman, M. D. Cotton, C. Fujii, S. Garland, B. Hatch, K. Horst, K. Roberts, L. Watthey, J. Weidman, H. O. Smith, and J. C. Venter. 1998. Complete genome sequence of *Treponema pallidum*, the syphilis spirochete. *Science* **281**:375–388.

20. Fromont-Racine, M., J.-C. Rain, and P. Legrain. 1997. Toward a functional analysis of the yeast genome through exhaustive two-hybrid screens. *Nat. Genet.* **16**:277–282.

21. Gari, E., L. Piedrafita, M. Aldea, and E. Herrero. 1997. A set of vectors with a tetracycline-regulatable promoter system for modulated gene expression in *Saccharomyces cerevisiae*. *Yeast* **13**:837–848.

22. Gietz, D., A. St. Jean, R. A. Woods, and R. H. Schiestl. 1992. Improved method for high efficiency transformation of intact yeast cells. *Nucleic Acids Res.* **20**:1425.

23. Gietz, R. D., and R. A. Woods. 1994. High efficiency transformation in yeast, p. 121–134. *In* J. A. Johnston (ed.), *Molecular Genetics of Yeast: Practical Approaches*. Oxford University Press, Oxford, United Kingdom.

24. Goebl, M. G., and T. D. Petes. 1986. Most of the yeast genomic sequences are not essential for cell growth and division. *Cell* **46**:983–992.

25. Goffeau, A., R. Aert, M. L. Agostini-Carbone, A. Ahmed, M. Aigle, L. Alberghina, K. Albermann, M. Albers, M. Aldea, D. Alexandraki, G. Aljinovic, E. Allen, R. Altmann, J. Alt-Mörbe, B. André, S. Andrews, W. Ansorge, G. Antoine, R. Anwar, A. Aparicio, R. Araujo, J. Arino, W. Arnold, J. Arroyo, E. Aviles, U. Backes, M. C. Baclet, K. Badcock, A. Bahr, J. P. G. Ballesta, A. T. Bankier, A. Banrevi, M. Bargues, L. Baron, T. Barreiros, B. G. Barrell, C. Barthe, A. B. Barton, A. Baur, A.-M. Bécam, A. Becker, I. Becker, J. Beinhauer, V. Benes, P. Benit, G. Berben, E. Bergantino, P. Bergez, A. Berno, I. Bertani, N. Biteau, A. J. Bjourson, H. Blöcker, C. Blugeon, C. Bohn, E. Boles, P. A. Bolle, M. Bolotin-Fukuhara, R. Bordonné, J. Boskovic, P. Bossier, D. Botstein, G. Bou, S. Bowman, J. Boyer, P. Brandt, T. Brandt, M. Brendel, T. Brennan, R. Brinkman, A. Brown, A. J. P. Brown, D. Brown, M. Brückner, C. V. Bruschi, J. M. Buhler, M. J. Buitrago, F. Bussereau, H. Bussey, A. Camasses, C. Carcano, G. Carignani, J. Carpenter, A. Casamayor, C. Casas, L. Castagnoli, H. Cederberg, E. Cerdan, N. Chalwatzis, R. Chanet, E. Chen, G. Chéret, J. M. Cherry, T. Chillingworth, C. Christiansen, J.-C. Chuat, E. Chung, C. Churcher, C. M. Churcher, M. W. Clark, M. L. Clemente, A. Coblenz, M. Coglievina, E. Coissac, L. Colleaux, R. Connor, R. Contreras, J. Cooper, T. Copsey, F. Coster, R. Coster, J. Couch, M. Crouzet, C. Cziepluch, B. Daignan-Fornier, F. Dal Paro, D. V. Dang, M. D'Angelo, C. J. Davies, K. Davis, R. W. Davis, A. De Antoni, S. Dear, K. Dedman, E. Defoor, M. de Haan, T. Delaveau, S. Del Bino, M. Delgado, H. Delius, D. Delneri, F. Del Rey, J. Demolder, N. Démolis, K. Devlin, P. de Wergifosse, F. S. Dietrich, H. Ding, C. Dion, T. Dipaolo, F. Doignon, C. Doira, H. Domdey, J. Dover, Z. Du, E. Dubois, B. Dujon, M. Duncan, P. Durand, A. Düsterhöft, S. Düsterhus, T. Eki, M. El Bakkoury, L. G. Eide, K.-D. Entian, P. Eraso, D. Erdmann, H. Erfle, V. Escribano, M. Esteban, L. Fabiani, F. Fabre, C. Fairhead, B. Fartmann, A. Favello, G. Faye, H. Feldmann, L. Fernandes, F. Feroli, M. Feuermann, T. Fiedler, W. Fiers, U. N. Fleig, M. Flöth, G. M. Fobo, N. Fortin, F. Foury, M. C. Francingues-Gaillard, L. Franco, A. Fraser, J. D. Friesen, C. Fritz, L. Frontali, H. Fukuhara, L. Fulton, L. J. Fuller, C. Gabel, C. Gaillardin, L. Gaillon, F. Galibert, F. Galisson, P. Galland, F.-J. Gamo, C. Gancedo, J. M. Garcia-Cantalejo, M. I. García-Gonzalez, J. J. Garcia-Ramirez, M. García-Saéz, H. Gassenhuber, M. Gatius, S. Gattung, C. Geisel, M. E. Gent, S. Gentles, M. Ghazvini, D. Gigot, V. Gilliquet, N. Glansdorff, A. Gómez-Peris, A. González, S. E. Goulding, C. Granotier, T. Greco, M. Grenson, P. Grisanti, L. A. Grivell, D. Grothues, U. Gueldener, P. Guerreiro, E. Guzman, M. Haasemann, B. Habbig, H. Hagiwara, J. Hall, K. Hallsworth, K. Hamberg, N. Hamlin, N. J. Hand, V. Hanemann, J. Hani, T. Hankeln, M. Hansen, D. Harris, D. E. Harris, G. Hartzell, D. Hatat, U. Hattenhorst, J. Hawkins, U. Hebling, J. Hegemann, C. Hein, A. Hennemann, K. Hennessy, C. J. Herbert, K. Hernandez, Y. Hernando, E. Herrero, K. Heumann, D. Heuss-Neitzel, N. Hewitt, R. Hiesel, H. Hilbert, F. Hilger, L. Hillier, K. Hinni, C. Ho, J. Hoenicka, B. Hofmann, J. Hoheisel, S. Hohmann, C. P. Hollenberg, K. Holmstrøm, O. Horaitis, T. S. Horsnell, M.-E. Huang, B. Hughes, S. Hunicke-Smith, S. Hunt, S. E. Hunt, K. Huse, R. W. Hyman, F. Iborra, K. J. Indge, I. Iraqui Houssaini, K. Isono, C. Jacq, M. Jacquet, A. Jacquier, K. Jagels, W. Jäger, C. M. James,

J. C. Jauniaux, Y. Jia, M. Jier, A. Jimenez, D. Johnson, L. Johnston, M. Johnston, L. Jones, M. Jones, J.-L. Jonniaux, D. B. Kaback, T. Kallesøe, S. Kalman, A. Kalogeropoulos, L. Karpfinger-Hartl, D. Kashkari, C. Katsoulou, A. Kayser, A. Kelly, T. Keng, H. Keuchel, P. Kiesau, L. Kirchrath, J. Kirsten, K. Kleine, U. Kleinhans, R. Klima, C. Komp, E. Kordes, S. Korol, P. Kötter, C. Krämer, B. Kramer, W. Kramer, P. Kreisl, T. Kucaba, H. Kuester, O. Kurdi, P. Laamanen, M. J. Lafuente, D. Landt, G. Lanfranchi, Y. Langston, D. Lashkari, P. Latreille, G. Lauquin, T. Le, P. Legrain, Y. Legros, A. Lepingle, H. Lesveque, H. Leuther, H. Lew, C. Lewis, Z. Y. Li, S. Liebl, A. Lin, D. Lin, M. Logghe, A. J. E. Lohan, E. J. Louis, G. Lucchini, K. Lutzenkirchen, R. Lyck, G. Lye, A. C. Maarse, M. J. Maat, C. Macri, A. Madania, M. Maftahi, A. Maia e Silva, E. Maillier, L. Mallet, G. Mannhaupt, V. Manus, R. Marathe, C. Marck, A. Marconi, E. Mardis, E. Martegani, R. Martin, A. Mathieu, K. T. C. Maurer, M. J. Mazón, C. Mazzoni, D. McConnell, S. McDonald, R. A. McKee, A. D. K. McReynolds, P. Melchioretto, S. Menezes, F. Messenguy, H. W. Mewes, G. Michaux, N. Miller, O. Minenkova, T. Miosga, S. Mirtipati, S. Möller-Rieker, D. Möstl, F. Molemans, A. Monnet, A.-L. Monnier, M. A. Montague, M. Moro, D. Mosedale, D. Möstl, S. Moule, L. Mouser, Y. Murakami, S. Müller-Auer, J. Mulligan, L. Murphy, M. Muzi Falconi, M. Naitou, K. Nakahara, A. Namath, F. Nasr, L. Navas, A. Nawrocki, J. Nelson, U. Nentwich, P. Netter, R. Neu, C. S. Newlon, M. Nhan, D. Niblett, J.-M. Nicaud, R. K. Niedenthal, C. Nombela, D. Noone, R. Norgren, B. Nußbaumer, B. Obermaier, C. Odell, P. Öfner, C. Oh, K. Oliver, S. G. Oliver, B. F. Ouellette, M. Ozawa, Y. Paces, C. Pallier, D. Pandolfo, L. Panzeri, S. Paoluzi, A. G. Parle-McDermott, S. Pascolo, N. Paricio, A. Pauley, L. Paulin, B. M. Pearson, D. Pearson, D. Peluso, J. Perea, M. Pérez-Alonso, J. E. Pérez-Ortin, A. Perrin, F. X. Petel, B. Pettersson, F. Pfeiffer, P. Philippsen, A. Piérard, E. Piravandi, R. J. Planta, P. Plevani, O. Poch, B. Poetsch, F. M. Pohl, T. M. Pohl, R. Pöhlmann, R. Poirey, D. Portetelle, F. Portillo, S. Potier, M. Proft, H. Prydz, A. Pujol, B. Purnelle, V. Puzos, M. A. Rajandream, M. Ramezani Rad, S. W. Rasmussen, A. Raynal, C. Rebischung, S. Rechmann, M. Remacha, J. L. Revuelta, P. Rice, G.-F. Richard, C. Richards, P. Richterich, M. Rieger, L. Rifkin, L. Riles, T. Rinaldi, M. Rinke, J. Robben, A. B. Roberts, D. Roberts, F. Rodriguez, E. Rodriguez-Belmonte, C. Rodrigues-Pousada, A. M. Rodriguez-Torres, M. Rose, R. Rossau, N. Rowley, T. Rupp, M. Ruzzi, W. Saeger, J. E. Saiz, M. Saliola, D. Salom, H. P. Saluz, M. Sánchez-Perez, M. A. Santos, E. Sanz, J. E. Sanz, A.-M. Saren, F. Sartorello, M. Sasanuma, S.-I. Sasanuma, T. Scarcez, I. Schaaf-Gerstenschläger, B. Schäfer, M. Schäfer, M. Scharfe, B. Scherens, N. Schroff, M. Sen-Gupta, T. Shibata, T. Schmidheini, E. R. Schmidt, C. Schneider, P. Scholler, S. Schramm, A. Schreer, M. Schröder, C. Schwager, S. Schwarz, C. Schwarzlose, B. Schweitzer, M. Schweizer, A.-M. Sdicu, P. Sehl, C. Sensen, J. G. Sgouros, T. Shogren, L. Shore, Y. Su, J. Skala, J. Skelton, P. P. Slonimski, P. H. M. Smits, V. Smith, H. Soares, E. Soeda, A. Soler-Mira, F. Sor, N. Soriano, J. L. Souciet, C. Soustelle, R. Spiegelberg, L. I. Stateva, H. Y. Steensma, J. Stegemann, S. Steiner, L. Stellyes, F. Sterky, R. K. Storms, H. St. Peter, R. Stucka, A. Taich, E. Talla, I. Tarassov, H. Tashiro, P. Taylor, C. Teodoru, H. Tettelin, A. Thierry, G. Thireos, E. Tobiasch, D. Tovan, E. Trevaskis, Y. Tsuchiya, M. Tzermia, M. Uhlen, A. Underwood, M. Unseld, J. H. M. Urbanus, A. Urrestarazu, S. Ushinsky, M. Valens, G. Valle, A. Van Broekhoven, M. Vandenbol, Q. J. M. Van der Aart, C. G. Van Der Linden, L. Van Dyck, M. Vanoni, J. C. van Vliet-Reedijk, A. Vassarotti, M. Vaudin, K. Vaughan, P. Verhasselt, I. Vetter, F. Vierendeels, D. Vignati, C. Vilela, S. Vissers, C. Vleck, D. T. Vo, D. H. Vo, M. Voet, G. Volckaert, D. Von Wettstein, H. Voss, P. Vreken, A. Wach, G. Wagner, S. V. Walsh, R. Wambutt, H. Wang, Y. Wang, J. R. Warmington, R. Waterston, M. D. Watson, N. Weber, E. Wedler, H. Wedler, Y. Wei, S. Whitehead, B. L. Wicksteed, S. Wiemann, L. Wilcox, C. Wilson, R. Wilson, A. Winant, E. Winnett, B. Winsor, P. Wipfli, S. Wölfl, P. Wohldman, K. Wolf, K. H. Wolfe, L. F. Wright, H. Wurst, G. Xu, M. Yamazaki, M. A. Yelton, K. Yokohama, A. Yoshikawa, S. Yuping, P. Zaccaria, M. Zagulski, F. K. Zimmermann, J. Zimmermann, M. Zimmermann, W.-W. Zhong, A. Zollner, and E. Zumstein. 1997. The Yeast Genome Directory. *Nature* **387(Suppl.):**1–105.

26. Goffeau, A., B. G. Barrell, H. Bussey, R. W. Davis, B. Dujon, H. Feldmann, F. Galibert, J. D. Hoheisel, C. Jacq, M., Johnston, E. J. Louis, H. W. Mewes, Y. Murakami, P. Philippsen, H. Tettelin, and S. G. Oliver. 1996. Life with 6000 genes. *Science* **274:**546–550.

27. Gold, L., D. Brown, Y.-Y. He, T. Shtatland, B. S. Singer, and Y. Wu. 1997. From oligonucleotide shapes to genomic SELEX: Novel biological regulatory loops. *Proc. Natl. Acad. Sci. USA* **94:**59–64.

28. Guarente, L. 1983. Yeast promoters and lacZ fusions designed to study expression of cloned genes in yeast. *Methods Enzymol.* **101:**181–191.

29. Guarente, L., R. R. Yocum, and P. Gifford. 1982. A GAL10-CYC1 hybrid yeast promoter identifies the GAL4 regulatory region as an upstream site. *Proc. Natl. Acad. Sci. USA* **79:**7410–7414.

30. Güldener, U., S. Heck, T. Fiedler, J. Beinhauer, and J. H. Hegemann. 1996. A new efficient gene disruption cassette for repeated use in budding yeast. *Nucleic Acids Res.* **24:**2519–2524.

31. Heiter, P., and M. Boguski. 1997. Functional genomics: it's all how you read it. *Science* **278:**601–602.

32. Heumann, K., C. Harris, and H. W. Mewes. 1996. A top-down approach to whole genome visualization. *Ismb.* **4:**98–108.

33. Hillenkamp, F., and M. Karas. 1990. Mass spectrometry of peptides and proteins by matrix-assisted ultraviolet laser desorption/ionization. *Methods Enzymol.* **193:**280–295.

34. Himmelreich, R., H. Hilbert, H. Plagens, E. Pirkl, B. C. Li, and R. Herrmann. 1996. Complete sequence analysis of the genome of the bacterium *Mycoplasma pneumoniae*. *Nucleic Acids Res.* **24:**4420–4449.

35. Itaya, M. 1995. An estimation of minimal genome size required for life. *FEBS Lett.* **362:**257–260.

36. Itaya, M., and T. Tanaka. 1990. Gene-directed mutagenesis on the chromosome of Bacillus subtilis 168. *Mol. Gen. Genet.* **223:**268–272.

37. Johnston, M. 1996. Towards a complete understanding of how a simple eukaryotic cell works. *Trends Genet.* **12:**242–243.

38. Kaback, D. B., P. W. Oeller, H. Yde Steensma, J. Hirschman, D. Ruezinsky, K. G. Coleman, and J. R. Pringle. 1984. Temperature-sensitive lethal mutations on yeast chromosome I appear to define only a small number of genes. *Genetics* **108:**67–90.

39. Kaneko, T., S. Sato, H. Kotani, A. Tanaka, E. Asamizu, Y. Nakamura, N. Miyajima, M. Hirosawa, M. Sugiura, S. Sasamoto, T. Kimura, T. Hosouchi, A. Matsuno, A. Muraki, N. Nakazaki, K. Naruo, S. Okumura, S. Shimpo, C. Takeuchi, T. Wada, A. Watanabe, M. Yamada, M. Yasuda, and S. Tabata. 1996. Sequence analysis of the genome of the unicellular cyanobacterium *Synechocystis sp.* strain PCC6803. II. Sequence determination of the entire genome and assignment of potential protein-coding regions. *DNA Res.* **3:**109–136.

40. Karas, M., and F. Hillenkamp. 1988. Laser desorption ionization of proteins with molecular masses exceeding 10,000 daltons. *Anal. Chem.* **60:**2299–2301.

41. Kawarabayasi, Y., M. Sawada, H. Horikawa, Y. Haikawa, Y. Hino, S. Yamamoto, M. Sekine, S. Baba, H. Kosugi, A. Hosoyama, Y. Nagai, M. Sakai, K. Ogura, R. Otsuka, H. Nakazawa, M. Takamiya, Y. Ohfuku, T. Funahashi, T. Tanaka, Y. Kudoh, J. Yamazaki, N. Kushida, A. Oguchi, K. Aoki, T. Yoshizawa, Y. Nakamura, F. T. Robb, K. Horikoshi, Y. Masuchi, H. Shizuya, and H. Kikuchi. 1998. Complete sequence and gene organization of the genome of a hyper-thermophilic archaebacterium, *Pyrococcus horikoshii* OT3 (supplement). *DNA Res.* **5:**147–155.

42. Klenk, H. P., R. A. Clayton, J. Tomb, O. White, K. E. Nelson, K. A. Ketchum, R. J. Dodson, M. Gwinn, E. K. Hickey, J. D. Peterson, D. L. Richardson, A. R. Kerlavage, D. E. Graham, N. C. Kyrpides, R. D. Fleischmann, J. Quackenbush, N. H. Lee, G. G. Sutton, S. Gill, E. F. Kirkness, B. A. Dougherty, K. McKenney, M. D. Adams, B. Loftus, S. Peterson, C. I. Reich, L. K McNeil, J. H. Badger, A. Glodek, L. Zhou, R. Overbeek, J. D. Gocayne, J. F. Weidman, L. McDonald, T. Utterback, M. D. Cotton, T. Spriggs, P. Artiach, B. P. Kaine, S. M. Sykes, P. W. Sadow, K. P. D'Andrea, C. Bowman, C. Fujii, S. A. Garland, T. M. Mason, G. J. Olsen, C. M. Fraser, H. O. Smith, C. R. Woese, and J. C. Venter. 1997. The complete genome sequence of the hyper-thermophilic sulphate-reducing archaeon *Archaeoglobus fulgidus*. *Nature* **390:**364–370.

43. Kunst, F., N. Ogasawara, I. Moszer, A. M. Albertini, G. Alloni, V. Azevedo, M. G. Bertero, P. Bessieres, A. Bolotin, S. Borchert, R. Borriss, L. Boursier, A. Brans, M. Braun, S. C. Brignell, S. Bron, S. Brouillet, C. V. Bruschi, B. Caldwell, V. Capuano, N. M. Carter, S. K. Choi, J. J. Codani, I. F. Connerton, N. J. Cummings, R. A. Daniel, F. Denizot, K. M. Devine, A. Dusterhoft, S. D. Ehrlich, P. T. Emmerson, K. D. Entian,

J. Errington, C. Fabret, E. Ferrari, D. Foulger, C Fritz, M. Fujita, Y. Fujita, S. Fuma, A. Galizzi, N. Galleron, S. Y. Ghim, P. Glaser, A. Goffeau, E. J. Golightly, G. Grandi, G. Guiseppi, B. J. Guy, K. Haga, J. Haiech, C. R. Harwood, A. Henaut, H. Hilbert, S. Holsappel, S. Hosono, M. F. Hullo, M. Itaya, L. Jones, B. Joris, D. Karamata, Y. Kasahara, M. Klaerr-Blanchard, C. Klein, Y. Kobayashi, P. Koetter, G. Koningstein, S. Krogh, M. Kumano, K. Kurita, A. Lapidus, S. Lardinois, J. Lauber, V. Lazarevic, S. M. Lee, A. Levine, H. Liu, S. Masuda, C. Mauel, C. Medigue, N. Medina, R. P. Mellado, M. Mizuno, D. Moestl, S. Nakai, M. Noback, D. Noone, M. O'Reilly, K. Ogawa, A. Ogiwara, B. Oudega, S. H. Park, V. Parro, T. M. Pohl, D. Portetelle, S. Porwollik, A. M. Prescott, E. Presecan, P. Pujic, B. Purnelle, G. Rapoport, M. Rey, S. Reynolds, M. Rieger, C. Rivolta, E. Rocha, B. Roche, M. Rose, Y. Sadaie, T. Sato, E. Scanlan, S. Schleich, R. Schroeter, F. Scoffone, J. Sekiguchi, A. Sekowska, S. J. Seror, P. Serror, B. S. Shin, B. Soldo, A. Sorokin, E. Tacconi, T. Takagi, H. Takahashi, K. Takemaru, M. Takeuchi, A. Tamakoshi, T. Tanaka, P. Terpstra, A. Tognoni, V. Tosato, S. Uchiyama, M. Vandenbol, F. Vannier, A. Vassarotti, A. Viari, R. Wambutt, E. Wedler, H. Wedler, T. Weitzenegger, P. Winters, A. Wipat, H. Yamamoto, K. Yamane, K. Yasumoto, K. Yata, K. Yoshida, H. F. Yoshikawa, E. Zumstein, H. Yoshikawa, and A. Danchin. 1997. The complete genome sequence of the Gram-positive bacterium *Bacillus subtilis. Nature* 390:364–370.

44. **Lafontaine, D., and D. Tollervey.** 1996. One-step PCR mediated strategy for the construction of conditionally expressed and epitope tagged yeast proteins. *Nucleic Acids Res.* **24:** 3469–3471.

45. **Link, A. J., D. Phillips, and G. M. Church.** 1997. Methods for generating precise deletions and insertions in the genome of wild-type *Escherichia coli*: application to open reading frame characterization. *J. Bacteriol.* 179:6228–6237.

46. **Lipshutz, R. J., D. Morris, M. Chee, E. Hubbell, M. J. Kozal, N. Shah, N. Shen, R. Yang, and S. P. Fodor.** 1995. Using oligonucleotide probe arrays to access genetic diversity. *Biotechniques* **19:**442–447.

47. **Lussier, M., A. M. White, J. Sheraton, I. di Paolo, J. Treadwell, S. B. Southard, C. I. Horenstein, J. Chen-Weiner, A. F. Ram, J. C. Kapteyn, T. W. Roemer, D. H. Vo, D. C. Bondoc, J. Hall, W. W. Zhong, A. M.** Sdicu, J. Davies, F. M. Klis, P. W. Robbins, and H. Bussey. 1997. Large scale identification of genes involved in cell surface biosynthesis and architecture in *Saccharomyces cerevisiae Genetics* **147:**435–450.

48. **Marshall, A., and J. Hodgson.** 1998. DNA chips: an array of possibilities. *Nat. Biotechnol.* **16:** 27–31.

49. **Miller, J. H.** 1992. A short course in bacterial genetics: a laboratory manual and handbook for *Escherichia coli* and related bacteria. Cold Spring Harbor Laboratory Press, Cold Spring Harbor, N.Y.

50. **Mullis. K., F. Faloona, S. Scharf, R. Saiki, G. Horn, and H. Erlich.** 1986. Specific enzymatic amplification of DNA in vitro: the polymerase chain reaction. Part 1. *Cold Spring Harbor Symp. Quant. Biol.* 51:263–273.

51. **Nagahashi, S., H. Nakayama, K. Hamada, H. Yang, M. Arisawa, and K. Kitada.** 1997. Regulation by tetracycline of gene expression in *Saccharomyces cerevisiae. Mol. Gen. Genet.* **255:** 372–375.

52. **Naitou, M., H. Hagiwara, F. Hanaoka, T. Eki, and Y. Murakami.** 1997. Expression profiles of transcripts from 126 open reading frames in the entire chromosome VI of *Saccharomyces cerevisiae* by systematic northern analyses. *Yeast* **13:** 1275–1290.

53. **Niedenthal, R. K., L. Riles, M. Johnston, and J. H. Hegemann.** 1996. Green fluorescent protein as a marker for gene expression and subcellular localization in budding yeast. *Yeast 1996.* **12:**773–786.

54. **Oliver, S.** 1996. A network approach to the systematic analysis of yeast gene function. *Trends Genet.* **12:**241–242.

55. **Oliver, S.G.** 1996. From DNA sequence to biological function. *Nature* **379:**597–600.

56. **Pearson, W. R.** 1998. Empirical statistical estimates for sequence similarity searches. *J. Mol. Biol.* **276:**71–84.

57. **Pearson, W. R., and D. J. Lipman.** 1988. Improved tools for biological sequence comparison. *Proc. Natl. Acad. Sci. USA* **85:**2444–2448.

58. **Rieger, K. J., A. Kaniak, J. Y. Coppee, G. Aljinovic, A. Baudin-Baillieu, G. Orlowska, R. Gromadka, O. Groudinsky, J. P. Rago, and P. P. Slonimski.** 1997. Large-scale phenotypic analysis—the pilot project on yeast chromosome. III. *Yeast* **13:**1547–1562.

59. **Riley, M., and B. Labedan.** 1996. *Escherichia coli* gene products: physiological functions and common ancestries, p. 2118–2202. *In* F. C. Neidhardt, R. Curtiss III, J. L. Ingraham, E. C. C. Lin, K. B. Low, B. Magasanik, W. S. Reznikoff, M. Riley, M. Schaechter, and H. E. Um-

barger (ed.), *Escherichia coli and Salmonella: Cellular and Molecular Biology*, 2nd ed. ASM Press, Washington, D.C.

60. **Rine, J.** 1991. Gene overexpression in studies of *Saccharomyces cerevisiae*. *Methods Enzymol.* **194:**239–251.

61. **Romanos, M. A., C. A. Scorer, and J. J. Clare.** 1992. Foreign gene expression in yeast: a review. *Yeast* **8:**423–488.

62. **Ronne, H.** 1995. Glucose repression in fungi. *Trends Genet.* 1995. **11:**12–17.

63. **Rothstein, R. J.** 1983. One-step gene disruption in yeast. *Methods Enzymol.* **101:**202–211.

64. **Schneider, J. C., and L. Guarente.** 1991. Vectors for expression of cloned genes in yeast: regulation, overproduction, and underproduction. *Methods Enzymol.* **194:**373–388.

65. **Seifert, H. S., E. Y. Chen, M. So, and F. Heffron.** 1986. Shuttle mutagenesis: a method of transposon mutagenesis for *Saccharomyces cerevisiae*. *Proc. Natl. Acad. Sci. USA* **83:**735–739.

66. **Shalon, D., S. J. Smith, and P. O. Brown.** 1996. A DNA microarray system for analyzing complex DNA samples using two-color fluorescent probe hybridization. *Genome Res.* **6:**639–645.

67. **Sherman, F.** 1991. Getting started with yeast. *Methods Enzymol.* **194:**3–21.

68. **Sherman, F., and J. Hicks.** 1991. Micromanipulation and dissection of asci. *Methods Enzymol.* **194:**21–37.

69. **Sherman, F., and P. Wakem.** 1991. Mapping yeast genes. *Methods Enzymol.* **194:**38–57.

70. **Shoemaker, D. D., D. A. Lashkari, D. Morris, M. Mittmann, and R. W. Davis.** 1996. Quantitative phenotypic analysis of yeast deletion mutants using a highly parallel molecular barcoding strategy. *Nat. Genet.* **14:**450–456.

71. **Smith, D. R., L. A. Doucette-Stamm, C. Deloughery, H.-M. Lee, J. Dubois, T. Aldredge, R. Bashirzadeh, D. Blakely, R. Cook, K. Gilbert, D. Harrison, L. Hoang, P. Keagle, W. Lumm, B. Pothier, D. Qiu, R. Spadafora, R. Vicare, Y. Wang, J. Wierzbowski, R. Gibson, N. Jiwani, A. Caruso, D. Bush, H. Safer, D. Patwell, S. Prabhakar, S. McDougall, G. Shimer, A. Goyal, S. Pietrovski, G. M. Church, C. J. Daniels, J.-I. Mao, P. Rice, J. Nolling, and J. N. Reeve.** 1997. Complete genome sequence of *Methanobacterium thermoautotrophicum* ΔH: functional analysis and comparative genomics *J. Bacteriol.* **179:**7135–7155.

72. **Smith, V., D. Botstein, and P. O. Brown.** 1995. Genetic footprinting: a genomic strategy for determining a gene's function given its sequence. *Proc. Natl. Acad. Sci. USA* **92:**6479–6483.

73. **Smith, V., K. N. Chou, D. Lashkari, D. Botstein, and P. O. Brown.** 1996. Functional analysis of the genes of yeast chromosome V by genetic footprinting. *Science* **274:**2069–2074.

74. **Stephens, R. S., S. Kalman, C. J. Lammel, J. Fan, R. Marathe, L. Aravind, W. P. Mitchell, L. Olinger, R. L. Tatusov, Q. Zhao, E. V. Koonin, and R. W. Davis.** 1998. Genome sequence of an obligate intracellular pathogen of humans: *Chlamydia trachomatis*. *Science* **282:**754–759.

75. **Tomb, J. F., O. White, A. R. Kerlavage, R. A. Clayton, G. G. Sutton, R. D. Fleischmann, K. A. Ketchum, H. P. Klenk, S. Gill, B. A. Dougherty, K. Nelson, J. Quackenbush, L. Zhou, E. F. Kirkness, S. Peterson, B. Loftus, D. Richardson, R. Dodson, H. G. Khalak, A. Glodek, K. McKenney, L. M. Fitzegerald, N. Lee, M. D. Adams, and J. C. Venter.** 1997. The complete genome sequence of the gastric pathogen *Helicobacter pylori*. *Nature* **388:**539–547.

76. **Tanaka, S., and K. Isono.** 1993. Correlation between observed transcripts and sequenced ORFs of chromosome III of *Saccharomyces cerevisiae*. *Nucleic Acids Res.* **21:**1149–1153.

77. **Tuerk, C.** 1997. Using the SELEX combinatorial chemistry process to find high affinity nucleic acid ligands to target molecules. *Methods Mol. Biol.* **67:**219–230.

78. **Tuerk, C., and L. Gold.** 1990. Systematic evolution of ligands by exponential enrichment: RNA ligands to bacteriophage T4 DNA polymerase. *Science* **249:**505–510.

79. **Ushinsky, S. C., H. Bussey, A. A. Ahmed, Y. Wang, J. Friesen, and R. K. Storms.** 1977. Functional analysis of a 38 kilobase region on chromosome XVI in *Saccharomyces cerevisiae*. *Genes Funct.* **4:**273–284.

80. **Velculescu, V. E., L. Zhang, B. Vogelstein, and K. W. Kinzler.** 1995. Serial analysis of gene expression. *Science* **270:**484–487.

81. **Velculescu, V.E., L. Zhang, W. Zhou, J. Vogelstein, M.A. Basrai, D. E. Bassett, Jr., P. Hieter, B. Vogelstein, and K. W. Kinzler.** 1997. Characterization of the yeast transcriptome. *Cell* **88:**243–251.

82. **Wach, A., A. Brachat, R. Pohlmann, and P. Philippsen.** 1994. New heterologous modules for classical or PCR-based gene disruptions in *Saccharomyces cerevisiae*. *Yeast* **10:**1793–1808.

83. **Witke, W., A. V. Podtelejnikov, A. Di Nardo, J. D. Sutherland, C. B. Gurniak, C. Dotti, and M. Mann.** 1998. In mouse brain profilin I and profilin II associate with regulators of the endocytic pathway and actin assembly. *EMBO J.* **17:**967–976.

84. **Wodicka, L., H. Dong, M. Mittmann, M. H. Ho, and D. J. Lockhart.** 1997. Genome-

wide expression monitoring in *Saccharomyces cerevisiae*. *Nat. Biotechnol.* **15:**1359–1367.

85. **Wolfe, K. H., and D. C. Shields.** 1997. Molecular evidence for an ancient duplication of the entire yeast genome. *Nature* **387:**708–713.

86. **Yoshikawa, A., and K. Isono.** 1990. Chromosome III of Saccharomyces cerevisiae: an ordered clone bank, a detailed restriction map and analysis of transcripts suggest the presence of 160 genes. *Yeast* **6:**383–401.

87. **Zachariae, W., A. Shevchenko, P. D. Andrews, R. Ciosk, M. Galova, M. J. Stark, M. Mann, and K. Nasmyth**. 1998. Mass spectrometric analysis of the anaphase-promoting complex from yeast: identification of a subunit related to cullins. *Science* **279:**1216–1219.

INDEX